# Illustrated A
## QUICK REFERENCE

Ralph Grabowski

**autode**

**THOMSON**

**DELMAR LEARNING**™

Australia ï Canada ï Mexico ï Singapore ı Spain ı UnitedeKingdom ı UnitedeStates

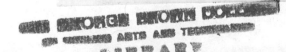

# THOMSON

## DELMAR LEARNING

# autodesk® Press

## The Illustrated AutoCAD® 2004 Quick Reference
### Ralph Grabowski

**Vice President, Technology and Trades SBU**
Alar Elken

**Editorial Director**
Sandy Clark

**Senior Acquisitions Editor**
James DeVoe

**Senior Development Editor**
John Fisher

**Production Editor**
Stacy Masucci

**Marketing Manager**
Maura Theriault

**Channel Manager**
Fair Huntoon

**Marketing Coordinator**
Sarena Douglass

**Production Director**
Mary Ellen Black

**Book Design and Typesetting**
Ralph Grabowski

**Production Manager**
Andrew Crouth

**Senior Art/Design Coordinator**
Mary Beth Vought

**Technology Project Manager**
David Porush

**Technology Project Specialist**
Kevin Smith

**Editorial Assistant**
Marry Ellen Martino

Library of Congress
Cataloging-in-Publication Data:
Card Number:
ISBN 0-4018-5014-6

## NOTICE TO THE READER

# About This Book

*The Illustrated AutoCAD 2004 Quick Reference* presents concise facts about all commands found in AutoCAD 2004 and earlier. The clear format of this reference book demonstrates each command, starting on its own page, illustrated by over 500 figures, plus these exclusive features:

- All variations of commands, such as the **View**, **-View**, and **+View** commands.
- Scores of AutoCAD 2004 commands not documented by Autodesk.
- A dozen "Quick Start" mini-tutorials that help you get started quicker.
- Over 100 definitions of acronyms and hard-to-understand terms.
- Nearly 1,000 context-sensitive tips.
- All system variables in Appendix A, including those not listed by the **SetVar** command.
- Obsolete commands that no longer work in AutoCAD 2004 in Appendix B.
- External commands that operate outside of AutoCAD in Appendix C.

The name of each command shown in mixed upper and lower case to help you understand the name, which is often condensed. For example, the name of the **VpClip** command is short for "ViewPort CLIP." Each command includes all alternative methods of command input:

- Alternate command spelling, such as **Donut** and **Doughnut**.
- ' (the apostrophe prefix) indicating transparent commands, such as **'Blipmode**.
- All aliases, such as **L** for the **Line** command.
- Pull-down menu picks, such as **Draw ⍦ Construction Line** for the **XLine** command.
- Control-key combinations, such as CTRL+E for the **Isoplane** toggle.
- Function keys, such as **F1** for the **Help** command.
- Alt-key combinations, such as ALT+TE for the **Spell** command.
- Table menu coordinates, such as **M2** for the **Hide** command.

The brief command description notes when the command is *renamed* from an earlier release of AutoCAD, or is *undocumented* by Autodesk.

The version or release number indicates when the command first appeared in AutoCAD, such as **Ver. 1.0**, **Rel. 14**, or **2004**. This is useful when working with older versions of AutoCAD. See Appendix B for the list of commands removed from every release of AutoCAD. Each command includes one or more of the following: command line options, dialog box options, shortcut menu options, related commands, related toolbar icons, and related system variables.

Many commands include one or more tips that help you use the command more efficiently or warn you of the command's limitations. Numrerous commands include a list of definitions of acronyms and jargon words.

Special thanks to to Stephen Dunning for his keen copy editing. *Soli Deo Gloria!*

**Ralph Grabowski**
Abbotsford, British Columbia, Canada
March 17, 2003
Contact: **editor@upfrontezine.com**

# Table of Contents

Indicates the command is new to AutoCAD 2004.

' Indicates the command is transparent.

# Dimensioning

# H

# I

# J

# L

# O

# P

# Q

# R

# S

# Appendices

# 'About

**Rel.12**  Displays the AutoCAD version and serial numbers, ownership information, and copyright notices.

| Command | Alias | Ctrl+ | F-key | Alt+ | Menu Bar | Tablet |
|---------|-------|-------|-------|------|----------|--------|
| 'about  | ...   | ...   | ...   | HO   | Help     | ...    |
|         |       |       |       |      | ⌐About   |        |

**Command:** about
  *Displays dialog box:*

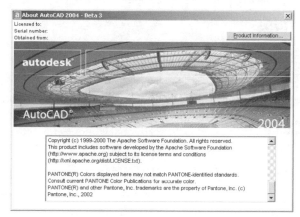

*Shown on the splash screen is the Stade de France stadium designed by Macary, Zublena et Regembal, Costantini - Architectes, Paris. www.stadefrance.com/presentation/vv.cfm*

## DIALOG BOX OPTIONS

  **x** dismisses dialog box, or press **ESC**.

  **Product Information** displays the Product Information dialog box:

  **License Agreement** opens the computer's default word processor and displays the Autodesk Software License Agreement document (*license.rtf*).

  **Save as Text File** displays the Save As dialog box to record the information displayed above.

## RELATED COMMANDS

  **List** displays information about selected objects.

  **Status** displays information about the drawing and environment.

## RELATED SYSTEM VARIABLES

  **_PkSer** displays the AutoCAD software serial number.

  **_Server** displays the network authorization code.

# AcisIn

<u>Rel. 13</u>  Imports SAT files into drawings, creating 3D solids, 2D regions, and bodies.

| Command | Alias | Ctrl+ | F-key | Alt+ | Menu Bar | Tablet |
|---------|-------|-------|-------|------|----------|--------|
| acisin | ... | ... | ... | IA | Insert | ... |
| | | | | | ⇘ACIS File | |

**Command:** acisin

*Displays* **Select ACIS File** *dialog box. Select a .sat file, and then click* **Open**.

## DIALOG BOX OPTIONS

**Cancel** dismisses the dialog box.

**Open** opens the selected *.sat* file.

## RELATED COMMANDS

**AcisOut** exports solid objects — 3D solids, 2D regions, and bodies — to *.sat* files for import into ACIS-aware CAD software.

**AmeConvert** converts AME v2.0 and v2.1 solid models and regions into solids.

## RELATED FILE

**\*.sat** is the ASCII format of ACIS model files.

## TIPS

- "Acis" comes from the first names of the original developers, "Andy, Charles, and Ian's System." In Greek mythology, Acis was the lover of the goddess Galatea; when Acis was killed by the jealous Cyclops, Galatea turned the blood of Acis into a river.

- ACIS is the solids modeling technology from the Spatial Technologies division of Dassault Systemes, and is used by numerous 3D CAD packages. As of AutoCAD 2004, Autodesk is using its own ACIS-derived solids modeler, called ShapeManager.

# AcisOut

Rel.13 Exports AutoCAD 3D solids, 2D regions, and bodies in SAT format.

| Command | Alias | Ctrl+ | F-key | Alt+ | Menu Bar | Tablet |
|---------|-------|-------|-------|------|----------|--------|
| acisout | ... | ... | ... | FE | File | ... |
| | | | | ⌂ACIS | ⌂Export | |
| | | | | | ⌂ACIS | |

**Command:** acisout
**Select objects:** *(Select one or more objects.)*
**Select objects:** *(Press **Enter** to end object selection.)*

*Displays **Create ACIS File** dialog box. Enter a name for the .sat file, and then click **Save**.*

## DIALOG BOX OPTIONS

**Cancel** dismisses the dialog box.

**Save** saves the selected objects as a *.sat* file.

## RELATED COMMANDS

**AcisIn** imports *.sat* files, and creates 3D solids, 2D regions, and bodies.

**StlOut** exports solid models in STL format for use by stereolithography devices.

**3dsOut** exports solid models as 3D faces for import into 3D Studio software.

## RELATED FILE

**\*.sat** is the ASCII format of ACIS model files; short for "save as text."

## TIPS

* **AcisOut** export objects that are 3D solids, 2D regions, and bodies only.

* The *.sat* files can be read by other ACIS-based CAD programs.

# 'AdcClose

**2000**  Closes the DesignCenter window.

| Command | Alias | Ctrl+ | F-key | Alt+ | Menu Bar | Tablet |
|---|---|---|---|---|---|---|
| 'adcclose | ... | 2 | ... | TG | Tools | X12 |
|  |  |  |  |  | ⬏ DesignCenter |  |

**Command:** adcclose

*The DesignCenter window is closed.*

## COMMAND LINE OPTIONS
*None.*

## RELATED COMMANDS
**AdCenter** opens the DesignCenter window.

**Close** closes the current drawing.

## TIPS
- The **AdcClose** command does not display a prompt.

- As an alternative to this command, click the small **x** button in the upper-right corner of the DesignCenter window.

- CTRL+**2** is a toggle keystroke: it opens and closes DesignCenter each time you press it.

# 'AdCenter

**2000**  Opens the DesignCenter window (*short for Autocad Design CENTER.*).

| Command | Aliases | Ctrl+ | F-key | Alt+ | Menu Bar | Tablet |
|---------|---------|-------|-------|------|----------|--------|
| 'adcenter | adc | 2 | ... | TG | Tools | X12 |
| | content | | | | ⮑DesignCenter | |

**Command:** adcenter

*Displays DesignCenter window:*

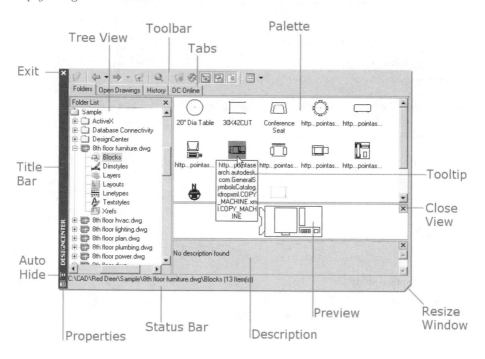

## TAB OPTIONS

**Folders** displays the content found on the computer.

**Open Drawings** displays the content found in drawings currently open in AutoCAD.

**History** displays the drawings previously opened.

**DC Online** displays content available from Autodesk via the Internet.

## TOOLBAR OPTIONS

**Open** displays the **Open** dialog box to open the following vector and raster file types: *.dwg, .bil, .bmp, .cal, .cg4, .dib, .flc, .fli, .gif, .gp4, .ig4, .igs, .jpg, .mil, .pat, .pcx, .png, .rlc, .rle, .rst, .tga,* and *.tif.*

**Back** returns to the previous view.

**Forward** goes to the next view.

**Up** moves up one folder level.

**Search** opens the **Search** dialog box to search for the following AutoCAD files (on the computer) and objects (in drawings): blocks, dimstyles, drawings, hatch patterns, layers, layouts, linetypes, text styles, and xrefs.

**Favorites** displays the files in the *\documents and settings\favorites\autodesk* folder.

**Home** displays the files in the *\autocad 2004\sample\designcenter* folder.

**Tree View Toggle** hides and displays the Folders and Open Drawings tree views.

**Preview** toggles the display of the preview image of *.dwg* and raster files.

**Description** toggles the display of the description area.

**View** changes the display format of the palette area.

## RELATED COMMANDS

**AdcClose** closes the DesignCenter window.

**AdcNavigate** specifies the initial path for DesignCenter.

## TIPS

- Use DesignCenter to keep track of drawings and parts of drawings, such as block libraries.

- You can drag blocks and other drawing parts from the DesignCenter into the drawing.

- The DesignCenter window can switch between floating and docked by right-clicking and selecting the option from the menu.

- CTRL+2 opens and closes the DesignCenter each time you press it.

# AdcNavigate

**2000** Specifies the initial path for the DesignCenter to access content.

| Command | Alias | Ctrl+ | F-key | Alt+ | Menu Bar | Tablet |
|---|---|---|---|---|---|---|
| adcnavigate | ... | ... | ... | ... | ... | ... |

**Command:** adcnavigate
**Enter pathname <>:** *(Enter a path, and then press **Enter**.)*

## COMMAND LINE OPTION

**Enter pathname** specifies the path, such as *c:\program files\acad2004\sample*.

## RELATED COMMAND

**AdCenter** opens the DesignCenter window.

## TIPS

- AutoCAD uses the path specified by **AdcNavigate** to locate content displayed by DesignCenter's Desktop option.

- You can enter the path to a file, folder, or network location:

  Example of a folder path:     `c:\program files\autocad 2004\sample`
  Example of a file path:     `c:\design center\welding.dwg`
  Example of a network path:  `\\downstairs\project`

 # Ai_Box

Draws 3D boxes as surface objects (*undocumented command*).

| Command | Alias | Ctrl+ | F-key | Alt+ | Menu Bar | Tablet |
|---------|-------|-------|-------|------|----------|--------|
| ai_box | ... | ... | ... | ... | ... | ... |

**Command:** ai_box
**Specify corner point of box:** *(Pick a point.)*
**Specify length of box:** *(Pick a point.)*
**Specify width of box or [Cube]:** *(Pick a point, or type **C**.)*
**Specify height of box:** *(Pick a point.)*
**Specify rotation angle of box about the Z axis or [Reference]:** *(Specify the rotation angle, or type **R**.)*

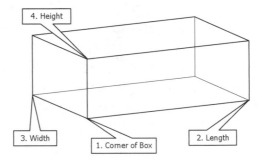

**COMMAND LINE OPTIONS**

**Corner of box** specifies the initial corner of the box.

**Length** specifies the length of one side of the box along the x-axis.

**Cube** creates a cube based on the Length.

**Width** specifies the width of the box, along the y-axis.

**Height** specifies the height of the box, along the z-axis.

**Rotation angle** specifies the angle the box rotates about the z-axis.

**Reference** prompts you to pick two points that represent the new angle.

**TIPS**

- This command creates 3D surface objects of rectangular boxes and cubes; to draw a solid model, use the **Box** command.

- When specifying the **Width** and **Height**, move the cursor back to the **Corner of box** point; otherwise the box may have a different size than you expect.

- The box is made of a single polymesh; **Explode** converts each side to an independent 3D face object.

- The base of the box is drawn at the current setting of the **Elevation** system variable.

*The following applies to the Ai_ surface model commands:*

**RELATED SYSTEM VARIABLES**

**SurfU** specifies the surface mesh density in the m-direction.

**SurfV** specifies the surface mesh density in the n-direction.

**TIPS**

- You *cannot* perform Boolean operations (such as intersect, subtract, and union) on 3D surface objects.

- You cannot convert 3D surface objects into 3D solid objects.

- To convert 3D solid objects into 3D surface objects, export with **3dsOut**; then import with the **3dsIn** command.

- Variants of 3D objects drawn by the **Ai_** series of commands:

| Command | Variant |
| --- | --- |
| **Ai_Box** | Rectangular boxes, and cubes. |
| **Ai_Cone** | Pointy cones, and truncated cones. |
| **Ai_Pyramid** | Pyramids, truncated pyramids, tetrahedrons, truncated tetrahedrons, and roof shapes. |
| **Ai_Torus** | Tori and donuts. |
| **Ai_Wedge** | Wedges. |

- Mesh m and n sizes are limited to values between 2 and 256.

- Each of these surface shapes is made of a single polymesh; use the **Explode** command to convert each facet to independent 3Dface objects.

- You can use the **Hide**, **Shade**, and **Render** commands on these surface model shapes.

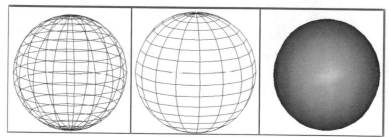

- The bases of boxes, cones, dishes, domes, meshes, and pyramids are drawn at the current setting of the **Elevation** system variable — unless you specify a z-coordinate.

- The centers of spheres and tori are at the current elevation.

# 'Ai_CircTan

Draws circles tangent to three points (*undocumented command*).

| Command | Alias | Ctrl+ | F-key | Alt+ | Menu Bar | Tablet |
|---------|-------|-------|-------|------|----------|--------|
| 'ai_circtan | ... | ... | ... | ... | ... | ... |

**Command:** ai_circtan
**Enter Tangent spec:** *(Pick a point.)*
**Enter second Tangent spec:** *(Pick a point.)*
**Enter third Tangent spec:** *(Pick a point.)*

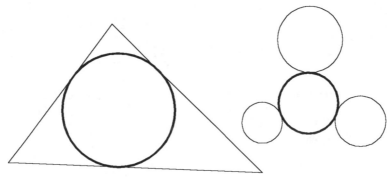

*A circle drawn tangent to three lines (left) and three circles (right).*

## COMMAND LINE OPTION

**Enter Tangent spec** picks the objects to which the circle will be made tangent.

## TIPS

• This command is meant for use in toolbar and menu macros.

• If the circle cannot be drawn between three tangents, the command complains, "Circle does not exist."

• This command is an alternative to the **Circle** command's **3P** option.

 # Ai_Cone

Draws 3D cones as surface objects (*undocumented command*).

| Command | Alias | Ctrl+ | F-key | Alt+ | Menu Bar | Tablet |
|---------|-------|-------|-------|------|----------|--------|
| ai_cone | ... | ... | ... | ... | ... | ... |

**Command:** ai_cone
**Specify center point for base of cone:** *(Pick a point.)*
**Specify radius for base of cone or [Diameter]:** *(Pick a point, or type **D**.)*
**Specify radius for top of cone or [Diameter] <0>:** *(Specify the radius, or type **D**; enter **0** for a pointy top.)*
**Specify height of cone:** *(Pick a point.)*
**Enter number of segments for surface of cone <16>:** *(Press **Enter**.)*

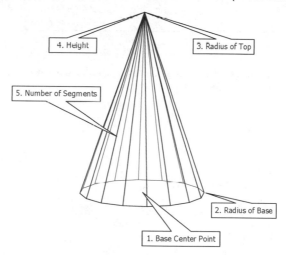

## COMMAND LINE OPTIONS

**Base center point** specifies the center point of the cone's base.

**Diameter of base** specifies the diameter of the base.

**Radius of base** specifies the radius of the base.

**Diameter of top** specifies the diameter of the top of the cone.

**Radius of top** specifies the radius of the top of the cone; 0 = cone with a point.

**Height** specifies the height of the cone.

**Number of segments** specifies the number of "lines" that define the curved surface of the cone; default = 16.

## TIPS

- The **Ai_Cone** command creates "pointy" and truncated cones; see figure.

- The base of the cone is drawn at the current setting of the **Elevation** system variable.

# 'Ai_Custom_Safe, *etc.*

Accesses Autodesk support Web pages (*undocumented command*).

| Commands | Alias | Ctrl+ | F-key | Alt+ | Menu Bar | Tablet |
|----------|-------|-------|-------|------|----------|--------|
| 'ai_custom_safe | ... | ... | ... | HRC | Help<br>⤷Online Resources<br>  ⤷Customization | ... |
| 'ai_product_support | | | | HRP | Help<br>⤷Online Resources<br>  ⤷Product Support | |
| 'ai_product_support_safe | | | | | | |
| 'ai_training_safe | | | | HRT | Help<br>⤷Online Resources<br>  ⤷Training | |

**Command:** ai_custom_safe
**_.browser Enter Web location (URL) <http://www.autodesk.com> http://
www.www.autodesk.com/developautocad**
*Opens Web browser.*

## COMMAND LINE OPTIONS
*None.*

## TIPS
• The commands access the following Web pages:

| Command | Web Page Accessed |
|---------|-------------------|
| ai_custom_safe | www.autodesk.com/developautocad |
| ai_product_support | www.autodesk.com/autocad-support |
| ai_product_support_safe | www.autodesk.com/autocad-support |
| ai_training_safe | www.autodesk.com/autocad-training |

• These commands are meant for use in menu macros and toolbars.

# AiDimPrec

Changes the precision displayed by existing dimensions (*undocumented command*).

| Command | Alias | Ctrl+ | F-key | Alt+ | Menu Bar | Tablet |
|---------|-------|-------|-------|------|----------|--------|
| aidimprec | ... | ... | ... | ... | ... | ... |

**Command:** aidimprec
**Enter option [0/1/2/3/4/5/6] <4>:** *(Enter a digit.)*
**Select objects:** *(Select one or more dimensions.)*
**Select objects:** *(Press **Enter**.)*

*Before (left) and after (right) applying **AiDimPrec** = 1 to a decimal dimension.*

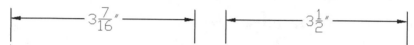

*Before and after applying **AiDimPrec** = 1 to a fractional dimension.*

## COMMAND LINE OPTIONS

**Enter option** specifies the precision (number of decimal places, or fractional equivalent); enter a number between 0 and 6.

**Select objects** selects one or more dimensions.

## TIPS

- This command allows you retroactively to change the precision displayed by selected dimensions. It is meant for use by menu and toolbar macros.

- Zero to six decimal places can be specified; fractional units are rounded to the nearest fraction:

| AiDim Prec | Architectural Units |
|------------|---------------------|
| 0 | Rounded to the nearest unit. |
| 1 | 1/2" |
| 2 | 1/4" |
| 3 | 1/8" |
| 4 | 1/16" |
| 5 | 1/32" |
| 6 | 1/64" |

- *Caution!* Since the **AiDimPrec** command rounds off dimensions, it can create false dimensions. For example, the dimension line below measures 3.4375", but setting **AiDimPrec** to 0 causes the dimension text to be rounded down to three inches.

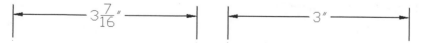

*Applying **AiDimPrec** = 0 to a 3 $^7/_{16}$" dimension.*

# AiDimStyle

Saves and applies preset dimension styles (*undocumented command*).

| Command | Alias | Ctrl+ | F-key | Alt+ | Menu Bar | Tablet |
|---------|-------|-------|-------|------|----------|--------|
| aidimstyle | ... | ... | ... | ... | ... | ... |

**Command:** aidimstyle
**Enter option [1/2/3/4/5/6/Other/Save] <1>:** *(Specify option, press **Enter**.)*
**Select objects:** *(Select one or more dimensions.)*
**Select objects:** *(Press **Enter**.)*

## COMMAND LINE OPTIONS

**Enter option** specifies a predefined dimension style, numbered 1 through 6.

**Other** applies a named dimension style to selected dimension(s).

**Save** saves the style of the selected dimension(s).

**Select objects** selects one or more dimensions.

Other option
**Enter option [1/2/3/4/5/6/Other/Save] <1>:** o

*Displays dialog box:*

**Select objects:** *(Select one or more dimensions, and then press **Enter**.)*
**Select objects:** *(Press **Enter**.)*

Save option
**Enter option [1/2/3/4/5/6/Other/Save] <1>:** s

*Displays dialog box:*

**OK** saves style to the selected dimension style name. AutoCAD warns:

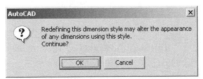

**Select objects:** *(Select one or more dimensions, and then press **Enter**.)*
**Select objects:** *(Press **Enter**.)*

## TIPS

• This command quickly applies and saves dimensions styles; it is meant for macros.

• *Caution!* The **Save** option overwrites existing dimstyles; it does not create new style names.

# Ai_Dim_TextAbove, Center, Home

Moves dimension text relative to dimension lines (*undocumented commands*).

| Command | Alias | Ctrl+ | F-key | Alt+ | Menu Bar | Tablet |
|---|---|---|---|---|---|---|
| ai_dim_textabove | ... | ... | ... | ... | ... | ... |
| ai_dim_textcenter | | | | | | |
| ai_dim_texthome | | | | | | |

**Command:** ai_dim_textabove
**Select objects:** *(Select one or more dimensions.)*
**Select objects:** *(Press **Enter**.)*

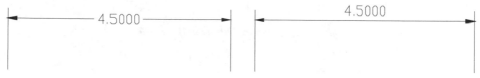

*Before (left) and after (right) applying **Ai_Dim_TextAbove** .*

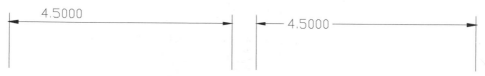

*Before (left) and after (right) applying **Ai_Dim_TextCenter**.*

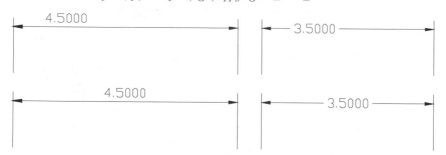

*Before (top) and after (bottom) applying **Ai_Dim_TextHome**.*

## COMMAND LINE OPTION

**Select objects** selects one or more dimensions.

## TIPS

- **Ai_Dim_TextAbove** quickly makes dimensions compliant with JIS dimensioning.

- **Ai_Dim_TextCenter** centers text vertically on the dimension line, but not horizontally.

- **Ai_Dim_TextHome** centers text horizontally on the dimension line, but not vertically.

- Use the **DimTEdit** command to align text to the left, center, or right on horizontal dimensions.

# AiDimTextMove

Moves the location of dimension text *(undocumented command)*.

| Command | Alias | Ctrl+ | F-key | Alt+ | Menu Bar | Tablet |
|---------|-------|-------|-------|------|----------|--------|
| aidimtextmove | ... | ... | ... | ... | ... | ... |

**Command:** aidimtextmove
**Enter option [0/1/2] <2>:** *(Enter an option, and then press **Enter**.)*
**Select objects:** *(Select one dimension.)*
**Select objects:** *(Press **Enter**.)*

*Before (left) and after (right) applying **AiDimTextMove** = 0 to dimension text.*

*Before and after applying **AiDimTextMove** = 1 to dimension text.*

*Before and after applying **AiDimTextMove** = 2 to dimension text.*

## COMMAND LINE OPTIONS

**Enter option** specifies the style of text movement:

| Option | Meaning |
|--------|---------|
| 0 | Moves text with dimension line. |
| 1 | Adds a leader to the moved text. |
| 2 | Moves text anywhere without leader line (default). |

**Select objects** selects one or more dimensions.

## TIPS

• This command allows you retroactively to change the position of dimension text. It is meant for use in menu and toolbar macros.

• Although the command allows you to select more than one dimension, it operates on the first-selected dimension only.

# Ai_Dish, Ai_Dome

Draws the bottom and top halves of 3D spheres as surface models (*undocumented command*).

| Command | Alias | Ctrl+ | F-key | Alt+ | Menu Bar | Tablet |
|---------|-------|-------|-------|------|----------|--------|
| ai_dish | ... | ... | ... | ... | ... | ... |
| ai_dome | | | | | | |

**Command:** ai_dish
**Specify center point of dish:** *(Pick a point.)*
**Specify radius of dish or [Diameter]:** *(Pick a point, or type **D**.)*
**Enter number of longitudinal segments for surface of dish <16>:** *(Press **Enter**.)*
**Enter number of latitudinal segments for surface of dish <8>:** *(Press **Enter**.)*

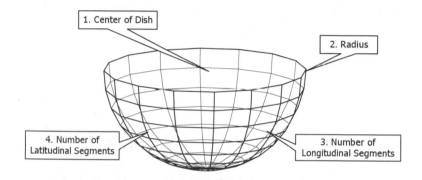

## COMMAND LINE OPTIONS

**Center of dish** specifies the center of the dish's base.

**Diameter** specifies the diameter of the dish.

**Radius** specifies the radius of the dish.

**Number of longitudinal segments** specifies the number of "lines" that define the curved surface in the vertical direction; default = 16.

**Number of latitudinal segments** specifies the number of "lines" that define the curved surface in the horizontal direction; default = 8.

## TIPS

- The **Ai_Dish** command draws the bottom half of a sphere; **Ai_Dome** draws the top half.

- The "base" of the dish is drawn at the current setting of the **Elevation** system variable; the dish is drawn downward (in the negative z-direction).

- The "base" of the dome is drawn at the current setting of the **Elevation** system variable; the dome is drawn upward (in the positive z-direction).

# Ai_Fms

Switches to layout mode, and then to floating model space (*short for Floating Model Space; undocumented command*).

| Command | Alias | Ctrl+ | F-key | Alt+ | Menu Bar | Tablet |
|---------|-------|-------|-------|------|----------|--------|
| ai_fms | ... | ... | ... | ... | ... | ... |

**Command:** ai_fms

*Switches to layout mode, then to the first floating model viewport.*

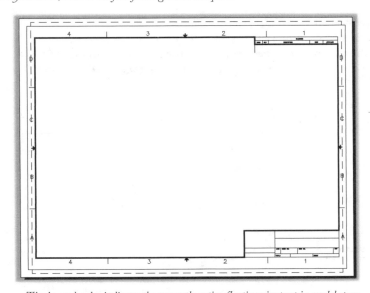

*The heavy border indicates the currently-active floating viewport in model space.*

## COMMAND LINE OPTIONS

*None.*

## TIPS

• This command combines two commands: the **Layout** command, followed by **MSpace**.

• The command is meant for use with menu and toolbar macros.

# Ai_Mesh

Draws non-planar meshes as surface models (*undocumented command*).

| Command | Alias | Ctrl+ | F-key | Alt+ | Menu Bar | Tablet |
|---------|-------|-------|-------|------|----------|--------|
| ai_mesh | ... | ... | ... | ... | ... | ... |

**Command:** ai_mesh
**Specify first corner point of mesh:** *(Pick a point.)*
**Specify second corner point of mesh:** *(Pick a point.)*
**Specify third corner point of mesh:** *(Pick a point.)*
**Specify fourth corner point of mesh:** *(Pick a point.)*
**Enter mesh size in the M direction:** *(Specify a number.)*
**Enter mesh size in the N direction:** *(Specify a number.)*

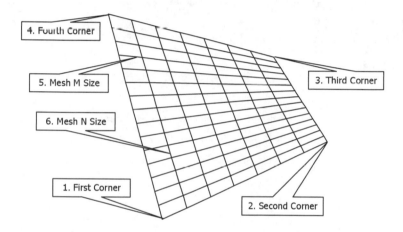

## COMMAND LINE OPTIONS

**First corner** specifies the location of the mesh's first corner.

**Second corner** specifies the location of the mesh's second corner.

**Third corner** specifies the location of the mesh's third corner.

**Fourth corner** specifies the location of the mesh's last corner.

**Mesh size M direction** specifies the number of horizontal "lines" that define the mesh's surface.

**Mesh size N direction** specifies the number of vertical "lines" that define the mesh's surface.

## TIP

- First use the **.xy** filter first to specify the x,y coordinate, followed by the z coordinate, as follows:

    **First corner:**.xy
    **of** *(Pick a point.)*
    **need Z:** *(Specify z.)*

# 'Ai_Molc

Changes the current layer to the one on which the selected object is located (*short for Make Object Layer Current; undocumented command*).

| Command | Alias | Ctrl+ | F-key | Alt+ | Menu Bar | Tablet |
|---------|-------|-------|-------|------|----------|--------|
| 'ai_molc | ... | ... | ... | ... | ... | ... |

**Command:** ai_molc
**Select object whose layer will become current:** *(Pick an object.)*

**COMMAND LINE OPTION**
    **Select object** selects a single object.

**RELATED COMMANDS**
    **Layer** displays the Layer Properties Manager dialog box.
    **LayerP** reverts to the previous layer.
    **MatchProp** matches the properties of one object to other objects.

**RELATED SYSTEM VARIABLE**
    **CLayer** holds the name of the current layer.

**TIPS**
- This command is activated by the **Make Object's Layer Current** button on the toolbar.
- This command is useful for determining on which layer an object resides.

# Ai_Pyramid

Draws 3D pyramids as surface models (*undocumented command*).

| Command | Alias | Ctrl+ | F-key | Alt+ | Menu Bar | Tablet |
|---|---|---|---|---|---|---|
| ai_pyramid | ... | ... | ... | ... | ... | ... |

**Command:** ai_pyramid
**Specify first corner point for base of pyramid:** *(Pick a point.)*
**Specify second corner point for base of pyramid:** *(Pick a point.)*
**Specify third corner point for base of pyramid:** *(Pick a point.)*
**Specify fourth corner point for base of pyramid or [Tetrahedron]:** *(Pick a point, or type T.)*

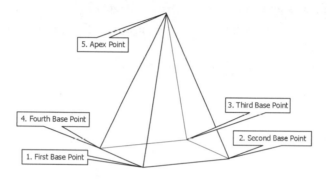

## COMMAND LINE OPTIONS

**First base point** specifies the location of the pyramid's first base point.

**Second base point** specifies the location of the pyramid's second base point.

**Third base point** specifies the location of the pyramid's third base point.

**Tetrahedron** draws a triangular pyramid with equal sides.

**Fourth base point** specifies the location of the pyramid's last base point.

**Ridge** specifies a ridge-top for the pyramid; see figure below.

**Top** specifies a flat-top for the pyramid; see figure below.

**Apex point** specifies a point for the pyramid's top.

## TIPS

• To draw a 2D pyramid, enter no z-coordinate for the **Ridge**, **Top**, and **Apex point** options.

• Use the **.xy** filter to specify the z-coordinate for the **Ridge**, **Top**, and **Apex** point option.

# 'Ai_SelAll

Selects all objects in drawings (*undocumented command*).

| Command | Alias | Ctrl+ | F-key | Alt+ | Menu Bar | Tablet |
|---------|-------|-------|-------|------|----------|--------|
| 'ai_selall | ... | A | ... | ... | ... | ... |

**Command:** ai_selall
**Selecting objects...done.**

## COMMAND LINE OPTIONS
*None.*

## TIPS
- This command is meant for use in menu macros and toolbars.

- Use the CTRL+A shortcut to select all objects in the drawing, other than those on frozen layers.

- The opposite command to un-select all is **(ai_deselect)**.

# Ai_Sphere

Draws 3D spheres as surface models (*undocumented command*).

| Command | Alias | Ctrl+ | F-key | Alt+ | Menu Bar | Tablet |
|---------|-------|-------|-------|------|----------|--------|
| ai_sphere | ... | ... | ... | ... | ... | ... |

**Command:** ai_sphere
**Specify center point of sphere:** *(Pick a point.)*
**Specify radius of sphere or [Diameter]:** *(Enter a radius, or type **D**.)*
**Enter number of longitudinal segments for surface of sphere <16>:** *(Enter a value.)*
**Enter number of latitudinal segments for surface of sphere <16>:** *(Enter a value.)*

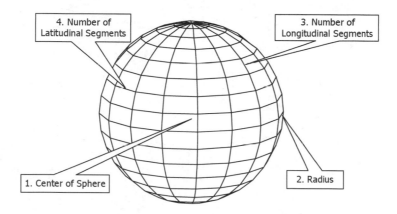

## COMMAND LINE OPTIONS

**Center of sphere** specifies the center point of the sphere.

**Diameter** specifies the diameter of the sphere.

**Radius** specifies the radius of the sphere.

**Number of longitudinal segments** specifies the number of "lines" that define the curved surface in the vertical direction; default = 16.

**Number of latitudinal segments** specifies the number of "lines" that define the curved surface in the horizontal direction; default = 16.

 # Ai_Torus

Draws 3D tori as surface models *(undocumented command)*.

| Command | Alias | Ctrl+ | F-key | Alt+ | Menu Bar | Tablet |
|---------|-------|-------|-------|------|----------|--------|
| ai_torus | ... | ... | ... | ... | ... | ... |

**Command:** ai_torus
**Specify center point of torus:** *(Pick a point.)*
**Specify radius of torus or [Diameter]:** *(Specify the radius, or type **D**.)*
**Specify radius of tube or [Diameter]:** *(Specify the radius, or type **D**.)*
**Enter number of segments around tube circumference <16>:** *(Press **Enter**.)*
**Enter number of segments around torus circumference <16>:** *(Press **Enter**.)*

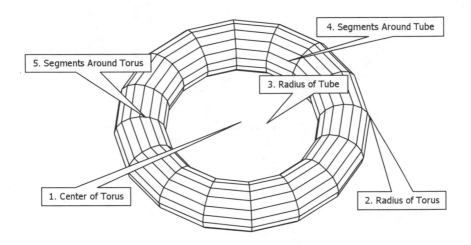

## COMMAND LINE OPTIONS

**Center of torus** specifies the center point of the torus.

**Diameter of torus** specifies the diameter of the torus.

**Radius of torus** specifies the radius of the torus.

**Diameter of tube** specifies the diameter of the tube.

**Radius of tube** specifies the radius of the tube.

**Segments around tube circumference <16>** specifies the number of "lines" defining the curved surface; default = 16.

**Segments around torus circumference <16>** specifies the number of "lines" defining the curved surface; default = 16.

## TIPS

- The **Ai_Torus** command draws donut shapes.

- The tube diameter cannot exceed the torus radius.

# Ai_Wedge

Draws 3D wedges as surface models (*undocumented command*).

| Command | Alias | Ctrl+ | F-key | Alt+ | Menu Bar | Tablet |
|---------|-------|-------|-------|------|----------|--------|
| ai_wedge | ... | ... | ... | ... | ... | ... |

**Command:** ai_wedge
**Specify corner point of wedge:** *(Pick a point.)*
**Specify length of wedge:** *(Pick a point.)*
**Specify width of wedge:** *(Pick a point.)*
**Specify height of wedge:** *(Pick a point.)*
**Specify rotation angle of wedge about the Z axis:** *(Enter a rotation angle, or type 0 for no rotation.)*

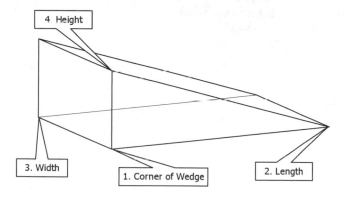

## COMMAND LINE OPTIONS

**Corner of wedge** specifies the corner of the wedge's base.

**Length** specifies the length of the wedge's base.

**Width** specifies the width of the wedge's base.

**Height** specifies the height of the wedge.

**Rotation angle** specifies the angle the box rotates about the z-axis.

## TIPS

- The point you pick for the **Corner of wedge** option determines the taller end of the wedge.

- When specifying the **Width** and **Height** options, move the cursor back to the **Corner of wedge** point; otherwise the wedge may have a different size than you expect.

- You can use the **.xy** point filter to specify the height.

# Align

<u>Rel.12</u>  Moves, transforms, and rotates objects in three dimensions.

| Command | Alias | Ctrl+ | F-key | Alt+ | Menu Bar | Tablet |
|---------|-------|-------|-------|------|----------|--------|
| align | al | ... | ... | M3L | Modify | X14 |
| | | | | | ↳3D Operation | |
| | | | | | ↳Align | |

**Command:** align
**Select objects:** *(Select one or more objects.)*
**Select objects:** *(Press* **Enter***.)*
**Specify first source point:** *(Pick a point.)*
**Specify first destination point:** *(Pick a point.)*
**Specify second source point:** *(Pick a point.)*
**Specify second destination point:** *(Pick a point.)*
**Specify third source point or <continue>:** *(Pick a point.)*
**Specify third destination point:** *(Pick a point.)*

### COMMAND LINE OPTIONS

**First point** moves object in 2D or 3D when one source and destination point are picked.

**Second point** moves, rotates, and scales object in 2D or 3D when two source and destination points are picked.

**Third point**  moves, rotates, and scales object in 3D when three source and destination points are picked.

Continue option
**Scale objects based on alignment points? [Yes/No] <N>:** *(Type* **Y** *or* **N***.)*

Specifies that the distance between the first and second destination points is used for the reference length by which to scale the object.

### RELATED COMMANDS

**Mirror3d** mirrors objects in three dimensions.

**Rotate3d** rotates objects in three dimensions.

### TIPS

• Enter the first pair of points to define the move distance:

   **Specify first source point:** *(Pick a point.)*
   **Specify first destination point:** *(Pick a point.)*
   **Specify second source point:** *(Press* **Enter***.)*

• Enter two pairs of points to define a 2D (or 3D) transformation, scaling, and rotation:

| Points | Alignment Defined |
|--------|-------------------|
| **First** | Base point for alignment. |
| **Second** | Rotation angle. |
| **Third** | Scale based on distance between first and second destination points. |

• The third pair defines the 3D transformation.

# AmeConvert

Rel.13 Converts PADL solid models and regions created by AME v2.0 and v2.1 (*created in AutoCAD Releases 11 and 12*) to ShapeManager solid models.

| Command | Alias | Ctrl+ | F-key | Alt+ | Menu Bar | Tablet |
|---------|-------|-------|-------|------|----------|--------|
| ameconvert | ... | ... | ... | ... | ... | ... |

**Command:** ameconvert
**Select objects:** *(Select one or more objects.)*
**Processing Boolean operations.**

## COMMAND LINE OPTION

**Select objects** selects AME objects to convert; ignores non-AME objects, such as the ACIS solids produced by AutoCAD Release 13 through 2002.

## RELATED COMMAND

**AcisIn** imports ACIS models from an *.sat* file.

## TIPS

* After conversion, the AME model remains in the drawing in the same location as the solid model. Erase, if necessary.

* AME holes may become blind holes in the solid model.

* AME fillets and chamfers may be placed higher or lower in the solid model.

* Once the Release 12 PADL drawings is converted to an AutoCAD 2004 solid model, it cannot be converted back to PADL format.

* This command ignores objects that are neither AME solids or regions.

* Old AME models are stored in AutoCAD as anonymous block references.

## DEFINITIONS

*ACIS* — name of the solids modeling technology used by AutoCAD in Release 13 through 2002.

*AME* — short for "Advanced Modeling Extension," the name of the solids modeling module used by AutoCAD in Releases 10 through 12.

*PADL* — short for "Parts and Description Language," the solids modeling technology used in AutoCAD Releases 10 through 12.

*ShapeManager* — name of the solids modeling technology used by AutoCAD 2004.

# 'Aperture

**V. 1.3** Sets the size (in pixels) of the object snap target height, or box cursor.

| Command | Alias | Ctrl+ | F-key | Alt+ | Menu Bar | Tablet |
|---------|-------|-------|-------|------|----------|--------|
| 'aperture | ... | ... | ... | ... | ... | ... |

**Command:** aperture
**Object snap target height (1-50 pixels) <10>:** *(Enter a value.)*

*Aperture size = 1 (left), 10 (center), and 50 pixels (right).*

## COMMAND LINE OPTION
**Height** specifies the height of the object snap cursor's target.

## RELATED COMMANDS
**Options** allows you to set the aperture size interactively (Drafting tab).

## RELATED SYSTEM VARIABLE
**Aperture** contains the current target height, in pixels:

| Aperture | Meaning |
|----------|---------|
| 1 | Minimum size. |
| 10 | Default size, in pixels. |
| 50 | Maximum size. |

## TIPS
- The *box* cursor appears only during object snap selection; to change the size of the *pick* cursor, use the **Pickbox** command.

- Use the **Options** command to change the size of the aperture visually.

# AppLoad

**Rel.12** Creates a list of LISP, VBA, ObjectARx, and other applications to load into AutoCAD (*short for APPlication LOADer*).

| Command | Alias | Ctrl+ | F-key | Alt+ | Menu Bar | Tablet |
|---------|-------|-------|-------|------|----------|--------|
| appload | ap | ... | ... | TL | Tools | V10 |
| | | | | | ⍓Load Applications | |

**Command:** appload

*Displays dialog box:*

## DIALOG BOX OPTIONS

**Look in** lists the names of drives and folders available to this computer.

**File name** specifies the name of the file to load.

**Files of type** displays a list of file types:

| Filetype | Meaning |
|----------|---------|
| **ARX** | objectARX |
| **DVB** | Visual Basic for Applications (VBA) |
| **DBX** | objectDBX |
| **FAS** | FASt load autolisp |
| **LSP** | autoLiSP |
| **VLX** | Visual Lisp eXecutable |

**Load** loads all or selected files into AutoCAD.

**Loaded Applications** displays the names of applications already loaded into AutoCAD.

**History List** displays the names of applications previously saved to this list.

**Add to History** adds the file to the history tab.

**Unload** unloads all or selected files out of AutoCAD.

**Close** exits the dialog box.

Startup Suite options

**Contents** displays dialog box:

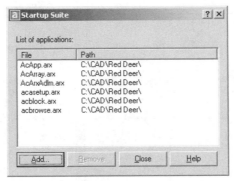

**List of applications** lists the file names and paths of applications automatically loaded each time AutoCAD starts.

**Add** displays the Add File to Startup Suite dialog box; allows you to select one or more application files.

**Remove** removes the application from the list.

**Close** returns to the Load/Unload Applications dialog box.

## RELATED COMMANDS

**Arx** lists ObjectARX programs currently loaded in AutoCAD.

**VbaLoad** loads VBA applications.

## RELATED AUTOLISP FUNCTIONS

**(load)** loads an AutoLISP program.

**(autoload)** predefines commands to load related AutoLISP programs.

## TIPS

• Use **AppLoad** when AutoCAD does not automatically load a command.

• ObjectARX, VBA, and DBX applications are loaded immediately; FAS, LSP, and VLX files are loaded after this dialog box closes.

• This command was a transparent command in earlier versions of AutoCAD.

• You can drag files from Windows Explorer into the **Loaded Applications** list.

• The *acad2004doc.lsp* file establishes autoloader and other utility functions, and is loaded automatically each time a drawing is opened; the *acad2004.lsp* file is loaded only once per AutoCAD session. Use the **AcadLspAsDoc** system variable to control whether these files are loaded with AutoCAD.

• To load LISP code into every drawing, add the code to *acad2004doc.lsp*.

 **Arc**

<u>**V. 1.0**</u>   Draws 2D arcs of less than 360 degrees, by eleven methods.

| Command | Alias | Ctrl+ | F-key | Alt+ | Menu Bar | Tablet |
|---------|-------|-------|-------|------|----------|--------|
| arc | a | ... | ... | DA | Draw | R10 |
| | | | | | ⏃Arc | |

**Command:** arc
**Specify start point of arc or [CEnter]:** *(Pick a point, or enter the **CE** option.)*
**Specify second point of arc or [CEnter/ENd]:** *(Pick a point, or enter an option.)*
**Specify end point of arc:** *(Pick a point.)*

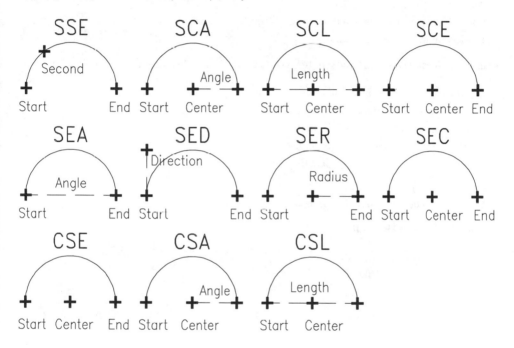

**COMMAND LINE OPTIONS**

*SSE (start, second, end) arc options*
    **Start point** indicates the start point of a three-point arc.
    **Second point** indicates a second point anywhere along the arc.
    **Endpoint** indicates the end point of the arc.

SCE (start, center, end), SCA (start, center, arc), and SCL (start, center, length) options
**Start point** indicates the start point of a two-point arc.

**Center** indicates the arc's center point.

**Angle** indicates the arc's included angle.

**Length of chord** indicates the length of the arc's chord.

**Endpoint** indicates the arc's endpoint.

SEA (start, end, angle), SED (start, end, direction), SER (start, end, radius), and SEC (start, end, center) options
**Start point** indicates the start point of a two-point arc.

**End** indicates the arc's end point.

**Center point** indicates the arc's center point.

**Angle** indicates the arc's included angle.

**Direction** indicates the tangent direction from the arc's start point.

**Radius** indicates the arc's radius.

CSE (center, start, end), CSA (center, start, angle), and CSL (center, start, length) options
**Center** indicates the center point of a two-point arc.

**Start point** indicates the arc's start point.

**Endpoint** indicates the arc's end point.

**Angle** indicates the arc's included angle.

**Length of chord** indicates the length of the arc's chord.

*Continued Arc option*
ENTER continues arc tangent from endpoint of last-drawn line or arc.

## RELATED TOOLBAR ICONS

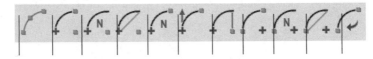

SSE    SCE    SCA    SCL    SEA    SED    SER    SEC    CSA    CSL    Continued

## RELATED COMMANDS
**Circle** draws an "arc" of 360 degrees.

**Ellipse** draws elliptical arcs.

**Polyline** draws connected polyline arcs.

**ViewRes** controls the roundness of arcs.

**RELATED SYSTEM VARIABLE**

**LastAngle** saves the included angle of the last-drawn arc (read-only).

**TIPS**

- To start an arc precisely tangent to the end point of the last line or arc, press ENTER at the 'Specify start point of arc or [CEnter]:' prompt.

- You can drag the arc only during the last-entered option.

- Specifying an x,y,z-coordinate as the starting point of the arc draws the arc at the z-elevation.

- In some cases, it may be easier to draw a circle, and then use **Break** and **Trim** to convert the circle into an arc.

- The components of an AutoCAD arc:

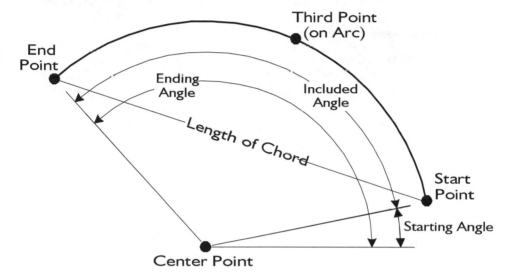

- When the chord length is positive, the minor arc is drawn counterclockwise from the start point; when negative, the major arc is drawn counterclockwise.

 # Area

**V. 1.0**  Calculates the area and perimeter of areas, closed objects, and polylines.

| Command | Alias | Ctrl+ | F-key | Alt+ | Menu Bar | Tablet |
|---------|-------|-------|-------|------|----------|--------|
| area | aa | ... | ... | TQA | Tools | T7 |
| | | | | | ⮡ Inquiry | |
| | | | | | ⮡ Area | |

**Command:** area
**Specify first corner point or [Object/Add/Subtract]:** *(Pick a point, or enter an option.)*
**Specify next corner point or press ENTER for total:** *(Pick a point, or press* **Enter***.)*
   *Sample response*
**Area = 1.8398, Perimeter = 6.5245**

## COMMAND LINE OPTIONS

**First point** indicates the first point to begin measurement.

**Object** indicates the object to be measured.

**Add** switches to add-area mode.

**Subtract** switches to subtract-area mode.

ENTER indicates the end of the area outline.

## RELATED COMMANDS

**Dist** returns the distance between two points.

**Id** lists the x,y,z coordinates of a selected point.

**MassProp** returns surface area, and so on of solid models.

## RELATED SYSTEM VARIABLES

**Area** contains the most recently-calculated area.

**Perimeter** contains the most recently-calculated perimeter.

## TIPS

• At least three points must be picked to calculate an area; AutoCAD "closes the polygon" automatically before measuring the area.

• You can specify 2D x,y coordinates or 3D x,y,z coordinates.

• The **Object** option returns the following information:

| Object | Measurement Returned |
|--------|----------------------|
| **Circle, ellipse** | Area and circumference. |
| **Planar closed spline** | Area and circumference. |
| **Closed polyline, polygon** | Area and perimeter. |
| **Open objects** | Area and length. |
| **Region** | Area of all objects in region. |
| **2D solid** | Area. |

• Areas of wide polylines are measured along center lines; closed polylines must have one closed area only.

• This command does not measure the area of solid objects; use the **MassProp** command.

# Array

**V. 1.3**   Creates 2D linear, rectangular, and polar arrays of objects.

| Commands | Aliases | Ctrl+ | F-key | Alt+ | Menu Bar | Tablet |
|----------|---------|-------|-------|------|----------|--------|
| array | ar | ... | ... | MA | Modify<br>⤷Array | V18 |
| -array | -ar | | | | | |

**Command:** array

*Displays dialog box:*

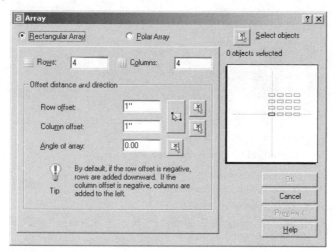

## DIALOG BOX OPTIONS

**Rectangular Array** displays the options for creating a rectangular array.

**Polar Array** displays the options for creating a polar array.

**Select objects** temporarily dismisses the dialog box, so that you can select the objects in the drawing:

> **Select objects:** *(Select one or more objects.)*
> **Select objects:** *(Press Enter.)*
> **Press Enter, or right-click to return to the dialog box.**

**Preview** temporarily dismisses the dialog box so that you can see what the array will look like.

### Rectangular Array options

**Rows** specifies the number of rows; minimum=1, maximum=100,000.

**Columns** specifies the number of columns; minimum=1, maximum=100,000.

**Row offset** specifies the distance between rows; use negative numbers to draw rows in the negative x-direction (to the right).

**Column offset** specifies the distance between columns; use negative numbers to draw rows in the negative y-direction (downward).

**Angle of array** specifies the angle of the array, which "tilts" the array.

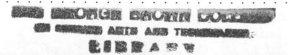

## Polar Array options

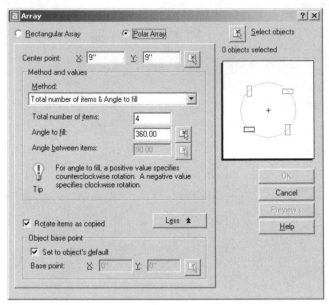

**Center point** specifies the center of the polar array.

**Method** specifies the method by which the array is constructed:

- Total number of items and angle to fill.
- Total number of items and angle between items.
- Angle to fill and angle between items.

**Total number of items** specifies the number of objects in the array; minimum=2.

**Angle to fill** specifies the angle of "arc" to construct the array; min=1 deg; max=360 deg.

**Angle between items** specifies the angle between each object in the array.

**Rotate items as copied**

☑ objects are rotated so that they face the center of the array.

☐ objects are not rotated.

**More** displays additional options for constructing a polar array.

### More options
**Set to the object's default:**

☑ uses the default base point of the object, as follows:

| Object | Default base point |
|---|---|
| Arc, circle, ellipse | Center point |
| Polygon, rectangle | First vertex |
| Line, polyline, 3D polyline, ray, spline | Start point |
| Donut | Start point |
| Block, mtext, text | Insertion point |
| Xline | Midpoint |
| Region | Grip point |

☐ allows you to specify the base point.

## -ARRAY Command

**Command:** -array
**Select objects:** *(Select one or more objects.)*
**Select objects:** *(Press* **Enter**.*)*
**Enter the type of array [Rectangular/Polar] <R>:** *(Type* **R** *or* **P**.*)*

Rectangular options
**Enter the number of rows (---) <1>:** *(Enter a value, or press* **Enter**.*)*
**Enter the number of columns (||||) <1>:** *(Enter a value, or press* **Enter**.*)*
**Enter the distance between rows or specify unit cell (---):** *(Enter a value.)*
**Specify the distance between columns (||||):** *(Enter a value.)*

Polar options
**Specify center point of array:** *(Select one or more objects.)*
**Enter the number of items in the array:** *(Enter a value.)*
**Specify the angle to fill (+=ccw, -=cw) <360>:** *(Enter a value, or press* **Enter**.*)*
**Rotate arrayed objects? [Yes/No] <Y>:** *(Enter* **Y** *or* **N**.*)*

### COMMAND LINE OPTIONS

**R** creates a rectangular array of the selected object.

**P** creates a polar array of the selected object.

**Center point** specifies the center point of the array.

**Rows** specifies the number of horizontal rows.

**Columns** specifies the number of vertical columns.

**Unit cell** specifies the vertical and horizontal spacing between objects.

### RELATED COMMANDS

**3dArray** creates a rectangular or polar array in 3D space.

**Copy** creates one or more copies of the selected object.

**MInsert** creates a rectangular block array of blocks.

### RELATED SYSTEM VARIABLE

**SnapAng** determines the angle of a rectangular array.

### TIPS

- To create a rectangular array at an angle, use the **Rotation** option of the **Snap** command.

- Rectangular arrays are drawn upward in the positive x-direction, and rightward in the positive y-direction; to draw the array in the opposite directions, specify negative row and column distances.

- Polar arrays are drawn counterclockwise; to draw the array clockwise, specify a negative angle.

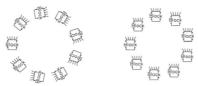

*Nine-item polar arrays — rotated (left) and unrotated (right).*

- For linear arrays, enter 1 for the number of rows or columns, or use **Divide** or **Measure**.

# Arx

<u>Rel.13</u> Displays information regarding currently loaded ObjectARX programs.

| Command | Alias | Ctrl+ | F-key | Alt+ | Menu Bar | Tablet |
|---------|-------|-------|-------|------|----------|--------|
| arx | ... | ... | ... | ... | ... | ... |

**Command:** arx
**Enter an option [?/Load/Unload/Commands/Options]:** *(Enter an option.)*

## COMMAND LINE OPTIONS

**?** lists the names of currently loaded ObjectARX programs.

**Load** loads the ObjectARX program into AutoCAD.

**Unload** unloads the ObjectARX program out of memory.

**Commands** lists the names of command associated with each ObjectARX program.

Options options

**CLasses** lists the class hierarchy for ObjectARX objects.

**Groups** lists the names of objects entered into the "system registry."

**Services** lists names of services entered in the ObjectARX "service dictionary."

## RELATED COMMAND

**AppLoad** loads LISP, VBA, ObjectDBX, and ObjectARX programs via a dialog box.

## RELATED AUTOLISP FUNCTIONS

**(arx)** lists currently loaded ObjectARX programs.

**(arxload)** loads an ObjectARX application.

**(autoarxload)** predefines commands that load the ObjectARX program.

**(arxunload)** unloads an ObjectARX application.

## RELATED FILE

***.arx** are objectARX program files.

## TIPS

• Use the **Load** option to load external commands that do not seem to work.

• Use the **Unload** option of the **Arx** command to remove ObjectARX programs from AutoCAD to free up memory.

. . . . . . . . . . . . . . . . . . . . . . . . . . . . . . . . . . . . . . . . . . . . . . . . . . .

## Removed Commands

The following ASE (AutoCAD SQL Extension) commands were removed from AutoCAD 2000: **AseAdmin**, **AseExport**, **AseLinks**, **AseRows**, **AseSelect**, and **AseSqlEd**. They were replaced by **DbConnect**.

. . . . . . . . . . . . . . . . . . . . . . . . . . . . . . . . . . . . . . . . . . . . . . . . . . .

**'Assist**

<u>2000i</u> Opens the Active Assistance window, which provides real-time assistance.

| Command | Alias | Ctrl+ | F-key | Alt+ | Menu Bar | Tablet |
|---------|-------|-------|-------|------|----------|--------|
| 'assist | ... | ... | ... | HA | Help<br>↳Active Assistance | ... |

**Command:** assist

*Displays window:*

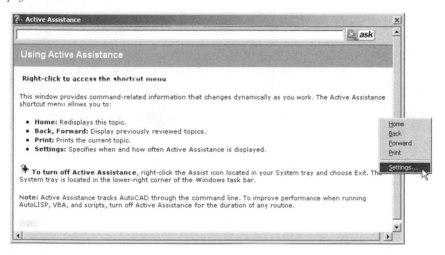

## WINDOW OPTIONS

**Ask** displays the Ask Me tab of AutoCAD's online help.

**x** closes the window.

## SHORTCUT MENU OPTIONS

**Home** displays the "home" page.

**Back** displays the previous topic.

**Forward** displays the next topic, if the Back option has been used.

**Print** prints the topic.

**Settings** displays the Active Assistance Settings dialog box.

## Active Assistance Settings dialog box

**Show on start** launches Active Assistance automatically when AutoCAD starts.

**Activation** determines when Active Assistance displays its help topics:

- **All commands** displays assistance for all commands.
- **New and enhanced commands** displays assistance for commands new to AutoCAD and for commands changed since the previous major release.
- **Dialogs only** displays assistance for dialog boxes only.
- **On demand** displays assistance only when you click the icon on the Windows taskbar, or enter the Assist command.

### RELATED COMMAND

**Help** displays online help in a window.

### TIPS

- To access Active Assistance on-demand, double-click the icon found on the Windows taskbar, or right-click and select an option from the shortcut menu:

- Not all commands are supported by Active Assistance. In those cases, the following error message is displayed:

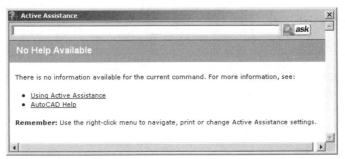

- To prevent Active Assistance from appearing, turn off the **Show on Start** option in the **Active Assistance Settings** dialog box.

# AttachURL

Rel.14 Attaches hyperlinks to objects and areas (*undocumented command*).

| Command | Alias | Ctrl+ | F-key | Alt+ | Menu Bar | Tablet |
|---------|-------|-------|-------|------|----------|--------|
| attachurl | ... | ... | ... | ... | ... | ... |

**Command:** attachurl
**Enter hyperlink insert option [Area/Object] <Object>:** *(Type **A** or **O**.)*
**Select objects:** *(Select one or more objects.)*
**Select objects:** *(Press **Enter**.)*
**Enter hyperlink <current drawing>:** *(Enter an address.)*

## COMMAND LINE OPTIONS

**Area** creates a 2D hyperlink by specifying two corners of a rectangle.

**Object** creates a 1D hyperlink by selecting one or more objects.

**Enter hyperlink** allows you to enter a valid hyperlink.

Area options
**First corner** picks the first corner of rectangle.

**Other corner** picks the second corner of rectangle.

## RELATED COMMANDS

**Hyperlink** displays a dialog box for adding a hyperlink to an object.

**SelectUrl** selects all objects with attached hyperlinks.

## TIPS

* The hyperlinks placed in the drawing can link to *any* other file: another AutoCAD drawing, an office document, or a file located on the Internet.

* Autodesk recommends that you use the following URL (uniform resource locator) formats:

| File Location | Example URL |
|---------------|-------------|
| **Web Site** | **http://**/*servername*/*pathname*/*filename*.**dwg** |
| **FTP Site** | **ftp://**/*servername*/*pathname*/*filename*.**dwg** |
| **Local File** | **file:///***drive:*/*pathname*/*filename*.**dwg** |
| *or* | **file:////***localPC*/*pathname*/*filename*.**dwg** |
| **Network File** | **file://***localhost*/*drive:*/*pathname*/*filename*.**dwg** |

* The URL (hyperlink) is stored as follows:

| Attachment | URL |
|------------|-----|
| **One object** | Stored as xdata (extended entity data). |
| **Multiple objects** | Stored as xdata in each object. |
| **Area** | Stored as xdata in a rectangle object on layer URLLAYER. |

* The **Area** option creates a layer named URLLAYER with the default color of red, and places a rectangle object on that layer; do not delete layer URLLAYER.

 # AttDef

**V. 2.0** Defines attribute modes and prompts (*short for ATTribute DEFinition*).

| Commands | Aliases | Ctrl+ | F-key | Alt+ | Menu Bar | Tablet |
|----------|---------|-------|-------|------|----------|--------|
| attdef | att | ... | ... | DKD | Draw | ... |
| | ddattdef | | | | ⬐Block | |
| | | | | | ⬐Define Attributes | |
| -attdef | -att | | | | | |

**Command:** attdef

*Displays dialog box:*

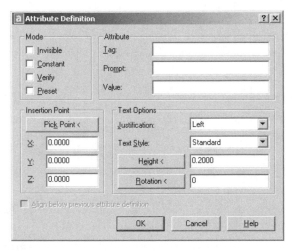

## DIALOG BOX OPTIONS

Mode options

**Invisible** makes the attribute text invisible.

**Constant** uses constant values for the attributes.

**Verify** verifies the text after input.

**Preset** presets the variable attribute text.

Attribute options

**Tag** identifies the attribute.

**Prompt** prompts the user for input.

**Value** sets the default value for the attribute.

Insertion Point options

**Pick point** picks insertion point with cursor.

**X** specifies the x coordinate insertion point.

**Y** specifies the y coordinate insertion point.

**Z** specifies the z coordinate insertion point.

Text options

Justification sets the text justification.

Text style selects a text style.

Height specifies the height.

Rotation sets the rotation angle.

*Additional option*

Align below previous attribute definition automatically places the text below the previous attribute.

. . . . . . . . . . . . . . . . . . . . . . . . . . . . . . . . . . . . . . . . . . . . . . . . . .

## -ATTDEF Command
**Command:** -attdef
**Current attribute modes: Invisible=N Constant=N Verify=N  Preset=N**
**Enter an option to change [Invisible/Constant/Verify/Preset] <done>:** *(Enter an option.)*
**Enter attribute tag:** *(Enter text, and then press* **Enter.***)*
**Enter attribute prompt:** *(Enter text, and then press* **Enter.***)*
**Enter default attribute value:** *(Enter text, and then press* **Enter.***)*
**Specify start point of text or [Justify/Style]:** *(Pick a point, or enter an option.)*
**Specify height <0.200>:** *(Enter a value.)*
**Specify rotation angle of text <0>:** *(Enter a value.)*

## COMMAND LINE OPTIONS

Attribute mode selects the mode(s) for the attribute:

- I toggles visibility of attribute text in drawing (short for Invisible).
- C toggles fixed or variable value of attribute (short for Constant).
- V toggles confirmation prompt during input (short for Verify).
- P toggles automatic insertion of default values (short for Preset).

Start point indicates the start point of the attribute text.

Justify selects the justification mode for the attribute text.

Style selects the text style for the attribute text.

Height specifies the height of the attribute text; not displayed if the style specifies a height other than 0.

Rotation angle specifies the angle of the attribute text.

## RELATED COMMANDS

AttDisp controls the visibility of attributes.

EAttEdit edits the values of attributes.

EAttExt extracts attributes to disk.

AttRedef redefines an attribute or block.

Block binds attributes to objects.

Insert inserts a block and prompts for attribute values.

. . . . . . . . . . . . . . . . . . . . . . . . . . . . . . . . . . . . . . . . . . . . . . . . . .

## RELATED SYSTEM VARIABLES

**AFlags** holds the value of modes in bit form:

| AFlags | Meaning |
|--------|---------|
| 0 | No attribute mode selected |
| 1 | Invisible |
| 2 | Constant |
| 4 | Verify |
| 8 | Preset |

**AttDia** toggles the use of dialog box during the Insert command:

| AttDia | Meaning |
|--------|---------|
| 0 | Uses command-line prompts |
| 1 | Uses dialog box |

**AttReq** toggles use of defaults or user prompts during the Insert command:

| AttReq | Meaning |
|--------|---------|
| 0 | Assumes default values of all attributes |
| 1 | Prompts for attributes |

## TIPS

• Constant attributes cannot be edited.

• Attribute tags cannot be null (have no value); attribute values may be null.

• You can enter any characters for the attribute tag, except a space or an exclamation mark. All characters are converted to uppercase.

• When you press **Enter** at 'Attribute Prompt,' AutoCAD uses the attribute *tag* as the prompt.

• When you press **Enter** at the 'Starting point:' prompt, **AttDef** automatically places the next attribute below the previous one.

*Block with attribute **value** (left) and attribute **tags** (right).*

• 'Attribute Prompt' and 'Default Attribute Value' are not displayed when constant mode is turned on. Instead, AutoCAD prompts 'Attribute Value.'

# 'AttDisp

**V. 2.0**   Controls the display of all attributes in the drawing (*short for ATTribute DISPlay*).

| Command | Alias | Ctrl+ | F-key | Alt+ | Menu Bar | Tablet |
|---------|-------|-------|-------|------|----------|--------|
| 'attdisp | ... | ... | ... | VLA | View | L1 |
| | | | | | ⬐Display | |
| | | | | | ⬐Attribute Display | |

**Command:** attdisp
**Enter attribute visibility setting [Normal/ON/OFF] <Normal>:** *(Enter an option.)*
**Regenerating drawing.**

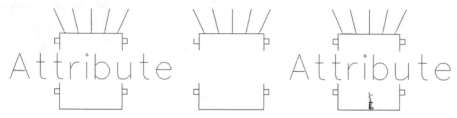

*Attribute display **Normal** (left), **Off** (center), and **On** (right).*

## COMMAND LINE OPTIONS

**Normal** displays attributes according to AttDef setting.

**ON** displays all attributes, regardless of AttDef setting.

**OFF** displays no attributes, regardless of AttDef setting.

## RELATED COMMAND

**AttDef** defines new attributes, including their default visibility.

## RELATED SYSTEM VARIABLE

**AttMode** holds the current setting of **AttDisp**:

| AttMode | Meaning |
|---------|---------|
| 0 | Off: no attributes are displayed. |
| 1 | Normal: invisible attributes are not displayed. |
| 2 | On: all attributes are displayed. |

## TIPS

• If necessary, use **Regen** after **AttDisp** to see change to attribute display.

• When you define invisible attributes, use **AttDisp** to view them.

• Use **AttDisp** to turn off the display of attributes, which increases display speed and reduces drawing clutter.

 **AttEdit**

**V. 2.0** Edits attributes in drawings (*short for ATTribute EDIT*).

| Commands | Aliases | Ctrl+ | F-key | Alt+ | Menu Bar | Tablet |
|----------|---------|-------|-------|------|----------|--------|
| attedit | ate | ... | ... | MOAS | Modify | Y20 |
| | ddatte | | | | ⬐Object | |
| | atte | | | | ⬐Attribute | |
| | | | | | ⬐Single | |
| -attedit | -ate | | | MOAG | Modify | |
| | | | | | ⬐Object | |
| | | | | | ⬐Attribute | |
| | | | | | ⬐Global | |

**Command:** attedit
**Select block reference:** *(Pick a block.)*

*Displays dialog box:*

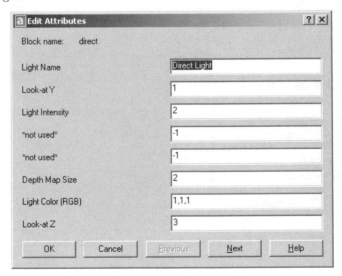

**DIALOG BOX OPTIONS**

**Block Name** names the selected block.

*Attribute-specific prompts* allow you to change attribute values.

*Buttons*

**OK** accepts the changes and closes the dialog box.

**Cancel** discards the changes and closes the dialog box.

**Previous** displays the previous list of attributes, if any.

**Next** displays the next list of attributes, if any.

## -ATTEDIT Command
**Command:** -attedit

One-at-time attribute editing options
**Edit attributes one at a time? [Yes/No] <Y>:** *(Enter Y.)*
**Enter block name specification <*>:** *(Press Enter to edit all.)*
**Enter attribute tag specification <*>:** *(Press Enter to edit all.)*
**Enter attribute value specification <*>:** *(Press Enter to edit all.)*
**Select Attributes:** *(Select one or more attributes.)*
**Select Attributes:** *(Press Enter.)*
**Enter an option [Value/Position/Height/Angle/Style/Layer/Color/Next] <N>:**
*(Enter an option, and then press Enter.)*

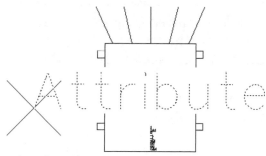

*During single attribute editing, **AttEdit** marks the current attribute with an 'X.'*

Global attribute editing options
**Edit attributes one at a time? [Yes/No] <Y>:** *(Enter N.)*
**Performing global editing of attribute values.**

**Edit only attributes visible on screen? [Yes/No] <Y>:** *(Press Enter.)*
**Enter block name specification <*>:** *(Press Enter.)*
**Enter attribute tag specification <*>:** *(Press Enter.)*
**Enter attribute value specification <*>:** *(Press Enter.)*
**Select Attributes:** *(Select one or more attributes.)*
**Select Attributes:** *(Press Enter.)*

**Enter string to change:** *(Enter existing string, and then press Enter.)*
**Enter new string:** *(Enter new string, and then press Enter.)*

### COMMAND LINE OPTIONS
**Value** changes or replaces the value of the attribute.

**Position** moves the text insertion point of the attribute.

**Height** changes the attribute text height.

**Angle** changes the attribute text angle.

**Style** changes the text style of the attribute text.

**Layer** moves the attribute to a different layer.

**Color** changes the color of the attribute text.

**Next** edits the next attribute.

## RELATED SYSTEM VARIABLE

**AttDia** toggles use of AttEdit during Insert command.

## RELATED COMMANDS

**AttDef** defines an attribute's original value and parameter.

**AttDisp** toggles an attribute's visibility.

**AttRedef** redefines attributes and blocks.

**Explode** reduces an attribute to its tag.

## TIPS

- Constant attributes cannot be edited with **AttEdit**.

- You can only edit attributes parallel to the current UCS.

- Unlike other text input to AutoCAD, attribute values are case-sensitive.

- To edit null attribute values, use **-AttEdit**'s global edit option, and enter \ (backslash) at the 'Enter attribute value specification' prompt.

- The wildcard characters **?** and **\*** are interpreted literally at the 'Enter string to change' and 'Enter new string' prompts.

  - To edit the different parts of an attribute, use the following commands:

| Command | Edit Attribute |
|---------|----------------|
| **Attedit** | Non-constant attribute *values* in one block. |
| **-AttEdit** | Attribute *values* and *properties* (such as position, height, and style) in one block or in all attributes. |

- When selecting attributes for global editing, you may pick the attributes, or use the following selection modes: **Window**, **Last**, **Crossing**, **BOX**, **Fence**, **WPolygon**, and **CPolygon**.

# AttExt

**V. 2.0** Extracts attribute data from drawings to files on disk (*short for ATTribute EXTract*).

| Commands | Alias | Ctrl+ | F-key | Alt+ | Menu Bar | Tablet |
|----------|-------|-------|-------|------|----------|--------|
| attext | ddattext | ... | ... | FE | File | ... |
| | | | | ⍓DXX | ⍓Export | |
| | | | | | ⍓DXX Extract | |
| -attext | | | | | | |

**Command:** attext

*Displays dialog box:*

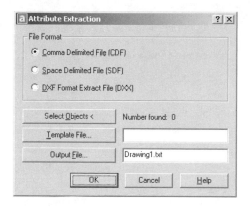

## DIALOG BOX OPTIONS

File Format options
- **Comma Delimited File (CDF)** creates a CDF text file, where commas separate fields.
- **Space Delimited File (SDF)** creates an SDF text file, where spaces separate fields.
- **DXF Format Extract File (DXX)** creates an ASCII DXF-format file.

*Additional options*

**Select Objects** returns to the graphics screen to select attributes for export.

**Template File** specifies the name of the TXT template file for CDF and SDF files.

**Output File** specifies the name of the attribute output file, *.txt* for CDF and SDF formats, or *.dxx* for DXF format.

# -ATTEXT Command

**Command:** -attext

**Enter extraction type or enable object selection [Cdf/Sdf/Dxf/Objects] <C>:**
*(Enter an option.)*

*Displays the* ***Select Template File*** *dialog box; select the template file,*

*Displays the* ***Create Extract File*** *dialog box.*

## COMMAND LINE OPTIONS

**Cdf** outputs attributes in comma-delimited format.

**Sdf** outputs attributes in space-delimited format.

**Dxf** outputs attributes in DXF format.

**Objects** selects objects from which to extract attributes.

## RELATED COMMAND

**AttDef** defines attributes.

## RELATED FILES

***\*.txt*** required extension for template file; extension for CDF and SDF files.

***\*.dxx*** extension for DXF extraction files.

## TIPS

- **CDF** is short for "Comma Delimited File"; it has one record for each block reference; a comma separates each field; single quotation marks delimit text strings.

- **SDF** is short for "Space Delimited File"; it has one record for each block reference; fields have fixed width padded with spaces; string delimiters are not used.

- **DXF** is short for "Drawing Interchange File"; it contains only block reference, attribute, and end-of-sequence DXF objects; no template file is required.

- CDF files use the following conventions:

  Specified field widths are the maximum width.

  Positive number fields have a leading blank.

  Character fields are enclosed in ' ' (single quotation marks).

  Trailing blanks are deleted.

  Null strings are " (two single quotation marks).

  Uses spaces; do not uses tabs.

  Use the C:DELIM and C:QUOTE records to change the field and string delimiters to another character.

- To output the attributes to the printer, specify:

  | Logical Filename | Meaning |
  |---|---|
  | **CON** | Displays on text screen. |
  | **PRN** *or* **LPT1** | Prints to parallel port 1. |
  | **LPT2** *or* **LPT3** | Prints to parallel ports 2 or 3. |

- Before you can specify the SDF or CDF option, you must create a template file.

# 'AttRedef

Rel.13 Redefines blocks and attributes (*short for ATTribute REDEFinition*).

| Command | Alias | Ctrl+ | F-key | Alt+ | Menu Bar | Tablet |
|---------|-------|-------|-------|------|----------|--------|
| 'attredef | ... | ... | ... | ... | ... | ... |

**Command:** redefine
**Name of Block you wish to redefine:** *(Enter name of block.)*
**Select objects for new Block...**
**Select objects:** *(Select one or more objects.)*
**Select objects:** *(Press **Enter**.)*
**Insertion base point of new block:** *(Pick a point.)*

## COMMAND LINE OPTIONS

**Name of Block you wish to redefine** specifies the name of the block to be redefined.

**Select objects** selects objects for the new block.

**Insertion base point of new block** picks the new insertion point.

## RELATED COMMANDS

**AttDef** defines an attribute's original value and parameter.

**AttDisp** toggles an attribute's visibility.

**EAttEdit** edits the attribute's values.

**Explode** reduces an attribute to its tag.

## TIPS

* Existing attributes retain their values.

* Existing attributes not included in the new block are erased.

* New attributes added to an existing block take on default values.

# AttSync

**2002** Updates blocks with new attribute definitions *(short for ATTribute SYNChronization).*

| Command | Alias | Ctrl+ | F-key | Alt+ | Menu Bar | Tablet |
|---------|-------|-------|-------|------|----------|--------|
| attsync | ... | ... | ... | ... | ... | ... |

**Command:** attsync
**Enter an option [?/Name/Select] <Select>:** *(Specify an option.)*
**Select a block:** *(Pick a block.)*
**ATTSYNC block name? [Yes/No] <Yes>:** *(Enter* **Y** *or* **N***.)*

### COMMAND LINE OPTIONS
**?** lists the names of all blocks in the drawing.

**Name** enters the name of the block.

**Select** selects a single block with the cursor.

### RELATED COMMANDS
**AttDef** defines an attribute.

**BattMan** edits the attributes in a block definition.

**EAttEdit** edits the attributes in block references.

### TIPS
- This command is used together with other attributed-related commands, in the following order:

  1. **AttDef** and **Block** defines the attributes, and attaches them to the block.
  2. **Insert** inserts the block, and gives values to the attributes.
  3. **EAttEdit** changes the attribute values in a specific block.
  4. **AttSync** reverts the attribute values to their original.
  5. **BattMan** changes the attributes in the original block definition.
  6. **AttSync** updates the attributes to the new definition.

- This command does not operate if the drawing lacks blocks with attributes. AutoCAD complains, "This drawing contains no attributed blocks."

# Audit

<u>Rel.11</u>  Examines drawing files for structural errors.

| Command | Alias | Ctrl+ | F-key | Alt+ | Menu Bar | Tablet |
|---------|-------|-------|-------|------|----------|--------|
| audit | ... | ... | ... | FUA | File | ... |
| | | | | | ↳Drawing Utilities | |
| | | | | | ↳Audit | |

**Command:** audit
**Fix any errors detected? [Yes/No] <N>:** *(Type Y or N.)*

*Sample output*

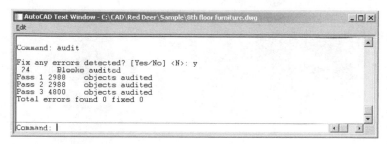

## COMMAND LINE OPTIONS

**N** reports errors found; does not fix errors.

**Y** reports and fixes errors found in the drawing file.

## RELATED COMMANDS

**Save** saves a recovered drawing to disk.

**Recover** recovers a damaged drawing file.

## RELATED SYSTEM VARIABLE

**AuditCtl** controls the creation of the *.adt* audit log file:

| AuditCtl | Meaning |
|----------|---------|
| 0 | No log file is written. |
| 1 | ADT log file is written to drawing directory. |

## RELATED FILE

**\*.adt** is the audit log file, which records the auditing process.

## TIPS

- The **Audit** command is a diagnostic tool for validating and repairing the contents of *.dwg* files.

- Objects with errors are placed in the **Previous** selection set. Use an editing command, such as **Copy**, to view the objects.

- If **Audit** cannot fix a drawing file, use the **Recover** command.

**Background**

Rel.14 Places a solid color, linear gradient, raster image, or the current view, in the background of renderings.

| Command | Alias | Ctrl+ | F-key | Alt+ | Menu Bar | Tablet |
|---|---|---|---|---|---|---|
| background | ... | ... | ... | VEB | View | Q2 |
| | | | | | ⮑Render | |
| | | | | | ⮑Background | |

**Command:** background

*Displays dialog box:*

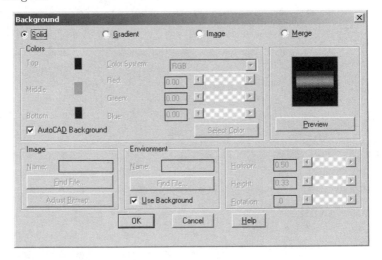

### DIALOG BOX OPTIONS

Environment options

**Environment** allows reflection and refraction effects on objects: mirror effect with the Photo Real renderer; or raytracing with the Photo Raytrace renderer.

**Use Background** specifies that objects reflect the background, whether a color, gradient, image, or merged image.

**Solid** view

**Colors** specifies a color from the RGB (red, green, blue) or HLS (hue, lightness, saturation) slider bars.

**Select Custom Color** displays the Windows Color dialog box.

**AutoCAD Background** sets the current AutoCAD background color (default = white).

## Gradient view

**Top** specifies the top color for two- and three-color gradients.

**Middle** specifies the middle color for three-color gradients.

**Bottom** specifies the bottom color for two- and three-color gradients.

**Horizon** determines the center of the gradient as a percent of the viewport's height.

**Height** determines the start of the second color of a three-color gradient; automatically set to 0 for two-color gradients.

**Rotation** rotates the angle of the gradient.

## Image view

**Image Name** specifies the name of the raster file to use as the background image.

**Find File** displays the file dialog box; allows selection of a *.bmp* (Windows bitmap), *.gif* (GIF), *.jpg* (JPEG), *.pcx* (PC Paintbrush), *.tga* (Targa), or *.tif* (TIFF) file.

**Adjust Bitmap** adjusts the position of a raster image; displays the Adjust Background Bitmap Placement dialog box.

## Merge view

*None. Displays the AutoCAD drawing as the background image.*

## Adjust Background Bitmap Placement dialog box

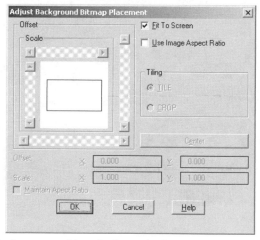

**Offset** uses the slider bars to position the image in the viewport.

**Fit to Screen** stretches the image to fit the viewport.

**Use Image Aspect Ratio** ensures the image is not distorted.

**Tiling** adjusts the size of the image when the image does not fit the viewport:

- **Tile** repeats the image.
- **Crop** cuts off the edges from the image to make it smaller.

**Center** centers the image in the viewport.

**X,Y Offset** changes the position of the image.

**X,Y scale** changes the size of the image.

## RELATED COMMANDS

**Fog** creates a fog-like effect.

**ImageAttach** loads a raster image as an xref file.

**Render** renders 3D objects in the drawing.

**Replay** displays a raster image in the current viewport.

## TIPS

- Gradient backgrounds are useful for simulating a sunset (cyan-pink-orange) or underwater view (green-blue-black).

- To create a 2-color gradient, set **Height** to 0.

- Image backgrounds are useful for placing the 3D rendered model in its environment, such as a rendered house on its building site.

- AutoCAD includes some background images in the \\*acad 2004*\\*textures* folder.

- The four types of background:

*Top:* **Solid** *option displays a uniform color (left);* **Gradient** *option displays a 2- or 3-color linear gradient (right).*

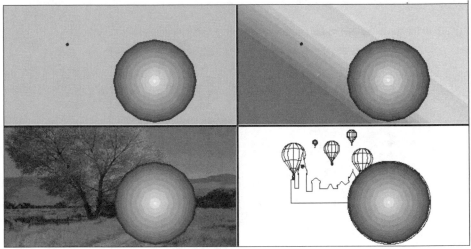

*Bottom:* **Image** *option displays a raster image (left);* **Merge** *option displays the current AutoCAD viewport (right).*

# 'Base

**V. 1.0** Changes the 2D or 3D insertion point of the current drawing, located by default at (0,0,0).

| Command | Alias | Ctrl+ | F-key | Alt+ | Menu Bar | Tablet |
|---------|-------|-------|-------|------|----------|--------|
| 'base | ... | ... | ... | DKB | Draw | ... |
| | | | | | ⇘Block | |
| | | | | | ⇘Base | |

**Command:** base
**Enter base point <0.0,0.0,0.0>:** *(Pick a point.)*

### COMMAND LINE OPTION
**Enter base point** specifies the x,y,z coordinates of the new insertion point.

### RELATED COMMANDS
**Block** specifies the insertion point of a new block.

**Insert** inserts another drawing into the current drawing.

**Xref** references another drawing.

### RELATED SYSTEM VARIABLE
**InsBase** contains the current setting of the drawing insertion point.

### TIPS
- Use this command to shift the insertion point of the current drawing.

- This command does not affect the current drawing. Instead, it comes into effect when you insert it or xref it into another drawing.

# BAttMan

<u>**2002**</u>  Edits all aspects of attributes in blocks; works with one block at a time
(*short for Block ATTribute MANager*).

| Command | Alias | Ctrl+ | F-key | Alt+ | Menu Bar | Tablet |
|---------|-------|-------|-------|------|----------|--------|
| battman | ... | ... | ... | MOAB | Modify | ... |
| | | | | | ⮑Object | |
| | | | | | ⮑Attribute | |
| | | | | | ⮑Block Attribute Manager | |

**Command:** battman

*When the drawing contains no blocks with attributes, displays error message, "This drawing contains no attributed blocks."*

*When drawing contains at least one block with attributes, displays dialog box:*

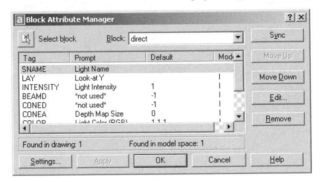

## DIALOG BOX OPTIONS

**Select block** dismisses the dialog box, and then prompts, "Select a block: ."

**Block** lists the names of blocks in the drawing, and displays the name of the selected block.

**Sync** changes the attributes in block insertions to match the changes made here.

**Move Up** moves the attribute tag up the list; constant attributes cannot be moved.

**Move Down** moves the attribute tag down the list.

**Edit** displays the Edit Attribute dialog box; see the EAttEdit command.

**Remove** removes the attribute tag and related data from the block; it does not operate when the block contains a single attribute.

**Settings** displays the Settings dialog box.

**Apply** applies the changes to the block definition.

**Settings** dialog box

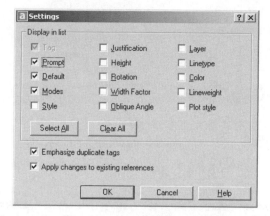

Display In List options

**Prompt** toggles (turns on and off) display of the column of attribute prompts.

**Default** toggles the display of the attribute's default value.

**Modes** toggles the display of the attribute's modes: invisible, constant, verify, and/or preset.

**Style** toggles the display of the attribute's text style name.

**Justification** toggles the display of the attribute's text justification.

**Height** toggles the display of the attribute's text height.

**Rotation** toggles the display of the attribute's text rotation angle.

**Width Factor** toggles the display of the attribute's text width factor.

**Oblique Angle** toggles the display of the attribute's text obliquing angle (slant).

**Layer** toggles the display of the attribute's layer.

**Linetype** toggles the display of the attribute's linetype.

**Color** toggles the display of the attribute's color.

**Lineweight** toggles the display of the attribute's lineweight.

**Plot style** toggles the display of the attribute's plot style name (available only when plot styles are turned on).

**Select All** selects all display options.

**Clear All** clears all display options, except tag name.

Additional options

**Emphasize duplicate tags:**

☑ highlights duplicate attribute tags in red.

☐ does not highlight duplicate tags.

**Apply changes to existing references:**

☑ applies changes to all block instances that reference this definition in the drawing.

☐ applies the new attribute definitions only to newly inserted-blocks.

## RELATED COMMANDS

**AttDef** defines attributes.

**Block** binds attributes to a symbol.

**Insert** inserts a block, and then allows you to specify the attribute data.

## TIPS

- Use this command to edit and remove attribute definitions, as well as to change the order in which attributes appear.

- The **Sync** option does not change the values you assigned to attributes.

- When an attribute has a mode of Constant, it cannot be moved up or down the list.

- Turning on all the options displays a lot of data. To see all the data columns, you can stretch the dialog box.

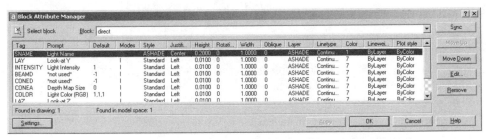

*With the cursor, grab the edge of the dialog box to make it larger and smaller.*

- An attribute definition cannot be changed to Constant via the **Edit Attribute** dialog box.

- The **Remove** option does not work when the block contains a single attribute.

# BHatch

**Rel.12** Automatically applies associative hatch pattern objects within boundaries (*short for Boundary HATCH*).

| Commands | Aliases | Ctrl+ | F-key | Alt+ | Menu Bar | Tablet |
|----------|---------|-------|-------|------|----------|--------|
| bhatch | bh | ... | ... | DH | Draw | P9 |
| | h | | | | ⌐Hatch | |
| -bhatch | | | | | | |

**Command:** bhatch

*Displays dialog box:*

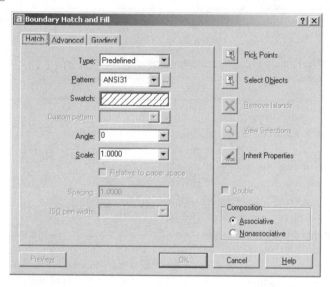

## DIALOG BOX OPTIONS

**Pick Points** detects a boundary, which will be filled with the hatch pattern.

**Select Objects** selects objects for hatching.

**Remove Islands** removes islands from the hatch pattern selection set.

**View Selections** views hatch pattern selection set.

**Inherit Properties** selects the hatch pattern parameters from an existing hatch pattern.

**Double:**

☑ hatch is applied a second time at 90 degrees to the first pattern.

☐ hatch is applied once.

**Composition** determines the associativity of the hatch pattern:

- **Associative** automatically updates the hatch when boundary or properties are modified; pattern is created as a hatch object.
- **Nonassociative** means the hatch cannot be updated; pattern is created as a block.

**Preview** displays a preview of the hatch pattern.

**Hatch** tab

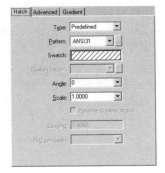

**Type** selects the pattern type:

| Type | Meaning |
|---|---|
| **Predefined** | Hatches predefined by AutoCAD; patterns are stored in *acad.pat* and *acadiso.pat* files. |
| **User Defined** | Hatches defined by you. |
| **Custom** | Hatches defined by *.pat* files are added to AutoCAD's search path. |

**Pattern** selects hatch pattern.

**...** displays **Hatch Pattern Palette** dialog box showing sample pattern types.

**Swatch** displays a non-scaled preview of the hatch pattern; click to display the **Hatch Pattern Palette** dialog box.

**Custom Pattern** lists the custom patterns, if any are available.

**Angle** specifies the hatch pattern angle; default = 0 degrees.

**Scale** specifies the hatch pattern scale; default = 1.0.

**Relative to Paper Space** specifies the scale of the hatch pattern relative to paper space units; available only in layout mode.

**Spacing** specifies the spacing between the lines of a user-defined hatch pattern.

**ISO Pen Width** scales pattern according to pen width.

**Advanced** tab

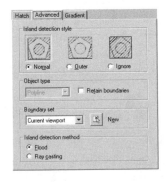

**Island Detection Style**

- **Normal** alternate areas are hatched; text is not hatched.
- **Outer** only the outermost areas are hatched; text is not hatched.
- **Ignore** everything within the boundary is hatched; text is hatched.

**Object Type** constructs the boundary from either a polyline object; or a region object.

**Retain Boundaries:**

  ☑ the boundary created during the hatching process is kept after BHatch finishes.

  ☐ the boundary is discarded.

**Boundary Set** defines the objects analyzed for defining a boundary; not available when Select Objects is used to define the boundary (default = current viewport).

**New** creates a new boundary set.

**Island Detection Method:**

  • **Flood** includes islands as boundary objects.

  • **Ray Casting** runs an imaginary line from the pick point to the nearest object, and then traces the boundary in a counter clockwise direction; it excludes islands.

## Gradient tab

*New to AutoCAD 2004.*

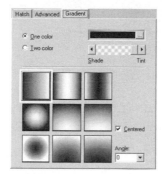

**One color** produces a color-shade gradient.

**Shade - Tint** (displayed only when One Color option is selected) changes the second color between white and black.

**Two color** produces a two-color gradient.

**Centered** centers the gradient in the hatch area.

**Angle** rotates the gradient.

**...** displays the Select Color dialog box; see Color command.

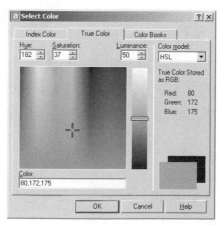

# Hatch Pattern Palette dialog box

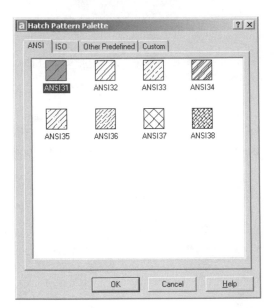

## ANSI Patterns:

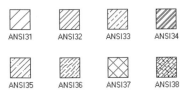

## ISO Patterns:

## Other Predefined Patterns:

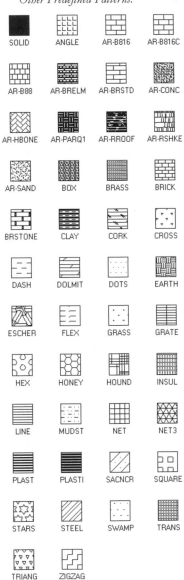

# -BHATCH Command

**Command:** -bhatch
**Current hatch pattern: ANSI31**
**Specify internal point or [Properties/Select/Remove islands/Advanced]:** *(Pick a point, or specify an option.)*

## COMMAND LINE OPTIONS

Internal Point option
  *Pick a point* creates a boundary inside the area of your pick point.

Property options
**Enter a pattern name or [?/Solid/User defined] <ANSI31>:** *(Enter a name, or select an option.)*
**Specify a scale for the pattern <1.0000>:** *(Enter a scale factor.)*
**Specify an angle for the pattern <0>:** *(Enter an angle.)*

  **Enter a pattern name** allows you to enter the name of the hatch pattern.

  **?** lists the names of available hatch patterns.

  **Solid** floods the area with a solid fill in the current color.

  **User defined** creates a simple, user-defined hatch pattern.

  **Specify a scale** specifies the hatch pattern angle (default = 0 degrees).

  **Specify an angle** specifies the hatch pattern scale (default: = 1.0).

Select option
**Select objects:** *(Select one or more objects.)*
**Select objects:** *(Press **Enter**.)*

  **Select objects** selects one or more objects to fill with hatch pattern.

Remove islands options
**Select island to remove:** *(Select one object.)*
**<Select island to remove>/Undo:** *(Press **Enter**, or type **U**.)*

  **Select island to remove** selects the island to remove, which is not filled with the hatch pattern.

  **Undo** adds the removed island.

Advanced options
**Enter an option [Boundary set/Retain boundary/Island detection/Style/Associativity]:** *(Specify an option.)*

  **Boundary set** defines the objects analyzed when a boundary is defined by a specified pick point.

  **Retain boundary:**
  • **On** the boundary created during the hatching process is kept after BHatch finishes.
  • **Off** the boundary is discarded.

  **Island detection:**
  • **On** objects within the outermost boundary are used as boundary objects.
  • **Off** all objects within outermost boundary are filled.

  **Style** selects the hatching style: ignore, outer, or normal.

**Associativity:**

- **On** hatch pattern is associative.
- **Off** hatch pattern is not associative.

## RELATED COMMANDS

**Boundary** traces a polyline automatically around a closed boundary.

**Convert** converts Release 13 (and earlier) hatch patterns into Release 14-2004 format.

**Hatch** creates a nonassociative hatch.

**HatchEdit** edits the hatch pattern.

**PsFill** floods a closed polyline with a PostScript fill pattern.

## RELATED SYSTEM VARIABLES

**DelObj** toggles whether boundary is erased after hatch is placed:

| DelObj | Meaning |
|--------|---------|
| 0 | Boundary is retained. |
| 1 | Boundary is deleted (default). |

**GfAng** specifies the angle of the gradient fill; ranges from 0 to 360 degrees.

**GfClr1** specifies the first gradient fill color in RGB format, such as "RGB 000, 128, 255."

**GfClr2** specifies the second gradient fill color in RGB format.

**GfClrLum** specifies the luminescence of a one-color gradient fill; ranges from 0.0 (black) to 1.0 (white).

**GfClrState** specifies whether the gradient fill is one-color or two-color.

**GfName** specifies the gradient fill pattern:

| GfName | Meaning |
|--------|---------|
| 1 | Linear. |
| 2 | Cylindrical. |
| 3 | Inverted cylindrical. |
| 4 | Spherical. |
| 5 | Inverted spherical. |
| 6 | Hemispherical. |
| 7 | Inverted hemispherical. |
| 8 | Curved. |
| 9 | Inverted curved. |

**GfShift** specifies whether the gradient fill is centered or is shifted to the upper-left.

**HpAng** specifies the current hatch pattern angle (default = 0).

**HpBound** specifies the hatch boundary object:

| HpBound | Meaning |
|---------|---------|
| 0 | Polyline object. |
| 1 | Region object (default). |

**HpDouble** specifies single or double hatching:

| HpDouble | Meaning |
|---|---|
| 0 | Single hatch (default). |
| 1 | Double hatch. |

**HpName** specifies the current hatch pattern name (up to 31 characters long):

| HpName | Meaning |
|---|---|
| **ANSI31** | Default pattern name. |
| "" | No current pattern name. |
| "." | Eliminate current pattern name. |

**HpScale** specifies the current hatch pattern scale factor (default = 1).

**HpSpace** specifies the current hatch pattern spacing factor (default = 1).

**PickStyle** controls the selection of hatch patterns:

| PickStyle | Meaning |
|---|---|
| 0 | Neither groups nor hatches selected. |
| 1 | Groups selected (default). |
| 2 | Associative hatches selected. |
| 3 | Both selected. |

**SnapBase** specifies the starting coordinates of hatch pattern (default = 0,0).

## RELATED FILES

*acad.pat* contains the ANSI and other hatch pattern definitions.

*acadiso.pat* contains the ISO hatch pattern definitions.

## TIPS

• The **BHatch** command first generates a boundary, and then hatches the inside area.

• Use the **Boundary** command to create just the boundary.

• **BHatch** stores hatching parameters in the pattern's extended object data.

# 'Blipmode

**V. 2.1**  Turns the display of pick-point markers, known as "blips," on and off.

| Command | Alias | Ctrl+ | F-key | Alt+ | Menu Bar | Tablet |
|---------|-------|-------|-------|------|----------|--------|
| 'blipmode | ... | ... | ... | ... | ... | ... |

**Command:** blipmode
**Enter mode [ON/OFF] <OFF>:** on

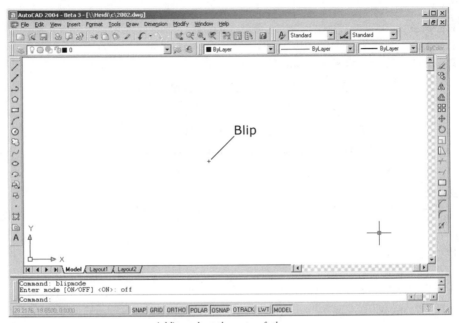

*A blipmark at the center of the screen.*

## COMMAND LINE OPTIONS
**ON** turns on the display of pick-point markers.

**OFF** turns off the display of pick-point markers.

## RELATED COMMANDS
**Options** allows blipmode toggling via a dialog box.

**Redraw** cleans blips off the screen.

## RELATED SYSTEM VARIABLE
**Blipmode** contains the current blipmode setting.

## TIPS
• You cannot change the size of the blipmark.

• Blipmarks are erased by any command that redraws the view, such as **Redraw**, **Regen**, **Zoom**, and **Vports**.

# Block

**V. 1.0** Defines a group of objects as a single named object; creates symbols.

| Commands | Aliases Ctrl+ | F-key | Alt+ | Menu Bar | Tablet |
|----------|---------------|-------|------|----------|--------|
| block | b ... | ... | DKM | Draw | N9 |
| | bmake | | | ⬐Block | |
| | | | | ⬐Make | |
| -block | -b | | | | |

**Command:** block

*Displays dialog box:*

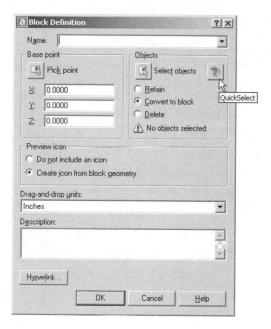

## DIALOG BOX OPTIONS

**Name** names the block (*maximum = 255 characters*).

Base point options

**Pick point** picks the block's insertion point from the drawing.

**X, Y,** and **Z** specify the x,y,z coordinates of the insertion point.

Objects options

**Select objects** selects the objects that make up the block:

- **Retain** leaves objects in place after the block is created.
- **Convert to Block** erases objects making up the block, and replaces them with the block.
- **Delete** erases the objects making up the block; the block is stored in drawing.

**Quick Select** displays the Quick Select dialog box; see the QSelect command.

Preview Icon options

**Do not include an icon** does not create an icon.

**Create icon from block geometry** creates a preview image of the block.

**Insert units** selects the units for the block when dragged from the DesignCenter.

**Description** allows you to enter a description of the block.

**Hyperlink** displays the Insert Hyperlink dialog box; see the Hyperlink command.

. . . . . . . . . . . . . . . . . . . . . . . . . . . . . . . . . . . . . . . . . . . . . . .

## -BLOCK Command

**Command:** -block
**Enter block name or [?]:** *(Enter a name, or type* **?***.)*
**Specify insertion base point:** *(Pick a point.)*
**Select objects:** *(Select one or more objects.)*
**Select objects:** *(Press* **Enter***.)*

### COMMAND LINE OPTIONS

**Block name** allows you to enter the name of the block.

**?** lists the names of blocks stored in the drawing.

**Insertion base point** specifies the x,y coordinates of the block's insertion point.

**Select objects** selects the objects and attributes that make up the block.

### RELATED COMMANDS

**Explode** reduces a block its original objects.

**Insert** adds a block or another drawing to the current drawing.

**Oops** returns objects to the screen after creating the block.

**WBlock** writes a block to a file on disk as a drawing.

**XRef** displays another drawing in the current drawing.

### RELATED SYSTEM VARIABLES

**InsName** default block name.

**InsUnits** drawing units for blocks dragged from the DesignCenter:

| InsUnits | Meaning | InsUnits | Meaning |
|----------|-------------|----------|---------------------|
| 0 | Unitless. | 11 | Angstroms. |
| 1 | Inches. | 12 | Nanometers. |
| 2 | Feet. | 13 | Microns. |
| 3 | Miles. | 14 | Decimeters. |
| 4 | Millimeters. | 15 | Decameters. |
| 5 | Centimeters. | 16 | Hectometers. |
| 6 | Meters. | 17 | Gigameters. |
| 7 | Kilometers. | 18 | Astronomical Units. |
| 8 | Microinches. | 19 | Light Years. |
| 9 | Mils. | 20 | Parsecs. |
| 10 | Yards. | | |

. . . . . . . . . . . . . . . . . . . . . . . . . . . . . . . . . . . . . . . . . . . . . . .

**RELATED FILE**

*\*.dwg* are all drawing files, which can be inserted as blocks.

**TIPS**

- A block consists of these parts:

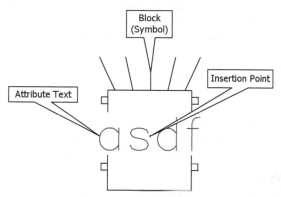

- A block name has up to 255 alphanumeric characters, including $ , -, and _.

- Use the **INSertion** object snap to select the block's insertion point.

- A block created on a layer other than layer 0 is always inserted on that other layer.

- A block created on layer 0 is inserted on the current layer.

- AutoCAD has five types of blocks:

| Block | Meaning |
|---|---|
| **User block** | Named blocks created by users. |
| **Nested block** | Blocks inside other blocks. |
| **Unnamed block** | Blocks created by AutoCAD. |
| **Xref** | Externally-referenced drawings. |
| **Dependent block** | Blocks in externally-referenced drawings. |

- AutoCAD creates the following unnamed blocks, also called "anonymous blocks":

| Name | Meaning |
|---|---|
| **\*A***n* | Group. |
| **\*D***n* | Associative dimension. |
| **\*U***n* | Created by AutoLISP or ObjectARx app. |
| **\*X***n* | Hatch pattern. |

- AutoCAD automatically purges unreferenced anonymous blocks when the drawing is first loaded.

- You cannot place an anonymous block with **Insert**.

# BlockIcon

**2000** Creates preview images for all blocks created with AutoCAD Release 14, or earlier, found in the drawing.

| Command | Alias | Ctrl+ | F-key | Alt+ | Menu Bar | Tablet |
|---------|-------|-------|-------|------|----------|--------|
| blockicon | ... | ... | ... | ... | ... | ... |

**Command:** blockicon
**Enter block names <\*>:** *(Enter a name, or press **Enter** for all names.)*
*n* **blocks updated.**

### COMMAND LINE OPTION

**Enter block names** specifies the blocks for which to create icons; press ENTER to add icons to all blocks in the drawing.

### RELATED COMMANDS

**Block** creates new blocks, as well as their icons.

**AdCenter** displays the icons created by this command:

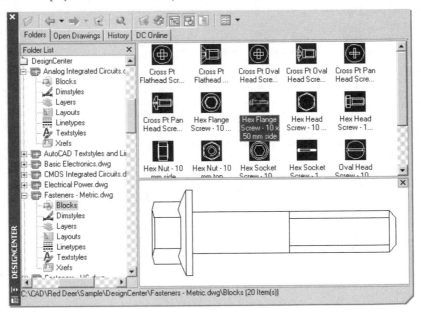

# BmpOut

**Rel.13** Exports the current viewport as a raster image in BMP bitmap format.

| Command | Alias | Ctrl+ | F-key | Alt+ | Menu Bar | Tablet |
|---------|-------|-------|-------|------|----------|--------|
| bmpout | ... | ... | ... | FE | File | ... |
| | | | | ⇘BMP | ⇘Export | |
| | | | | | ⇘BMP | |

**Command:** bmpout
  *Displays **Create BMP File** dialog box. Enter a filename, and then click **Save**.*

## DIALOG BOX OPTION
**Save** saves the drawing as a BMP file.

## RELATED COMMANDS
**JpgOut** exports objects and viewports in JPEG format.

**PngOut** exports objects and viewports in PNG format.

**TifOut** exports objects and viewports in TIFF format.

**WmfOut** exports selected objects in WMF format.

## RELATED WINDOWS COMMANDS
PRT SCR saves screen to the Clipboard.

ALT+PRT SCR saves the topmost window to the Clipboard.

## TIPS
• The *.bmp* extension is short for "bitmap," a raster file standard for Windows.

• The **BmpOut** command exports all objects visible in the current viewport.

• This command creates a compressed *.bmp* file, which cannot be read by most Windows programs.

# Boundary

**Rel.12**  Creates boundaries as polylines or 2D regions.

| Command | Aliases | Ctrl+ | F-key | Alt+ | Menu Bar | Tablet |
|---------|---------|-------|-------|------|----------|--------|
| boundary | bo ... | | ... | DB | Draw | Q9 |
| | bpoly | | | | ⤷Boundary | |
| -boundary | -bo | | | | | |

**Command:** boundary

*Displays dialog box:*

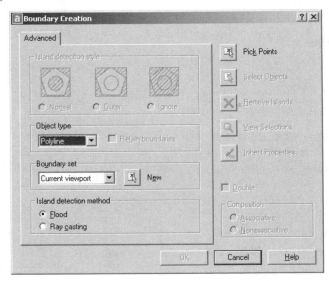

## DIALOG BOX OPTIONS

Object Type options
- Polyline object.
- Region object.

Boundary Set options
**Boundary Set** defines the objects analyzed for defining a boundary; not available when Select Objects is used to define the boundary (default = current viewport).
**New** creates a new boundary set.

Island Detection options
- **Flood** includes islands as boundary objects.
- **Ray Casting** runs an imaginary line from the pick point to the nearest object, and then traces the boundary in a counter clockwise direction; excludes islands.

*Buttons*
**Pick Points** picks points inside of closed areas.

## -BOUNDARY Command
**Command:** -boundary
**Specify internal point or [Advanced options]:** *(Pick a point, or type **A**.)*

### COMMAND LINE OPTIONS
**Specify internal point** creates a boundary based on the point you pick.

Advanced Options options
**Enter an option [Boundary set/Island detection/Object type]:** *(Enter an option.)*
**Boundary set** defines the objects BHatch analyzes when defining a boundary from a specified pick point: a new set of objects, or all objects visible in the current viewport.
**Island detection:**
- **On** uses objects within the outermost boundary are used as boundary objects.
- **Off** sues all objects within outermost boundary are filled.

**Object type** specifies polyline or region as the boundary object.

### RELATED COMMANDS
**PLine** draws a polyline.
**PEdit** edits a polyline.
**Region** creates a 2D region from a collection of objects.

### RELATED SYSTEM VARIABLE
**HpBound** object used to create boundary:

| HpBound | Meaning |
|---------|---------|
| 0 | Draw as region. |
| 1 | Draw as polyline (default). |

### TIPS
- Although the **Boundary Creation** dialog box looks similar to the **BHatch** command's **Advanced Options** dialog box, be aware of differences indicated by the grayed out sections.
- Use this command to measurement of irregular areas:
    1. Apply the **Boundary** command to an irregular area.
    2. Use the **Area** command to find the area and perimeter of the boundary.
- Use **Boundary** together with the **Offset** command to help create poching.

 # Box

**Rel.13**   Draws a 3D box as a solid model.

| Command | Alias | Ctrl+ | F-key | Alt+ | Menu Bar | Tablet |
|---------|-------|-------|-------|------|----------|--------|
| box | ... | ... | ... | DIB | Draw | J7 |
| | | | | | ⌐Solids | |
| | | | | | ⌐Box | |

**Command:** box
**Specify corner of box or [CEnter] <0,0,0>:** *(Pick a point, or type the CE option.)*
**Specify corner or [Cube/Length]:** *(Pick a point, or enter an option.)*
**Specify height:** *(Pick a point.)*
**Second point:** *(Pick a point.)*

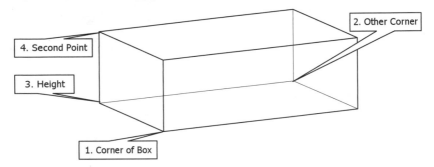

### COMMAND LINE OPTIONS
**Corner of box** specifies one corner for the base of box.

**Center** draws the box about a center point.

**Cube** draws a cube box — all sides are the same length.

**Length** specifies the x, y, z lengths.

**Height** specifies the height of the box.

### RELATED COMMANDS
**Ai_Box** draws a 3D wireframe (surface) box.

**Cone** draws a 3D solid cone.

**Cylinder** draws a 3D solid tube.

**Sphere** draws a 3D solid ball.

**Torus** draws a 3D solid donut.

**Wedge** draws a 3D solid wedge.

### RELATED SYSTEM VARIABLE
**DispSilh** displays 3D objects as silhouettes after hidden-line removal and shading.

 # Break

**V. 1.4** Removes portions of objects.

| Command | Alias | Ctrl+ | F-key | Alt+ | Menu Bar | Tablet |
|---------|-------|-------|-------|------|----------|--------|
| break | br | ... | ... | MK | Modify | W17 |
| | | | | | ↳Break | |

**Command:** break
**Select object:** *(Select one object.)*
**Specify second break point or [First point]:** *(Pick a point, or type **F**.)*

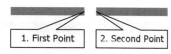

*Breaking a line at two points.*

## COMMAND LINE OPTIONS

**Select object** selects one object to break; pick point becomes first break point, unless the **F** option is used at the next prompt.

**@** uses the first break point's coordinates for the second break point.

 *First Point options*

**Enter first point:** *(Pick a point.)*
**Enter second point:** *(Pick a point.)*

**First point** specifies the first break point.

**Second point** specifies the second break point.

## RELATED COMMANDS

**Change** changes the length lines.

**PEdit** removes and relocates vertices of polylines.

**Trim** shortens the lengths of open objects.

## TIPS

• Use this command to convert a circle into an arc.

• This command can erase a portion of an object (as shown in the figure) or remove the end of an open object.

• The second point does not need to be on the object; AutoCAD breaks the object at the point nearest to the pick point.

• The **Break** command works on the following objects: lines, arcs, circles, polylines, ellipses, rays, xlines, and splines, as well as objects made from polylines, such as donuts and polygons.

# Browser, Browser2

**Rel.14**  Launches a Web browser.

| Commands | Alias | Ctrl+ | F-key | Alt+ | Menu Bar | Tablet |
|----------|-------|-------|-------|------|----------|--------|
| browser | ... | ... | ... | HD | Help | Y8 |
| | | | | | ↳Autodesk On The Web | |

| browser2 |
|----------|

**Command:** browser
**Enter Web location (URL) <http://www.autodesk.com>:** *(Enter a Web address.)*
*Launches browser:*

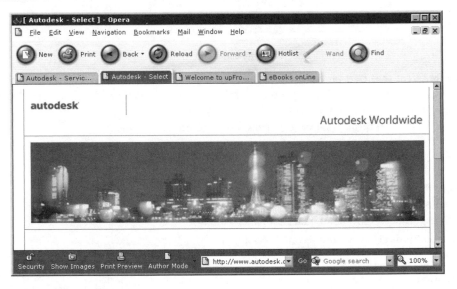

## COMMAND LINE OPTION

**Enter Web location** specifies the URL; see the AttachURL command for information about URLs.

## RELATED SYSTEM VARIABLE

**InetLocation** contains the name of the last-accessed URL.

## TIPS

- **URL** is short for "uniform resource locator," the universal file-naming system used on the Internet; also called a link or hyperlink.

- An example of a URL is http://www.autodeskpress.com, the Autodesk Press Web site.

- Many file dialog boxes also give you access to the Web browser; see the **Open** command.

- The undocumented **Browser2** command opens a double URL.

# 'Cal

**Rel.12** Command-line algebraic and vector geometry calculator (*short for CALculator*).

| Command | Alias | Ctrl+ | F-key | Alt+ | Menu Bar | Tablet |
|---------|-------|-------|-------|------|----------|--------|
| 'cal | ... | ... | ... | ... | ... | ... |

**Command:** cal
**>>Expression:** *(Enter an expression, and then press **Enter**.)*

## COMMAND LINE OPTIONS

| | | | |
|---|---|---|---|
| ( ) | Grouping of expressions. | pi | The value PI. |
| [ ] | Vector expressions. | xyof | x,y coordinates of a point. |
| + | Addition. | xzof | x,z coordinates of a point. |
| - | Subtraction. | yzof | y,z coordinates of a point. |
| * | Multiplication. | xof | x coordinate of a point. |
| / | Division. | yof | y coordinate of a point. |
| ^ | Exponentiation. | zof | z coordinate of a point. |
| & | Vector product of vectors. | rxof | Real x coordinate of a point. |
| sin | Sine. | ryof | Real y coordinate of a point. |
| cos | Cosine. | rzof | Real z coordinate of a point. |
| tang | Tangent. | cur | x,y,z coordinates of a picked point. |
| asin | Arc sine. | rad | Radius of object. |
| acos | Arc cosine. | pld | Point on line, distance from. |
| atan | Arc tangent. | plt | Point on line, using parameter t. |
| ln | Natural logarithm. | rot | Rotated point through angle about origin. |
| log | Logarithm. | ill | Intersection of two lines. |
| exp | Natural exponent. | ilp | Intersection of line and plane. |
| exp10 | Exponent. | dist | Distance between two points. |
| sqr | Square. | dpl | Distance between point and line. |
| sqrt | Square root. | dpp | Distance between point and plane. |
| abs | Absolute value. | ang | Angle between lines. |
| round | Round off. | nor | Unit vector normal. |
| trunc | Truncate. | ESC | Exits Cal mode. |
| cvunit | Converts units using *acad.unt*. | | |
| w2u | WCS to UCS conversion. | | |
| u2w | UCS to WCS conversion. | | |
| r2d | Radians-to-degrees conversion. | | |
| d2r | Degrees-to-radians conversion. | | |

## RELATED COMMANDS

*All.*

## RELATED SYSTEM VARIABLES

**UserI1 — UserI5** holds user-definable variables, which can be used to store integers.

**UserR1 — UserR5** holds user-definable variables, which can be used to store real numbers.

## TIPS

- To use **Cal**, type an expression at the >> prompt. For example, to find the area of a circle with radius of 1.2 units:

   **Expression >>** (1.2^2)*pi
   **4.52389**

- **Cal** understands the following prefixes:

   **\*** Scalar product of vectors.

   **&** Vector product of vectors.

- And the following suffixes:

   **r** Radian (degrees is the default).

   **g** Grad.

   **'** Feet (unitless distance is the default).

   **"** Inches.

- Since **'Cal** is a transparent command, it can be used to perform a calculation in the middle of another command.

 # Camera

<u>**2000**</u>  Sets the camera and target coordinates to create 3D viewpoints.

| Command | Alias | Ctrl+ | F-key | Alt+ | Menu Bar | Tablet |
|---------|-------|-------|-------|------|----------|--------|
| camera | ... | ... | ... | ... | ... | ... |

**Command:** camera
**Specify new camera position <7.6,4.5,21.1>:** *(Pick a point.)*
**Specify new camera target <7.6,4.5,0.0>:** *(Pick a point.)*
**Regenerating model.**

### COMMAND LINE OPTIONS

**Specify new camera position** specifies the x, y, z coordinates for the "look from" point.

**Specify new camera target** specifies the x, y, z coordinates for the "look at" point.

*Camera position: 1,1,1*          *Camera position: -1,-1,-1*
*Camera target: -1,-1,-1*          *Camera target: 1,1,1*

### RELATED COMMANDS

**DView** has the CAmera option, which interactively sets a new 3D viewpoint.

**VPoint** sets a 3D viewpoint through x, y, z coordinates or angles.

**3dOrbit** sets a new 3D viewpoint interactively.

### TIPS

• Following the **Camera** command, you may need to use **Zoom Extents** to see the model.

• Use **Hide** to check whether you are looking at the model from above or below.

# Chamfer

**V. 2.1** Bevels the intersection of two lines, all vertices of 2D polylines, and the faces of 3D solid models.

| Command | Alias | Ctrl+ | F-key | Alt+ | Menu Bar | Tablet |
|---------|-------|-------|-------|------|----------|--------|
| chamfer | cha | ... | ... | MC | Modify<br>↳Chamfer | W18 |

**Command:** chamfer
**(TRIM mode) Current chamfer Dist1 = 0.0, Dist2 = 0.0**
**Select first line or [Polyline/Distance/Angle/Trim/Method/mUltiple]:** *(Pick an object, or enter an option.)*
**Select second line:** *(Pick an object.)*

**COMMAND LINE OPTIONS**

**Select first line** selects the first line, arc, face, or edge.

**Select second line** selects the second line, arc, face, or edge.

**mUltiple** allows more than one pair of lines to be chamfered (new to AutoCAD 2004).

Polyline option
**Select 2D polyline:** *(Pick a polyline.)*
**n lines were chamfered**

**2D polyline** chamfers *all* segments of a 2D polyline; if the polyline is not closed with the Close option, the first and last segments are not chamfered.

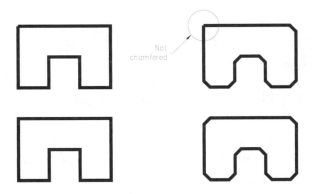

Distance options
**Specify first chamfer distance <0.5000>:** *(Enter a value.)*
**Specify second chamfer distance <0.5000>:** *(Enter a value.)*

**First distance** specifies the chamfering distance along the line picked first.
**Second distance** specifies the chamfering distance along the line picked second.

Angle options
**Specify chamfer length on the first line <1.0000>:** *(Enter a distance.)*
**Specify chamfer angle from the first line <0>:** *(Enter an angle.)*

**Chamfer length** specifies the chamfering distance along the line picked first.
**Chamfer angle** specifies the chamfering angle by an angle from the line picked first.

Trim options
**Enter Trim mode option [Trim/No trim] <Trim>:** *(Type* **T** *or* **N***.)*

**Trim** trims lines, edges, and faces are after chamfer.

**No trim** does not trim lines, edges, and faces after chamfer.

*Intersecting lines before (at right) and after a no-trim chamfer (at left).*

Method option
**Enter trim method [Distance/Angle] <Angle>:** *(Type* **D** *or* **A***.)*

**Distance** determines chamfer by two specified distances.

**Angle** determines chamfer by the specified angle and distance.

- - - - - - - - - - - - - - - - - - - - - - - - - - - - - - - - - - - - -

Chamfering 3D Solids — **Edge Mode** options
**Command:** chamfer
**(TRIM mode) Current chamfer Dist1 = 0.0, Dist2 = 0.0**
**Polyline/.../<Select first line>:** *(Pick a 3D solid model.)*
**Select base surface:** *(Pick surface edge on the solid model — see step 1, below.)*
**Next/<OK>:** *(Enter* **N** *or select the next surface, or* **OK** *to end selection.)*
**Enter base surface distance <0.0>:** *(Enter a value.)*
**Enter other surface distance <0.0>:** *(Enter a value.)*
**Loop/<Select edge>:** *(Select an edge — see step 2, below.)*
**Loop/<Select edge>:** *(Press* **Enter***.)*

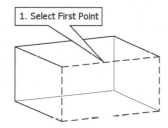
1. Select First Point

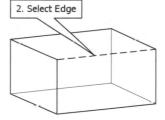
2. Select Edge

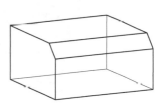

- - - - - - - - - - - - - - - - - - - - - - - - - - - - - - - - - - - - -

Chamfering 3D Solids — **Loop Mode** options
**Command:** chamfer
**(TRIM mode) Current chamfer Dist1 = 0.0, Dist2 = 0.0**
**Polyline/.../<Select first line>:** *(Pick a 3D solid model.)*
**Select base surface:** *(Pick a surface edge — see step 1, below.)*
**Next/<OK>:** *(Enter* **N** *or select the next surface, or* **OK** *to end selection.)*
**Enter base surface distance <0.0>:** *(Enter a value.)*
**Enter other surface distance <0.0>:** *(Enter a value.)*
**Loop/<Select edge>:** *(Enter* **L***.)*
**Edge/<Select edge loop>:** *(Pick a surface edge — see step 2, below.)*

- - - - - - - - - - - - - - - - - - - - - - - - - - - - - - - - - - - -

**Edge/<Select edge loop>:** *(Press* **Enter.***)*

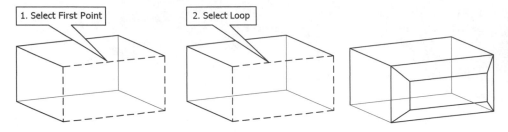

1. Select First Point

2. Select Loop

## COMMAND LINE OPTIONS

**Select first line** selects the 3D solid.

**Next** selects the adjacent face, or press ENTER to accept face.

**Enter base surface distance** specifies the first chamfer distance (default = 0.5).

**Enter other surface distance** specifies the second chamfer distance (default = 0.5).

**Loop** selects all edges of the face.

**Select edge** selects a single edge of the face.

## RELATED COMMANDS

**Fillet** rounds the intersection with a radius.

**SolidEdit** edits the faces and edges of solids.

## RELATED SYSTEM VARIABLES

**ChamferA** is the first chamfer distance (default = 0.5).

**ChamferB** is the second chamfer distance (default = 0.5).

**ChamferC** is the length of chamfer (default = 1).

**ChamferD** is the chamfer angle (default = 0).

**ChamMode** is the toggles chamfer measurement:

| ChamMode | Meaning |
|---|---|
| 0 | Chamfer by two distances (default). |
| 1 | Chamfer by distance and angle. |

**TrimMode** determines whether lines/edges are trimmed after chamfer:

| TrimMode | Meaning |
|---|---|
| 0 | Do not trim selected edges. |
| 1 | Trim selected edges (default). |

## TIPS

* *Caution!* The associativity of a hatch pattern is only maintained when the pattern boundary is a polyline; when the pattern boundary consists of lines, associativity is lost.

* When **TrimMode** is set to 1 and when lines do not intersect, **Chamfer** extends or trims the lines to intersect.

* When the two objects are not on the same layer, **Chamfer** places the chamfer line on the current layer; the chamfer line takes on the layer, or current color and linetype.

# Change

**V. 1.0** Modifies the color, elevation, layer, linetype, linetype scale, lineweight, plot style, and thickness of most objects, and certain other properties of lines, circles, blocks, text, and attributes.

| Command | Alias | Ctrl+ | F-key | Alt+ | Menu Bar | Tablet |
|---------|-------|-------|-------|------|----------|--------|
| change | -ch | ... | ... | ... | ... | ... |

**Command:** change
**Select objects:** *(Pick one or more objects.)*
**Select objects:** *(Press **Enter**.)*
**Specify change point or [Properties]:** *(Pick a point, an object, or type **P**.)*
**Enter property to change [Color/Elev/LAyer/LType/ltScale/LWeight/ Thickness/Plotstyle]:** *(Enter an option.)*

## COMMAND LINE OPTIONS

**Specify change point** selects an object to change:

*(pick a line)* indicates the new length of a line.

*(pick a circle)* indicates the new radius of a circle.

*(pick a block)* indicates the new insertion point or rotation angle of a block.

*(pick text)* indicates the new location of text.

*(pick an attribute)* indicates an attribute's new text insertion point, text style, height, rotation angle, text, tag, prompt, or default value.

ENTER changes the insertion point, style, height, rotation angle, and text of a text string.

*Properties options*

**Color** changes the color of the objects.

**Elev** changes the elevation of the objects.

**LAyer** moves the object to a different layers.

**LType** changes the linetype of the objects.

**ltScale** changes the scale of the linetypes.

**LWeight** changes the lineweight of the objects.

**Thickness** changes the thickness of any object, except blocks and 3D solids.

**Plotstyle** changes the plotstyle of the objects; available only when plotstyles are turned on.

## RELATED COMMANDS

**AttRedef** changes a block or attribute.

**ChProp** contains the properties portion of the Change command.

**Color** changes the current color setting.

**Elev** changes the working elevation and thickness.

**LtScale** changes the linetype scale.

**Properties** changes most aspects of all objects.

**PlotStyle** sets the plotstyle.

## RELATED SYSTEM VARIABLES

**CeColor** specifies the current color setting.

**CeLType** specifies the current linetype setting.

**CelWeight** specifies the current lineweight.

**CircleRad** contains the current circle radius.

**CLayer** contains the name of the current layer.

**CPlotstyle** contains the names of the current plotstyle.

**Elevation** contains the current elevation setting.

**LtScale** contains the current linetype scale.

**TextSize** contains the current height of text.

**TextStyle** contains the current text style.

**Thickness** contains the current thickness setting.

## TIPS

• The **Change** command cannot change the size of donuts, the radius or length of arcs, the length of polylines, or the justification of text.

• Use this command to change the endpoints of a group of lines to a common vertex:

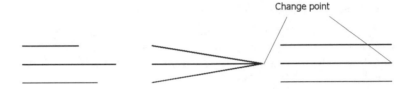

*Change with ortho mode turned off (center) and ortho mode turned on (at left).*

• Turn on ortho mode to extend or trim a group of lines, without needing a cutting edge (as do the **Extend** and **Trim** commands).

• The **PlotStyle** option is not displayed when plotstyles are not turned on.

# 'CheckStandards

2001 Checks the drawing for adherence to standards previously specified by the Standards command.

| Command | Alias | Ctrl+ | F-key | Alt+ | Menu Bar | Tablet |
|---------|-------|-------|-------|------|----------|--------|
| 'checkstandards | chk | ... | ... | TSK | Tools | ... |
| | | | | | ⤷CAD Standards | |
| | | | | | ⤷Check | |

**Command:** checkstandards

*When the **Standards** command has not set up standards for the drawing, AutoCAD displays this error message:*

*When standards have been set up for the drawing, this command checks the CAD standards, and then displays this dialog box:*

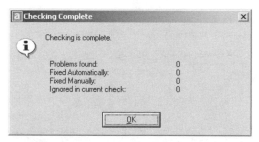

*Click **OK**.*

*To change settings in the **Configure Standards** dialog box, click **Settings**.*

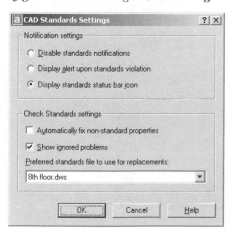

## DIALOG BOX OPTIONS

### Check Standards Settings dialog box

**Notification settings:**

- Disable standards notifications.

- Display alert upon standards violation.

- Display standards status bar icon.

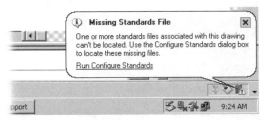

**Automatically fix non-standard properties:**

☑ fixes properties not matching the CAD standard automatically.

☐ allows you to step manually through the properties not matching the CAD standard.

**Show ignored problems:**

☑ displays problems marked as ignored.

☐ does not display ignored problems.

**Preferred standards file to use for replacements** selects the default *.dws* file.

### Check Standards dialog box

*When an object does not match the standards, AutoCAD displays this dialog box:*

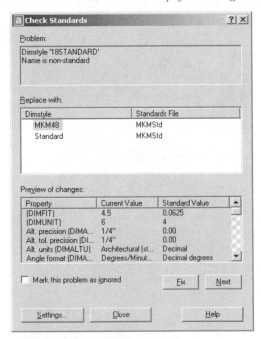

**Problem** describes the property in the drawing that does not match the standard; this dialog box displays one problem at a time.

**Replace With** lists linetypes, text styles, etc. found in the *.dws* standards file.

**Mark this problem as ignored**:

    ☑ ignores non-standard properties, and marks them with the user's login name; some errors are always ignored by AutoCAD, such as layer 0 and DefPoints.

    ☐ does not ignore non-standard properties.

**Fix** replaces the non-standard property with the selected standard; the color checkmark icon means a fix is available.

**Next** displays the next non-standard property.

**Settings** displays the Check Standards Settings dialog box.

**Close** closes the dialog box; displays warning dialog box if standards are not fully checked:

## RELATED COMMANDS

**Standards** selects the *.dws* standards file.

**LayTrans** translates layers.

## RELATED SYSTEM VARIABLES

*None.*

## RELATED PROGRAM

**DwgCheckStandards.Exe** is the Batch Standards Checker program that checks one or more drawings at a time.

## RELATED FILES

*\*.dws* drawing standards file; stored in DWG format.

*\*.chs* standard check file; stored in XML format.

## TIPS

• While this dialog box is open, you can use the following shortcut keys:

| Shortcut | Meaning |
|----------|---------------|
| F4 | Fix problem. |
| F5 | Next problem. |

• AutoCAD includes a single drawing standards file, *mkmstds.dws* found in the \\*support* folder.

# ChProp

**Rel.10** Modifies the color, layer, linetype, linetype scale, lineweight, plotstyle, and thickness of most objects.

| Command | Alias | Ctrl+ | F-key | Alt+ | Menu Bar | Tablet |
|---------|-------|-------|-------|------|----------|--------|
| chprop | ... | ... | ... | ... | ... | ... |

**Command:** chprop
**Select objects:** *(Select one or more objects.)*
**Select objects:** *(Press Enter.)*
**Enter property to change [Color/LAyer/LType/ltScale/LWeight/Thickness/Plotstyle]:** *(Enter an option.)*

## COMMAND LINE OPTIONS

**Color** changes the color of the object.

**LAyer** moves the object to a different layer.

**LType** changes the linetype of the object.

**ltScale** changes the linetype scale.

**LWeight** changes the lineweight of the object.

**Thickness** changes the thickness of all objects, except blocks.

**Plotstyle** changes the plotstyle of the object; available only when plotstyles are turned on.

## RELATED COMMANDS

**Change** changes lines, circles, blocks, text and attributes.

**Color** changes the current color setting.

**Elev** changes the working elevation and thickness.

**LtScale** changes the linetype scale.

**LWeight** sets the lineweight options.

**Properties** changes most aspects of all objects.

**PlotStyle** sets the plotstyle.

## RELATED SYSTEM VARIABLES

**CeColor** specifies the current color setting.

**CeLType** specifies the current linetype setting.

**CelWeight** specifies the current lineweight.

**CLayer** contains the name of the current layer.

**CPlotstyle** contains the names of the current plotstyle.

**LtScale** contains the current linetype scale.

**Thickness** specifies the current thickness setting.

## TIPS

• Use the **Change** command to change the elevation of an object.

• The **Plotstyle** option is not displayed when plotstyles are not turned on.

# Circle

**V. 1.0**  Draws 2D circles by five different methods.

| Command | Alias | Ctrl+ | F-key | Alt+ | Menu Bar | Tablet |
|---------|-------|-------|-------|------|----------|--------|
| circle | c | ... | ... | DC | Draw<br>⬫Circle | J9 |

**Command:** circle
**Specify center point for circle or [3P/2P/Ttr (tan tan radius)]:** *(Pick a point, or enter an option.)*
**Specify radius of circle or [Diameter]:** *(Pick a point, or type D.)*

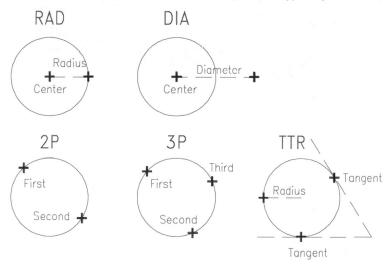

## COMMAND LINE OPTIONS

Center and Radius or Diameter circle options
**Specify center point for circle or [3P/2P/Ttr (tan tan radius)]:** *(Pick a point, or enter an option.)*
**Specify radius of circle or [Diameter] <0.5>:** *(Pick a point, or type D.)*
**Specify diameter of circle <0.5>:** *(Enter a value.)*

    **Center point** indicates the circle's center point.

    **Radius** indicates the circle's radius.

    **Diameter** indicates the circle's diameter.

3P (three-point) circle options
**Specify first point on circle:** *(Pick a point.)*
**Specify second point on circle:** *(Pick a point.)*
**Specify third point on circle:** *(Pick a point.)*

    **First point** indicates the first point on the circle.

    **Second point** indicates the second point on the circle.

    **Third point** indicates the third point on the circle.

2P (two-point) circle options
**Specify first end point of circle's diameter:** *(Pick a point.)*
**Specify second end point of circle's diameter:** *(Pick a point.)*

First end point indicates the first point on the circle.

Second end point indicates the second point on the circle.

TTR (tangent-tangent-radius) circle options
**Specify point on object for first tangent of circle: (Pick** *a point.)*
**Specify point on object for second tangent of circle:** *(Pick a point.)*
**Specify radius of circle <0.5>:** *(Enter a radius.)*

First tangent indicates the first point of tangency.

Second tangent indicates the second point of tangency.

Radius indicates the first point of radius.

## RELATED TOOLBAR ICONS

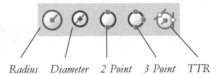

Radius    Diameter    2 Point    3 Point    TTR

## RELATED COMMANDS

**Ai_CircTan** draws a circle tangent to three objects.

**Arc** draws an arc.

**Donut** draws a solid-filled circle or donut.

**Ellipse** draws an elliptical circle or arc.

**Sphere** draws a 3D solid ball.

**ViewRes** controls the visual roundness of circles.

## RELATED SYSTEM VARIABLE

**CircleRad** specifies the current circle radius.

## TIPS

• The 3P (three-point) circle defines three points on the circle's circumference.

• When drawing a TTR (tangent, tangent, radius) circle, AutoCAD draws the circle with tangent points closest to the pick points; note that more than one circle placement is possible.

• Selecting **Draw | Circle | Tan, Tan, Radius** from the menu bar automatically turns on the **TANgent** object snap.

• Giving a circle thickness turns it into a cylinder.

# CleanScreenOn, CleanScreenOff

<u>2004</u> Maximizes the drawing area.

| Command | Alias | Ctrl+ | F-key | Alt+ | Menu Bar | Tablet |
|---|---|---|---|---|---|---|
| cleanscreenon | ... | 0 | ... | VC | View | ... |
| | | | | | ⅋Clean Screen | |
| cleanscreenoff | ... | 0 | ... | VC | View | ... |
| | | | | | ⅋Clean Screen | |

**Command:** cleanscreenon

*AutoCAD turns off the title bar, toolbars, and window edges:*

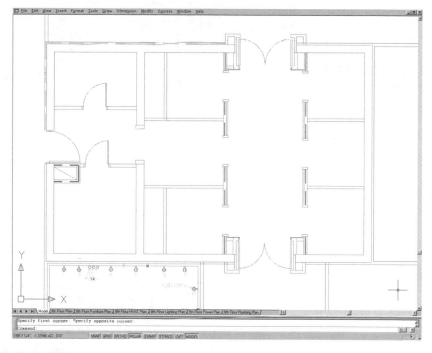

## COMMAND LINE OPTIONS
*None.*

## RELATED COMMANDS
**CleanScreenOff** returns AutoCAD to normal display.

**GraphScr** returns AutoCAD from text window.

## TIPS
• For an even larger drawing area, turn off the scroll bars and layout tabs through the **Options | Display** dialog box, and drag the command prompt area into a window.

• To toggle this command, you can press CTRL+0 (zero).

# Close

**2000**   Closes the current drawing.

| Command | Alias | Ctrl+ | F-key | Alt+ | Menu Bar | Tablet |
|---------|-------|-------|-------|------|----------|--------|
| close | ... | | F4 | ... | FC | File | ... |
| | | | | | ↳Close | |

**Command:** close

*Displays dialog box if the drawing has not been saved since the last change:*

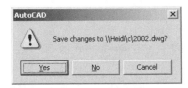

### DIALOG BOX OPTIONS

**Yes** displays the Save Drawing As dialog box, or saves the drawing if it has been previously saved.

**No** exits the drawing without saving it.

**Cancel** returns to the drawing.

### RELATED COMMANDS

**Quit** exits AutoCAD.

**Open** opens additional drawings, each in its own window.

**CloseAll** closes all drawings.

### TIP

• As an alternative to this command, you can click the **x** button on the title bar:

# CloseAll

<u>2000i</u>  Closes all drawings; does not exit AutoCAD.

| Command | Alias | Ctrl+ | F-key | Alt+ | Menu Bar | Tablet |
|---------|-------|-------|-------|------|----------|--------|
| closeall | ... | ... | ... | WL | Window<br>↳Close All | ... |

**Command:** closeall

*Displays dialog box if one or more drawings have not been saved since the last change:*

## DIALOG BOX OPTIONS

**Yes** displays the Save Drawing As dialog box, or saves the drawing if it has been previously saved.

**No** exits the drawing without saving it.

**Cancel** returns to the drawing.

## RELATED COMMANDS

**Quit** exits AutoCAD.

**Open** opens additional drawings, each in its own window.

**Close** closes the current drawing only.

# 'Color

**V. 2.5** Sets the new working color.

| Commands | Aliases | Ctrl+ | F-key | Alt+ | Menu Bar | Tablet |
|----------|---------|-------|-------|------|----------|--------|
| 'color | ddcolor | ... | ... | OC | Format | U4 |
| | col | | | | ⬚Color | |
| | colour | | | | | |
| -color | ... | ... | ... | ... | ... | ... |

**Command:** color

*Displays dialog box:*

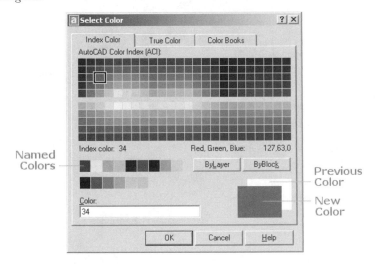

## DIALOG BOX OPTIONS

**Index Color** tab

**AutoCAD Color Index (ACI)** selects one of AutoCAD's colors (short for AutoCAD Color Index).

**Bylayer** sets color to BYLAYER.

**Byblock** sets color to BYBLOCK.

**Color** sets the color by number or name.

**True Color** tab

**Hue** selects the hue (the color) in the range from 0 (red) to 360 (violet).

**Saturation** selects the saturation (intensity of color) in the range from 0 (gray) to 100 (color).

**Luminance** selects the luminance (brightness of color) in the range from 0 (white) to 100 (black).

**Color** specifies the color number as hue, saturation, luminance.

**Color model** selects the type of color model:
- **HSL** is hue, saturation, luminance.
- **RGB** is red, green, blue.

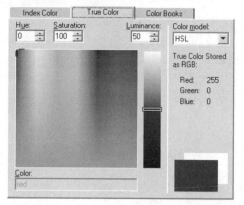 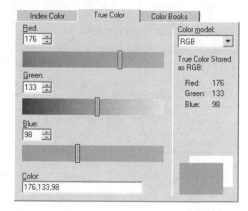

**Red** select the range of red from 0 to 256.

**Green** select the range of green from 0 to 256.

**Blue** select the range of blue from 0 to 256.

**Color** specifies the color number as red, green, blue.

## Color Book tab

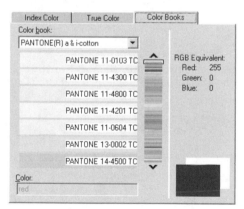

**Color book** selects a predefined collection of colors.

**Color** specifies the name of the Pantone and RAL colors.

*Historical notes:* The Pantone Color System was designed in 1963 to specify color for graphic arts, textiles, and plastics, based on the assumption that colors are seen differently by individuals. Designers typically work with Pantone's fan-format book of standardized colors <www.pantone.com> .

The RAL color system was designed in 1927 to standardize colors by limiting the number of color gradations, first just 30 and now over 1,600. RAL (Reichs Ausschuß für Lieferbedingungen – German for "Imperial Committee for Supply Conditions") is adminis-tered by the German Institute for Quality Assurance and Labeling <www.ral.de>.

# -COLOR Command

**Command:** -color

**Enter default object color [Truecolor/COlorbook] <BYLAYER>:** *(Enter a color number or name, or enter an option.)*

## COMMAND LINE OPTIONS

**BYLAYER** sets the working color to the color of the current layer.

**BYBLOCK** sets the working color of inserted blocks.

*color number* sets working color using number (1 through 255), name, or abbreviation:

| Color Number | Color Name | Abbreviation | Comments |
|---|---|---|---|
| 1 | Red | R | |
| 2 | Yellow | Y | |
| 3 | Green | G | |
| 4 | Cyan | C | |
| 5 | Blue | B | |
| 6 | Magenta | M | |
| 7 | White | W | Or black. |
| 8 - 249 | | | Other colors. |
| 250 - 255 | | | Shades of gray. |

Truecolor options

**Red, Green, Blue:** *(Specify color values separated by commas.)*

**Red, Green, Blue** specifies a color by the amount of red, green, and blue, each in the range from 0 to 255; for example to produce orange:

**Red, Green, Blue:** 255,128,0

COlorbook options

**Enter Color Book name:** *(Specify a name, such as **Pantone**.)*

**Enter color name:** *(Enter a color name, such as **11-0103TC**.)*

**Enter Color Book name** specifies the name of a color book.

**Enter color name** specifies the name of the color.

## RELATED COMMANDS

**ChProp** changes the color via the command line in fewer keystrokes.

**Properties** changes the color of objects via a dialog box.

## RELATED SYSTEM VARIABLES

**CeColor** is the current object color setting.

## TIPS

- 'BYLAYER' means that objects take on the color assigned to that layer; 'BYBLOCK' means objects take on the color in effect at the time the block is inserted.

- White objects display as black when the background color is white.

- When more than one method is used to assign color to objects in a block, unpredictable results occur when the block is inserted or has its color changed.

- Color "0" cannot be specified; AutoCAD uses it internally as the background color.

- The *colorwh.dwg* drawing has 255-color and 156.7 million color wheels.

# Compile

Rel.12 Compiles SHP shape and font files, and PFB font files into SHX files.

| Command | Alias | Ctrl+ | F-key | Alt+ | Menu Bar | Tablet |
|---------|-------|-------|-------|------|----------|--------|
| compile | ... | ... | ... | ... | ... | ... |

**Command:** compile

*Displays **Select Shape or Font File** dialog box. Select a .shp or .pfb file, and then click **Open**.*

## DIALOG BOX OPTIONS

**Open** opens the *.shp* or *.pfb* file for compiling.

**Cancel** closes the dialog box without loading the file.

## RELATED COMMANDS

**Load** loads compiled *.shx* shape files into the current drawing.

**Style** loads *.shx* and *.ttf* font files into the current drawing.

## RELATED SYSTEM VARIABLE

**ShpName** is the current *.shp* filename.

## RELATED FILES

***.shp*** are AutoCAD font and shape source files.

***.shx*** are AutoCAD compiled font and shape files.

***.pfb*** are PostScript Type B font files.

## TIPS

- As of AutoCAD Release 12, **Style** converts *.shp* font files on-the-fly; it is only necessary to use the **Compile** command to obtain an *.shx* font file.

- As of Release 14, AutoCAD no longer direcetly supports PostScript font files. Instead, use the **Compile** command to convert *.pfb* files to *.shx* format.

- TrueType fonts are not compiled.

 # Cone

<u>Rel.11</u>   Draws a 3D solid cone with a circular or elliptical base.

| Command | Alias | Ctrl+ | F-key | Alt+ | Menu Bar | Tablet |
|---|---|---|---|---|---|---|
| cone | ... | ... | ... | DIO | Draw | M7 |
| | | | | | ⮡Solids | |
| | | | | | ⮡Cone | |

**Command:** cone

Cone with circular base
**Current wire frame density: ISOLINES=4**
**Specify center point for base of cone or [Elliptical] <0,0,0>:** *(Pick center point.)*
**Specify radius for base of cone or [Diameter]:** *(Specify radius, or type **D**.)*
**Specify height of cone or [Apex]:** *(Specify height, or type **A**.)*

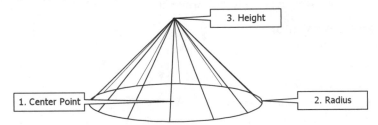

Cone with elliptical base and apex
**Specify center point for base of cone or [Elliptical] <0,0,0>:** *(Type **E**.)*
**Specify axis endpoint of ellipse for base of cone or [Center]:** *(Pick a point, or type **C**.)*
**Specify second axis endpoint of ellipse for base of cone:** *(Pick a point.)*
**Specify length of other axis for base of cone:** *(Pick a point.)*
**Specify height of cone or [Apex]:** *(Specify height, or type **A**.)*

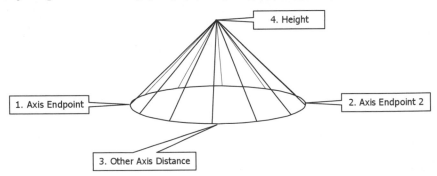

## COMMAND LINE OPTIONS

**Elliptical** draws a cone with an elliptical base.

**Center point** specifies the center point of the cone's base.

**Axis endpoint** picks the endpoint of the axis defining the elliptical base.

**Center of ellipse** picks the center of the elliptical base.

**Height** specifies the height of the cone.

**Apex** determines the height and orientation of the cone's tip.

## RELATED COMMANDS

**Ai_Cone** draws a 3D wireframe cone.

**Box** draws a 3D solid box.

**Cylinder** draws a 3D solid tube.

**Sphere** draws a 3D solid ball.

**Torus** draws a 3D solid donut.

**Wedge** draws a 3D solid wedge.

## RELATED SYSTEM VARIABLES

**DispSilh** toggles display of 3D objects as silhouettes after hidden-line removal and shading.

**IsoLines** specifies the number of isolines on solid surfaces:

| IsoLines | Meaning |
| --- | --- |
| 0 | Minimum; no isolines. |
| 4 | Default. |
| 12 *or* 16 | A reasonable value. |
| 2047 | Maximum value. |

## TIPS

* You define the elliptical base in two ways: by the length of the major and minor axes, or by the center point and two radii.

* To draw a cone at an angle, use the **Apex** option.

## Changed Command

The **Config** command now displays the **Options** dialog box; see the **Options** command.

# Convert

Rel.14  Converts 2D polylines and associative hatches to optimized formats.

| Command | Alias | Ctrl+ | F-key | Alt+ | Menu Bar | Tablet |
|---------|-------|-------|-------|------|----------|--------|
| convert | ... | ... | ... | ... | ... | ... |

**Command:** convert
**Enter type of objects to convert [Hatch/Polyline/All] <All>:** *(Enter an option.)*
**Enter object selection preference [Select/All] <All>:** *(Type **S** or **A**.)*

## COMMAND LINE OPTIONS

**Hatch** converts associative hatch patterns from anonymous blocks to hatch objects; displays warning dialog box:

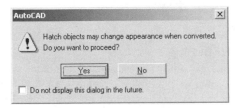

**Yes** converts hatch patterns to hatch objects.

**No** does not convert hatch patterns.

**Polyline** converts 2D polylines to Lwpolyline objects.

**All** converts all polylines and hatch patterns.

**Select** selects the hatch patterns and polylines to convert.

## RELATED COMMANDS

**BHatch** creates associative hatch patterns.

**PLine** draws 2D polylines.

## RELATED SYSTEM VARIABLE

**PLineType** determines whether pre-Release 14 polylines are converted in AutoCAD 2004.

| PlineType | Meaning |
|-----------|---------|
| 0 | Not converted; **PLine** creates old-format polylines. |
| 1 | Not converted; **PLine** creates lwpolylines. |
| 2 | Converted; **PLine** creates lwpolylines (default). |

## TIPS

• When a Release 13 or earlier drawing is opened in AutoCAD 2004, it automatically converts most (not all) 2D polylines to lwpolylines; hatch patterns are not automatically updated.

• A hatch pattern is automatically updated the first time the **HatchEdit** command is applied, or when its boundary is changed.

• Polylines are not converted when they contain curve fit segments, splined segments, extended object data in their vertices, or 3D polylines.

• **PLineType** affect the following commands: **Boundary** (when a polyline), **Donut, Ellipse** (**PEllipse** = 1), **PEdit** (converts a line or arc), **Polygon**, and **Sketch** (**SkPoly** = 1).

# ConvertCTB

<u>**2001**</u>  Converts a plotstyle file from CTB color-dependent format to STB named format (*short for CONVERT Color TaBle*).

| Command | Alias | Ctrl+ | F-key | Alt+ | Menu Bar | Tablet |
|---------|-------|-------|-------|------|----------|--------|
| convertctb | ... | ... | ... | ... | ... | ... |

**Command:** convertctb

*Displays **Select File** dialog box.*

*1. Select a .ctb file, and then click **Open**. AutoCAD displays the **Create File** dialog box.*

*2. Specify the name of an .stb file, and then click **Save**.*

*When AutoCAD has completed the conversion, it displays dialog box:*

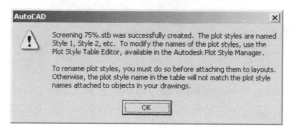

## DIALOG BOX OPTIONS

**Open** opens the *.ctb* file.

**Save** saves as an *.stb* file.

## RELATED COMMANDS

**ConvertPStyles** converts a drawing between color-dependent and named plotstyles.

**PlotStyle** sets the plotstyle for a drawing.

## RELATED FILES

*****.ctb** are color-dependent plotstyle table files.

*****.stb** are named plotstyle tables.

## TIPS

- "Color-dependent plotstyles" are used by older versions of AutoCAD, where the color of the object controls the pen selection.

- "Named plotstyles" is the alternative introduced with AutoCAD 2000, which allows plotter-specific information to be assigned to layers and objects.

# ConvertPStyles

**2001** Converts a drawing between color-dependent and named plotstyles (*short for Convert Plot STYLES*).

| Command | Alias | Ctrl+ | F-key | Alt+ | Menu Bar | Tablet |
|---------|-------|-------|-------|------|----------|--------|
| convertpstyles | ... | ... | ... | ... | ... | ... |

**Command:** convertpstyles

*Displays warning dialog box:*

*When AutoCAD has completed the conversion, it displays the message:*

**Drawing converted from Named plotstyle mode to Color Dependent mode.**

## DIALOG BOX OPTIONS

**OK** proceeds with the conversion.

**Cancel** prevents the conversion.

## RELATED COMMANDS

**ConvertCTB** converts a plotstyle file from CTB color-dependent format to STB named format.

**PlotStyle** sets the plot style for a drawing.

## TIPS

- "Color-dependent plot styles" was used by older versions of AutoCAD, where the color of the object controlled the pen selection.

- "Named plot styles" is the alternative introduced with AutoCAD 2000, which allows plotter-specific information to be assigned to layers and objects.

 # Copy

**V. 1.0** Creates one or more copies of an object.

| Command | Alias | Ctrl+ | F-key | Alt+ | Menu Bar | Tablet |
|---------|-------|-------|-------|------|----------|--------|
| copy | co | ... | ... | MY | Modify | V15 |
| | cp | | | | ⸂Copy | |

**Command:** copy
**Select objects:** *(Select one or more objects.)*
**Select objects:** *(Press* **Enter***.)*
**Specify base point or displacement, or [Multiple]:** *(Pick a point, or type* **M***.)*
**Specify second point of displacement or <use first point as displacement>:**
*(Pick a point.)*

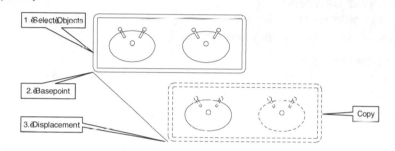

## COMMAND LINE OPTIONS

**Base point** indicates the starting point.

**Multiple** allows the object(s) to be copied more than once.

**Second point** indicates the point to move to.

**Displacement** uses the selection point as the base point when you press ENTER.

ESC cancels multiple object copying.

## RELATED COMMANDS

**Array** draws a rectangular or polar array of objects.

**MInsert** places an array of blocks.

**Move** moves an object to a new location.

**Offset** draws parallel lines, polylines, circles, and arcs.

## TIPS

- Use the **M** (*multiple*) option to quickly place several copies of the original object.

- Inserting a block multiple times is more efficient than placing multiple copies.

- Turn on ortho mode to copy objects in a precise horizontal and vertical direction; turn snap mode on to copy objects in precise increments; use object snap modes to copy objects precisely from one geometric feature to another.

- To copy an object by a known distance, enter 0,0 as the 'Base point.' Then enter the known distance as the 'Second point.'

# CopyBase

2000 Copies selected objects to the Clipboard with a specified base point (*short for COPY with BASEpoint*).

| Command | Alias | Ctrl+ | F-key | Alt+ | Menu Bar | Tablet |
|---|---|---|---|---|---|---|
| copybase | ... | Shift+C | ... | EB | Edit | ... |
| | | | | | ↳Copy with Basepoint | |

**Command:** copybase
**Specify base point:** *(Pick a point.)*
**Select objects:** *(Select one or more objects.)*
**Select objects:** *(Press **Enter**.)*

## COMMAND LINE OPTIONS

**Specify base point** specifies the base point.

**Select objects** selects the objects to copy to the Clipboard.

## RELATED COMMANDS

**CopyClip** copies selected objects to the Clipboard without a specified basepoint.

**PasteBlock** pastes from the Clipboard into the drawing as a block.

**PasteClip** pastes from the Clipboard into the drawing.

## TIPS

• The purpose of this command is to allow you to place the copied objects with the **PasteClip** command.

• When **PasteClip** pastes objects selected with the **CopyBase** command, it prompts you 'Specify insertion point' and pastes the objects as a block with a name similar to A$C7E1B27BE.

• When specifying the **All** option at the 'Select objects' prompt, **CopyBase** selects all objects visible only in the current viewport.

• As of AutoCAD 2004, you can use the CTRL+SHIFT+C shortcut for this command.

# CopyClip

**Rel.12**  Copies selected objects from the drawing to the Clipboard (*short for COPY to CLIPboard*).

| Command | Alias | Ctrl+ | F-key | Alt+ | Menu Bar | Tablet |
|---------|-------|-------|-------|------|----------|--------|
| copyclip | ... | C | ... | EC | Edit | T14 |
| | | | | | ⤷Copy | |

**Command:** copyclip
**Select objects:** *(Select one or more objects.)*
**Select objects:** *(Press* **Enter.***)*

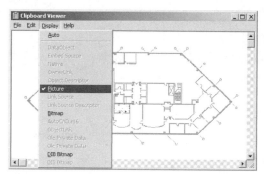

*The Clipboard Viewer shows an object copied from AutoCAD to the Clipboard.*

## COMMAND LINE OPTION

**Select objects** selects the objects to copy to Clipboard.

## RELATED COMMANDS

**CopyBase** copies objects to the Clipboard with a specified basepoint.

**CopyHist** copies Text window text to the Clipboard.

**CopyLink** copies the current viewport to the Clipboard.

**CutClip** cuts selected objects to the Clipboard.

**PasteBlock** pastes from the Clipboard into the drawing as a block.

**PasteClip** pastes from the Clipboard into the drawing.

## RELATED WINDOWS COMMANDS

PRT SCR copies the entire screen to the Clipboard

ALT+PRT SCR copies the topmost window to the Clipboard.

## TIPS

• Contrary to the AutoCAD *Command Reference*, text objects are *not* copied to the Clipboard in text format; instead, text is copied as an AutoCAD object.

• When the **All** option is specified at the 'Select objects' prompt, **CopyClip** selects objects visible only in the current viewport.

# CopyHist

Rel.13 Copies Text window text to the Clipboard (*short for COPY HISTory*).

| Command | Alias | Ctrl+ | F-key | Alt+ | Menu Bar | Tablet |
|---------|-------|-------|-------|------|----------|--------|
| copyhist | ... | ... | ... | EH | Edit | ... |
| | | | | | ⌐Copy History | |

**Command:** copyhist

## COMMAND LINE OPTIONS
*None.*

## RELATED COMMAND
**CopyClip** copies selected text from the drawing to the Clipboard.

## RELATED WINDOWS COMMAND
ALT+PRT SCR copies the Text window to the Clipboard in graphics format.

## TIPS
- To copy a selected portion of Text window text to the Clipboard, highlight the text first, then select **Edit | Copy** from the Text window's menu bar.

- To paste text to the command line, select **Edit | Paste to Cmdline** from the menu bar. However, this only works when the Clipboard contains text — not graphics.

- As an alternative, you can right-click in the **Text** window to bring up the cursor menu.

# Getting the Right Clipboard Result

You can use the Clipboard to display AutoCAD drawings in other Windows applications, such as word processing, desktop publishing, and paint programs. AutoCAD has two primary commands for copying objects from the drawing to the Clipboard: **CopyLink** and **CopyClip**. The result of using these commands, however, may surprise you, since each command produces a different result, depending on whether AutoCAD is in model or layout mode.

- **CopyClip** (**Edit | Copy**) copies selected objects from the drawing, as follows:

> **Command:** copyclip
> **Select objects:** all
> **Select objects:** ENTER

*Warning*: when you select **All** objects, AutoCAD selects only those object visible in the current viewport.

- **CopyLink** (**Edit | Copy Link**) copies everything visible in the current viewport. Use this command to capture a layout view with *all* viewports.

The two commands take on different meanings, as described below.

## Model Mode

In *model mode*, AutoCAD copies only objects visible in the current viewport. If the drawing contains more than one viewport, you must select the correct viewport before copying objects to the Clipboard. **CopyClip** and **CopyLink** have the same effect:

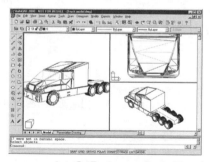 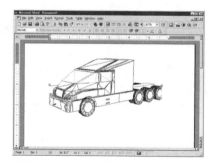

*AutoCAD viewports in model mode.*    *Drawing pasted in Word after either*
*CopyClip or CopyLink.*

## Layout Mode

In layout mode's **PAPER** space, **CopyClip** copies objects drawn only in paper space — *nothing* drawn in model space is copied to the Clipboard, even if it is visible.

**CopyLink** copies *all* visible objects, whether drawn in paper space or model space. Notice that this command also copies the margin lines and gray background.

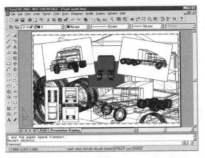

*AutoCAD in layout mode's PAPER space.*

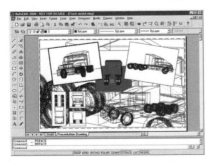 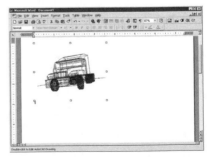

*Result in Word after using AutoCAD's **ClipClip** (left) and **CopyLink** (right).*

In layout mode's **MODEL** space, AutoCAD only copies objects drawn in model space that are visible in the selected viewport. **CopyClip** and **CopyLink** produce the same result when pasted in Word and other Windows applications.

 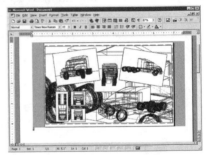

*A viewport in layout mode's MODEL space.*  *Drawing pasted in Word after either*
*CopyClip or CopyLink.*

# CopyLink

Rel.13 Copies the current viewport to the Clipboard; optionally allows you to link the drawing back to AutoCAD.

| Command | Alias | Ctrl+ | F-key | Alt+ | Menu Bar | Tablet |
|---------|-------|-------|-------|------|----------|--------|
| copylink | ... | ... | | EL | Edit | ... |
| | | | | | ↳Copy Link | |

**Command:** copyclip

## COMMAND LINE OPTIONS
*None.*

## RELATED COMMANDS
**CopyClip** copies selected objects to the Clipboard.

**CopyEmbed** copies selected objects to the Clipboard.

**CopyHist** copies Text window text to the Clipboard.

**CutClip** cuts selected objects to the Clipboard.

**PasteClip** pastes from the Clipboard into the drawing.

## RELATED WINDOWS COMMANDS
PRT SCR copies the entire screen to the Clipboard

ALT+PRT SCR copies the topmost window to the Clipboard.

## TIPS
- In the other application, use the **Edit | Paste Special** commands to paste the AutoCAD image into the document; to link the drawing back to AutoCAD, select the **Paste Link** option.

- AutoCAD does not let you link a drawing to itself.

- This command copies everything in the current viewport (if in model space) or the entire drawing (if in paper space).

- **CopyEmbed** is identical to **CopyLink**, except that **CopyEmbed** prompts you to select objects.

# Customize

<u>2000i</u>  Creates and changes toolbars, toolbar macros and icons, and keyboard shortcuts.

| Command | Aliases Ctrl+ | F-key | Alt+ | Menu Bar | Tablet |
|---------|---------------|-------|------|----------|--------|
| customize | toolbar ...<br>to<br>tbconfig | ... | TC | Tools<br>⤷Customize | T13 |

**Command:** customize

*Displays dialog box:*

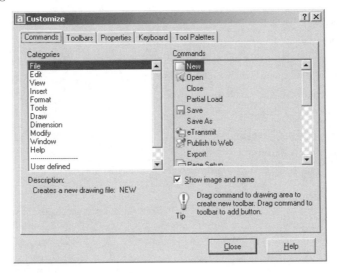

## DIALOG BOX OPTIONS

**Close** closes the dialog box, and saves the changes.

### Commands tab

**Categories** lists names of primary menu items.

**Commands** lists command names and related icons, if any; drag a command name out of the dialog box to start creating a new toolbar.

**Show image and name**

☑ lists all command names and related icons.

☐ lists icons only; commands without an icon are shown as a gray square.

**Toolbars** tab

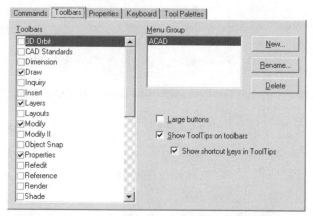

**Toolbars** lists the names of defined toolbars, and toggles their visibility.

**Menu Group** lists the names of menu groups loaded in AutoCAD.

**New** creates a new toolbar; displays New Toolbar dialog box, which prompts you to select the menu group to which the new toolbar should be added.

**Rename** changes the name of the selected toolbar; displays the Rename Toolbar dialog box.

**Delete** deletes the selected toolbar; displays a warning dialog box: "Are you sure you want to delete the toolbar from the menu group?"

**Large buttons**

☑ toolbar buttons and icons are displayed at a larger size for better visibility on very-high resolution screens, or for visually-impaired users (40 pixels wide).

☐ toolbar buttons and icons displayed at standard size (24 pixels wide).

*Normal buttons (left) and large buttons (right).*

**Show tooltips on toolbars**

☑ displays a small yellow tag with the command name.

☐ does not display tooltips.

**Show shortcut keys in Tooltips**

☑ displays shortcut keystrokes, such as CTRL+N, in the tooltip.

☐ does not display shortcut keystrokes.

*Tooltip with shortcut key CTRL+N.*

**Button Properties** tab

*When you first click the **Button Properties** tab, AutoCAD prompts:*

Tip: Select a toolbar item to view or modify its properties.

*Select any toolbar button; AutoCAD displays the **Button Properties** tab:*

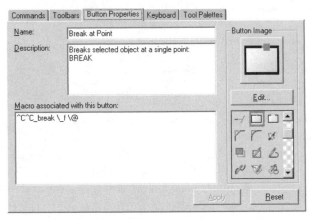

**Name** specifies the wording for the tooltip.

**Description** specifies the help text displayed on the status line.

**Macro associated with this button** specifies the macro executed when the button is clicked.

**Edit** edits the icon; displays the Button Editor dialog box.

**Apply** applies the changes to the button.

**Reset** changes the parameters back.

**Keyboard** tab

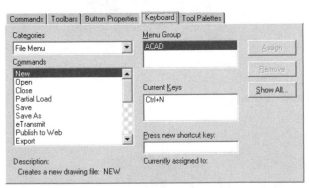

**Categories** allows you to select the name of a toolbar, menu item, or all AutoCAD commands.

**Commands** lists the commands, as specified by Categories.

**Menu Group** lists the names of menu groups loaded in AutoCAD.

**Current Keys** lists the shortcut key(s) assigned to the command; this option appears to not work with pre-assigned keys, such as CTRL+Q.

**Press new shortcut key** selects the shortcut key: hold down CTRL or CTRL+SHIFT, plus a letter or number key between A and Z, and 1 and 0.

**Assign** assigns the shortcut keystroke to the command.

**Remove** removes the shortcut keystroke from the command.

**Show All** lists all assigned shortcut keys; displays the Shortcut Keys dialog box.

## Tool Palettes tab

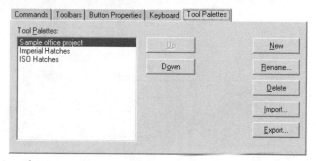

**Tool Palettes** lists the names of tool palettes.

**Up** moves the selected palette up one position.

**Down** moves the selected palette down one position.

**New** create a new, blank palette.

**Rename** renames the selected palette.

**Delete** removes the selected palette.

**Import** imports palettes (*.xtp* files).

**Export** exports the selected palette in an XML-like *.xtp* file.

## Shortcut Keys dialog box

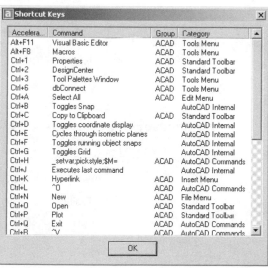

**OK** closes the dialog box.

**Button Editor** dialog box

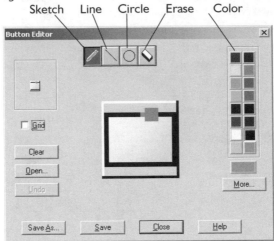

Sketch  Line  Circle  Erase  Color

**Grid**:

    ☑ displays a grid in the icon drawing area.

    ☐ hides the grid.

**Clear** erases the icon.

**Open** opens a *.bmp* (bitmap) file, which can be used as an icon.

**Undo** undoes the last operation.

**Save As** saves the icon as a *.bmp* file.

**Save** saves the changes to the icon in AutoCAD's internal storage.

**Close** closes the dialog box.

**More** displays the Select Color dialog box.

## RELATED COMMANDS

**BmpOut** exports selected objects in the current view to a BMP file.

**MenuLoad** loads partial menus into AutoCAD.

**-Toolbar** opens and closes toolbars via the command line.

## TIP

• Elements of a toolbar:

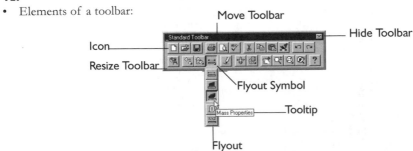

Move Toolbar

Hide Toolbar

Icon

Resize Toolbar

Flyout Symbol

Mass Properties — Tooltip

Flyout

 # CutClip

**Rel.12**   Cuts the selected objects from the drawing to the Clipboard (*short for CUT to CLIPboard*).

| Command | Alias | Ctrl+ | F-key | Alt+ | Menu Bar | Tablet |
|---------|-------|-------|-------|------|----------|--------|
| cutclip | ... | X | ... | ET | Edit<br>↳Cut | T13 |

**Command:** cutclip
**Select objects:** *(Select one or more objects.)*
**Select objects:** *(Press **Enter**.)*

## COMMAND LINE OPTION
**Select objects** selects the objects to cut to Clipboard.

## RELATED COMMANDS
**BmpOut** exports selected objects in the current view to a *.bmp* file.
**CopyClip** copies selected objects to the Clipboard.
**CopyHist** copies Text window text to the Clipboard.
**CopyLink** copies current viewport to the Clipboard.
**PasteClip** pastes from the Clipboard into the drawing.
**WmfOut** exports selected objects to a *.wmf* file.

## RELATED WINDOWS COMMANDS
RRT SCR copies the entire screen to the Clipboard
ALT+PRT SCR copies the topmost window to the Clipboard.

## TIPS
- When the **All** option is specified at the 'Select objects:' prompt, **CutClip** selects all objects only visible in the current viewport.

- In the other application, use the **Edit | Paste** or **Edit | Paste Special** commands to paste the AutoCAD image into the document; the **Paste Special** command lets you specify the pasted format.

- You can use the **Undo** command to return the "cut" objects to the drawing.

 # Cylinder

**Rel.12** Draws a 3D solid cylinder with a circular or elliptical cross section.

| Command | Alias | Ctrl+ | F-key | Alt+ | Menu Bar | Tablet |
|---------|-------|-------|-------|------|----------|--------|
| cylinder | ... | ... | ... | DIC | Draw | L7 |
| | | | | | ⤷Solids | |
| | | | | | ⤷Cylinder | |

**Command:** cylinder
**Current wire frame density: ISOLINES=4**
**Specify center point for base of cylinder or [Elliptical] <0,0,0>:** *(Pick center point, or type **E**.)*
**Specify radius for base of cylinder or [Diameter]:** *(Specify radius, or type **D**.)*
**Specify height of cylinder or [Center of other end]:** *(Specify height, or type **C**.)*

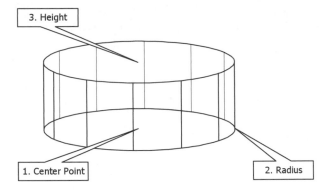

3. Height

1. Center Point

2. Radius

## COMMAND OPTIONS

**Center point** specifies the center point of the cylinder's base.

**Radius** specifies the cylinder's radius.

**Diameter** specifies the cylinder's diameter.

**Height** specifies the cylinder's height.

**Center of other end** specifies the z orientation of the cylinder.

Elliptical options
**Specify axis endpoint of ellipse for base of cylinder or [Center]:** *(Pick point, or type **C**.)*
**Specify second axis endpoint of ellipse for base of cylinder:** *(Pick point.)*
**Specify length of other axis for base of cylinder:** *(Specify length.)*
**Specify height of cylinder or [Center of other end]:** *(Specify height, or type **C**.)*

**Axis endpoint** specifies one end of the elliptical axis.

**Center** specifies the center point of the elliptical base.

**Second axis endpoint** specifies the other end of the axis.

**Length of other axis** specifies the length of the other elliptical axis.

**RELATED COMMANDS**

**Box** draws a 3D solid box.

**Cone** draws a 3D solid cone.

**Elevation** turns a circle into a wireframe cylinder.

**Extrude** creates a cylinder with an arbitrary cross-section and sloped walls.

**Sphere** draws a 3D solid ball.

**Torus** draws a 3D solid donut.

**Wedge** draws a 3D solid wedge.

**RELATED SYSTEM VARIABLES**

**DispSilh** displays 3D objects as silhouettes after hidden-line removal and shading.

**IsoLines** determines the number of isolines on solid surfaces:

| IsoLines | Meaning |
|----------|---------|
| 0 | Minimum (no isolines). |
| 4 | default. |
| 12 *or* 16 | A reasonable value. |
| 2047 | Maximum value. |

**TIPS**

- The **Ellipse** option draws a cylinder with an elliptical cross-section.

- Use the **Intersect** command remove solid portions from the cylinder; use the **Union** command to add solid portions to the cylinder.

# DbcClose

<u>2000</u>  Closes the dbConnect Manager window.

| Command | Alias | Ctrl+ | F-key | Alt+ | Menu Bar | Tablet |
|---------|-------|-------|-------|------|----------|--------|
| dbcclose | ... | 6 | ... | TD | Tools | ... |
| | | | | | ⬚dbConnect | |

**Command:** dbcclose

**COMMAND LINE OPTIONS**
*None.*

**RELATED COMMAND**
**DbConnect** opens the **dbConnect Manager** window.

**TIP**
- Pressing CTRL+6 toggles on and off the display of the **dbConnect Manager** window; see the **DbConnect** command.

# DbConnect

<u>**2000**</u>  Opens the dbConnect Manager window to connect objects with external database tables (*short for Data Base CONNECTion*).

| Command | Alias | Ctrl+ | F-key | Alt+ | Menu Bar | Tablet |
|---------|-------|-------|-------|------|----------|--------|
| dbconnect | dbc | 6 | ... | TD | Tools | W12 |
| | | | | | ⮑dbConnect | |

**Command:** dbconnect

*Displays window. If a red x appears, the database is disconnected from the drawing:*

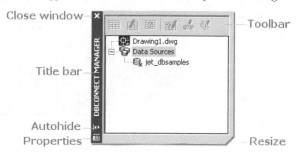

*To connect the drawing with the database, right-click the database icon, and then select* **Connect**.
*The icons have the following meaning:*

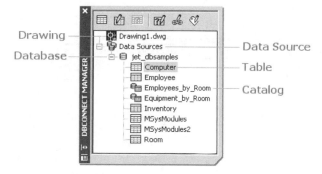

## TOOLBAR OPTIONS

**View Table** opens an external database table in *read-only* mode; select a table, link template, or label template to make this button available.

**Edit Table** opens an external database table in *edit* mode; select a table, link template, or label template to make this button available.

**Execute Query** executes a query; select a previously-defined query to make this button available.

**New Query** displays the New Query dialog box when a table or link template is selected; displays the Query Editor when a query is selected.

**New Link Template** displays the New Link Template dialog box when a table is selected; displays the Link Template dialog box when a link template is selected; not available for link templates with links already defined in a drawing.

**New Label Template** displays the New Label Template dialog box when a table or link template is selected; displays the Label Template dialog box when a label template is selected.

### View Table and Edit Table windows

*The **View Table** and **Edit Table** windows are identical, with the exception that **View Table** is read-only; hence all text in columns is grayed-out.*

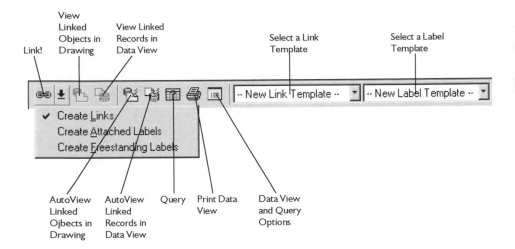

**Link!** creates a link or a label; click the drop list to select an option:

- **Create Links** turns on link creation mode.
- **Create Attached Labels** turns on the attached label creation mode.
- **Create Freestanding Labels** turns on freestanding label creation mode.

**View Link Objects in Drawing** highlights objects linked to selected records.

**View Linked Records in Data View** highlights records that are linked to selected objects in the drawing.

**AutoView Linked Objects in Drawing** automatically highlights objects that are linked to selected records.

**AutoView Linked Records in Data View** highlights records that are linked to selected objects in the drawing.

**Query** displays the New Query dialog box.

**Print Data View** prints the data in this window.

**Data View and Query Options** displays the Data View and Query Options dialog box.

**Select a Link Template** lists the names of previously-defined link templates.

**Select a Label Template** lists the names of previously-defined label templates.

**New Query** dialog box

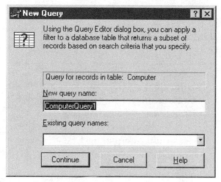

**New query name** specifies the name of the query.

**Existing query names** uses an existing query.

**Continue** displays the Query Editor dialog box.

**Query Editor** dialog box

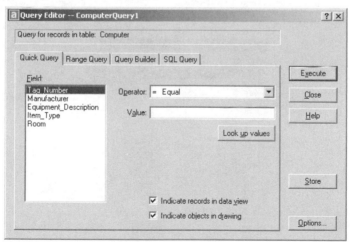

**Execute** executes the query.

**Close** closes the dialog box.

**Store** saves the settings.

**Options** displays the Data View and Query Options dialog box.

**Quick Query** tab

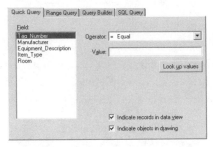

**Field** selects a field name.

**Operator** specifies a conditional operator:

| Operator | Meaning |
|----------|---------|
| = | **Equal** — Exactly match the value (default). |
| <> | **Not equal** — Does not match the value. |
| > | **Greater than** — Greater than the value. |
| < | **Less than** — Less than the value. |
| >= | **Greater than or equal** — Greater than or equal to the value. |
| <= | **Less than or equal** — Less than or equal to the value. |
| **Like** | Contains the value; use the % wild-card character (equivalent to * in DOS). |
| **In** | Matches two values separated by a comma. |
| **Is null** | Does not have a value; used for locating records that are missing data. |
| **Is not null** | Has a value; used for excluding records that are missing data. |

**Value** specifies the value for which to search.

**Look up values** displays a list of existing values:

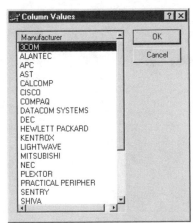

**Range Query** tab

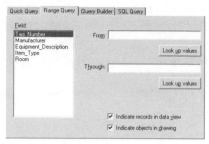

**Field** lists the names of fields in the current table.

**From** specifies the first value of the range.

**Look Up Values** displays the **Column Values** dialog box.

**Through** specifies the second value of the range.

**Indicate Records in Data View** highlights records that match the search criteria in the Data View window (default = on).

**Indicate Objects in Drawing** highlights objects that match the search criteria in the drawing (default = on).

**Query Builder** tab

**(** groups search criteria with parentheses; up to four sets can be nested.

**Field** specifies a field name; double-click the cell to display the list of fields in the current table.

**Operator** specifies a logical operator; double-click to display the list of operators.

**Value** specifies a value for the query; click **...** to display a list of current values.

**)** specifies a closing parenthesis.

**Logical** specifies an And or Or operator; click once to add And; click again to change to Or.

**Fields in table** displays the fields in the current table; when no fields are selected, the query displays all fields from the table; double-click a field to add it to the list.

**Show fields** specifies the fields displayed by the Data View window; drag the field out of the list to remove it.

**Add** adds a field from the Fields in Table list to the Show Fields list.

**Sort By** specifies the sort order: the first field is the primary sort; to change the sort order, drag the field to another location in the list; press Delete to remove a field from the list.

**Add** adds a field from the Fields in table list to the Sort by list (default = ascending).

⋏⋎ reverses the sort order.

**Indicate Records in Data View** highlights records that match the search criteria in the Data View window (default = on).

**Indicate Objects in Drawing** highlights objects that match the search criteria in the drawing (default = on).

**SQL Query** tab

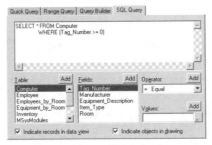

**Table** lists the names of all database tables available in the current data source.

**Add** adds the selected table to the SQL text editor.

**Fields** displays a list of field names in the selected database table.

**Add** adds the selected field to the SQL text editor.

**Operator** specifies the logical operator, which is added to the query (default = Equal).

**Add** adds the selected operator to the SQL text editor.

**Values** specifies a value for the selected field.

**Add** adds the value to the SQL text editor.

**...** lists available values for the field.

**Indicate Records in Data View** highlights records that match the search criteria in the Data View window (default = on).

**Indicate Objects in Drawing** highlights objects that match the search criteria in the drawing (default = on).

## Data View and Query Options dialog box

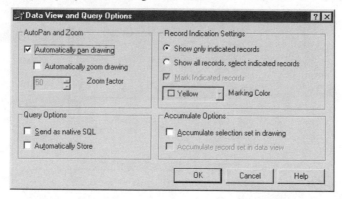

### AutoPan and Zoom options

**Automatically Pan Drawing** causes AutoCAD to pan the drawing automatically to display associated objects (default = on).

**Automatically Zoom Drawing** causes AutoCAD to zoom the drawing automatically to display associated objects (default = off).

**Zoom Factor** specifies the zoom factor as a percentage of the viewport area:

| Zoom Factor | Meaning |
|---|---|
| 20 | Minimum. |
| 50 | Default. |
| 90 | Maximum. |

### Query Options options

**Send as Native SQL** makes queries to database tables in:

☑ the format of the source table.

☐ SQL 92 format (default).

**Automatically Store** automatically stores queries when they are executed (default = off).

**Record Indication Settings**:

- **Show Only Indicated Records** displays the records associated with the current AutoCAD selections in the Data View window (default).

- **Show All Records, Select Indicated Records** displays all records in the current database table.

**Mark Indicated Records** colors linked records to differentiate them from unlinked records.

**Marking Color** specifies the marking color (default = yellow).

### Accumulate Options options

**Accumulate Selection Set in Drawing**

☑ adds objects to the selection set as data view records are added.

☐ replaces the selection set each time data view records are selected (default).

**Accumulate Record Set in Data View**

☑ adds records to the selection set as drawing objects are selected.

☐ replaces the selection set each time drawing objects are selected (default).

**New Link Template** dialog box

**New link template name** specifies the name of the link template.

**Start with template** reuses an existing template.

**Continue** displays the Link Template dialog box.

**Link Template** dialog box

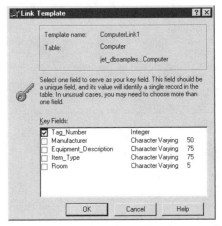

**Key Fields** selects one field name; you may select more than one field name, but AutoCAD warns you that too many key fields may slow performance.

**New Label Template** dialog box

**New label template name** specifies the name of the label template.

**Start with template** reuses an existing template.

**Continue** displays the Label Template dialog box.

**Label Template** dialog box

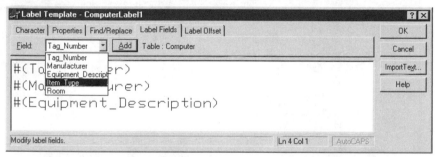

**Field** specifies the names of the fields.

**Add** adds the field to the label.

**Label Offset** displays the Label Offset tab:

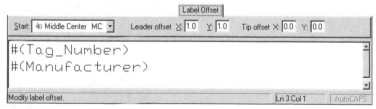

**Start** specifies the justification of the label starting point.

**Leader offset** specifies the x,y distance between the label's starting point and the leader line.

**Tip offset** specifies the offset distance to the leader tip or label text.

*See MText command for more information.*

## RELATED COMMANDS

**AttDef** creates an attribute definition, akin to an internal database.

**EAttExt** exports attributes.

## TIPS

- Query searches are case sensitive: "Computer" is not the same as "computer."

- OLE DB v2.0 must be installed before you use the **dbConnect Manager**.

- Leaders must have a length; to get rid of a leader, use a freestanding label.

- SQL is short for "structured query language."

- The properties of a link template can only be edited if it contains no links, and if the drawing is fully loaded (cannot be partially loaded).

- Before you can edit a record with an SQL Server table, you must define a primary key.

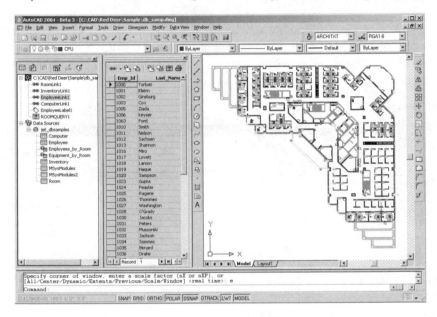

# Constructing Your First Query

### Step 1

From the **Tools** menu, select **dbConnect**. AutoCAD adds the **dbConnect** item to the menu.

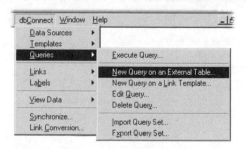

### Step 2

From the **dbConnect** menu, select **Queries | New Query on an External Table**. AutoCAD displays the **Select Data Object** dialog box.

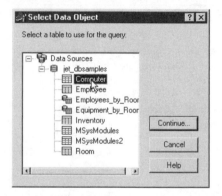

### Step 3

Select a table, and then click **Continue**. AutoCAD displays the **New Query** dialog box.

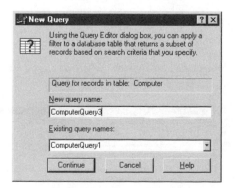

**Step 4**

Enter a name for the query in **New query name** text field, or select an existing name.

**Step 5**

Click **Continue**. AutoCAD opens the **Query Editor** dialog box.

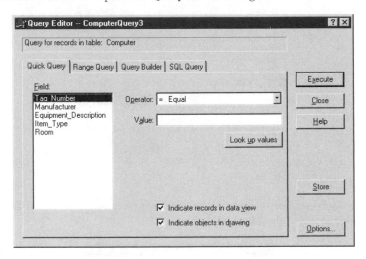

**Step 6**

Ensure the **Quick Query** tab is showing. Also, make sure that **Indicate records in data view** and **Indicate objects in drawing** are both checked.

**Step 7**

Select a field name from the **Field** list by highlighting it.

**Step 8**

Select an operator from the **Operator** list. For example, to match a value, select = **Equal**.

**Step 9**

Enter a value in the **Value** field, or click **Look up values** and select a value from the list of values already in the database.

**Step 10**

Click **Store** to save the query for reuse in the future.

**Step 11**

Click **Execute** to run the query. AutoCAD closes the dialog box, and then displays the records that match your selection in the **Data View** window.

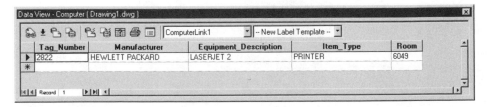

# DblClkEdit

**2000i**   Determines whether objects are edited by double-clicking the mouse button (*short for DouBLe CLicK EDITing*).

| Command | Alias | Ctrl+ | F-key | Alt+ | Menu Bar | Tablet |
|---|---|---|---|---|---|---|
| dblclkedit | ... | ... | ... | ... | ... | ... |

**Command:** dblclkedit
**Enter double-click editing mode [ON/OFF] <ON>:** *(Enter **ON** or **OFF**.)*

## COMMAND LINE OPTIONS

**ON** enables double-click editing.

**OFF** disables double-click editing for compatibility with earlier versions of AutoCAD.

## RELATED COMMANDS

**[Right-click]** displays a shortcut menu with editing commands.

**Properties** displays the Properties dialog box.

## TIPS

- When this command is turned on, double-clicking objects is the equivalent of entering a related editing command:

| Object | Dialog Box | Equivalent Command |
|---|---|---|
| Attribute | Edit Attribute | **DdEdit** or **EAttEdit** |
| Block | Reference Edit | **RefEdit** |
| Hatch | Hatch Edit | **HatchEdit** |
| Leader | Multiline Text Editor | **MtEdit** |
| Mline | Multiline Edit Tools | **MlEdit** |
| Mtext | Multiline Text Editor | **MtEdit** |
| Text | Edit Text | **DdEdit** |
| Xref | Reference Edit | **RefEdit** |

- In other cases, double-clicking an object displays the **Properties** window, and highlights the object with grips:

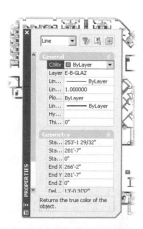

# DbList

**V. 1.0** Lists information on all objects in the drawing (*short for Data Base LISTing*).

| Command | Alias | Ctrl+ | F-key | Alt+ | Menu Bar | Tablet |
|---------|-------|-------|-------|------|----------|--------|
| dblist | ... | ... | ... | ... | ... | ... |

**Command:** dblist

*Sample listing:*

```
AutoCAD Text Window - C:\CAD\Red Deer\Sample\db_samp.dwg        _□ x

Edit

Command: dblist

               .LWPOLYLINE  Layer: "E-B-MULL"
                        Space: Model space
                  Handle = 3A
         Closed
Constant width    0'-0"
         area     11.25 square in. (0.0781 square ft.)
    perimeter     1'-2 1/2"

    at point  X=233'-4 33/64"  Y=270'-10 27/64"  Z=    0'-0"
    at point  X=233'-0 63/64"  Y=270'-6 7/8"   Z=    0'-0"
    at point  X=233'-2 9/16"   Y=270'-5 19/64"  Z=    0'-0"

Press ENTER to continue: |
```

## COMMAND LINE OPTIONS

ENTER continues display after pause.

ESC cancels database listing.

## RELATED COMMANDS

**Area** lists the area and perimeter of objects.

**Dist** lists the 3D distance and angle between two points.

**Id** lists the 3D coordinates of a point.

**List** lists information about selected objects in the drawing.

. . . . . . . . . . . . . . . . . . . . . . . . . . . . . . . . . . . . . . . . . . . . . . . . . . . . . . . . . . . . . .

## Removed Commands

**DdAttDef** was removed from AutoCAD 2000; it was replaced by **AttDef**.

**DdAttE** was removed from AutoCAD 2000; it was replaced by **AttEdit**.

**DdAttExt** was removed from AutoCAD 2000; it was replaced by **AttExt**.

**DdChProp** was removed from AutoCAD 2000; it was replaced by **Properties**.

**DdColor** was removed from AutoCAD 2000; it was replaced by **Color**.

. . . . . . . . . . . . . . . . . . . . . . . . . . . . . . . . . . . . . . . . . . . . . . . . . . . . . . . . . . . . . .

# DdEdit

**Rel.11** Edits a single line of text; launches the text editor for editing multiline text (*short for Dynamic Dialog EDITor*).

| Command | Alias | Ctrl+ | F-key | Alt+ | Menu Bar | Tablet |
|---------|-------|-------|-------|------|----------|--------|
| ddedit | ed | ... | ... | MOTE | Modify | Y21 |
| | | | | | ⤷Object | |
| | | | | | ⤷Text | |
| | | | | | ⤷Edit | |

**Command:** ddedit
**Select an annotation object or [Undo]:** *(Select a text object, or type **U**.)*
    *For single-line text placed by the **Text** and **DText** commands, it displays **Edit Text** dialog box.*
    *For paragraph text placed by the **MText** command, it displays the **Multiline Text Editor** dialog box.*
**Select an annotation object or [Undo]:** *(Press **Esc** to exit dialog box.)*

## COMMAND LINE OPTIONS
**Undo** undoes editing.

ESC ends the command.

## DIALOG BOX OPTIONS

**Text** displays the selected text; editing options:

| Keystroke | Meaning |
|-----------|---------|
| **Left** or **Up** | Moves cursor one character left. |
| **Right** or **Down** | Moves cursor one character right. |
| HOME | Moves cursor to beginning of line. |
| END | Moves cursor to end of line. |
| DELETE | Erases character to right of cursor. |
| BACKSPACE | Erases character to left of cursor. |

**OK** accepts editing changes, and dismisses dialog box.

**Cancel** ignores editing changes, and dismisses dialog box.

## Multiline Text Editor dialog box
*See **MText** command for **Multiline Text Editor** options.*

## RELATED COMMANDS

**AttEdit** edits all text attributes connected with a block.

**MtEdit** edits paragraph text.

**Properties** edits all text *properties*, including the text itself.

## RELATED SYSTEM VARIABLE

**MTextEd** specifies the name of the text editor used for editing multiline text.

## TIPS

• **DdEdit** automatically repeats; press ESC to cancel the command.

• As of AutoCAD 2000, this command no longer edits attribute text; use **AttEdit** instead.

## Removed Commands

**DdGrips** was removed from AutoCAD 2000; it was replaced by the **Selection** tab of the **Options** command.

**DDim** was removed from AutoCAD 2000; it was replaced by **DimStyle**.

**DdInsert** was removed from AutoCAD 2000; it was replaced by **Insert**.

**DdModify** was removed from AutoCAD 2000; it was replaced by **Properties**.

# 'DdPtype

Rel.12 Sets the style and size of points (*short for Dynamic Dialog Point TYPE*).

| Command | Alias | Ctrl+ | F-key | Alt+ | Menu Bar | Tablet |
|---------|-------|-------|-------|------|----------|--------|
| 'ddptype | ... | ... | ... | OP | Format | U1 |
| | | | | | ⬐ Point Style | |

**Command:** ddptype

*Displays dialog box:*

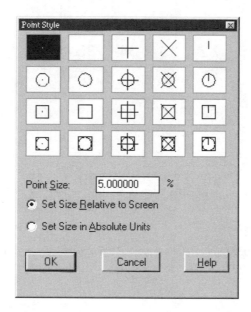

## DIALOG BOX OPTIONS

**Point size** sets the size in percentage or pixels.

**Set Size Relative to Screen** sets the size as a percentage of the total viewport height.

**Set Size in Absolute Units** sets the size in pixels.

## RELATED COMMANDS

**Divide** draws points along an object.

**Point** draws points.

**Measure** draws points a measured distance along an object.

**Regen** displays the new point format with a regeneration.

## RELATED SYSTEM VARIABLES

**PdSize** contains the size of the point:

| PdSize | Meaning |
|--------|---------|
| 0 | Point is 5% of viewport height (default). |
| *positive* | Absolute size in pixels. |
| *negative* | Percentage of the viewport size. |

**PdMode** determines the look of a point:

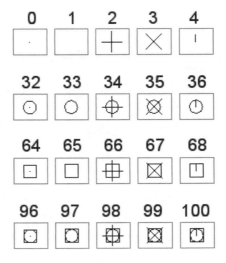

## TIP

- Points often cannot be seen in the drawing. To make them visible, change its mode and size.

## Removed Commands

**DdEModes** was removed from AutoCAD Release 14; it was replaced by the **Object Properties** toolbar.

**DdLModes** was removed in AutoCAD 2000; is was replaced by **Layer**.

**DdLtype** was removed in AutoCAD 2000; it was replaced by **Linetype**.

**DdRename** was removed from AutoCAD 2000; it was replaced by **Rename**.

**DdRModes** was removed from AutoCAD 2000; it was replaced by **DSettings**.

**DdSelect** was removed from AutoCAD 2000; it was replaced by the **Selection** tab of the **Options** command.

**DdUcs** and **DdUcsP** have been removed from AutoCAD 2000; they have been replaced by **UcsMan**.

**DdUnits** was removed from AutoCAD 2000; it was replaced by **Units**.

**DdView** was removed from AutoCAD 2000; it was replaced by **View**.

# DdVPoint

**Rel.12**  Changes the 3D viewpoint through a dialog box (*short for Dynamic Dialog ViewPOINT*).

| Command | Alias | Ctrl+ | F-key | Alt+ | Menu Bar | Tablet |
|---------|-------|-------|-------|------|----------|--------|
| ddvpoint | vp | ... | ... | V3I | View | N5 |
| | | | | | ⌐3D Views | |
| | | | | | ⌐Viewpoint Presets | |

**Command:** ddvpoint

*Displays dialog box:*

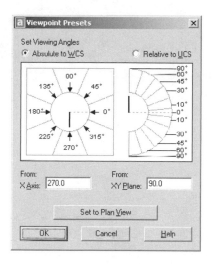

## DIALOG BOX OPTIONS

Set Viewing Angles options

**Absolute to WCS** sets the view direction relative to the WCS.

**Relative to UCS** sets the view direction relative to the current UCS.

**From X Axis** measures the view angle from the x axis.

**From XY Plane** measures the view angle from the x,y plane

**Set to Plan View** changes the view to plan view.

## RELATED COMMANDS

**VPoint** adjusts the viewpoint from the command line.

**3dOrbit** changes the 3D viewpoint interactively.

## RELATED SYSTEM VARIABLE

**WorldView** determines whether viewpoint coordinates are in WCS or UCS.

## TIPS

- After changing the viewpoint, AutoCAD performs an automatic **Zoom** extents.

- In the image tile, the black arm indicates the new angle.

- In the image tile, the red arm indicates the current angle.

- To select an angle with your mouse:

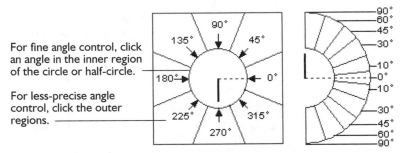

For fine angle control, click an angle in the inner region of the circle or half-circle.

For less-precise angle control, click the outer regions.

- WCS is short for "world coordinate system."

- UCS is short for "user-defined coordinate system."

# 'Delay

**v. 1.4**  Delays the next command, in milliseconds.

| Command | Alias | Ctrl+ | F-key | Alt+ | Menu Bar | Tablet |
|---------|-------|-------|-------|------|----------|--------|
| 'delay  | ...   | ...   | ...   | ...  | ...      | ...    |

**Command:** delay
**Delay time in milliseconds:** *(Enter the number of milliseconds.)*

## COMMAND LINE OPTION

**Delay time** specifies the number of milliseconds by which to delay the next command.

## RELATED COMMAND

**Script** initiates scripts.

## TIPS

- Use **Delay** to slow down the execution of a script file.
- The maximum delay is 32767, just over 32 seconds.

# DetachURL

<u>Rel.14</u> Removes URLs from objects and areas *(undocumented command)*.

| Command | Alias | Ctrl+ | F-key | Alt+ | Menu Bar | Tablet |
|---------|-------|-------|-------|------|----------|--------|
| detachurl | ... | ... | ... | ... | ... | ... |

**Command:** detachurl
**Select objects:** *(Select one or more objects.)*
**Select objects:** *(Press Enter.)*

## COMMAND LINE OPTION

**Select objects** selects the objects from which to remove URL(s).

## RELATED COMMANDS

**AttachUrl** attaches a hyperlink to an object or an area.

**Hyperlink** attaches and removes hyperlinks via a dialog box.

**SelectUrl** selects all objects with attached hyperlinks.

## TIPS

- When you select a hyperlinked area to detach, AutoCAD reports:

      1. hyperlink ()
      Remove, deleting the Area.
      1 hyperlink deleted...

- When you select an object with no hyperlink attached, AutoCAD reports nothing.

- A URL (short for "uniform resource locator") is the universal file naming convention of the Internet; also called a link or hyperlink.

# Dim

**V. 1.2**  Changes the prompt from 'Command' to 'Dim', allowing access to AutoCAD's old dimensioning mode (*short for DIMensions*).

| Command | Alias | Ctrl+ | F-key | Alt+ | Menu Bar | Tablet |
|---------|-------|-------|-------|------|----------|--------|
| dim | ... | ... | ... | ... | ... | ... |

**Command:** dim
**Dim:** *(Enter a dimension command from the list below.)*

## COMMAND LINE OPTIONS
*Aliases for the dimension commands are shown in UPPERCASE.*

**ALigned** draws linear dimensions aligned with objects (*first introduced with AutoCAD version 2.0*); replaced by DimAligned.

**ANgular** draws angular dimensions that measure angles (*ver. 2.0*); replaced by DimAngular.

**Baseline** continues dimensions from basepoints (*ver. 1.2*); replaced by QDim and DimBaseline.

**CEnter** draws '+' center marks on circles and arcs centers (*ver. 2.0*); replaced by DimCenter.

**COntinue** continues dimensions from previous dimensions' extension lines (*ver. 1.2*); replaced by the QDim and DimContinue commands.

**Diameter** draws diameter dimensions on circles, arcs, and polyarcs (*ver. 2.0*); replaced by the QDim and DimDiameter commands.

**Exit** returns to 'Command' prompt from 'Dim' prompt (*ver. 1.2*).

**HOMetext** returns dimension text to its original position (*ver. 2.6*); replaced by the DimEdit command's Home option.

**HORizontal** draws horizontal dimensions (*ver. 1.2*); replaced by DimLinear.

**LEAder** draws leaders (*ver. 2.0*); replaced by the Leader and QLeader commands.

**Newtext** edits text in associative dimensions (*ver. 2.6*); replaced by the DimEdit command's New option.

**OBlique** changes the angle of extension lines in associative dimensions (*rel. 11*); replaced by the DimEdit command's Oblique option.

**ORdinate** draws x- and y-ordinate dimensions (*Rel. 11*); replaced by QDim and DimOrdinate.

**OVerride** overrides current dimension variables (*Rel. 11*); replaced by DimOverride.

**RAdius** draws radial dimensions on circles, arcs, and polyline arcs (*ver. 2.0*); replaced by the QDim and DimRadius commands.

**REDraw** redraws the current viewport (same as 'Redraw; *ver. 2.0*).

**REStore** restores dimensions to the current dimension style (*Rel. 11*); replaced by the -DimStyle command's Restore option.

**ROtated** draws linear dimensions at any angle (*ver. 2.0*); replaced by DimLinear.

**SAve** saves the current settings of dimension styles (*Rel. 11*); replaced by the -DimStyle command's Save option.

**STAtus** lists the current settings of dimension variables (*ver. 2.0*); replaced by the -DimStyle command's Status option.

**STYle** defines styles for dimensions (*ver. 2.5*); replaced by DimStyle.

**TEdit** changes locations and orientation of text in associative dimensions (*Rel. 11*); replaced by DimTEdit.

**TRotate** changes the rotation angle of text in associative dimensions (*Rel. 11*); replaced by the DimTEdit command's Rotate option.

**Undo** undoes the last dimension action (*ver. 2.0*); replaced by the Undo command.

**UPdate** updates selected associative dimensions to the current dimvar setting (*ver. 2.6*); replaced by the -DimStyle command's Apply option.

**VAriables** lists the values of variables associated with dimension styles, *not* dimvars (*Rel. 11*); replaced by the -DimStyle command's Variables option.

**VErtical** draws vertical linear dimensions (*ver. 1.2*); replaced by DimLinear.

## RELATED DIM VARIABLES

**Dim**xxx specifies system variables for dimensions; see the DimStyle command.

**DimAso** determines whether dimensions are drawn associatively.

**DimScale** determines the dimension scale.

## TIPS

- The 'Dim' prompt dimension commands are included for compatibility with AutoCAD Release 12 and earlier.

- Only transparent commands and dimension commands work at the 'Dim' prompt. To use other commands, you must exit the 'Dim' prompt and return to the 'Command' prompt with the **Exit** command.

- The defpoint (short for "definition point") is used by earlier versions of AutoCAD to locate the extension lines. Defpoints appear as a small dot on the **DefPoints** layer. When stretching a dimension, make sure you include the defpoints, otherwise the dimension will not be updated automatically.

- Most dimensions consist of four basic components, as shown below:

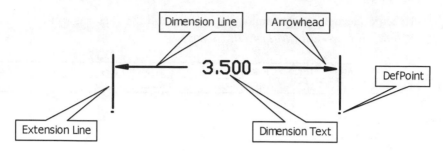

- As of AutoCAD 2002, defpoints are not used when **DimAssoc**=2 (the default); dimensions are attached directly to objects instead.

- All the components of an *associative dimension* are treated as a single object; components of a nonassociative dimension are treated as individual objects.

- For additional control over dimensions, see the following undocumented commands:

  AiDimPrec
  AiDimStyle
  Ai_Dim_TextAbove
  Ai_Dim_TextCenter
  Ai_Dim_TextHome
  AiDimTextMove

- When changing dimension text, AutoCAD recognizes the <> metacharacters to represent the dimension text. For example, if AutoCAD measures a dimension as 25.4, then **<>mm** means 25.4mm.

# Dim1

**V. 2.5**
Executes a single dimensioning command, and then returns to the 'Command' prompt (*short for DIMension once*).

| Command | Alias | Ctrl+ | F-key | Alt+ | Menu Bar | Tablet |
|---------|-------|-------|-------|------|----------|--------|
| dim1 | ... | ... | ... | ... | ... | ... |

**Command:** dim1
**Dim:** *(Enter a dimension command.)*

## COMMAND LINE OPTIONS
*All "old" dimension commands; see the **Dim** command for the complete list.*

## RELATED COMMANDS
**DimStyle** displays the dialog box for setting dimension variables.

**Dim** switches to the "old" dimensioning mode and remains there.

## RELATED DIM VARIABLES
**DimAso** determines whether dimensions are drawn associatively.

**DimTxt** determines the height of text.

**DimScale** determines the dimension scale.

## TIP
- Use **Dim1** when you require just a single "old" dimension command.

# DimAligned

**Rel 13**   Draws linear dimensions aligned with objects.

| Command | Aliases | Ctrl+ | F-key | Alt+ | Menu Bar | Tablet |
|---------|---------|-------|-------|------|----------|--------|
| dimaligned | dal | ... | ... | NG | Dimension | W4 |
| | dimali | | | | ⇧Aligned | |

**Command:** dimaligned
**Specify first extension line origin or <select object>:** *(Pick a point, or select object.)*
**Select object to dimension:** *(Select an object.)*
**Specify dimension line location or [Mtext/Text/Angle]:** *(Pick a point, or enter an option.)*
**Dimension text =** *nnn*

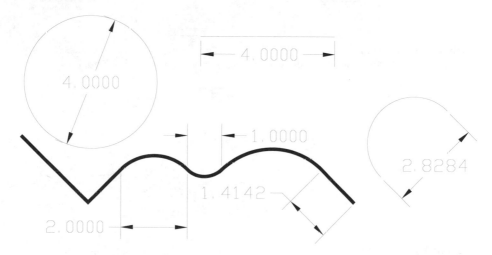

## COMMAND LINE OPTIONS

**Specify first extension line origin** picks a point for the origin of the first extension line.

**Specify second extension line origin** picks a point for the origin of the second extension line.

**Select object** selects an object to dimension after pressing ENTER.

**Select object to dimension** picks a line, arc, polyline, and so on; the individual segments of a polyline are dimensioned.

**Specify dimension line location** picks a point from which to locate the dimension line and text.

**Mtext** changes the wording of the dimension text.

**Text** changes the position of the text.

**Angle** changes the angle of dimension text.

## RELATED DIM COMMAND

**DimRotated** draws an angular dimension line with a perpendicular extension line.

# Editing Dimensions with Grips

Dimensions may be edited directly without first invoking an editing command. Click the dimension once to display grips (small blue squares). Double-click the dimension to display the **Properties** window; see the **Properties** command.

## Linear Dimensions

Grips change the location of the dimension line, text, and extension lines.

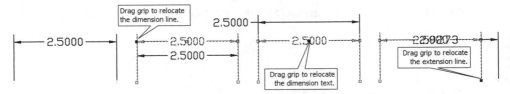

## Aligned Dimensions

Grips change the location of the dimension line, text, and angle of the entire dimension.

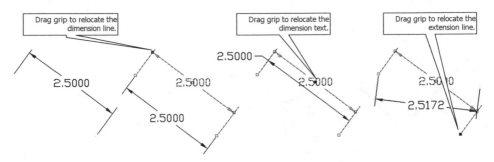

## Angular Dimensions

Grips change the location of the dimension line, text, arrowhead location, and extension lines.

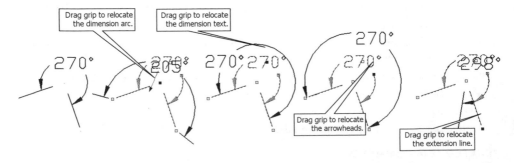

## Diameter and Radius Dimensions

Grips change the location of the dimension line, text, mark.

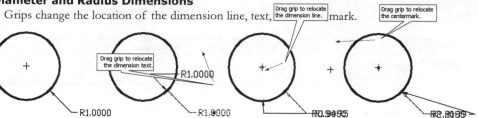

## Ordinate Dimensions

Grips change the location of the ordinate line, text, and endpoint.

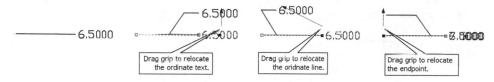

## Leader Dimensions

Grips allow you to change the location of the dimension line, text, and extension lines.

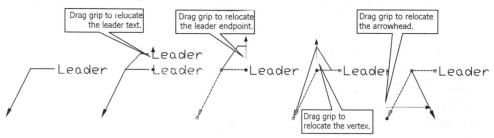

# DimAngular

**Rel 13**  Draws dimensions that measure angles.

| Command | Aliases | Ctrl+ | F-key | Alt+ | Menu Bar | Tablet |
|---------|---------|-------|-------|------|----------|--------|
| dimangular | dan | ... | ... | NA | Dimension | X3 |
| | dimang | | | | ↳Angular | |

**Command:** dimangular
**Select arc, circle, line, or <specify vertex>:** *(Select an object, or pick a vertex.)*
**Specify second angle endpoint:** *(Pick a point.)*
**Specify dimension arc line location or [Mtext/Text/Angle]:** *(Pick a point, or enter an option.)*
**Dimension text =** *nnn*

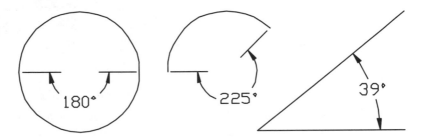

## COMMAND LINE OPTIONS

**Select arc** measures the angle of the arc.

**Circle** prompts you to pick two points on the circle.

**Line** prompts you to pick two lines.

**Specify vertex** prompts you to pick points to make an angle.

**Specify dimension arc/line location** specifies the location of the angular dimension.

**Mtext** changes the wording of the dimension text.

**Text** changes the position of the text.

**Angle** changes the angle of the dimension text.

## RELATED DIM COMMANDS

**DimCenter** places a center mark at the center of an arc or circle.

**DimRadius** dimensions the radius of an arc or circle.

# DimBaseline

<u>Rel 13</u>  Draws linear dimensions based on previous starting points.

| Command | Aliases | Ctrl+ | F-key | Alt+ | Menu Bar | Tablet |
|---|---|---|---|---|---|---|
| dimbaseline | dba | ... | ... | NB | Dimension | ... |
| | dimbase | | | | ⍦Baseline | |

**Command:** dimbaseline
**Specify a second extension line origin or [Undo/Select] <Select>:** *(Pick a point, or enter an option.)*
**Dimension text = *nnn***
**Specify a second extension line origin or [Undo/Select] <Select>:** *(Pick a point, enter an option, or press **Esc** to exit command.)*

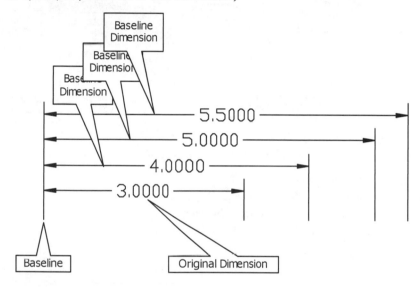

## COMMAND LINE OPTIONS

**Specify a second extension line origin** positions the extension line of the next baseline dimension.

**Select** prompts you to select the originating dimension.

**Undo** undoes the previous baseline dimension.

**Esc** exits the command.

## RELATED DIM COMMANDS

**Continue** continues linear dimensioning from the last extension point.

**QDim** creates a continuous or baseline dimension quickly.

## RELATED DIM VARIABLES

**DimDli** specifies the distance between baseline dimension lines.

**DimSe1** suppresses the first extension line.

**DimSe2** suppresses the second extension line.

# DimCenter

__Rel 13__  Draws centermarks and lines on arcs and circles.

| Command | Alias | Ctrl+ | F-key | Alt+ | Menu Bar | Tablet |
|---------|-------|-------|-------|------|----------|--------|
| dimcenter | dce | ... | ... | NM | Dimension | X2 |
| | | | | | ⮑ Center Mark | |

**Command:** dimcenter
**Select arc or circle:** *(Select an arc or a circle.)*

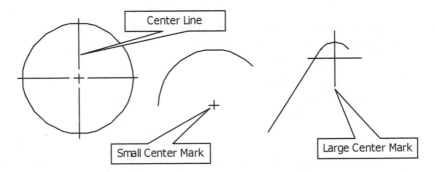

**COMMAND LINE OPTION**

**Select arc or circle** places the center mark at the center of the selected arc, circle, or polyarc.

**RELATED DIM COMMANDS**

**DimAngular** dimensions arcs and circles.

**DimDiameter** dimensions arcs and circles by diameter value.

**DimRadius** dimensions arcs and circles by radius value.

**RELATED DIM VARIABLE**

**DimCen** specifies the size and type of the center mark:

| DimCen | Meaning |
|--------|---------|
| *negative value* | Draws center marks and lines. |
| 0 | Does not draw center mark or center lines. |
| *positive value* | Draws center marks. |
| 0.09 | Default value. |

# DimContinue

**Rel 13**   Continues dimension from the second extension line of previous
dimensions.

| Command | Aliases | Ctrl+ | F-key | Alt+ | Menu Bar | Tablet |
|---------|---------|-------|-------|------|----------|--------|
| dimcontinue | dco | ... | ... | NC | Dimension | ... |
| | dimcont | | | | ↳Continue | |

**Command:** dimcontinue
**Specify a second extension line origin or [Undo/Select] <Select>:** *(Pick a point, or enter an option.)*
**Dimension text = ***nnn*
**Specify a second extension line origin or [Undo/Select] <Select>:** *(Pick a point, enter an option, or press **Esc** to exit command.)*

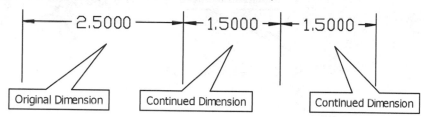

## COMMAND LINE OPTIONS

**Specify a second extension line origin** positions the extension line of the next continued dimension.

**Select** prompts you to select the originating dimension.

**Undo** undoes the previous continued dimension.

**Esc** exits the command.

## RELATED DIM COMMANDS

**DimBaseline** continues dimensioning from the first extension point.

**QDim** creates a continuous or baseline dimension quickly.

## RELATED DIM VARIABLES

**DimDli** sets the distance between continuous dimension lines.

**DimSe1** suppresses the first extension line.

**DimSe2** suppresses the second extension line.

# DimDiameter

**Rel 13** Draws diameter dimensions on arcs, circles, and polyline arcs.

| Command | Aliases | Ctrl+ | F-key | Alt+ | Menu Bar | Tablet |
|---------|---------|-------|-------|------|----------|--------|
| dimdiameter | ddi | ... | ... | ND | Dimension | X4 |
| | dimdia | | | | ⌐Diameter | |

**Command:** diameter
**Select arc or circle: (Select** *an arc or a circle.)*
**Dimension text =** *nnn*
**Specify dimension line location or [Mtext/Text/Angle]:** *(Pick a point, or else enter an option.)*

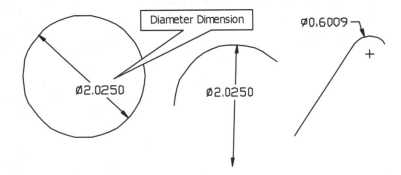

## COMMAND LINE OPTION

**Select arc or circle** selects an arc, circle, or polyarc.

**Specify dimension line location** specifies the location of the angular dimension.

**Mtext** changes the wording of the dimension text.

**Text** changes the position of the text.

**Angle** changes the angle of the dimension text.

## RELATED DIM COMMANDS

**DimCenter** marks the center point of arcs and circles.

**DimRadius** draws the radius dimension of arcs and circles.

## TIP

• To include the diameter symbol, use the **%%d** code or the Unicode **\U+2205** .

# DimDisassociate

2002 Converts associative dimensions to non-associative (*the AutoCAD command that's most difficult to spell correctly*).

| Command | Alias | Ctrl+ | F-key | Alt+ | Menu Bar | Tablet |
|---|---|---|---|---|---|---|
| dimdisassociate | ... | ... | ... | ... | ... | ... |

**Command:** dimdisassociate
**Select dimensions to disassociate...**
**Select objects:** *(Select one or more dimensions, or enter All to select all dimensions.)*
**Select objects:** *(Press Enter to end object selection.)*
**36 disassociated.**

## COMMAND LINE OPTION

**Select objects** selects dimensions to convert to non-associative type.

## RELATED DIM COMMANDS

**DimReassociate** converts dimensions from non-associative to associative.

**DimRegen** makes all dimensions associative automatically.

## RELATED SYSTEM VARIABLE

**DimAssoc** determines whether newly-created dimensions are associative:

| DimAssoc | Meaning |
|---|---|
| 0 | Dimension is created "exploded," where all parts (such as dimension lines, arrowheads) are individual, ungrouped objects. |
| 1 | Dimension is created as a single object, but is not associative. |
| 2 | dimension is created as a single object, and is associative (default). |

## TIPS

• As you select objects, this command ignores non-dimensions, dimensions on locked layers, and those not in the current space (model or paper).

• The command displays a report of filtered and disassociated dimensions.

• The effect of this command can be reversed with the **U** and the **DimReassociate** commands.

# DimEdit

**Rel 13**  Applies editing changes to the dimension text.

| Command | Aliases Ctrl+ | F-key | Alt+ | Menu Bar | Tablet |
|---------|---------------|-------|------|----------|--------|
| dimedit | ded  ... | ... | NQ | Dimension | Y1 |
|         | dimed |  |  | ⤷Oblique |  |

**Command:** dimedit
**Enter type of dimension editing [Home/New/Rotate/Oblique] <Home>:** *(Enter an option.)*
**Select objects:** *(Select one or more objects.)*
**Select objects:** *(Press* **Enter.***)*

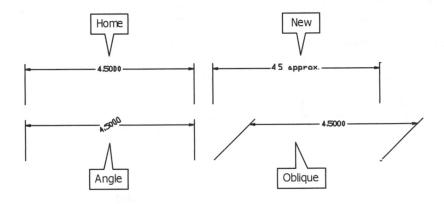

## COMMAND LINE OPTIONS
**Angle** rotates the dimension text.

**Home** returns the dimension text to its original position.

**Oblique** rotates the extension lines.

**New** allows editing of the dimension text.

## RELATED DIM COMMANDS
*All.*

## RELATED DIM VARIABLES
*Most.*

## TIPS
• When you enter dimension text with the **DimEdit** command's **New** option, AutoCAD recognizes the <> metacharacters represent existing text.

• Use the **Oblique** option to angle dimension lines by 30 degrees, suitable for isometric drawings; use **Style** command to oblique text by 30 degrees. See the **Isometric** command.

• **DimEdit** operates differently when used from the command line than from the menu bar: the **Oblique** option is the default when selected from the menu bar.

# DimHorizontal

<u>**2004**</u>   Draw horizontal dimensions *(undocumented command).*

| Command | Aliases | Ctrl+ | F-key | Alt+ | Menu Bar | Tablet |
|---|---|---|---|---|---|---|
| dimhorizontal | ... | ... | ... | ... | ... | ... |

**Command:** dimhorizontal
**Specify first extension line origin or <select object>:** *(Pick a point, or press* **Enter** *to select one object.)*
**Specify second extension line origin:** *(Pick a point.)*
**Specify dimension line location or [Mtext/Text/Angle]:** *(Pick a point, or select an option.)*
**Dimension text =** *nnn*

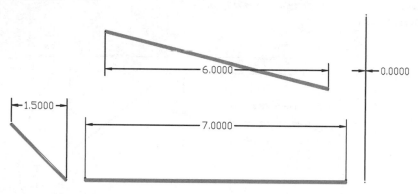

## COMMAND LINE OPTIONS

**Specify first extension line origin** specifies the location of the first extension line's origin.

**Select object** dimensions a line, arc, or circle automatically.

**Specify second extension line location** specifies the location of the second extension line.

**Specify dimension line location** specifies the location of the dimension line.

**Mtext** displays the Multiline Text Editor dialog box, which allows you to modify the dimension text; see the MText command.

**Text** prompts you to replace the dimension text on the command line: 'Enter dimension text <*nnn*>'.

**Angle** changes the angle of the dimension text: 'Specify angle of dimension text:'.

## RELATED DIM COMMANDS
*All.*

## RELATED DIM VARIABLES
*Most.*

# DimLinear

**Rel 13** Draws horizontal dimensions.

| Command | Aliases | Ctrl+ | F-key | Alt+ | Menu Bar | Tablet |
|---------|---------|-------|-------|------|----------|--------|
| dimlinear | dli | ... | ... | NL | Dimension | W5 |
| | dimlin | | | | ↳Linear | |

**Command:** dimlinear
**Specify first extension line origin or <select object>:** *(Pick a point, or press* **Enter** *to select an object.)*
**Specify second extension line origin:** *(Pick a point.)*
**Specify dimension line location or [Mtext/Text/Angle/Horizontal/Vertical/Rotated]:** *(Pick a point, or else enter an option.)*
**Dimension text =** *nnn*

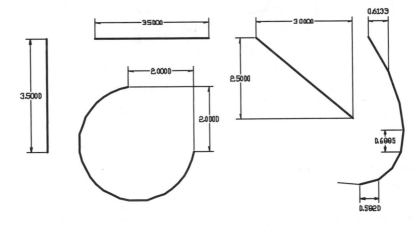

## COMMAND LINE OPTIONS

**Specify first extension line origin** specifies the location of the origin of the first extension line.

**Select object** dimensions a line, arc, or circle automatically.

**Specify second extension line location** specifies the location of the second extension line.

**Specify dimension line location** specifies the location of the dimension line.

**Mtext** displays the Multiline Text Editor dialog box, which allows you to modify the dimension text; see the MText command.

Text option
**Enter dimension text <*nnn*>:** *(Enter dimension text, or press* **Enter** *to accept default value.)*

**Enter dimension text** prompts you to replace the dimension text on the command line.

Angle option
**Specify angle of dimension text:** *(Enter an angle.)*

**Specify angle** changes the angle of dimension text.

Horizontal option
**Specify dimension line location or [Mtext/Text/Angle]:** *(Pick a point or enter an option.)*

Horizontal option forces dimension to be horizontal.

Vertical option
**Specify dimension line location or [Mtext/Text/Angle]:** *(Pick a point or enter an option.)*

Vertical forces dimension to be vertical.

Rotated option
**Specify angle of dimension line <0>:** *(Enter an angle.)*

Rotated forces dimension to be rotated.

Select Object options

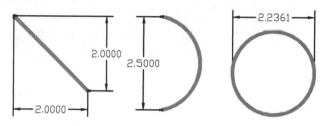

**Command:** dimlinear
**Specify first extension line origin or <select object>:** *(Press **Enter**.)*
**Select object to dimension:** *(Select one object.)*
**Specify dimension line location or**
**[Mtext/Text/Angle/Vertical/Rotated]:** *(Pick a point, or enter an option.)*
**Dimension text** = *nnn*

## RELATED DIM COMMANDS

**DimAligned** draws linear dimensions aligned with objects.

**QDim** dimensions an object quickly.

# DimOrdinate

**Rel 13**   Draws x- and y-ordinate dimensions.

| Command | Aliases | Ctrl+ | F-key | Alt+ | Menu Bar | Tablet |
|---------|---------|-------|-------|------|----------|--------|
| dimordinate | dor | ... | ... | NO | Dimension | W3 |
|  | dimord |  |  |  | ⌖Ordinate |  |

**Command:** dimordinate
**Specify feature location:** *(Pick a point.)*
**Specify leader endpoint or [Xdatum/Ydatum/Mtext/Text/Angle]:** *(Pick a point, or else enter an option.)*
**Dimension text =** *nnn*

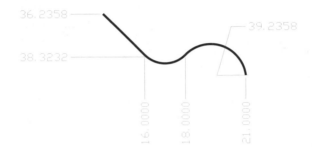

### COMMAND LINE OPTIONS

**Xdatum** forces x ordinate dimension.

**Ydatum** forces y ordinate dimension.

**Mtext** displays the Multiline Text Editor dialog box, which allows you to modify the dimension text; see the MText command.

**Text** prompts you to replace the dimension text on the command line: 'Enter dimension text *<nnn>*'.

**Angle** changes the angle of dimension text: 'Specify angle of dimension text:'.

### RELATED DIM COMMANDS

**Leader** draws leader dimensions.

**Tolerance** draws geometric tolerances.

### RELATED TOOLBAR ICONS

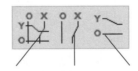

DimOrdinate   XDatum   YDatum

### TIP

• AutoCAD misuses the term "ordinate," which by definition is the distance from the x-axis only.

# DimOverride

<u>Rel 13</u>  Overrides the current dimension variables.

| Command | Aliases | Ctrl+ | F-key | Alt+ | Menu Bar | Tablet |
|---|---|---|---|---|---|---|
| dimoverride | dov | ... | ... | NV | Dimension | Y4 |
| | dimover | | | | ⌦Override | |

**Command:** dimoverride
**Enter dimension variable name to override or [Clear overrides]:** *(Enter the name of a dimension variable, or type* **C**.*)*
**Enter new value for dimension variable** *<nnn>*: *(Enter a new value.)*
**Select objects:** *(Select one or more dimensions.)*
**Select objects:** *(Press* **Enter**.*)*

## COMMAND LINE OPTIONS

**Dimension variable to override** requires you to enter the name of the dimension variable.

**Clear** removes the override.

**New Value** specifies the new value of the dimvar.

**Select objects** selects the dimension objects to which the change applies.

## RELATED DIM COMMAND

**DimStyle** creates and modifies dimension styles.

## RELATED DIM VARIABLES

*All dimension variables.*

# DimRadius

<u>Rel 13</u>  Draws radial dimensions on circles, arcs, and polyline arcs.

| Command | Aliases | Ctrl+ | F-key | Alt+ | Menu Bar | Tablet |
|---------|---------|-------|-------|------|----------|--------|
| dimradius | dra | ... | ... | NR | Dimension | X5 |
| | dimrad | | | | ↳Radius | |

**Command:** dimradius
**Select arc or circle:** *(Select an arc or a circle.)*
**Dimension text =** *nnn*
**Specify dimension line location or [Mtext/Text/Angle]:** *(Pick a point, or else enter an option.)*

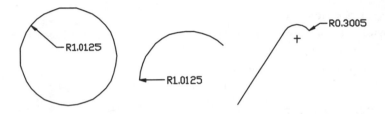

## COMMAND LINE OPTIONS

**Select arc or circle** selects the arc, circle, or polyarc to dimension.

**Specify dimension line location** specifies the location of the angular dimension.

**Mtext** displays the **Multiline Text Editor** dialog box, which allows you to modify the dimension text; see the **MText** command.

**Text** prompts you to replace the dimension text at the command line: "Enter dimension text *<nnn>*".

**Angle** changes the angle of dimension text, and prompts: "Specify angle of dimension text:".

## RELATED DIM COMMANDS

**DimCenter** draws center mark on arcs and circles.

**DimDiameter** draws diameter dimensions on arcs and circles.

## RELATED DIM VARIABLE

**DimCen** determines the size of the center mark.

# DimReassociate

2002 Associates dimensions with objects.

| Command | Alias | Ctrl+ | F-key | Alt+ | Menu Bar | Tablet |
|---------|-------|-------|-------|------|----------|--------|
| dimreassociate | ... | ... | ... | NN | Dimension | X4 |
| | | | | | ↳ Reassociate Dimensions | |

**Command:** dimreassociate

**Select dimensions to reassociate...**

**Select objects:** *(Select one or more dimensions, or enter **All** to select all dimensions.)*

**58 found. 4 were on a locked layer. 18 were not in current space.**

**Select objects:** *(Press **Enter** to end object selection.)*

**Specify first extension line origin or [Select object]:** *(Pick a point, or select an object.)*

**Select are or circle <next>:** *(Press **Enter** to end object selection.)*

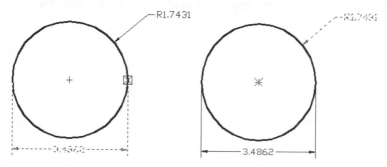

*The boxed X indicates that the linear dimension is associated with the circle; the X indicates the radius dimension is not associated.*

## COMMAND LINE OPTIONS

**Select objects** selects the dimensions to reassociate.

**Specify first extension line origin** picks a point to which the dimension should be associated.

**Select arc or circle** picks a circle or arc to which the dimension should be associated.

**Next** goes to the next circle or arc.

## RELATED DIM COMMANDS

**DimDisassociate** converts dimensions from associative to non-associative.

**DimRegen** makes all dimensions associative automatically.

## TIPS

- As you select objects, this command ignores non-dimensions, dimensions on locked layers, and those not in the current space (model or paper).

- AutoCAD displays a boxed X to indicate the object to which the dimension is associated; an unboxed X indicates an unassociated dimension. The markers disappear when a wheelmouse performs a zoom or pan.

# DimRegen

Updates the locations of associative dimensions.

| Command | Alias | Ctrl+ | F-key | Alt+ | Menu Bar | Tablet |
|---------|-------|-------|-------|------|----------|--------|
| dimregen | ... | ... | ... | ... | ... | ... |

**Command:** dimregen

## COMMAND LINE OPTIONS
*None.*

## RELATED DIM COMMANDS
**DimDisassociate** converts dimensions from associative to non-associative.

**DimReassociate** converts dimensions from non-associative to associative.

## TIPS
- This command is meant for use after three conditions:

    The drawing has been edited by a version of AutoCAD prior to 2004.

    The drawing contains externally-referenced dimensions, and the xref has been edited.

    The drawing is in layout mode, model space is active, and a wheelmouse has been used to pan or zoom.

# DimRotated

2004 Draw rotated dimensions (*undocumented command*).

| Command | Alias | Ctrl+ | F-key | Alt+ | Menu Bar | Tablet |
|---|---|---|---|---|---|---|
| dimrotated | ... | ... | ... | ... | ... | ... |

**Command:** dimrotated
**Specify angle of dimension line <0>:** *(Enter a rotation angle.)*
**Specify first extension line origin or <select object>:** *(Pick a point, or press* **Enter** *to select one object.)*
**Specify second extension line origin:** *(Pick a point.)*
**Specify dimension line location or [Mtext/Text/Angle]:** *(Pick a point, or select an option.)*
**Dimension text =** *nnn*

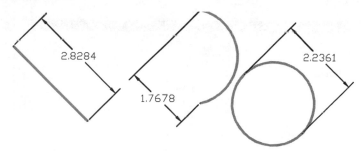

*Rotated dimensions placed at -45 degrees.*

## COMMAND LINE OPTIONS

**Specify angle of dimension line** specifies the angle at which the dimension line is rotated.

**Specify first extension line origin** specifies the location of the first extension line's origin.

**Select object** dimensions a line, arc, or circle automatically.

**Specify second extension line location** specifies the location of the second extension line.

**Specify dimension line location** specifies the location of the dimension line.

**Mtext** displays the Multiline Text Editor dialog box, which allows you to modify the dimension text; see the MText command.

**Text** prompts you to replace the dimension text on the command line: 'Enter dimension text <*nnn*>'.

**Angle** changes the angle of dimension text: 'Specify angle of dimension text:'.

## RELATED DIM COMMANDS
*All.*

## RELATED DIM VARIABLES
*Most.*

# DimStyle

**Rel 13**   Creates and edits dimstyles *(short for DIMension STYLE)*.

| Commands | Aliases | Ctrl+ | F-key | Alt+ | Menu Bar | Tablet |
|----------|---------|-------|-------|------|----------|--------|
| dimstyle | d | ... | ... | NS | Dimension | Y5 |
|          | dst |   |   |   | ⓑStyle |   |
|          | dimsty |   |   | OD | Format |   |
|          | ddim |   |   |   | ⓑDimension Style |   |
| -dimstyle | ... | ... | ... | NU | Dimension |   |
|          |   |   |   |   | ⓑUpdate |   |

## Command: dimstyle

*Displays dialog box:*

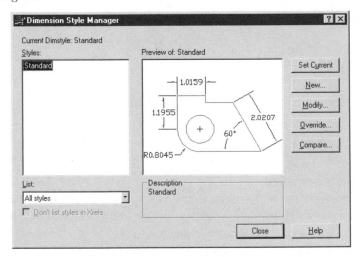

## DIALOG BOX OPTIONS

**Styles** lists the names of dimension styles in the drawing.

**List** modifies the style names listed under Styles: All styles or Styles in use.

**Don't list styles in Xrefs**

   ☑ Dimension styles found in externally-referenced drawings are not listed.

   ☐ Xref dimension styles are listed under **Styles** (default).

**Preview** displays a preview of the elements of the current dimension style.

**Description** names the current dimstyle; shows changes to dimension variables, if any.

**Set Current** sets the selected style as the current dimension style.

**New** creates a new dimension style via Create New Dimension Style dialog box.

**Modify** modifies an existing dimension style; displays the Modify Dimension Style dialog box.

**Override** allows temporary changes to a dimension style; displays the Override Dimension Style dialog box.

**Compare** lists the differences between dimension variables of two styles; displays the Compare Dimension Styles dialog box.

## CURSOR MENU OPTIONS

*Right-click a dimension style name under **Styles:***

**Set Current** sets the selected dimension style as the current style.

**Rename** renames dimensions styles.

**Delete** erases selected dimension styles from the drawing; you cannot erase the STANDARD style, or styles that are in use.

## Create New Dimension Style dialog box

*When creating a new dimension style, typically you start by making changes to an existing style.*

**New Style Name** specifies the name of the new dimension style.

**Start With** lists the names of the current dimension style(s), which are used as the template for the new dimension style.

**Use for** creates a substyle that applies to a specific type of dimension type: linear, angular, radius, diameter, ordinate, leaders and tolerances.

**Continue** continues to the next dialog box, New Dimension Style.

**Cancel** dismisses this dialog box, and returns to the Dimension Style Manager dialog box.

## New Dimension Style dialog box

**OK** records the changes made to dimension properties, and returns to the Dimension Manager dialog box.

**Cancel** cancels the changes, and returns to the Dimension Manager dialog box.

## Lines and Arrows tab

*Sets the format of dimension lines, extension lines, arrowheads, and center marks.*

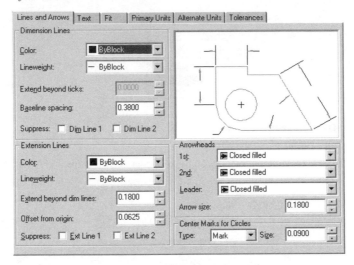

### Dimension Lines section

**Color** specifies the color of the dimension line; select Other to display the Select Color dialog box (stored in dimension variable DimClrD; default = ByBlock).

**Lineweight** specifies the lineweight of the dimension line (DimLwD; default = ByBlock).

**Extend beyond ticks** specifies the distance the dimension line extends beyond the extension line; used with oblique, architectural, tick, integral, and no arrowheads (DimDlE; default = 0).

**Baseline spacing** specifies the spacing between the dimension lines of a baseline dimension (DimDlI; default = 0.38).

**Suppress** suppresses the first and second dimension lines when outside the extension lines (DimSd1 and DimSd1; default = off).

### Extension Lines section

**Color** specifies the color of the extension line; select Other to display the **Select Color** dialog box (stored in dimension variable DimClrE ; default=ByBlock).

**Lineweight** specifies the lineweight of the extension line (DimLwE; default=ByBlock).

**Extend beyond dim lines** specifies the distance the extension line extends beyond the dimension line; used with oblique, architectural, tick, integral, and no arrowheads (**DimExe**; default = 0.18).

**Offset from origin** specifies the distance from the origin point to the start of the extension lines (**DimExO**; default = 0.0625).

**Suppress** suppresses the first and second extension lines (DimSe1 and DimSde1; default = off).

## Arrowheads options

**1st** specifies the name of the arrowhead to use for the first end of the dimension line (DimBlk1; default = closed filled).

*To use a custom arrowhead, select **User Arrow** to display the **Select Custom Arrow Block** dialog box:*

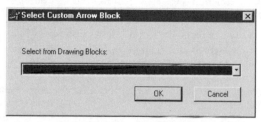

**2nd** specifies the arrowhead for the second dimension line; select User Arrow to display the Select Custom Arrow Block dialog box (DimBlk2, default = closed filled).

**Leader** specifies the arrowhead for the leader; select User Arrow to display the Select Custom Arrow Block dialog box (DimLdrBlk; default = closed filled).

**Arrow Size** specifies the size of arrowheads (DimASz; default = 0.18).

## Center Marks for Circles options

**Type** specifies the type of center mark (DimCen; default = 0.09):

| Type | Meaning |
| --- | --- |
| **Mark** | Places a center mark (**DimCen** > 0). |
| **Line** | Places a center mark and centerlines (**DimCen** < 0). |
| **None** | Places no center mark or centerline (**DimCen**=0). |

**Size** specifies the size of the center mark or centerline (**DimCen**; default = 0.09).

## Text tab

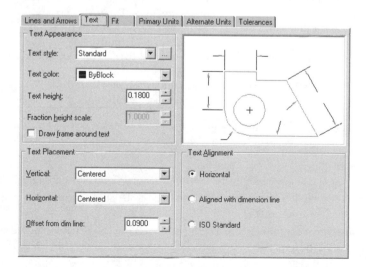

## Text Appearance options

**Text style** specifies the text style name for dimension text (DimTxSty; default = Standard). ... displays the Text Style dialog box; see the Style command.

**Text color** specifies the color of the dimension line; select Other to display the Select Color dialog box (DimClrT; default = ByBlock).

**Text height** specifies the height of the dimension text, when the height defined by the text style is 0 (DimTxt; default = 0.18).

**Fraction height scale** scales fraction text height relative to dimension text; AutoCAD multiples this value by the text height (DimTFac; default = 1.0).

**Draw frame around text** draws a rectangle around dimension text; when on, dimension variable **DimGap** is set to a negative value (DimGap; default = off).

## Text Placement options

**Vertical** specifies the vertical justification of dimension text relative to the dimension line (DimTad; default = 1):

| Vertical | Meaning |
| --- | --- |
| **Centered** | Centers dimension text in the dimension line (**DimTad** = 0). |
| **Above** | Places text above dimension line (**DimTad** = 1). |
| **Outside** | Places text on the side of the dimension line farthest from the first defining point (**DimTad** = 2). |
| **JIS** | Places text in conformity with JIS **DimTad**= 3). |

**Horizontal** specifies the horizontal justification of dimension text along the dimension and extension lines (DimJust; default = 0):

| Horizontal | Meaning |
| --- | --- |
| **Centered** | Centers dimension text along the dimension line between the extension lines (**DimJust** = 0). |
| **1st Extension Line** | Left-justifies the text with the first extension line (**DimJust** = 1). |
| **2nd Extension Line** | Right-justifies the text with the second extension line (**DimJust** = 2). |
| **Over 1st Extension Line** | Places the text over the first extension line (**DimJust** = 3). |
| **Over 2nd Extension Line** | Places the text over the second extension line (**DimJust** = 4). |

**Offset from dimension line** specifies the text gap, the distance between dimension text and the dimension line (DimGap; default = 0.09).

## Text Alignment options

**Horizontal** forces dimension text to be always horizontal (DimTih = on; DimToh = on).

**Aligned with dimension line** forces dimension text to be aligned with the dimension line (DimTih = off; DimToh = off).

**ISO Standard** forces text to be aligned with the dimension line when inside the extension lines; forces text to be horizontal when outside the extension lines (DimTih = off; **DimToh** = on).

**Fit** tab

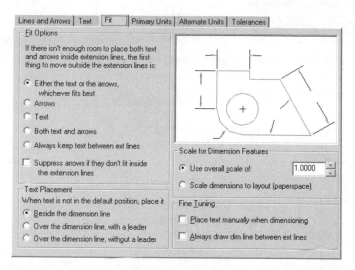

Fit Options options

**If there isn't enough room to place both text and arrows inside extension lines, the first thing to move outside the extension lines is:**

- **Either the text or the arrows, whichever fits best** places dimension text and arrowheads between the extension lines when space is available; when space is not available for both, the text or the arrowheads are placed outside the extension lines, whichever fits best; if there is room for neither, both are placed outside the extension lines (DimAtFit = 3; default).

- **Arrows** places arrowheads between the extension lines when there is not enough room for arrowheads and dimension text (DimAtFit = 2).

- **Text** places text between extension lines when there is not enough room for arrowheads and dimension text (DimAtFit = 1).

- **Both text and arrows** places both outside the extension lines when there is not enough room for dimension text and arrowheads (DimAtFit = 0).

- **Always keep text between ext lines** forces text between the extension lines (DimTix; default = off).

**Suppress arrows if they don't fit inside the extension lines** suppresses arrowheads when there is not enough room between the extension lines.

**Text Placement:**

- **Beside the dimension line** places dimension text beside the dimension line (DimTMove =0; default).

- **Over the dimension line, with a leader** draws a leader when dimension text is moved away from the dimension line (DimTMove = 1).

- **Over the dimension line, without a leader:** does not draw a leader when dimension text is moved away from the dimension line (DimTMove = 2).

### Scale for Dimension Features

- **Use overall scale of** specifies the scale factor for all dimensions in the drawing; affects text and arrowhead sizes, distances, and spacings (DimScale; default = 1.0).

- **Scale dimension to layout (paper space)** determines the scale factor of dimensions in layout mode; based on the scale factor between the current model space viewport and the layout (DimScale = 0; default = off).

## Fine Tuning options

**Place text manually when dimensioning** places text at the position picked at the 'Dimension line location' prompt (DimUpt; default = off).

**Always draw dim line between ext lines** forces the dimension line between the extension lines (DimTofl; default = off).

## Primary Units tab

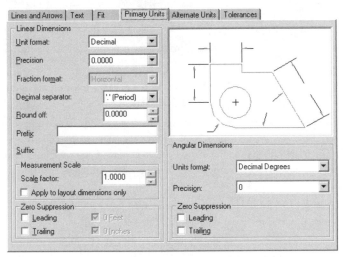

## Linear Dimensions options

**Unit Format** specifies the linear units format; does not apply to angular dimensions (DimLUnit; default = Decimal):

| DimLUnit | Meaning |
|----------|---------|
| 1 | Scientific. |
| 2 | Decimal (default). |
| 3 | Engineering. |
| 4 | Architectural. |
| 5 | Fractional. |
| 6 | Windows desktop setting. |

**Precision** specifies the number of decimal places (or fractional accuracy) for linear dimensions (DimDec; default = 4).

**Fraction Format** specifies the stacking format of fractions (DimFrac; default = 0):

| DimFrac | Meaning |
|---------|---------|
| 0 | Horizontal stacked: $\frac{1}{2}$ (default). |
| 1 | Diagonal stacked: ½ |
| 2 | Not stacked: 1/2. |

**Decimal Separator** specifies the separator for decimal formats (DimDSep; default = .).

**Round Off** specifies the format for rounding dimension values; does not apply to angular dimensions (DimRnd; default = 0).

**Prefix** specifies a prefix for dimension text (DimPost; default = nothing); you can use the following control codes to show special characters:

| Control Code | Meaning |
| --- | --- |
| **%%nnn** | Character specified by ASCII number nnn. |
| **%%o** | Turns on and off overscoring. |
| **%%u** | Turns on and off underscoring. |
| **%%d** | Degrees symbol (°). |
| **%%p** | Plus/minus symbol (±). |
| **%%c** | Diameter symbol (Ø). |
| **%%%** | Percentage sign (%). |

**Suffix** specifies a suffix for dimension text (DimPost; default = nothing); you can use the control codes listed above to show special characters.

### Measurement Scale options

**Scale factor** specifies a scale factor for linear measurements, except for angular dimensions (DimLFac; default = 1.0); use this, for example, to change dimension values from imperial to metric.

**Apply to layout dimensions only** specifies that the scale factor is applied only to dimensions created in layout mode or paper space (stored as a negative value in DimLFac; default = off).

**Zero Suppression** options:

- **Leading** suppresses leading zeros in all decimal dimensions (DimZin = 4).
- **Trailing** suppresses trailing zeros in all decimal dimensions (DimZin = 8).
- **0 Feet** suppresses zero feet of feet-and-inches dimensions (DimZin = 0).
- **0 Inches** suppresses zero inches of feet-and-inches dimensions (DimZin = 2).

| DimZin | Meaning |
| --- | --- |
| 0 | Suppresses zero feet and precisely zero inches (default). |
| 1 | Includes zero feet and precisely zero inches. |
| 2 | Includes zero feet and suppresses zero inches. |
| 3 | Includes zero inches and suppresses zero feet. |
| 4 | Suppresses leading zeros in decimal dimensions. |
| 8 | Suppresses trailing zeros in decimal dimensions . |
| 12 | Suppresses leading and trailing zeros. |

### Angular Dimensions options:

**Units Format** specifies the format of angular dimensions (DimAUnit; default = 0):

| DimAUnit | Meaning |
| --- | --- |
| 0 | Decimal degrees (default). |
| 1 | Degrees/minutes/seconds. |
| 2 | Grads. |
| 3 | Radians. |

**Precision** specifies the precision of angular dimensions (DimADec; default = 0).

## Zero Suppression options
*Same as for linear dimensions (DimAZin; default = 0).*

## **Alternate Units** tab

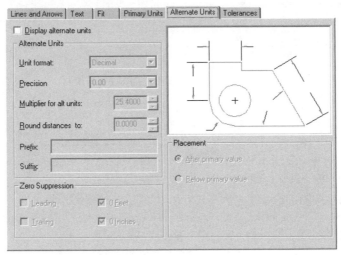

**Display alternate units** adds alternate units to dimension text (DimAlt; default = off).

## Alternate Units options

**Unit Format** specifies the alternate units formats (DimAltU; default = Decimal).

**Precision** specifies the number of decimal places or fractional accuracy (DimAltD; default=2).

**Multiplier for alt units** specifies the conversion factor between primary and alternate units (DimAltF; default = 25.4).

**Round distances to** specifies the format for rounding dimension values; does not apply to angular dimensions (DimAltRnd; default = 0.0000).

**Prefix** specifies a prefix for dimension text (DimAPost; default = nothing); you can use control codes to show special characters.

**Suffix** specifies a suffix for dimension text (DimAPost; default = nothing).

## Zero Suppression options

**Leading** suppresses leading zeros in all decimal dimensions (DimAltZ = 4).

**Trailing** suppresses trailing zeros in all decimal dimensions (DimAltZ = 8).

**0 Feet** suppresses zero feet of feet-and-inches dimensions (DimAltZ = 0).

**0 Inches** suppresses zero inches of feet-and-inches dimensions (DimAltZ = 2).

**Placement**:

- **After primary units** places alternate units behind the primary units (DimAPost).
- **Below primary units** places alternate units below the primary units.

# Tolerances tab

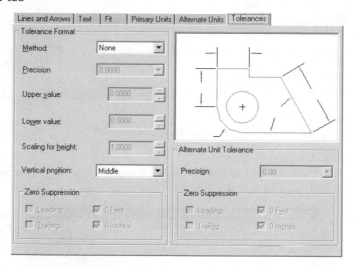

## Tolerance Format options

**Method** specifies the tolerance format:

- **None** does not display tolerances (DimTol = 0; default).
- **Symmetrical** places a ± after the dimension (DimTol=0; DimLim=0).
- **Deviation** places + and – symbols (DimTol = 1; DimLim = 1).
- **Limits** places maximum over minimum value (DimTol=0; DimLim=1):
  Maximum value = dimension value + upper value.
  Minimum value = dimension value - lower value.
- **Basic** boxes the dimension text (DimGap=negative value).

**Precision** specifies the number of decimal places for tolerance values (DimTDec; default=4).

**Upper value** specifies the upper tolerance value (DimTp; default = 0).

**Lower value** specifies the lower tolerance value (DimTm; default = 0).

**Scaling for height** specifies the scale factor for tolerance text height (DimTFac; default = 1.0).

**Vertical position** specifies the vertical text position for symmetrical and deviation tolerances (DimTolJ; default = 1):

| Vertical | Meaning |
| --- | --- |
| Top | Aligns the tolerance text with the top of the dimension text (**DimTolJ** = 2). |
| Middle | Aligns the tolerance text with the middle of the dimension text (**DimTolJ** = 1). |
| Bottom | Aligns the tolerance text with the bottom of the dimension text (**DimTolJ** = 0). |

## Zero Suppression options

**Leading** suppresses leading zeros in all decimal dimensions (DimTZin = 4).

**Trailing** suppresses trailing zeros in all decimal dimensions (DimTZin = 8).

**0 Feet** suppresses zero feet of feet-and-inches dimensions (DimTZin = 0).

**0 Inches** suppresses zero inches of feet-and-inches dimensions (DimTZin = 2).

Alternate Unit Tolerance options

**Precision** specifies the precision — the number of decimal places — of tolerance text (DimAltTd; default = 2).

*Zero Suppression options:* the same as for tolerance format; stored in **DimAltTz**.

## Modify Dimension Style dialog box

*This dialog box is identical to the New Dimension Style dialog box.*

## Override Dimension Style dialog box

*This dialog box is identical to the New Dimension Style dialog box.*

## Compare Dimension Styles dialog box

*The list is blank when AutoCAD finds no differences. When **With** is set to **<none>** or the same style as Compare, AutoCAD displays all dimension variables.*

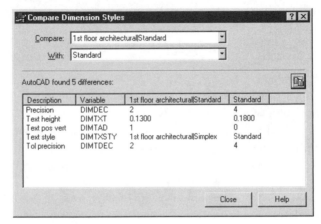

**Compare** displays the name of one dimension style.

**With** displays the name of the second dimension style.

**Copy to Clipboard** copies the style comparison text to the Clipboard, which can be pasted in another Windows application.

**Description** describes the dimension variable.

**Variable** names the dimension variable.

**Close** closes the dialog box.

# -DIMSTYLE Command

**Command:** -dimstyle
**Current dimension style: Standard**
**Enter a dimension style option**
**[Save/Restore/STatus/Variables/Apply/?] <Restore>:** *(Enter an option.)*
**Current dimension style: Standard**
**Enter a dimension style name, [?] or <select dimension>:** *(Enter an option.)*
**Select dimension:** *(Select a dimension in the drawing.)*
**Current dimension style: Standard**

## COMMAND LINE OPTIONS

**Save** saves current dimvar (dimension variables) settings as a named dimstyle (dimension style).

**Restore** sets dimvar settings from a named dimstyle.

**STatus** lists dimvars and current settings.

**Variables** lists dimvars and their current settings.

**Apply** updates selected dimension objects with current dimstyle settings.

**?** lists names of dimstyles stored in drawing.

## INPUT OPTIONS

**~dimvar** *(tilde prefix)* lists the differences between current and selected dimstyle.

**ENTER** lists the dimvar settings for the selected dimension object.

## RELATED DIM COMMANDS

**DDim** changes dimvar settings.

**DimScale** determines the scale of dimension text.

## RELATED DIM VARIABLES

*All*

**DimStyle** contains the name of the current dimstyle.

## TIPS

- At the 'Dim' prompt, the **Style** command sets the text style for the dimension text and does *not* select a dimension style.

- Dimstyles cannot be stored to disk, except in a drawing.

- Read dimstyles from other drawings with the **XBind Dimstyle** command.

# DimTEdit

<u>**Rel 13**</u>  Dynamically changes the location and orientation of text in dimensions.

| Command | Alias | Ctrl+ | F-key | Alt+ | Menu Bar | Tablet |
|---------|-------|-------|-------|------|----------|--------|
| dimtedit | ... | ... | ... | NX | Dimension | Y2 |
| | | | | | ⌐Align Text | |

**Command:** dimtedit
**Select dimension:** *(Select a dimension in the drawing.)*
**Specify new location for dimension text or [Left/Right/Center/Home/Angle]:**
*(Pick a point, or enter an option.)*

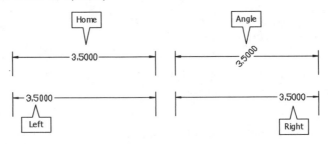

## COMMAND LINE OPTIONS

**Select dimension** selects the dimension to edit.

**Angle** rotates the dimension text.

**Center** centers the text on the dimension line.

**Home** returns the dimension text to original position.

**Left** moves the dimension text to the left.

**Right** moves the dimension text to the right.

## RELATED DIM VARIABLES

**DimSho** specifies whether dimension text is updated dynamically while dragged.

**DimTih** specifies whether dimension text is drawn horizontally or aligned with the dimension line.

**DimToh** specifies whether dimension text is forced inside the dimension lines.

## RELATED TOOLBAR ICONS

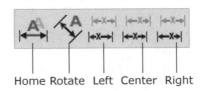

Home  Rotate  Left  Center  Right

## TIPS

• This command works only with associative dimensions; use the **DdEdit** command to edit text in non-associative dimensions.

• An angle of 0 returns dimension text to its default orientation.

# DimVertical

2004 Draw vertical dimensions (*undocumented command*).

| Command | Alias | Ctrl+ | F-key | Alt+ | Menu Bar | Tablet |
|---------|-------|-------|-------|------|----------|--------|
| dimvertical | ... | ... | ... | ... | ... | ... |

**Command:** dimvertical
**Specify first extension line origin or <select object>:** *(Pick a point, or press* **Enter** *to select one object.)*
**Specify second extension line origin:** *(Pick a point.)*
**Specify dimension line location or [Mtext/Text/Angle]:** *(Pick a point, or select an option.)*
**Dimension text =** *nnn*

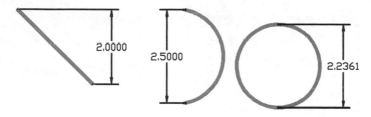

## COMMAND LINE OPTIONS

**Specify first extension line origin** specifies the location of the first extension line's origin.

**Select object** dimensions a line, arc, or circle automatically.

**Specify second extension line location** specifies the location of the second extension line.

**Specify dimension line location** specifies the location of the dimension line.

**Mtext** displays the Multiline Text Editor dialog box, which allows you to modify the dimension text; see the MText command.

**Text** prompts you to replace the dimension text on the command line: 'Enter dimension text <*nnn*>'.

**Angle** changes the angle of dimension text: 'Specify angle of dimension text:'.

## RELATED DIM COMMANDS
*All.*

## RELATED DIM VARIABLES
*Most.*

 **'Dist**

**V. 1.0** Lists the 3D distances and angles between two points *(short for DISTance)*.

| Command | Alias | Ctrl+ | F-key | Alt+ | Menu Bar | Tablet |
|---------|-------|-------|-------|------|----------|--------|
| 'dist | di | ... | ... | TYD | Tools | T8 |
| | | | | | ⤷Inquiry | |
| | | | | | ⤷Distance | |

**Command:** dist
**Specify first point:** *(Pick a point.)*
**Specify second point:** *(Pick another point.)*

*Sample result:*

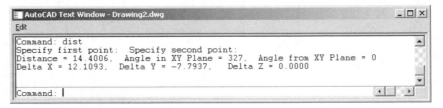

**COMMAND LINE OPTIONS**

**Specify first point** determines the start point of distance measurement.

**Specify second point** determines the end point.

**RELATED COMMANDS**

**Area** calculates the area and perimeter of objects.

**Id** lists the 3D coordinates of a point.

**Length** reports the length of open objects.

**List** lists information about selected objects.

**MassProp** reports on 2D regions and 3D solid models.

**RELATED SYSTEM VARIABLE**

**Distance** is the last calculated distance.

**TIPS**

• Use object snaps to measure precisely the distance between two geometric features.

• When the z-coordinate is left out, **Dist** assumes the current elevation for z.

# Divide

**V. 2.5**   Places points or blocks at equally-divided distances along an object.

| Command | Alias | Ctrl+ | F-key | Alt+ | Menu Bar | Tablet |
|---------|-------|-------|-------|------|----------|--------|
| divide | div | ... | ... | DOD | Draw | V13 |
| | | | | | ⌐Point | |
| | | | | | ⌐Divide | |

**Command:** divide
**Select object to divide:** *(Select one object.)*
**Enter the number of segments or [Block]:** *(Enter a number, or enter **B**.)*

*Polyline (left) divided by ten points (right).*

## COMMAND LINE OPTIONS

**Select object to divide** selects a single open or closed object.

**Enter the number of segments** specifies the number of segments; must be a number between 2 and 32767.

Block options
**Enter name of block to insert:** *(Enter the name of a block.)*
**Align block with object? [Yes/No] <Y>:** *(Enter **Y** or **N**.)*
**Enter the number of segments:** *(Enter a number.)*

**Enter name of block to insert** specifies the name of a block to insert in the current drawing.

**Align block with object** aligns the block's x axis with the object.

## RELATED COMMANDS

**Block** creates the block to use with the Divide command.

**Insert** places a single block in the drawing.

**MInsert** places an array of blocks in the drawing.

**Measure** divides an entity into measured distances.

## RELATED SYSTEM VARIABLES

**PdMode** sets the style of the point drawn.

**PdSize** sets the size of the point, in pixels.

## TIPS

• The first dividing point on a closed polyline is its initial vertex; on circles, the first dividing point is in the 0-degree direction from the center.

• The points or blocks are placed in the **Previous** selection set, so that you can select them with the next 'Select Objects' prompt.

 # Donut

**V. 2.5** Draws solid circles with a pair of wide polyline arcs.

| Commands | Alias | Ctrl+ | F-key | Alt+ | Menu Bar | Tablet |
|----------|-------|-------|-------|------|----------|--------|
| donut | do | ... | ... | DD | Draw | K9 |
| | | | | | ⸗Donut | |
| | | | | | | |
| doughnut | | | | | | |

**Command:** donut
**Specify inside diameter of donut <0.5000>:** *(Enter a value.)*
**Specify outside diameter of donut <1.0000>:** *(Enter a value.)*
**Specify center of donut or <exit>:** *(Pick a point.)*
**Specify center of donut or <exit>:** *(Press **Enter** to exit command.)*

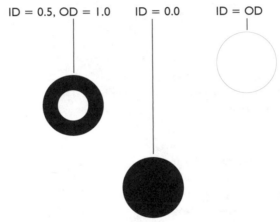

*Default donut (left), solid donut (center), and polyline circle (right).*

## COMMAND LINE OPTIONS

**Inside diameter** indicates the inner diameter by entering a number or picking two points.

**Outside diameter** indicates the outer diameter.

**Center of donut** indicates the donut's center point by entering coordinates, or by picking a point.

**Exit** exits the command.

## RELATED COMMAND

**Circle** draws a circle.

## RELATED SYSTEM VARIABLES

**DonutId** specifies the current donut internal diameter.

**DonutOd** specifies the current donut outside diameter.

**Fill** toggles the filling of the donut.

## TIP

• This command automatically repeats itself until cancelled.

# 'Dragmode

**V. 2.0**  Controls the display of objects during dragging operations.

| Command | Alias | Ctrl+ | F-key | Alt+ | Menu Bar | Tablet |
|---------|-------|-------|-------|------|----------|--------|
| 'dragmode | ... | ... | ... | ... | ... | ... |

**Command:** dragmode
**Enter new value [ON/OFF/Auto] <Auto>:** *(Enter an option.)*

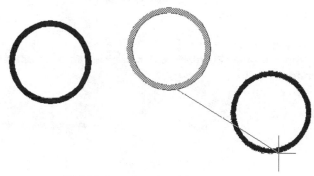

*Highlight image (center) and drag image (right).*

## COMMAND LINE OPTIONS

**ON** enables dragging display only with the Drag option.

**OFF** turns off all dragging displays.

**Auto** allows AutoCAD to determine when to display drag image.

## COMMAND MODIFIER

**Drag** displays drag images when DragMode = on.

## RELATED SYSTEM VARIABLES

**DragMode** is the current drag setting:

| DragMode | Meaning |
|----------|---------|
| 0 | No drag image. |
| 1 | On if required. |
| 2 | Automatic. |

**Drag1** drag regeneration rate (default = 10).

**Drag2** drag redraw rate (default = 25).

## TIP

• Turn off **DragMode** and **Highlight** in very large drawings to help speed up editing.

# DrawOrder

**Rel. 14** Controls the display of overlapping objects.

| Command | Alias | Ctrl+ | F-key | Alt+ | Menu Bar | Tablet |
|---------|-------|-------|-------|------|----------|--------|
| draworder | dr | ... | ... | TO | Tools | T9 |
| | | | | | ↳Display Order | |

**Command:** draworder
**Select objects:** *(Select one or more objects.)*
**Select objects:** *(Press* **Enter.***)*
**Enter object ordering option**
**[Above object/Under object/Front/Back] <Back>:** *(Enter an option.)*
**Regenerating model.**

*Text **under** solid (left) and text **above** solid (right).*

## COMMAND LINE OPTIONS

**Select objects** selects the objects to be moved.

**Above object** forces selected objects to appear above the reference object.

**Under object** forces selected objects to appear below the reference object.

**Front** forces selected objects to the top of the display order.

**Back** forces selected objects to the bottom of the display order.

## TIPS

• When you pick more than one object for reordering, AutoCAD maintains the relative display order of the selected objects.

• The order in which you select objects has no effect on drawing order.

• *Caution!* In the past, AutoCAD has been known not to maintain the display order correctly; this failing is corrected with AutoCAD 2004.

# 'DSettings

2000

Controls the most-common settings for drafting operations (*short for Drafting SETTINGS; replaces the DdRModes command*).

| Commands | Aliases | Ctrl+ | F-key | Alt+ | Menu Bar | Tablet |
|---|---|---|---|---|---|---|
| +dsettings | ds | ... | ... | TF | Tools | W10 |
| | se | | | | ⌐Drafting Settings | |
| | ddrmodes | | | | | |

'dsettings

**Command:** dsettings

*Displays dialog box:*

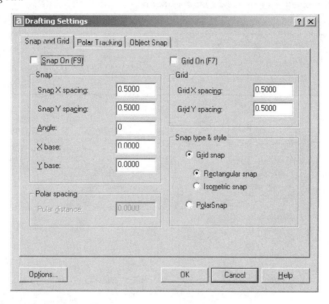

## DIALOG BOX OPTIONS

**Options** displays the Options dialog box; see the Options command.

**OK** saves the changes to settings, and closes the dialog box.

**Cancel** discards the changes to settings, and closes the dialog box.

### Snap and Grid tab

**Snap On (F9)** turns on and off snap mode (the setting is stored in system variable SnapMode; default = off).

**Grid On (F7)** turns on and off the grid display (GridMode; default = off).

Snap options

**Snap X spacing** specifies the snap spacing in the x direction (SnapUnit; default = 0.5).

**Snap Y spacing** specifies the snap spacing in the y direction (SnapUnit; default = 0.5).

**Angle** specifies the snap rotation angle (SnapAng; default = 0).

**X base** specifies the x coordinate for the snap origin (SnapBase; default = 0).

**Y base** specifies the y coordinate for the snap origin (SnapBase; default = 0).

. . . . . . . . . . . . . . . . . . . . . . . . . . . . . . . . . . . . . . . . . . . . . . .

Polar Spacing options
**Polar distance** specifies the snap distance, when Snap type & style is set to Polar snap; when 0, the polar snap distance is set to the value of Snap X spacing (PolarDist; default = 13).

Grid options
**Grid X spacing** specifies the spacing of grid dots in the x direction; when 0, the grid spacing is set to the value of Snap X spacing (GridUnit; default = 0.5).

**Grid Y spacing** specifies the spacing of grid dots in the y direction; when 0, the grid spacing is set to the value of Snap Y spacing (GridUnit; default = 0.5).

Snap Type & Style options
**Grid snap** specifies non-polar snap (SnapType; default = on).

**Rectangular snap** specifies rectangular snap (SnapStyl; default = on).

**Isometric snap** specifies isometric snap mode (SnapStyl; default = off).

**Polar snap** specifies polar snap (SnapType; default = on).

| SnapType | SnapStyl | Meaning |
|----------|----------|---------|
| 0 (off)  | 0 (off)  | Rectangular snap. |
| 0 (off)  | 1 (on)   | Isometric snap. |
| 1 (on)   | 0 (off)  | Polar snap. |

**Polar Tracking** tab

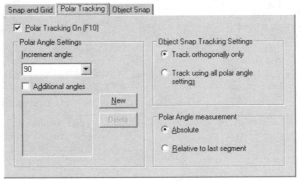

**Polar Tracking On (F10)** turns on and off polar tracking (**AutoSnap**; default = off).

Polar Angle Settings options
**Increment angle** specifies the increment angle displayed by the polar tracking alignment path; select a preset angle — 90, 60, 45, 30, 22.5, 18, 15, 10, or 5 degrees — or enter a value (PolarAng).

**Additional angles** allows additional polar tracking angles to be set (PolarMode; default=off).

**New** adds up to ten polar tracking alignment angles (PolarAddAng; default = 0;15;23;45).

**Delete** deletes added angles.

Object Snap Tracking Settings options
**Track orthogonally only** displays orthogonal tracking paths when object snap tracking is on (PolarMode).

**Track using all polar angle settings** tracks cursor along polar angle tracking path when object snap tracking is turned on (PolarMode).

Polar Angle Measurement options

**Absolute** forces polar tracking angle along the current user coordinate system (UCS).

**Relative to Last Segment** forces polar tracking angles on the last-created object.

**Object Snap** tab

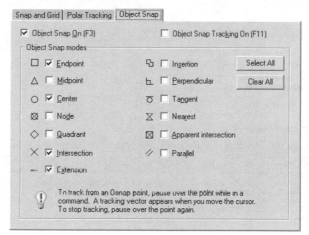

**Object Snap On (F3)** turns on and off running object snaps (OsMode; default = off).

**Object Snap Tracking On (F11)** toggles object snap tracking (AutoSnap; default = on).

Object Snap Modes options

**ENDpoint** snaps to the nearest endpoint of a line, multiline, polyline segment, ray, arc, and elliptical arc; and to the nearest corner of a trace, solid, and 3D face.

**MIDpoint** snaps to the midpoint of a line, multiline, polyline segment, solid, spline, xline, arc, ellipse, and elliptical arc.

**CENter** snaps to the center of an arc, circle, ellipse, and elliptical arc.

**NODe** snaps to a point.

**QUAdrant** snaps to a quadrant point (90 degrees) of an arc, circle, ellipse, and elliptical arc.

**INTersection** snaps to the intersection of a line, multiline, polyline, ray, spline, xline, arc, circle, ellipse, and elliptical arc; edges of regions; it does not snap to the edges and corners of 3D solids.

**EXTension** displays an extension line from the endpoint of objects; snaps to the point where two objects would intersect if they were infinitely extended; does not work with the edges and corners of 3D solids; automatically turns on intersection mode (do not turn on apparent intersection at the same time as extended intersection).

**INSertion** snaps to the insertion point of text, block, attribute, or shape.

**PERpendicular** snaps to the perpendicular of a line, multiline, polyline, ray, solid, spline, xline, arc, circle, ellipse, and elliptical arc; snaps from a line, arc, circle, polyline, ray, xline, multiline, and 3D solid edge; in which case deferred perpendicular mode is automatically turned on.

**TANgent** snaps to the tangent of an arc, circle, ellipse, or elliptical arc; deferred tangent snap mode is automatically turned on when more than one tangent snap is required.

**NEArest** snaps to the nearest point on a line, multiline, point, polyline, spline, xline, arc, circle, ellipse, and elliptical arc.

**APParent intersection** snaps to the apparent intersection of two objects that do not actually intersect but appear to intersect in 3D space; works with a line, multiline, polyline, ray, spline, xline, arc, circle, ellipse, and elliptical arc; does not work with edges and corners of 3D solids.

**PARallel** snaps to a parallel point when AutoCAD prompts for a second point.

**Clear All** turns off all object snap modes.

**Select All** turns on all object snap modes.

## +DSETTINGS Command
**Command:** +dsettings
**Tab Index <0>:** *(Enter a digit.)*

### COMMAND LINE OPTION
**Tab Index** displays the **Drafting Settings** dialog box with the associated tab:

| Tab Index | Meaning |
|-----------|---------|
| 0 | Displays the **Snap and Grid** tab (default). |
| 1 | Displays the **Polar Tracking** tab. |
| 2 | Displays the **Object Snap** tab. |

### RELATED COMMANDS
**Grid** sets the grid spacing and toggles visibility.

**Isoplane** selects the working isometric plane.

**Ortho** toggles orthographic mode.

**Snap** sets the snap spacing and isometric mode.

### RELATED SYSTEM VARIABLES
**AutoSnap** controls AutoSnap, polar tracking, and object snap tracking.

**GridMode** indicates the current grid visibility:

| GridMode | Meaning |
|----------|---------|
| 0 | Off (default). |
| 1 | On. |

**GridUnit** indicates the current grid spacing (default = 0.0).

**PolarAddAng** specifies user-defined polar angles, separated by semicolons.

**PolarAng** specifies the increments of the polar angle.

**PolarDist** specifies the polar snap distance.

**SnapAng** specifies the current snap rotation angle (default = 0).

**SnapBase** sets the base point of the snap rotation angle (default = 0,0).

**OsMode** holds the current object snap modes:

| OsMode | Meaning | OsMode | Meaning |
|--------|---------|--------|---------|
| 0 | NONe | 128 | PERpendicular |
| 1 | ENDpoint | 256 | TANgent |
| 2 | MIDpoint | 512 | NEArest |
| 4 | CENter | 1024 | QUIck |
| 8 | NODe | 2048 | APParent Intersection |
| 16 | QUAdrant | 4096 | EXTension |
| 32 | INTersection | | |
| 64 | INSertion | | |
| 8192 | PARallel | | |

**PolarMode** holds the settings for polar and object snap tracking:

| PolarMode | Meaning |
|-----------|---------|
| 0 | Measures polar angles based on current UCS (absolute); tracks orthogonally; doesn't use additional polar tracking angles; and acquires object tracking points automatically. |
| 1 | Measures polar angles from selected objects (relative). |
| 2 | Uses polar tracking settings in object snap tracking. |
| 4 | Uses additional polar tracking angles (via **PolarAng**). |
| 8 | Acquires object snap tracking points when SHIFT is pressed. |

**SnapIsoPair** is the current isoplane.

**SnapMode** sets the current snap mode setting.

**SnapStyl** specifies the snap style setting.

**SnapUnit** sets the current snap spacing (default = 1,1).

## TIPS

• Use snap to set the cursor movement increment.

• Use the grid as a visual display to help you better gauge distances.

• Use these CTRL and function keys to change modes during commands:

| Mode | Ctrl Key | Function Key |
|------|----------|--------------|
| **Grid** | CTRL+G | F7 |
| **Isoplane** | CTRL+E | F5 |
| **Object Snap** | CTRL+F | F3 |
| **Object Snap Tracking** | ... | F11 |
| **Ortho** | CTRL+L | F8 |
| **Polar Tracking** | ... | F10 |
| **Snap** | CTRL+B | F9 |

• Use object snaps to draw precisely to geometric features.

# DsViewer

<u>Rel.13</u>   Displays the bird's-eye view window; provides real-time pan and zoom
*(short for DiSplay VIEWer).*

| Command | Alias | Ctrl+ | F-key | Alt+ | Menu Bar | Tablet |
|---------|-------|-------|-------|------|----------|--------|
| dsviewer | av | ... | ... | VW | View | K2 |
| | | | | | ↳Aerial View | |

**Command:** dsviewer

*Displays the **Aerial View** window:*

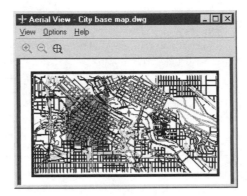

## MENU BAR OPTIONS

View menu

**Zoom In** increases centered zoom by a factor of 2.

**Zoom Out** decreases centered zoom by a factor of 2.

**Global** displays entire drawing in Aerial View window.

Options menu

**Auto Viewport** updates the **Aerial View** automatically with the current viewport.

**Dynamic Update** updates the **Aerial View** automatically with editing changes in the current viewport.

**Realtime Zoom** updates the drawing in real time as you zoom in the **Aerial View** window.

## TOOLBAR ICONS

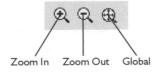

Zoom In   Zoom Out   Global

## RELATED COMMANDS

**Pan** moves the drawing view.

**View** creates and displays named views.

**Zoom** makes the view larger or smaller.

**TIPS**

• The purposes of the **Aerial View** are to let you see the entire drawing at all times, and to zoom and pan without entering the **Zoom** and **Pan** commands, or selecting items from the menu.

• The parts of the **Aerial View** window:

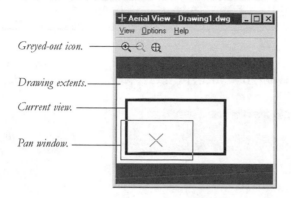

*Greyed-out icon.*

*Drawing extents.*

*Current view.*

*Pan window.*

• *Warning!* When in paper space, the **Aerial View** window shows only paper space objects.

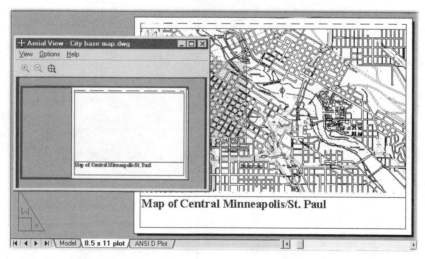

• To switch quickly between **Pan** (default) and **Zoom** modes, click on the **Aerial View** window.

**Removed Command**

**DText** was removed from AutoCAD 2000; it was combined with the **Text** command.

# DView

**Rel.10** Dynamically zooms and pans 3D drawings; turns on perspective mode (*short for Dynamic VIEW*).

| Command | Alias | Ctrl+ | F-key | Alt+ | Menu Bar | Tablet |
|---------|-------|-------|-------|------|----------|--------|
| dview | dv | ... | ... | ... | ... | ... |

**Command:** dview
**Select objects or <use DVIEWBLOCK>:** *(Select objects, or press **Enter**.)*
**Enter option [CAmera/TArget/Distance/POints/PAn/Zoom/TWist/CLip/Hide/Off/Undo]:** *(Enter an option.)*

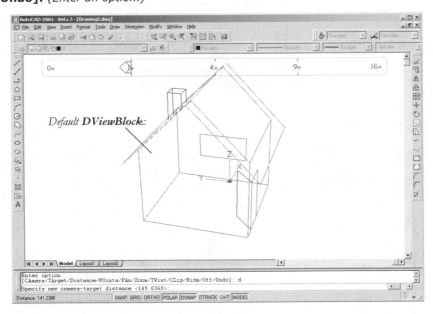

Default *DViewBlock:*

## COMMAND LINE OPTIONS

**CAmera** indicates the camera angle relative to the target.

**Toggle** switches between input angles.

**TArget** indicates the target angle relative to the camera.

**Distance** indicates the camera-to-target distance; turns on perspective mode.

**POints** indicates both the camera and target points.

**PAn** pans the view dynamically.

**Zoom** zooms the view dynamically.

**TWist** rotates the camera.

CLip options

*Back clip options*

**ON** turns on the back clipping plane.

**OFF** turns off the back clipping plane.

**Distance from target** indicates the location of the back clipping plane.

*Front clip options*
      **Eye** positions the front clipping plane at the camera.

      **Distance from target** indicates the location of the front clipping plane.

      **Off** turns off view clipping.

**Hide** removes hidden lines.

**Off** turns off the perspective view.

**Undo** undoes the most recent **DView** action.

**eXit** exits **DView**.

## RELATED COMMANDS

**Hide** removes hidden-lines from a non-perspective view.

**Pan** pans a non-perspective view.

**VPoint** selects a non-perspective viewpoint of a 3D drawing.

**Zoom** zooms a non-perspective view.

**3dOrbit** creates a 3D view interactively.

## RELATED SYSTEM VARIABLES

**BackZ** specifies the back clipping plane offset.

**FrontZ** specifies the front clipping plane offset.

**LensLength** specifies the perspective view lens length, in millimeters.

**Target** specifies the UCS 3D coordinates of target point.

**ViewCtr** specifies the 2D coordinates of current view center.

**ViewDir** specifies the WCS 3D coordinates of camera offset from target.

**ViewMode** specifies the perspective and clipping settings.

**ViewSize** specifies the height of view.

**ViewTwist** specifies the rotation angle of current view.

## RELATED SYSTEM BLOCK

**DViewBlock** is the alternate viewing object displayed during DView.

## TIPS

* The view direction is from the camera to target.

* Press ENTER at the 'Select objects' prompt to display the **DViewBlock** house.

* You can replace the house block with your own by redefining the **DViewBlock** block.

* To view a 3D drawing in one-point perspective, use the **Zoom** option.

* Menus and transparent zoom and pan are not available during **DView**.

* Once the view is in perspective mode, you cannot use **Sketch**, **Zoom**, and **Pan**.

. . . . . . . . . . . . . . . . . . . . . . . . . . . . . . . . . . . . . . . . . . . . .

## Removed Commands

**DwfOut** was removed from AutoCAD 2004; it was made part of **Plot**.

**DwfOutD** was removed from AutoCAD 2000; it was combined with **DwfOut**.

. . . . . . . . . . . . . . . . . . . . . . . . . . . . . . . . . . . . . . . . . . . . .

# Two- and 3-point Perspectives

In two-point perspective, the camera and the target are at the same height. Vertical lines remain vertical. In three-point perspective, the camera and target are at different heights.

### Step 1: Two-Point Perspective

1. Start **DView** and select all objects:

   **Command:** dview
   **Select objects:** all
   **Select objects:** *(Press* **Enter.***)*

### Step 2: Place Camera and Target

1. The **POints** option combines the **TArget** and **CAmera** options into one step:

   **CAmera/TArget/Distance/POints/.../Undo/<eXit>:** po

2. Use the **.xy** filter to pick the target point on the floorplan:

   **Enter target point <0.4997, 0.4999, 0.4997>:** .xy
   **of** *(Pick target point.)*

3. Enter a number for your eye height, such as 5'10" or 180cm:

   **(need Z):** *(Enter height.)*

4. Use the **.xy** filter to pick the camera point:

   **Enter camera point <0.4997, 0.4999, 1.4997>:** .xy
   **of** *(Pick camera point.)*

5. Type the same z coordinate for the camera height:

   **(need Z):** *(Enter same height as in #3, above.)*

### Step 3: Turn On Perspective Mode

1. The **Distance** option turns on perspective mode:

   **CAmera/TArget/Distance/POints/.../Undo/<eXit>:** d

2. In perspective mode, the UCS icon becomes a perspective icon. Use the slider bar to set the distance while in **Distance** mode:

   **New camera/target distance <1.0943>:** *(Move slider bar.)*

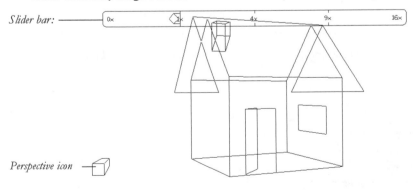

*Slider bar:* ——

*Perspective icon* —

## Step 4: Three-Point Perspective

In three-point perspective, the target and camera heights differ. Most commonly, the camera is higher than the target, so that you look down on the 3D scene.

1. Follow the earlier steps, but change the camera and target heights, as follows:

   **Command:** dview
   **Select objects:** all
   **1 found Select objects:** *(Press* **Enter.***)*
   **CAmera/TArget/Distance/POints/.../Undo/<eXit>:** po

2. For target height, enter the height of an object you are looking at, such as a window or table:

   **Enter target point <0.4997, 0.4999, 0.4997>:** .xy
   **of** *(Pick target point.)*
   **(need Z):** *(Enter a height.)*

3. For the camera height, enter your eye height or a larger number for a bird's-eye view:

   **Enter camera point <0.4997, 0.4999, 1.4997>:** .xy
   **of** *(Pick camera point.)*
   **(need Z):** *(Enter a height greater than in #2, above.)*
   **CAmera/TArget/Distance/POints/.../Undo/<eXit>:** d
   **New camera/target distance <1.0943>:** *(Adjust distance.)*

4. Use the **Hide** option to create a hidden-line view:

   **CAmera/TArget/Distance/POints/.../Hide/Off/Undo/<eXit>:** h

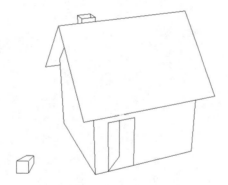

*Hidden-line view in three-point perspective mode.*

## Step 5: Exit Dview
1. Exit **DView**:

   **CAmera/TArget/Distance/POints/.../Undo/<eXit>:** *(Press* **Enter.***)*

   The view remains in perspective mode. While in perspective mode, the **Zoom**, **Pan**, and **DsViewer** commands and scroll bars do not work.

2. To exit perspective mode, use the **Plan** command.

# DwgProps

**2000**  Records and reports information about drawings (*short for DraWinG PROPertieS*).

| Command | Alias | Ctrl+ | F-key | Alt+ | Menu Bar | Tablet |
|---------|-------|-------|-------|------|----------|--------|
| dwgprops | ... | ... | ... | FI | File ⮑Drawing Properties | ... |

**Command:** dwgprops

*Displays tabbed dialog box; see below.*

## DIALOG BOX OPTIONS

**OK** records the changes, and exits the dialog box.

**Cancel** discards the changes, and exits the dialog box.

## General tab

*Displays information about the drawing obtained from the operating system:*

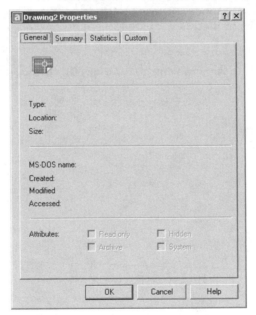

**File Type** indicates the type of the file.

**Location** indicates the location of the file.

**Size** indicates the size of the file.

**MS-DOS Name** indicates the MS-DOS filename; usually truncated to eight characters.

**Created** indicates the date and time the file was first saved.

**Modified** indicates the date and time the file was last saved.

**Accessed** indicates the date and time the file was last opened.

## Attributes options

**Read-Only** indicates the file cannot edited or erased.

**Archive** indicates the file has been changed since it was last backed up.

**Hidden** indicates the file cannot be seen in file listings.

**System** indicates the file is a system file; DWG drawing files never have this attribute turned on.

### Summary tab

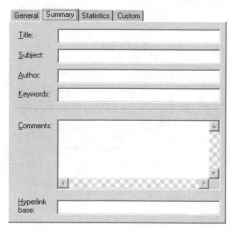

**Title** specifies a title for this drawing; is usually different from the filename.

**Subject** specifies a subject for this drawing.

**Author** specifies the name of the drafter of this drawing.

**Keywords** specifies keywords used by the operating system's Find command to locate the drawing.

**Comments** contains comments on this drawing.

**Hyperlink Base** specifies the base address for relative links in the drawing, such as http://www.upfrontezine.com; may be an operating system path name, such as *c:\autocad 2004*, or a network drive name. Stored in system variable HyperlinkBase.

### Statistics tab

*Displays information about the drawing obtained from the drawing:*

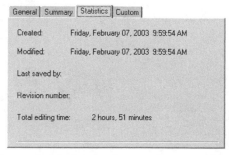

**Created** indicates the date and time the drawing was first opened, as stored in system variable TdCreate.

**Modified** indicates the date and time the drawing was last opened or modified, as stored in system variable TdUpdate.

**Last saved by** indicates, by the name stored in the LoginName system variable, who last accessed this drawing file.

**Revision number** indicates the revision number; usually blank.

**Total editing time** indicates the total amount of time that the drawing has been open, as stored in system variable TdInDwg.

### Custom tab

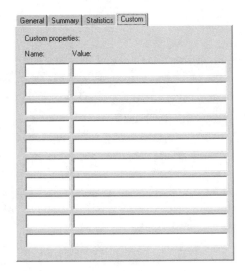

Custom Properties options

**Name** specifies the name of a customized field.

**Value** specifies the value of the customized field.

### RELATED COMMANDS

**Properties** lists information about objects in the drawing.

**Status** lists information about the drawing.

### TIP

• Use the **Find** button in **DesignCenter** to search for drawings containing values in the **Custom Properties** tab.

# DxbIn

**Ver. 2.1** Imports DXB files into drawings (*short for Drawing eXchange Binary INput*).

| Command | Alias | Ctrl+ | F-key | Alt+ | Menu Bar | Tablet |
|---------|-------|-------|-------|------|----------|--------|
| dxbin | ... | ... | ... | IE | Insert | ... |
| | | | | | ⮡Drawing Exchange Binary | |

**Command:** dxbin

*Displays **Select DXB File** dialog box. Select a .dxf file, and then click **Open**.*

## DIALOG BOX OPTION

**Open** opens the .*dxb* file, and inserts it in the drawing.

## RELATED COMMANDS

**DxfIn** reads .*dxf*-format files.

**Plot** writes .*dxb*-format files when configured for an ADI plotter.

## TIPS

• Configure AutoCAD with the ADI plotter driver to produce a .*dxb* file. After starting the **PlotterManager** command's **Add-a-Plotter Wizard**, select "AutoCAD DXB File" as the manufacturer.

• This command was created for an early Autodesk software product called CAD/camera, which converted raster scans into the .*dxb* vector format.

# DxfIn

**V. 2.0** Imports DXF files into drawings (*short for Drawing interchange Format INput; undocumented command*).

| Command | Alias | Ctrl+ | F-key | Alt+ | Menu Bar | Tablet |
|---------|-------|-------|-------|------|----------|--------|
| dxfin | ... | ... | ... | FO | File | ... |
| | | | | ⬥DXF | ⬥Open | |
| | | | | | ⬥DXF | |

**Command:** dxfin

*Displays **Select File** dialog box. Select a .dxf file, and then click **Open**.*

## DIALOG BOX OPTION

**Open** opens the *.dxf* file.

## RELATED COMMANDS

**DxbIn** reads a *.dxb*-format file.

**DxfOut** writes a *.dxf*-format file.

## TIPS

* The *.dxf* file comes in two styles: *complete* and *partial*:

    A complete *.dxf* file contains all data required to reproduce a complete drawing file.

    A partial *.dxf* file must be imported into an existing drawing.

* To load a complete *.dxf* file, AutoCAD requires the current drawing to be empty; a partial *.dxf* file can be imported into any drawing, empty or not.

* When you try to import a complete *.dxf* file but the drawing is not new, some versions of AutoCAD complain, "DXFIN requires a new drawing." To create an empty drawing, use **New** with the **Start from Scratch** option.

* If you need to import the complete *.dxf* file into a non-empty drawing (named, for example, *first.dwg*), take these steps:

    1. Use **DxfIn** to import the complete *.dxf* file into an empty drawing.
    2. Save the drawing with the **SaveAs** command, using the name of, for example, *second.dwg*.
    3. Open the non-empty *first.dwg* drawing, and use the **Insert** command with the * option to place the contents of the *second.dwg*.

* Autodesk documents the DXF format at usa.autodesk.com/adsk/servlet/ item?siteID=123112&id=752569

# DxfOut

**V. 2.0** Writes DXF files of part of (or entire) AutoCAD drawings for exchange with other programs (*short for Drawing interchange Format OUTput; undocumented command*).

| Command | Alias | Ctrl+ | F-key | Alt+ | Menu Bar | Tablet |
|---------|-------|-------|-------|------|----------|--------|
| dxfout | ... | ... | ... | FA | File | ... |
| | | | | ⌐DXF | ⌐Save As | |
| | | | | | ⌐DXF | |

**Command:** dxfout

*Displays **Save Drawing As** dialog box. Enter a filename, and then click **Save**.*

## DIALOG BOX OPTIONS

**Save** saves the drawing as a *.dxf* file.

**Files of type** creates a *.dxf* file compatible with these versions of AutoCAD:

- AutoCAD 2004.
- AutoCAD 2000 (includes 2000, 2000i, and 2002, and AutoCAD LT).
- AutoCAD Release 12 and AutoCAD LT Release 2.

**Tools | Options** dialog box

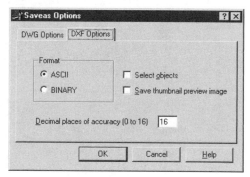

Format options
- **ASCII** creates a file in text format, which is human-readable and importable by most applications (default).
- **Binary** creates a binary file with a smaller filesize, but it cannot be read by all applications.

Additional options

**Select objects** selects objects to export, instead of the entire drawing.

**Save thumbnail preview image** includes a preview image in the *.dxf* file.

**Decimal places of accuracy (0 to 16)** specifies the decimal places of accuracy.

## RELATED COMMANDS

**DxfIn** reads *.dxf*-format files.

**AcisOut** saves solid model objects in the drawing as ACIS-compatible *.sat* format.

**SaveAs** writes drawings in *.dwg* format.

## TIPS

- Use the ASCII *.dxf* format to exchange drawings with other CAD and graphics programs.

- A binary *.dxf* file is much smaller, and is created much faster, than an ASCII binary file; few applications, however, read a binary *.dxf* file.

- The AutoCAD Release 12 dialect of *.dxf* is the most compatible with other applications.

- Drawing files created by AutoCAD 2000 are compatible with AutoCAD 2000i and 2002.

- *Caution!* When saving as an earlier release of DXF format, AutoCAD 2004 erases or converts some objects into simpler objects.

# EAttEdit

<u>2002</u>  Edits attribute values and properties in a selected block *(short for Enhanced ATTribute EDITor).*

| Command | Alias | Ctrl+ | Key | Alt+ | Menu Bar | Tablet |
|---------|-------|-------|-----|------|----------|--------|
| eattedit | ... | ... | ... | MOAS | Modify | ... |
| | | | | | ⮮Object | |
| | | | | | ⮮Attribute | |
| | | | | | ⮮Single | |

**Command:** eattedit

*If drawing contains no blocks with attributes, AutoCAD complains, "This drawing contains no attributed blocks."*

*When attributes exist, the command continues:*

**Select a block:** *(Select a single block.)*

*Displays dialog box:*

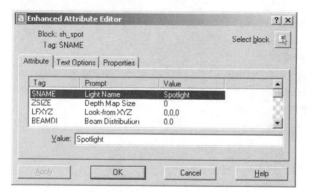

## DIALOG BOX OPTIONS

**Select block** selects another block for attribute editing.

**Apply** applies the changes to the attributes.

### Attribute tab

**Value** modifies the value of the selected attribute; neither the tag nor the prompt can be modified by this command.

### Text Options tab

**Text Style** selects a text style name from the list; text styles are defined by the Style command; default is Standard.

**Justification** selects a justification mode from the list; default is left justification.

**Height** specifies the text height; can be changed only when height is set to 0.0 in the text style.

**Rotation** specifies the rotation angle of the attribute text; default = 0 degrees.

**Backwards** displays the text backwards.

**Upside Down** displays the text upside-down.

**Width Factor** specifies the relative width of characters; default = 1.

**Oblique Angle** specifies the slant of characters; default = 0 degrees.

**Properties** tab

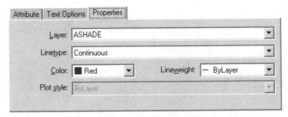

**Layer** selects a layer name from the list; layers are defined by the Layer command.

**Linetype** selects a linetype name from the list; linetypes are loaded into the drawing with the **Linetype** command.

**Color** selects a color from the list; to select from the full 255-color spectrum, select Other.

**Lineweight** selects a lineweight from the list; to turn on the display of lineweights, click LWT on the status bar.

**Plot style** selects a plot style name from the list; available only if plot styles are enabled in the drawing.

### RELATED COMMANDS

**AttDef** creates attribute definitions.

**Block** attaches attributes to objects.

**BAttMan** manages attributes.

**EAttExt** extracts attributes to a file.

### TIPS

- This command edits only attribute values and their properties; to edit all aspects of an attribute, use the **BAttMan** command.

- If you select a block with no attributes, AutoCAD complains, "The selected block has no editable attributes."

- When you select an object that isn't a block, AutoCAD complains, "Error selecting entity."

# EAttExt

**2002**

Extracts attribute data to file via a step-by-step guide procedure *(short for Enhanced ATTribute EXTraction)*.

| Command | Alias | Ctrl+ | Key | Alt+ | Menu Bar | Tablet |
|---------|-------|-------|-----|------|----------|--------|
| eattext | ... | ... | ... | ... | ... | ... |

**Command:** eattext

*Displays dialog box.*

## DIALOG BOX OPTIONS

**Back** goes back to the previous step.

**Next** proceeds to the next step.

**Cancel** exits the dialog box.

**Help** displays helpful information.

**Select Drawing** page

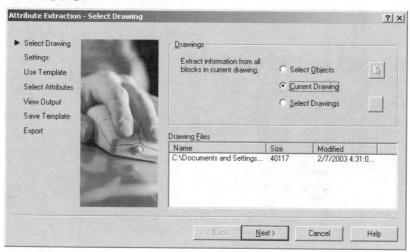

**Drawings** specifies the location of the blocks containing attributes:

- **Select objects** selects the specific blocks in the current drawing.
- **Current drawing** selects all blocks containing attributes in the current drawing.
- **Select drawings** selects drawings located on your computer or network.

**Drawing Files** lists the names of drawing files from which attributes will be extracted.

**Settings** page

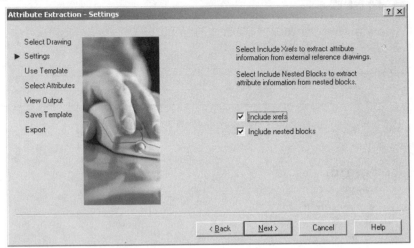

**Include xrefs** specifies that attributes in xrefs should be extracted.

**Include nested blocks** specifies that nested blocks (blocks within blocks) should be searched for attribute data.

**Use Template** page

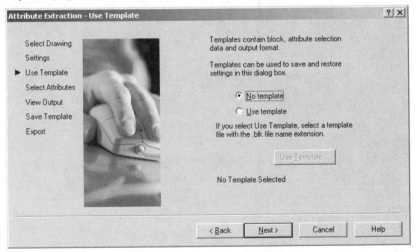

- **No template** means you will specify the data output format.
- **Use template** means that AutoCAD uses the existing output format, previously stored in a *.blk* file.

**Use Template** displays the Open dialog box; select a Block Template (*.blk*) file. AutoCAD does not include any *.blk* files; you can create one later during the Save Template step.

## Select Attributes page

*List of Blocks Found in Drawing(s)*    *List of Attributes for each Block*

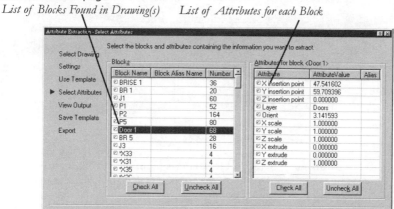

☑ *or* ☐ click the box next to each block (in left hand column) and each attribute (in right hand column) either to include or exclude them.

**Check All** selects all blocks or attributes.

**Uncheck All** unselects all blocks or attributes.

**Alias** allows you to enter an alias for each block and attribute; an *alias* is an alternative name.

| Block Information | Attribute Information |
|---|---|
| **Block Name** | Attribute. |
| **Block Alias** | Attribute Value. |
| **Number** | Alias. |

## View Output page

*Note*: This page can take a long time to complete on slow computers with drawings containing a large number of blocks.

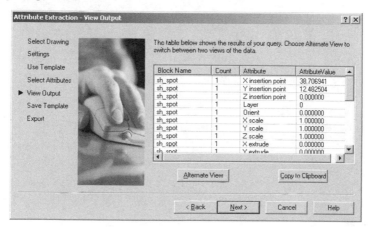

*Alternate view 1: Block Name, Count, and attribute data.*

| Block Name | C.. | X insertion .. | Y in.. | Z i.. | Layer | Orient |
|---|---|---|---|---|---|---|
| BRISE 1 | 4 | 35.316713 | 24.1.. | 0.0.. | Interior ... | 0.0015 |
| BR 1 | 4 | 62.408729 | 21.7.. | 0.0.. | 0 | 0.0000 |
| J1 | 4 | 67.008472 | 26.7.. | 0.0.. | 0 | 3.9269 |
| J1 | 4 | 40.322257 | 22.4.. | 0.0.. | 0 | 2.3561 |
| BRISE 1 | 4 | 48.593551 | 10.8.. | 0.0.. | Interior ... | 0.0015 |
| BRISE 1 | 4 | 45.938184 | 13.5.. | 0.0.. | Interior ... | 0.0015 |
| P1 | 4 | 46.223056 | 16.7.. | 0.0.. | 0 | 2.3561 |
| P2 | 52 | 0.000000 | 0.00.. | 0.0.. | Interior ... | 0.0000 |
| J1 | 4 | 50.928832 | 11.8.. | 0.0.. | 0 | 2.3561 |
| BRISE 1 | 4 | 37.972081 | 21.4.. | 0.0.. | Interior ... | 0.0015 |

*Alternate view 2: Block Name, Count, and block data.*

**Block Name** lists the name of each block; block names prefixed with *X are hatch patterns created by AutoCAD. Click the header to list block names in alphabetical order.

**Count** specifies the number of times the block appears.

**Alternate View** displays the data in two formats:

- Information about each block.
- Information about each attribute.

| Alternate View 1 | Alternate View 2 |
|---|---|
| Block Name. | Block Name. |
| Count. | Count. |
| Attribute . | X, Y, Z Insertion. |
| Attribute Value. | Layer. |
| | Orient. |
| | X, Y, Z scale. |
| | X, Y, Z Extrusion. |
| | User Defined. |

**Copy to Clipboard** copies the displayed data to the Windows Clipboard in tab-delimited format (each field is separated by a tab), which can be pasted into a spreadsheet or document with the Edit | Paste (CTRL+v) command.

*A field:*

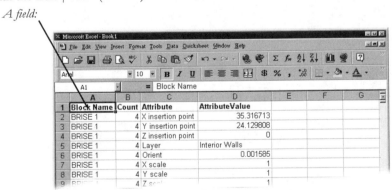

**Save Template** page

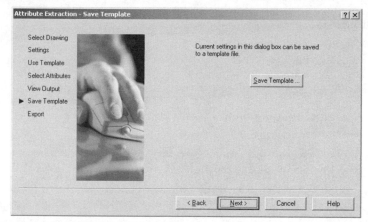

**Save Template** saves the block and attribute selections (made in the Select Attributes step) to a *.blk* (Block template) file; this *.blk* file can be reused during the Use Template step.

**Export** page

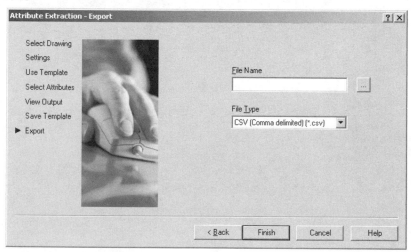

**File Name** specifies the name of the file that holds the extracted attribute data.

**File Type** selects the format of the data:

- **CSV** (comma separated values) separates fields with a comma.
- **TXT** (tab delimited values) separates fields with a tab.
- **XLS** (Excel spreadsheet format) saves the data in a proprietary format.

**Finish** completes the attribute data extraction process.

## RELATED COMMANDS

**AttExt** exports to DXF format, as well as comma- and tab-delimited formats; it is an older method of attribute extraction.

**AttDef** creates attribute definitions.

**Block** attaches attributes to objects.

**BAttMan** manages attributes.

## TIPS

- This command does not export attribute data in *.dxf* format; if you require this format, use the **AttExt** command.

- When the attribute data has been exported, you may open the file in another program, such as WordPerfect (word processing), Lotus 1-2-3 (spreadsheet), or Access (database).

- A drawing (and its xrefs) can contain many attributes. For example, the *1st Floor.dwg* sample drawing contains nearly a thousand blocks, which take up nearly 8,000 rows in a spreadsheet.

- You can use this command to create a crude BOM (bill of material). During the **View Output** step, click the **Copy to Clipboard** button, and paste the data in the AutoCAD drawing.

 # Edge

**Rel.12** Toggles the visibility of 3D faces.

| Command | Alias | Ctrl+ | F-key | Alt+ | Menu Bar | Tablet |
|---------|-------|-------|-------|------|----------|--------|
| edge | ... | ... | ... | DFE | Draw | ... |
| | | | | | ⤷Surfaces | |
| | | | | | ⤷Edge | |

**Command:** edge
**Specify edge of 3dface to toggle visibility or [Display]:** *(Pick an edge, or type D.)*
**Enter selection method for display of hidden edges [Select/All] <All>:** *(Type S or A.)*
**Select objects:** *(Select one or more objects.)*
**Specify edge of 3dface to toggle visibility or [Display]:** *(Press Esc to end the command.)*

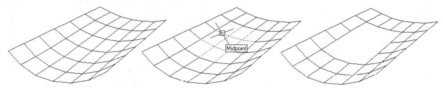

*3d faces (left), edges selected with **Edge** (center), and invisible edges (right).*

## COMMAND LINE OPTIONS
**Specify edge** selects edge to make invisible.

*Display options:*
**Select** highlights invisible edges.

**All** selects all hidden edges, and regenerates them.

## RELATED COMMAND
**3dFace** creates 3D faces.

## RELATED SYSTEM VARIABLE
**SplFrame** toggles visibility of 3D face edges.

## TIPS
* Make edges invisible to improve the appearance of 3D objects.

* **Edge** applies only to objects made of 3d faces; it does not work with polyface meshes or solid models.

* Use the **Explode** command to convert meshed objects into 3D faces.

* Re-execute **Edge** to display an edge that has been made invisible.

* Command repeats itself until you press ENTER or ESC at the 'Specify edge of 3dface to toggle visibility or [Display]:' prompt.

# EdgeSurf

**Rel.10**  Draws 3D polygon meshes as Coons surface patches between four boundaries (*short for EDGE-defined SURFace*).

| Command | Alias | Ctrl+ | F-key | Alt+ | Menu Bar | Tablet |
|---------|-------|-------|-------|------|----------|--------|
| edgesurf | ... | ... | ... | DFD | Draw | R8 |
| | | | | | ⮡Surfaces | |
| | | | | | ⮡Edge Surface | |

**Command:** edgesurf
**Current wire frame density:  SURFTAB1=6  SURFTAB2=6**
**Select object 1 for surface edge:** *(Pick an object.)*
**Select object 2 for surface edge:** *(Pick an object.)*
**Select object 3 for surface edge:** *(Pick an object.)*
**Select object 4 for surface edge:** *(Pick an object.)*

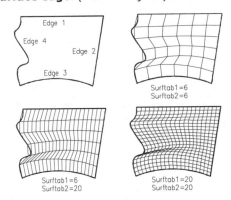

## COMMAND LINE OPTION
**Select object** picks an edge.

## RELATED COMMANDS
**3dMesh** creates a 3D mesh by specifying every vertex.
**3dFace** creates a 3D mesh of irregular vertices.
**PEdit** edits the mesh created by EdgeSurf.
**TabSurf** creates a tabulated 3D surface.
**RuleSurf** creates a ruled 3D surface.
**RevSurf** creates a 3D surface of revolution.

## RELATED SYSTEM VARIABLES
**SurfTab1** is the current m-density of meshing; the maximum mesh density is 32767.
**SurfTab2** is the current n-density of meshing.

## TIP
• The four boundary edges can be made from lines, arcs, and open 2D and 3D polylines; the edges must meet at their endpoints.

# 'Elev

V. 2.1 Sets elevation and thickness to created extruded 3D objects (*short for ELEVation*).

| Command | Alias | Ctrl+ | F-key | Alt+ | Menu Bar | Tablet |
|---------|-------|-------|-------|------|----------|--------|
| 'elev | ... | ... | ... | ... | ... | ... |

**Command:** elev
**Specify new default elevation <0.0000>:** *(Enter a value for elevation.)*
**Specify new default thickness <0.0000>:** *(Enter a value for thickness.)*

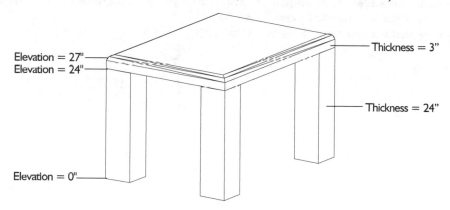

## COMMAND LINE OPTIONS
**Elevation** changes the base elevation from z = 0.
**Thickness** extrudes new 2D objects in the z-direction.

## RELATED COMMANDS
**Change** changes the thickness and z coordinate of objects.
**Move** moves objects, even in the z direction.
**Properties** changes the thickness of objects.

## RELATED SYSTEM VARIABLES
**Elevation** stores the current elevation setting.
**Thickness** stores the current thickness setting.

## TIPS
• The current value of elevation is used whenever a z coordinate is not supplied.
• Thickness is measured up from the current elevation in the positive z-direction.

# Ellipse

**V. 2.5** Draws ellipses by four different methods, as well as elliptical arcs and isometric circles.

| Command | Alias | Ctrl+ | F-key | Alt+ | Menu Bar | Tablet |
|---------|-------|-------|-------|------|----------|--------|
| ellipse | el | ... | ... | DE | Draw | M9 |
| | | | | | ⤷Ellipse | |

**Command:** ellipse
**Specify axis endpoint of ellipse or [Arc/Center]:** *(Pick a point, or enter an option.)*
**Specify other endpoint of axis:** *(Pick a point.)*
**Specify distance to other axis or [Rotation]:** *(Pick a point, or type **R**.)*

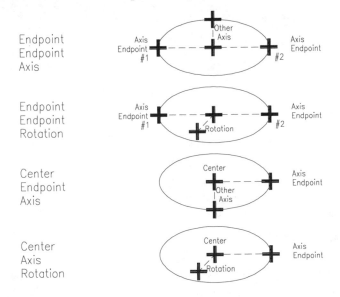

Endpoint
Endpoint
Axis

Endpoint
Endpoint
Rotation

Center
Endpoint
Axis

Center
Axis
Rotation

## COMMAND LINE OPTIONS

**Specify axis endpoint of ellipse** indicates the first endpoint of the major axis.

**Specify other endpoint of axis** indicates the second endpoint of the major axis.

**Specify distance to other axis** indicates the half-distance of the minor axis.

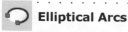 **Elliptical Arcs**

**Command:** ellipse
**Specify axis endpoint of elliptical arc or [Center]:** *(Pick a point, or type* **C.***)*
**Specify other endpoint of axis:** *(Pick a point.)*
**Specify distance to other axis or [Rotation]:** *(Pick a point, or type* **R.***)*
**Specify start angle or [Parameter]:** *(Enter an angle, or type* **P.***)*
**Specify end angle or [Parameter/Included angle]:** *(Enter an angle, or enter an option.)*

## COMMAND LINE OPTIONS

**Specify start angle** indicates the starting angle of the elliptical arc.

**Specify end angle** indicates the ending angle of the elliptical arc.

**Parameter** indicates the starting angle of the elliptical arc; draws the arc with this formula:

$$p(u)=c+(a*\cos(u))+(b*\sin(u))$$

| Parameter | Meaning |
|---|---|
| a | Major axis. |
| b | Minor axis. |
| c | Center of ellipse. |

**Included angle** indicates an angle measured relative to the start angle, rather than 0 degrees.

Center option
**Specify center of ellipse:** *(Pick a point.)*

**Specify center of ellipse** indicates the center point of the ellipse.

Rotation option
**Specify rotation around major axis:** *(Enter an angle.)*

**Specify rotation around major axis** indicates a rotation angle around the major axis:

| Rotation | Meaning |
|---|---|
| 0 degrees | Minimum rotation: creates a round ellipse, like a circle. |
| 89.4 degrees | Maximum rotation: creates a very thin ellipse. |

## Isometric Circles

*This option appears only when system variable **SnapStyl** is set to **1** (isometric snap mode).*
**Command:** ellipse
**Specify axis endpoint of ellipse or [Arc/Center/Isocircle]:** *(Type* **I.***)*
**Specify center of isocircle:** *(Pick a point.)*
**Specify radius of isocircle or [Diameter]:** *(Pick a point, or type* **D.***)*

## COMMAND LINE OPTIONS

**Specify center** indicates the center point of the isocircle.

**Specify radius** indicates the radius of the isocircle.

**Diameter** indicates the diameter of the isocircle.

## RELATED COMMANDS

**IsoPlane** sets the current isometric plane.

**PEdit** edits ellipses (when drawn with a polyline).

**Snap** controls the setting of isometric mode.

## RELATED SYSTEM VARIABLES

**PEllipse** determines how the ellipse is drawn:

| PEllipse | Meaning |
|----------|---------|
| 0 | Draws ellipse with the ellipse object (default). |
| 1 | Draws ellipse as a series of polyline arcs. |

**SnapIsoPair** is the current isometric plane:

| SnapIsoPair | Meaning |
|-------------|---------|
| 0 | Left (default). |
| 1 | Top. |
| 2 | Right. |

**SnapStyl** specifies regular or isometric drawing mode:

| SnapStyl | Meaning |
|----------|---------|
| 0 | Standard (default). |
| 1 | Isometric. |

## TIPS

- Previous to AutoCAD Release 13, **Ellipse** constructed the ellipse as a series of short polyline arcs.

- When **PEllipse** = 1, the **Arc** option is not available.

- Use ellipses to draw circles in isometric mode. When **Snap** is set to isometric mode, **Ellipse**'s isocircle option projects a circle into the working isometric drawing plane. Use CTRL+E to toggle isoplanes.

- The **Isocircle** option only appears in the option prompt when **Snap** is set to isometric mode.

These isometric circles were drawn with the **Ellipse** command's **Isocircle** option.

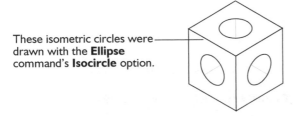

- See the **Isoplane** command for a tutorial on creating isometric objects.

. . . . . . . . . . . . . . . . . . . . . . . . . . . . . . . . . . . . . . . . . . . . . . . .

## Removed Commands

**End** was removed from AutoCAD Release 14; it was replaced by **Quit**.

**EndToday** was removed from AutoCAD 2004.

. . . . . . . . . . . . . . . . . . . . . . . . . . . . . . . . . . . . . . . . . . . . . . . .

# Erase

**V. 1.0**  Erases objects from drawings.

| Command | Alias | Ctrl+ | Key | Alt+ | Menu Bar | Tablet |
|---------|-------|-------|-----|------|----------|--------|
| erase | e | ... | Del | ME | Modify<br>⍾Erase | V14 |
| | | | | | Edit<br>⍾Clear | U14 |

**Command:** erase
**Select objects:** *(Select one or more objects.)*
**Select objects:** *(Press Enter to end object selection.)*

## COMMAND LINE OPTION

**Select objects** selects the objects to erase.

## RELATED COMMANDS

**Break** erases a portion of a line, circle, arc, or polyline.

**Oops** returns the most-recently erased objects to the drawing.

**Trim** cuts off the end of a line, arc, and other objects.

**Undo** returns the erased objects to the drawing.

## TIPS

- The **Erase L** command combination erases the last-drawn item visible in the current viewport.

- **Oops** brings back the most-recently erased objects; use **U** to bring back other erased objects.

- *Warning!* The **Erase All** command erases all objects in the drawing, except on locked, frozen, and off layers.

# eTransmit

<u>2000i</u> Transmits drawings and related files as email messages *(short for Electronic TRANSMITtal).*

| Commands | Alias | Ctrl+ | Key | Alt+ | Menu Bar | Tablet |
|----------|-------|-------|-----|------|----------|--------|
| etransmit | ... | ... | ... | FT | File | ... |
| | | | | | ⍽eTransmit | |

-etransmit

**Command:** etransmit

*If the drawing has not been saved, displays error dialog box:*

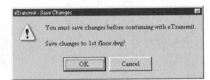

**OK** saves drawing and displays the Create Transmittal dialog box.

**Cancel** cancels the eTransmit command.

## DIALOG BOX OPTIONS

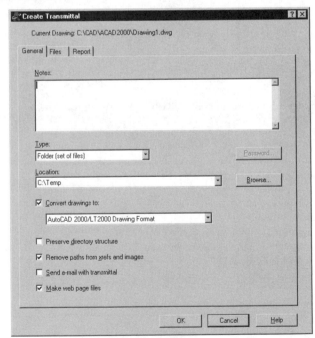

### General tab

**Notes** allows you to enter a note for the recipient of the transmittal.

**Type** selects how the files will be bundled:

- In a folder (set of files).
- In a self-extracting executable program (*.exe).
- In a Zip file (*.zip).

**Password** displays the **Set Password** dialog box.

**Location** specifies the drive and folder in which to place the transmittal files.

**Browse** displays the file dialog box for selecting the drive and folder.

**Convert drawings to:**

- ☑ converts the drawings to AutoCAD 2000/LT 2000 format.
- ☐ keeps the drawings in AutoCAD 2004 format.

**Preserve directory structure:**

- ☑ keeps the folder structure; useful when the destination computer has the same folder structure as the sending computer.
- ☐ collects files into a single folder.

**Remove paths from xrefs and images:**

- ☑ strips the path name from externally-referenced drawings and attached images.
- ☐ preserves the paths.

**Send email with transmittal:**

- ☑ starts up the Windows Messaging software.
- ☐ does not start the mail software.

**Make Web page files:**

- ☑ generates a Web page (found in the \*my documents* folder), which can be used to download the transmittal with a Web browser; see sample below.
- ☐ does not generate a Web page.

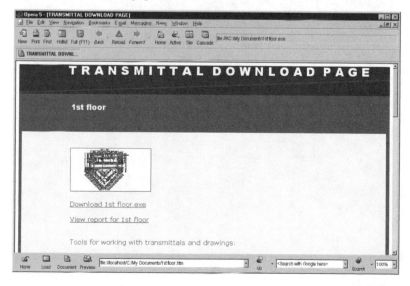

**Files** tab

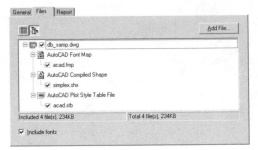

**Add File** displays the file dialog box so that you can select additional files to include with the transmittal.

**Include fonts**

    ☑ includes *.shx* (AutoCAD) and *.ttf* (Microsoft) font files with the transmittal.

    ☐ does not include fonts.

**Report** tab

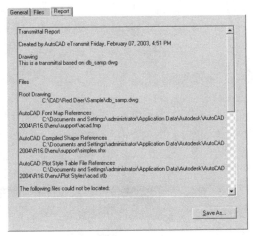

**Save As** saves the report as a text (*.txt*) file.

## -ETRANSMIT command

**Command:** -etransmit

**Transmittal type [Folder/self-extracting Executable/Zip/Report only] <Report only>:** *(Enter an option.)*

> *Displays* **Save Report File As** *dialog box.*
> *Provide a file name, and then click* **Save.**

**Transmittal note:** *(Enter text, and then press* **Enter** *to continue.)*

**Include fonts? [Yes/No] <Yes>:** *(Enter* **Y** *or* **N.***)*

**Convert all drawings? [No/R14(LT98/LT97)/2000(LT2000)] <No>:** *(Enter an option.)*

**Preserve directory structure? [Yes/No] <No>:** *(Type* **Y** *or* **N.***)*

**Remove xref and image paths? [Yes/No] <Yes>:** *(Type* **Y** *or* **N.***)*

**Make web page files? [Yes/No] <No>:** *(Type* **Y** *or* **N.***)*

**Password (press ENTER for none):** *(Enter a password, or press* **Enter.***)*

**Transmittal created: C:\filename.zip.**

## COMMAND LINE OPTIONS

**Folder** saves the drawing and related files in a folder (subdirectory).

**self-extracting Executable** saves the drawing and related files in a compressed self-extracting file (*.exe*).

**Zip** saves the drawing and related files in a compressed PkZIP (*.zip*) file.

**Report only** saves the report to a txt *(.txt)* file.

**Transmittal note** allows you to include a note to the recipient; press ENTER twice to end the note function.

**Include fonts?** includes *.shx* (AutoCAD) and *.ttf* (Microsoft) fonts.

**Convert all drawings?** converts the drawings to AutoCAD 2000 or Release 14 format.

**Preserve directory structure?** preserves the folder structure, so that files are restored into the same folders.

**Remove xref and image paths?** removes the path from externally-referenced drawings and attached images.

**Make web page files?** generates a Web page for accessing the transmittal by Web browser.

**Password** specifies a password to prevent unauthorized opening of the transmittal; press ENTER for no password.

## RELATED COMMANDS

**Publish** creates a *.dwf* file with drawing sheets.

**PublishToWeb** saves the drawing as an HTML file.

**TIPS**

- The **eTransmit** command is an expanded version of the **Pack'n Go** command, a bonus command first included with AutoCAD Release 14.

- The **Password** button is *not* available when you select the **Folder** option.

- Including a TrueType font (*.ttf*) is touchy, because sending a copy of the file might infringe on its copyright. All *.shx* and *.ttf* files included with AutoCAD may be transmitted. For this reason, the dialog box has the **Include Fonts** option, which you should turn off when the fonts cannot be copied legally, or when the recipient already has the fonts. In addition, not including the fonts can save a significant amount of file space; recall that a smaller file takes less time to transmit via the Internet.

- "Self-extracting executable" means that the files are compressed into a single file with the *.exe* extension. The email recipient double-clicks the file to extract (uncompress) the files. The benefit is that the recipient does not need to have a copy of PkUnzip or WinZip on their computer; the drawback is that a virus could hide in the *.exe* file.

- The benefit to converting the drawing to AutoCAD 2000 or Release 14 format is that clients with older versions of the software can read the file; the drawback is that R2000-specific objects are either erased or modified to a simpler format. Two problems that can occur include: (1) lineweights are no longer displayed (lineweights are restored when the drawing is opened again in AutoCAD 2004); and (2) database links and freestanding labels are converted to Release 14 links and displayable attributes.

- The **Make Web page files** option is useful when recipients do not have email access; for example, when out in the field, or when they want the transmittal available on a public basis (the file can be locked via the password, if necessary).

- If your computer cannot compress the files (or if your recipient cannot uncompress them), you are probably lacking the software needed to uncompress the *.zip* file. (Do not confuse *.zip* files with Iomega's ZIP disk drive; the two having nothing in common, except the name.) You can obtain the PkUnzip and WinZIP utilities as freeware or shareware.

- Files occasionally become corrupted when sent by email; the transmittal may need to be resent. If you continue to have problems, you may need to change a setting in your email software. Try changing the attachment encoding method from BinHex or Uuencode to MIME.

# Explode

**V. 2.5**  Explodes polylines, blocks, associative dimensions, hatches, multilines, 3D solids, regions, bodies, and polyface meshes into their constituent objects.

| Command | Alias | Ctrl+ | F-key | Alt+ | Menu Bar | Tablet |
|---------|-------|-------|-------|------|----------|--------|
| explode | x | ... | ... | MX | Modify<br>⬚Explode | Y22 |

**Command:** explode
**Select objects:** *(Select one or more objects.)*
**Select objects:** *(Press **Enter** to end object selection.)*

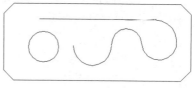

*Polylines (left) exploded into lines and arcs (right).*

## COMMAND LINE OPTION

**Select objects** selects the objects to explode.

## RELATED COMMANDS

**Block** recreates a block after an explode.

**PEdit** converts a line into a polyline.

**Region** converts 2D objects into a region.

**Undo** reverses the effects of explode.

**Xplode** provides control over the explosion process.

## RELATED SYSTEM VARIABLE

**ExplMode** toggles whether non-uniformly scaled blocks can be exploded:

| ExplMode | Meaning |
|----------|---------|
| 0 | Does not explode (compatible with AutoCAD Release 12 and earlier). |
| 1 | Explodes (default). |

**TIPS**

- As of Release 13, you *can* explode blocks inserted with unequal scale factors, mirrored blocks, and blocks created by the **MInsert** command.

- You cannot explode xrefs and dependent blocks (blocks from xref drawings).

- Parts making up exploded blocks and associative dimensions may change their color and linetype, most commonly to color White (or black when the background color is white) and to linetype Continuous.

- The **Explode** command reduces:

| Objects | Exploded Into |
|---------|---------------|
| **Blocks** | Constituent parts. |
| **Circles in non-uniformly scaled block** | Ellipse. |
| **Arcs in non-uniformly scaled block** | Elliptical arc. |
| **Associative dimensions** | Lines, solids, and text. |
| **3D polylines** | Lines. |
| **Multilines** | Lines. |
| **Polygon meshes** | 3D faces. |
| **Polyface meshes** | 3D faces, lines, and points. |
| **3D solids** | Regions and bodies. |
| **Regions** | Lines, arcs, ellipses, and splines. |
| **Bodies** | Single bodies, regions, and curves. |
| **2D polylines** | Lines and arcs; width and tangency information are lost. |

- Resulting objects become the previous selection set.

# Export

<u>Rel.13</u>   Saves drawings in formats other than DWG and DXF.

| Command | Alias | Ctrl+ | F-key | Alt+ | Menu Bar | Tablet |
|---------|-------|-------|-------|------|----------|--------|
| export | ... | ... | ... | FE | File | W24 |
| | | | | | ⇘Export | |

**Command:** export

*Displays dialog box:*

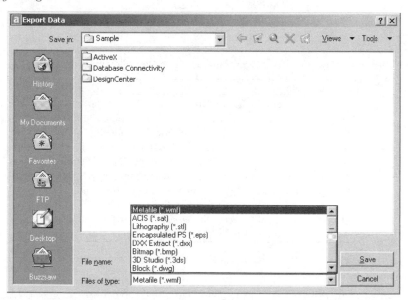

## DIALOG BOX OPTIONS

**Save in** selects the folder (subdirectory) and drive into which to export the file.

**Back** returns to the previous folder (**ALT+1**).

**Up One Level** moves up one level in the folder structure (**ALT+2**).

**Search the Web** displays a simple Web browser that accesses the Autodesk Web site (**ALT+3**).

**Delete** erases the selected file(s) or folder (**DEL**).

**Create New Folder** creates a new folder (**ALT+5**).

**Views** displays files and folders in a list or with details.

**Tools** lists several additional commands, including the Options dialog box — available only with encapsulated PostScript; see the PsOut command.

**File name** specifies the name of the file, or accepts the default.

**Save as type** selects the file format in which to save the drawing.

**Save** saves the drawing.

**Cancel** dismisses the dialog box, and returns to AutoCAD.

## RELATED COMMANDS

**AttExt** exports attribute data in the drawing in *.cdf*, *.sdf*, or *.dxf* formats.

**CopyClip** exports the drawing to the Clipboard.

**CopyHist** exports text from the text screen to the Clipboard.

**Import** imports several vector and raster formats.

**LogFileOn** saves the command line text as ASCII text in the *acad.log* file.

**MassProp** exports the mass property data as ASCII text in an *.mpr* file.

**MSlide** exports the current viewport as an *.sld* slide file.

**Plot** exports the drawing in a many vector and raster formats.

**SaveAs** saves the drawing in AutoCAD *.dwg* format.

**SaveImg** exports the rendering in TIFF, Targa, or BMP formats.

## TIPS

- The **Export** command exports the current drawing in the following formats:

| Extension | Meaning |
|-----------|---------|
| **3DS** | 3D Studio file: **3dsOut** command. |
| **BMP** | Device-independent bitmap file: **BmpOut** command. |
| **DWG** | AutoCAD drawing file : **WBlock** command. |
| **DWF** | Drawing Web format file: **DwfOut** command. |
| **DXF** | AutoCAD drawing interchange file: **DxfOut** command. |
| **DXX** | Attribute extract DXF file: **AttExt** command. |
| **EPS** | Encapsulated PostScript file: **PsOut** command. |
| **SAT** | ACIS solid object file: **AcisOut** command. |
| **STL** | Solid object stereo-lithography file: **StlOut** command. |
| **WMF** | Windows metafile: **WmfOut** command |

- This command acts as a "shell": it launches other AutoCAD commands that perform the actual export function, as noted above.

- When drawings are exported in formats other than *.dwg* or *.dxf*, information is lost, such as layers and attributes.

## Removed Command

The **ExpressTools** command (also known as "bonus CAD tools" in earlier versions of AutoCAD) was removed from AutoCAD 2002, but added back in AutoCAD 2004.

 # Extend

**V. 2.5** Extends the length of lines, rays, open polylines, arcs, and elliptical arcs to boundary objects.

| Command | Alias | Ctrl+ | F-key | Alt+ | Menu Bar | Tablet |
|---------|-------|-------|-------|------|----------|--------|
| extend | ex | ... | ... | MD | Modify<br>⮑Extend | W16 |

**Command:** extend
**Current settings: Projection=UCS Edge=None**
**Select boundary edges ...**
**Select objects:** *(Select one or more objects.)*
**Select objects:** *(Press **Enter**.)*
**Select object to extend or shift-select to trim or [Project/Edge/Undo]:** *(Select an object, or enter an option.)*
**Select object to extend or shift-select to trim or [Project/Edge/Undo]:** *(Press **Enter** to end the command.)*

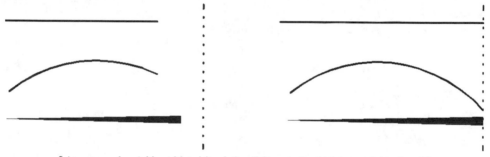

*Line, arc, and variable-width polyline before (left) and after (right) extended to dotted line.*

## COMMAND LINE OPTIONS

**Select objects** selects the objects to be used for the extension boundary.

**Select objects to extend** selects the objects that will be extended.

**Shift-select to trim** trims objects when you hold down the SHIFT key.

**Undo** undoes the most recent extend operation.

Project options
**Enter a projection option [None/Ucs/View] <Ucs>:** *(Enter an option.)*

**None** extends objects to boundary (Release 12-compatible).

**Ucs** extends objects in the x,y-plane of the current UCS.

**View** extends objects in the current view plane.

Edge options
**Enter an implied edge extension mode [Extend/No extend] <No extend>:** *(Enter an option.)*

**Extend** extends to implied boundary.

**No extend** extends only to actual boundary (Release 12-compatible).

## RELATED COMMANDS

**Change** changes the length of lines.

**Lengthen** changes the length of open objects.

**SolidEdit** extends the face of a solid object.

**Stretch** stretches objects wider and narrower.

**Trim** reduces the length of lines, polylines and arcs.

## RELATED SYSTEM VARIABLES

**EdgeMode** toggles boundary mode for the **Extend** and **Trim** commands:

| EdgeMode | Meaning |
|----------|---------|
| 0 | Use actual edges; Release 12 compatible (default). |
| 1 | Use implied edge. |

**ProjMode** toggles projection mode for the **Extend** and **Trim** commands:

| ProjMode | Meaning |
|----------|---------|
| 0 | None; Release 12 compatible. |
| 1 | Current UCS (default). |
| 2 | Current view plane. |

## TIPS

• The following objects can be used as a boundary:

| | |
|---|---|
| 2D polyline | Line |
| 3D polyline | Ray |
| Arc | Region |
| Circle | Spline |
| Ellipse | Text |
| Floating viewport | Xline |

• When a wide polyline is the edge, **Extend** extends to the polyline's centerline.

• Pick the object a second time to extend it to a second boundary line.

• Circles and other closed objects are valid edges: the object is extended in the direction nearest to the pick point.

• Extending a variable-width polyline widens it proportionately; extending a splined polyline adds a vertex.

# Extrude

**Rel.11**  Creates 3D solids by extruding 2D objects, optionally with tapered sides.

| Command | Alias | Ctrl+ | F-key | Alt+ | Menu Bar | Tablet |
|---------|-------|-------|-------|------|----------|--------|
| extrude | ext | ... | ... | DIX | Draw | P7 |
| | | | | | ⤷Solids | |
| | | | | | ⤷Extrude | |

**Command:** extrude
**Select objects:** *(Select one or more objects.)*
**Select objects:** *(Press **Enter**.)*
**Specify height of extrusion or [Path]:** *(Enter an value, or type **P**.)*
**Specify angle of taper for extrusion <0>:** *(Enter an angle, or press **Enter**.)*

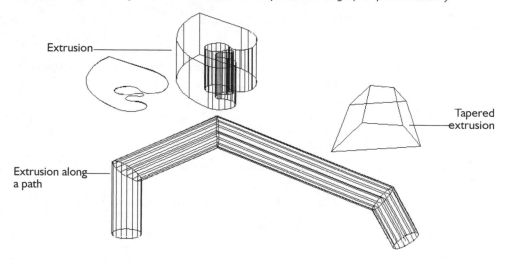

Extrusion

Tapered extrusion

Extrusion along a path

## COMMAND LINE OPTIONS

**Select objects** selects the 2D objects to extrude.

**Specify height of extrusion** specifies the extrusion height.

**Specify angle of taper for extrusion** specifies the taper angle; ranges from -90 to +90 degrees.

Path option
**Select extrusion path or [Taper angle]:** *(Select an object, or type **T**.)*

**Select extrusion path** specifies the path for the extrusion.

## RELATED COMMANDS

**Revolve** creates a 3D solid by revolving a 2D object.

**Elev** gives thickness to non-solid objects to extrude them.

**TIPS**

- This command extrudes the following objects:

  > Circles
  > Ellipses
  > Donuts
  > Closed polylines
  > Polygons
  > Closed splines
  > Regions

- This command can use the following objects as extrusion paths:

  > Lines
  > Polylines
  > Arcs
  > Elliptical arcs
  > Circles
  > Ellipses

- You *cannot* extrude polylines with less than 3 or more than 500 vertices; similarly, you cannot extrude crossing or self-intersecting polylines.

- Objects within a block cannot be extruded; use the **Explode** command first.

- The taper angle must be between 0 (default) and 90 degrees.

- Positive angles taper in from base; negative angles taper out.

- This command does not work when a taper angle is less than -90 degrees or more than +90 degrees.

- **Extrude** also does not work if the combination of angle and height makes the object's extrusion walls intersect.

# FileOpen

**Rel.12**  Opens drawing files without dialog boxes (*undocumented command*).

| Command | Alias | Ctrl+ | F-key | Alt+ | Menu Bar | Tablet |
|---------|-------|-------|-------|------|----------|--------|
| fileopen | ... | ... | ... | ... | ... | ... |

**Command:** fileopen
**Enter name of drawing to open <filename.dwg>:** *(Enter a file name.)*

## COMMAND LINE OPTIONS
**Enter name of drawing** specifies the name of the *.dwg* file to open.

## RELATED COMMANDS
**Close** closes the current drawing.

**Open** opens multiple drawing files.

## RELATED SYSTEM VARIABLES
**DbMod** detects whether the drawing was changed since being opened.

**SDI** allows only one drawing to be open in AutoCAD at a time.

## TIPS
* Use this command in menu and toolbar macros to open a drawing file when you don't want to display a dialog box.

* This command can only be used in SDI (single drawing interface) mode; if you use this command when two or more drawings are open, AutoCAD complains, "The SDI variable cannot be reset unless there is only one drawing open. Cannot run FILEOPEN if SDI mode cannot be established."

* Use **QSave** before using **FileOpen**; otherwise, **FileOpen** displays the following dialog box:

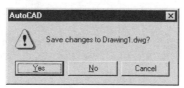

When you click **Yes**, AutoCAD closes the current drawing, and displays the **Select Drawing File** dialog box.

- - - - - - - - - - - - - - - - - - - - - - - - - - - - - - - - - - - - - -

## Removed Command
The **Files** command was removed from Release 14. In its place, use Windows Explorer.

- - - - - - - - - - - - - - - - - - - - - - - - - - - - - - - - - - - - - -

# 'Fill

<u>V. 1.4</u>  Toggles whether wide objects — traces, multilines, solids, and polylines —
are displayed and plotted as solid-filled or as outlines.

| Command | Alias | Ctrl+ | F-key | Alt+ | Menu Bar | Tablet |
|---------|-------|-------|-------|------|----------|--------|
| 'fill   | ...   | ...   | ...   | ...  | ...      | ...    |

**Command:** fill
**Enter mode [ON/OFF] <ON>:** *(Type **ON** or **OFF**.)*

*Fill on (left) and off (right) with a wide polyline, donut, and 2D solid.*

## COMMAND LINE OPTIONS

**ON** turns on fill after the next regeneration.

**OFF** turns off fill after the next regeneration.

## RELATED SYSTEM VARIABLE

**FillMode** holds the current setting of fill status:

| FillMode | Meaning |
|----------|---------|
| 0 | Fill mode is off. |
| 1 | Fill mode is on (default). |

## RELATED COMMAND

**Regen** changes the display to reflect the current fill or no-fill status.

## TIPS

• The state of fill (or no fill) does not come into effect until the next regeneration:

   **Command:** regen
   **Regenerating model.**

• Traces, solids, and polylines are only filled in plan view, regardless of the setting of **Fill**.

• Since filled objects take longer to regenerate, redraw, and plot, consider leaving fill off during
   editing and plotting. During plotting, use a wide pen for filled areas.

• **Fill** affects objects derived from polylines, including donuts, polygons, rectangles, and
   ellipses, when created with **PEllipse** = 1.

• **Fill** does *not* affect TrueType fonts, which have their own system variable — **TextFill** — that
   toggles their fill-no fill status.

• **Fill** does not toggle in rendered mode.

 # Fillet

**V. 1.4**  Joins intersecting lines, polylines, arcs, circles, and 3D solids with a radius.

| Command | Alias | Ctrl+ | F-key | Alt+ | Menu Bar | Tablet |
|---------|-------|-------|-------|------|----------|--------|
| fillet | f | ... | ... | MF | Modify<br>↳Fillet | W19 |

**Command:** fillet
**Current settings: Mode = TRIM, Radius = 0.0**
**Select first object or [Polyline/Radius/Trim/mUltiple]:** *(Select an object, or enter an option.)*
**Select second object:** *(Select another object.)*

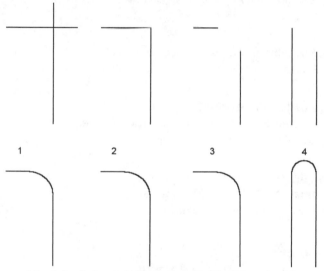

*Lines before (top) and after (bottom) applying the **Fillet** command:*
*1. Crossing lines.   2. Touching lines.   3. Non-intersecting lines.   4. Parallel lines.*

## COMMAND LINE OPTIONS

**Select first object** selects the first object to be filleted.

**Select second object** selects the second object to be filleted.

**mUltiple** prompts you to select additional pairs of objects to fillet (new in AutoCAD 2004).

Polyline option
**Select 2D polyline:** *(Pick a polyline.)*

**Select 2D polyline** fillets all vertices of a 2D polyline; 3D polylines cannot be filleted.

Radius option
**Specify fillet radius <0.0>:** *(Enter a value.)*

**Specify fillet radius** specifies the filleting radius.

Trim option
**Enter Trim mode option [Trim/No trim] <Trim>:** *(Type **T** or **N**.)*

**Trim** trims objects when filleted.

**No trim** does not trim objects.

· · · · · · · · · · · · · · · · · · · · · · · · · · · · · · · · · · · · · · · · · · · · · ·

## Filleting 3D Solids

**Select an edge or [Chain/Radius]:** *(Pick an edge, or enter an option.)*
**Select an edge or [Chain/Radius]:** *(Press **Enter**.)*
*n* **edge(s) selected for fillet.**

*Box before (left) and after (right) applying **Fillet** and **Render** to a 3D solid.*

Edge option
**Select an edge or [Chain/Radius]:** *(Pick an edge.)*

**Select edge** selects a single edge.

Chain option
**Select an edge or [Chain/Radius]:** *(Type **C**.)*
**Select an edge chain or [Edge/Radius]:** *(Pick an edge.)*

**Select an edge chain** selects all tangential edges.

Radius option
**Select an edge or [Chain/Radius]:** *(Type **R**.)*
**Enter fillet radius <1.0000>:** *(Enter a value.)*

**Enter fillet radius** specifies the fillet radius.

### RELATED COMMANDS

**Chamfer** bevels intersecting lines or polyline vertices.

**SolidEdit** edits 3D solid models.

### RELATED SYSTEM VARIABLES

**FilletRad** specifies the current filleting radius.

**TrimMode** toggles whether objects are trimmed.

**TIPS**

- Pick the end of the object you want filleted; the other end will remain untouched.

- The lines, arcs, or circles need not touch.

- As a faster substitute for the **Extend** and **Trim** commands, use the **Fillet** command with the radius of zero.

- If the lines to be filleted are on two different layers, the fillet is drawn on the current layer.

- The fillet radius must be smaller than the length of the lines. For example, if the lines to be filleted are 1.0m long, the fillet radius must be less than 1.0m.

- Use the **Close** option of the **PLine** command to ensure a polyline is filleted at all vertices.

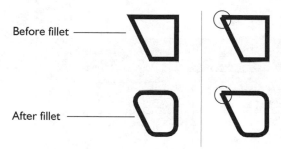

*Polyline closed with **Close** option (left) and without (right).*

- You cannot fillet polyline segments from different polylines.

- Filleting a pair of circles does not trim them.

- As of AutoCAD Release 13, the **Fillet** command fillets a pair of parallel lines; the radius of the fillet is automatically determined as half the distance between the lines.

- The **mUltiple** option is new with AutoCAD 2004.

# 'Filter

**Rel.12**  Creates filter lists that are applied to create selection sets.

| Command | Alias | Ctrl+ | F-key | Alt+ | Menu Bar | Tablet |
|---------|-------|-------|-------|------|----------|--------|
| 'filter | fi | ... | ... | ... | ... | ... |

**Command:** filter

*Displays dialog box:*

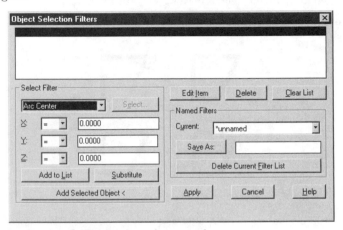

*After clicking **Apply**, AutoCAD continues in the command area:*

**Applying filter to selection.**
**Select objects:** *(Select one or more objects.)*
**Select objects:** *(Press **Enter** to end object selection.)*
**Exiting filtered selection.**

*AutoCAD highlights filtered objects with grips.*

## DIALOG BOX OPTIONS

Select Filter options

**Select** displays all items of the specified type in the drawing.

**X, Y, Z** specifies the object's coordinates.

**Add to List** adds the current select-filter option to the filter list.

**Substitute** replaces a highlighted filter with the selected filter.

**Add Selected Object** selects the object to be added from the drawing.

**Edit Item** edits the highlighted filter item.

**Delete** deletes the highlighted filter item.

**Clear list** clears the entire filter list.

Named Filter options

**Current** selects the named filter from the list.

**Save As** saves the filter list with a name and the *.nfl* extension.

**Delete Current Filter List** deletes the named filter.

**Apply** closes the dialog box, and applies the filter operation.

## COMMAND LINE OPTION

**Select options** selects the objects to be filtered; use the All option to select all non-frozen objects in the drawing.

## RELATED COMMANDS

*Any AutoCAD command with a 'Select objects' prompt.*

**QSelect** creates a selection set quickly, via a dialog box.

**Select** creates a selection set via the command line.

## RELATED FILE

***.nfl**is a named filter list.

## TIPS

- The selection set created by **Filter** is accessed via the **P** (previous) selection option.

- Alternatively, **'Filter** is used transparently at the 'Select objects' prompt.

- **Filter** cannot find objects when the color and linetype are set to BYLAYER.

- Save selection sets by name to an *.nfl* (short for *named filter*) file on disk for use in other drawings or editing sessions.

- **Filter** uses the following grouping operators:

| | | |
|---|---|---|
| **Begin OR | *with* | **End OR |
| **Begin AND | *with* | **End AND |
| **Begin XOR | *with* | **End XOR |
| **Begin NOT | *with* | **End NOT |

- **Filter** uses the following relational operators:

| Operator | Meaning |
|---|---|
| < | Less than. |
| <= | Less than or equal to. |
| = | Equal to. |
| != | Not equal to. |
| > | Greater than. |
| >= | Greater than or equal to. |
| * | All values. |

# Filtering Selection Sets

In this tutorial, you erase all construction lines from a drawing:

## Step 1

Start the **Erase** command, and then invoke the **Filter** command transparently:

**Command:** erase
**Select objects:** 'filter

AutoCAD displays the **Object Selection Filters** dialog box:

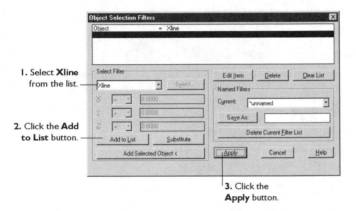

1. Select **Xline** from the list.

2. Click the **Add to List** button.

3. Click the **Apply** button.

## Step 2

In the **Select Filter** section:

Select **Xline** from the drop list.

Click the **Add to List** button.

Click **Apply**. Notice that the **Filter** command continues by displaying prompts on the command line: "Applying filter to selection."

Specify that **Filter** should search the entire drawing with the **All** option:

**Select objects:** all
*n* **found** *n* **were filtered out.**

## Step 3

Exit the **Filter** command by pressing ENTER:

**Select objects:** *(Press **Enter** to end object selection.)*
**Exiting filtered selection. <Selection set: 7>**
*n* **found**

AutoCAD resumes the **Erase** command. Press ENTER to end it:

**Select objects:** *(Press **Enter** to end the command.)*

AutoCAD uses the selection set created by the **Filter** command to erase the xlines.

 # Find

**2000** Finds, and optionally replaces, text in drawings.

| Command | Alias | Ctrl+ | F-key | Alt+ | Menu Bar | Tablet |
|---------|-------|-------|-------|------|----------|--------|
| find | ... | ... | ... | EF | **Edit** | **X10** |
| | | | | | ⃕**Find** | |

**Command:** find

*Displays dialog box:*

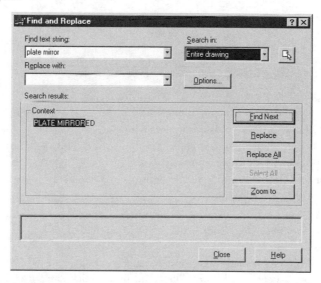

## DIALOG BOX OPTIONS

**Find text string** specifies the text find; enter text or click the down arrow to select one of the most recent lines of text searched for.

**Replace with** specifies the text to be replaced (only if replacing found text); enter text or click the down arrow to select one of the most recent lines of text searched for.

**Search in**:

- **Current Selection** searches the current selection set for text; click the Select objects button to create a selection set, if necessary.
- **Entire Drawing** searches the entire drawing.

**Select objects** allows you to select objects in the drawing; press ENTER to return to dialog box.

**Options** displays the Find and Replace Options dialog box.

### Search Results options

**Context** displays the found text in its context.

**Find Next** finds the text entered in the Find text string field.

**Replace** replaces a single found text with the text entered in the Replace with field.

**Replace All** replaces all instances of the found text.

. . . . . . . . . . . . . . . . . . . . . . . . . . . . . . . . . . . . . . . . . . . . . . . . . .

**Select All** selects all objects containing the text entered in the **Find text string** field; AutoCAD displays the message "AutoCAD found and selected *n* objects that contain "...".""

**Zoom to** zooms in to the area of the current drawing containing the found text.

**Find and Replace Options** dialog box

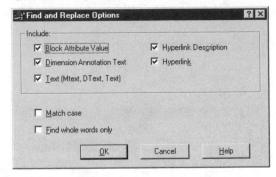

Include options

**Block Attribute Value** finds text in attributes.

**Dimension Annotation Text** finds text in dimensions.

**Text (Mtext, DText, Text)** finds text in paragraph text placed by the MText command, and single-line text placed by the Text command.

**Hyperlink Description** finds text in the description of a hyperlink.

**Hyperlink** finds text in a hyperlink.

**Match case** finds text that exactly matches the uppercase and lowercase pattern; for example, when searching for "Quick Reference," AutoCAD would find "Quick Reference" but not "quick reference."

**Find whole words only** finds text that exactly matches whole words; for example, when searching for "Quick Reference," AutoCAD would find "Quick Reference" but not "Quickly Reference."

### RELATED COMMANDS

**AttDef** creates attribute text.

**DdEdit** edits text.

**Dim***xxx* creates dimension text.

**Hyperlink** creates hyperlinks and hyperlink descriptions.

**MText** creates paragraph text.

**Properties** edits selected text.

**QSelect** finds text objects.

**Text** creates single-line text.

# Fog

**Rel.14** Creates fog-like effects in renderings to add visual distance cues that suggest the apparent distance of objects from cameras.

| Command | Alias | Ctrl+ | F-key | Alt+ | Menu Bar | Tablet |
|---------|-------|-------|-------|------|----------|--------|
| fog | ... | ... | ... | VEF | View | P2 |
| | | | | | ⤷Render | |
| | | | | | ⤷Fog | |

**Command:** fog

*Displays dialog box:*

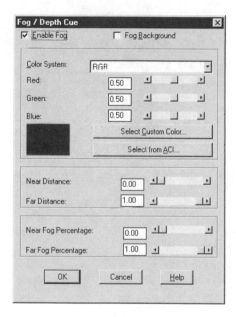

## DIALOG BOX OPTIONS

**Enable Fog**   turns fog effect on; allows turning fog off without affecting parameters.

**Fog Background** applies fog effect to background; see the Background command.

**Color System** selects the color of the fog by either RGB (red, green blue) or HLS (hue, lightness, saturation) methods.

**Near Distance** defines where the fog effect begins; value is the percentage distance between camera and the back clipping plane.

**Far Distance** defines where the fog effect ends.

**Near Fog** specifies the percentage of fog effect at the near distance; ranges between 0 and 100%.

**Far Fog** specifies the percentage of fog effect at the far distance; ranges between 0 and 100%.

**RELATED COMMAND**

**Render** creates a rendering with optional fog effect.

**TIPS**

- Apply the **Fog** command, and then use the **Render** command to see the effect.

- The fog can be any color:

| Color | RGB | HLS | Effect |
|-------|-----|-----|--------|
| White | 1,1,1 | 0,1,0 | Fog. |
| Black | 0,0,0 | 0,0,0 | Distance. |
| Green | 0,1,0 | 0.33,0.5,0 | Alien mist. |

- The effect of using White as the fog color:

- It can be tricky getting the fog effect to work. Use the **3dOrbit** command to set the back clipping plane at the back of the model, or where you want the fog to have its full effect. Then use **Fog** to set up the following parameters:

> Near distance: 0.70
>
> Far distance: 1.00
>
> Near fog percentage: 0.00
>
> Far fog percentage: 1.00

**Removed Command**

**GiffIn** was removed from Release 14. In its place, use **Image**.

# 'GotoUrl

2000 Goes to hyperlinks contained by objects (*undocumented command*).

| Command | Alias | Ctrl+ | F-key | Alt+ | Menu Bar | Tablet |
|---------|-------|-------|-------|------|----------|--------|
| 'gotourl | ... | ... | ... | ... | ... | ... |

**Command:** gotourl
**Select objects:** *(Select one or more objects.)*
**Select objects:** *(Press Enter to end object selection.)*
**browser Enter Web location (URL) <http://www.autodesk.com>: http://www.upfrontezine.com**
*AutoCAD launches your computer's default Web browser, and attempts to access the URL.*

## COMMAND LINE OPTION

**Select objects** selects one or more objects containing a hyperlink.

## RELATED COMMANDS.

**Browser** launches the Web browser.

**Hyperlink** attaches, edits, and removes hyperlinks from objects.

## TIPS

- This command is meant for use by macros and menus.

- AutoCAD uses this command for the shortcut menu's **Hyperlink | Open** option.

# 'GraphScr

**V. 2.1** Switches the text window back to the graphics window.

| Command | Alias | Ctrl+ | F-key | Alt+ | Menu Bar | Tablet |
|---------|-------|-------|-------|------|----------|--------|
| 'graphscr | ... | ... | F2 | ... | ... | ... |

**Command:** graphscr

*AutoCAD displays the drawing window:*

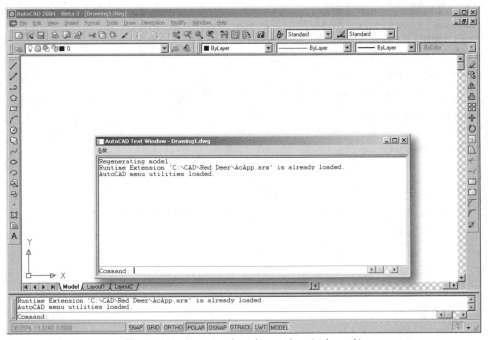

*Text window (on top) and graphics windows (underneath).*

## COMMAND LINE OPTIONS
*None.*

## RELATED COMMANDS
**CopyHist** copies text from the Text window to the Clipboard.

**TextScr** switches from the graphics window to the Text window.

## RELATED SYSTEM VARIABLE
**ScreenMode** indicates whether the current screen is in text or graphics mode:

| ScreenMode | Meaning |
|------------|---------|
| 0 | Text window. |
| 1 | Graphics window (default). |
| 2 | Dual screen displaying both text and graphics. |

## TIP
• The Text window appears to be frozen when a dialog box is active. Click the dialog box's **OK** or **Cancel** buttons to regain access to the Text window.

# 'Grid

<u>V. 1.0</u>  Displays a grid of reference dots within the current drawing limits.

| Command | Alias | Ctrl+ | F-key | Alt+ | Status Bar | Tablet |
|---------|-------|-------|-------|------|------------|--------|
| 'grid | ... | G | F7 | ... | GRID | ... |

**Command:** grid
**Specify grid spacing(X) or [ON/OFF/Snap/Aspect] <0.5000>:** *(Enter a value, or enter an option.)*

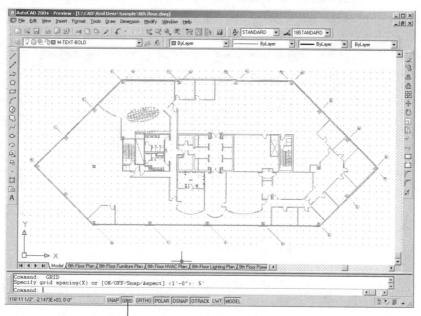

*Click **GRID** to toggle the grid display; right-click for menu.*

## COMMAND LINE OPTIONS

**Specify grid spacing(X)** sets the x and y direction spacing; an *X* following the value sets the grid spacing to a multiple of the current snap setting.

**ON** turns on grid markings.

**OFF** turns off grid markings.

**Snap** makes the grid spacing the same as the snap spacing.

Aspect options
**Specify the horizontal spacing(X) <0.0000>:** *(Enter a value.)*
**Specify the vertical spacing(Y) <0.0000>:** *(Enter a value.)*

**Horizontal** sets the spacing in the x direction.

**Vertical** sets the spacing in the y direction.

## RELATED COMMANDS

**Options** sets the grid via a dialog box.

**Limits** sets the limits of the grid in WCS.

**Snap** sets the snap spacing.

## RELATED SYSTEM VARIABLES

**GridMode** is the current grid visibility:

| GridMode | Meaning |
|----------|---------|
| 0 | Grid is off (default). |
| 1 | Grid is on. |

**GridUnit** specifies the current grid x,y spacing.

**LimMin** specifies the x,y coordinates of the lower-left corner of the grid display.

**LimMax** specifies the x,y coordinates of the upper-right corner of the grid display.

**SnapStyl** displays a normal or isometric grid:

| SnapStyl | Meaning |
|----------|---------|
| 0 | Normal (default). |
| 1 | Isometric grid. |

## TIPS

- The grid is most useful when set to the snap spacing, or to a multiple of the snap spacing.

- When the grid spacing is set to 0, it matches the snap spacing.

- You can set a different grid spacing in each viewport, and a different grid spacing in the x and y directions.

- Rotate the grid with the **Snap** command's **Rotate** option.

- The **Snap** command's **Isometric** option displays an isometric grid.

- If a very dense grid spacing is selected, the grid will take a long time to display; press **Esc** to cancel the display.

- AutoCAD will not display a grid that is too dense, and returns the message, "Grid too dense to display."

- Grid markings are not plotted; to create a plotted grid, use the **Array** command to place an array of points.

# Group

Rel.13 Creates named selection sets of objects.

| Commands | Aliases | Ctrl+ | F-key | Alt+ | Menu Bar | Tablet |
|----------|---------|-------|-------|------|----------|--------|
| group | g | ... | ... | ... | ... | X8 |
| -group | -g | | | | | |

**Command:** group

*Displays dialog box:*

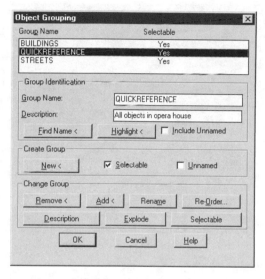

## DIALOG BOX OPTIONS

**Group Name** lists the names of groups in the drawing.

Group Identification options

**Group Name** displays the name of the current group.

**Description** describes the group; may be up to 64 characters long.

**Find Name** lists the name(s) of group(s) that a selected object belongs to.

**Highlight** highlights the objects included in the current group.

**Include Unnamed** lists unnamed groups in the dialog box.

Create Group options

**New** selects objects for the new group.

**Selectable** toggles selectability: picking one object picks the entire group.

**Unnamed** creates an unnamed group; AutoCAD gives the name **\*A***n*, where *n* is a number that increases with each group.

## Change Group options

**Remove** removes objects from the current group.

**Add** adds objects to the current group.

**Rename** renames the group.

**Re-order** changes the order of objects in the group; displays Order Group dialog box.

**Description** changes the description of the group.

**Explode** removes the group description; does not erase group members.

**Selectable** toggles selectability.

## Order Group dialog box

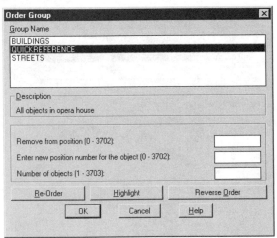

**Group Name** lists the names of groups in the current drawing.

**Description** describes the selected group.

**Remove from position (0 - *n*)** selects the object to move.

**Replace at position (0 - *n*)** moves the group name to a new position.

**Number of objects (1 - *n*)** lists the number of objects to reorder.

**Re-Order** applies the order changes.

**Highlight** highlights the objects in the current group.

**Reverse Order** reverses the order of the groups.

## -GROUP Command

**Command:** -group

**Enter a group option**
**[?/Order/Add/Remove/Explode/REName/Selectable/Create] <Create>:**
*(Enter an option.)*

### COMMAND LINE OPTIONS

**?** lists the names and descriptions of currently-defined groups.

**Order** changes the order of objects within the group.

**Add** adds objects to the group.

**Remove** removes objects from the group.

**Explode** removes the group definition from the drawing.

**REName** renames the group.

**Selectable** toggles whether the group is selectable.

**Create** creates a newly named group from the objects selected.

### RELATED COMMANDS

**Block** creates a named symbol from a group of objects.

**Select** creates a selection set.

### RELATED SYSTEM VARIABLE

**PickStyle** toggles whether groups are selected by the usual selection process:

| PickStyle | Meaning |
|-----------|---------|
| 0 | Groups and associative hatches are not selected. |
| 1 | Groups are included in selection sets. |
| 2 | Associate hatches are included in selection sets. |
| 3 | Both are selected (default). |

### TIPS

- As of AutoCAD 2002, you can no longer toggle groups on and off with the CTRL+A shortcut keystroke.

- Consider a group as a named selection set; unlike a regular selection set, a group is not "lost" when the next group is created.

- Group descriptions are up to 64 characters long.

- Anonymous groups are unnamed; AutoCAD refers to them as **\*A***n*.

# Hatch

<u>**V. 1.4**</u>  Draws non-associative crosshatch patterns within closed boundaries.

| Command | Alias | Ctrl+ | F-key | Alt+ | Menu Bar | Tablet |
|---------|-------|-------|-------|------|----------|--------|
| hatch | -h | ... | ... | ... | ... | ... |

**Command:** hatch
**Enter a pattern name [?/Solid/User defined] <ANSI31>:** *(Enter a pattern name, or enter an option.)*
**Specify a scale for the pattern <1.0000>:** *(Enter a scale factor.)*
**Specify an angle for the pattern <0>:** *(Enter a rotation angle.)*
**Select objects:** *(Press **Enter**, or select one or more objects.)*
**Select objects:** *(Press **Enter** to end object selection.)*

## COMMAND LINE OPTIONS

**Enter a pattern name** specifies the valid name of a hatch pattern; include one of the following optional, undocumented style parameters, such as:

**Enter a pattern name [?/Solid/User defined]:** ansi31,o

| Style | Meaning |
|-------|---------|
| N | Hatches alternate boundaries (Normal). |
| O | Hatches only outermost boundary (Outermost). |
| I | Hatches everything within outermost boundary (Ignore). |

**Specify a scale** specifies the hatch pattern scale.

**Specify an angle** specifies the hatch pattern angle.

**Select objects** selects the objects that make up the hatch pattern boundary.

*Pattern Name options*
**?** lists the hatch pattern names: 'Enter pattern(s) to list <*>:'.

**Solid** specifies a solid fill drawn with the current color.

*User Defined options*
**Specify angle for crosshatch lines <0>:** *(Enter an angle.)*
**Specify spacing between the lines <1.0000>:** *(Enter a spacing distance.)*
**Double hatch area? [Yes/No] <N>:** *(Type **Y** or **N**.)*
**Select objects to define hatch boundary or <direct hatch>,**
**Select objects:** *(Select one or more objects.)*
**Select objects:** *(Press **Enter** to end object selection.)*

**Specify angle for crosshatch lines** specifies the hatching angle.

**Specify spacing between the lines** specifies the distance between lines.

**Double hatch area?** determines whether a second set of hatch lines is drawn at 90 degrees to the first.

**Select objects to define hatch boundary** applies the hatch within objects.

**Direct hatch** specifies points that bound the hatch area.

Direct Hatch options

*Press ENTER at the first 'Select objects' prompt.*

**Retain polyline boundary? [Yes/No] <N>:** *(Enter **Y** or **N**.)*
**Specify start point:** *(Pick a point.)*
**Specify next point or [Arc/Close/Length/Undo]:** *(Pick a point, or enter option.)*
**Specify next point or [Arc/Close/Length/Undo]:** *(Pick a point, or enter option.)*
**Specify next point or [Arc/Close/Length/Undo]:** *(Enter **C** to close.)*
**Specify start point for new boundary or <apply hatch>:** *(Press **Enter** to apply.)*

**Retain polyline boundary?**

**Yes** leaves boundary in place after the hatch is complete.

**No** erases boundary after the hatch is complete.

**Specify start point** begins drawing the hatch boundary.

**Specify next point** draws straight lines.

**Arc** draws arc hatch boundaries; prompts 'Enter an arc boundary option [Angle/ CEnter/ CLose/Direction/Line/Radius/Second pt/Undo/Endpoint of arc] <Endpoint>"; see the Arc command.

**Close** closes the hatch boundary.

**Length** continues the boundary by a specified distance; prompts 'Specify length of line.'

**Undo** undoes the last-drawn segment.

## RELATED COMMANDS

**BHatch** places automatic, associative hatching.

**Boundary** automatically creates polyline or region boundary.

**Explode** reduces hatch pattern to its constituent lines.

**Snap** changes the hatch pattern's origin.

## RELATED SYSTEM VARIABLES

**HpAng** specifies the current hatch pattern angle.

**HpDouble** specifies doubled hatch pattern.

**HpName** specifies the current hatch pattern name.

**HpScale** specifies the current hatch pattern scale.

**HpSpace** specifies the current hatch pattern spacing.

**SnapBase** controls the origin of the hatch pattern.

**SnapAng** controls the angle of the hatch pattern.

## RELATED FILES

*acad.pat* holds hatch definition file for ANSI patterns.

*acadiso.pat* holds hatch definition file for ISO patterns.

**TIPS**

- The **Hatch** command draws non-associative hatch patterns; the pattern remains in place when its boundary is edited.

- For complex hatch areas, you may find it easier to outline the area with a polyline — using object snap — or with the **Boundary** or **BHatch** commands.

- To create the hatch as lines, precede the pattern name with * (asterisk).

- This command fills areas with solid color, but does not apply gradient fills; use the **BHatch** or **-BHatch** commands instead.

- For a listing of hatch patterns included with AutoCAD, see the **BHatch** command.

# HatchEdit

**Rel.13**  Edits associative hatch objects.

| Commands | Alias | Ctrl+ | F-key | Alt+ | Menu Bar | Tablet |
|----------|-------|-------|-------|------|----------|--------|
| hatchedit | he | ... | ... | ... | Modify | Y16 |
| | | | | | ⤷Hatch | |
| -hatchedit | | | | | | |

**Command:** hatchedit
**Select associative hatch object:** *(Select one hatch object.)*
  *Displays dialog box:*

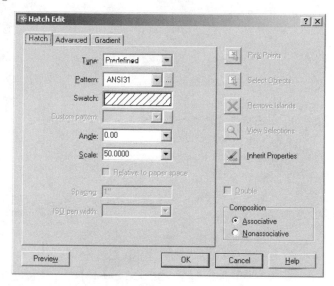

## DIALOG BOX OPTIONS

*Most options are grayed-out because they are unavailable.*

**Preview** dismisses the dialog box temporarily, so that you can see the effect of the editing changes on the hatch pattern.

**Inherit Properties** allows you to pick another hatch pattern whose properties will apply to the selected hatch pattern.

### Quick tab

**Pattern Type** selects the source of the hatch pattern:

| Source | Meaning |
|--------|---------|
| **Predefined** | *acad.pat* (default). |
| **User-defined** | Create hatch pattern on the fly. |
| **Custom** | Select hatch from another *.pat* file. |

**Pattern** selects a pattern name from the drop list.

**...** selects a pattern from a tabbed dialog box.

**Custom Pattern** names the custom hatch pattern.

**Scale** specifies the hatch pattern scale.

**Angle** specifies the hatch pattern angle.

**Spacing** specifies the spacing between pattern lines.

## Advanced tab

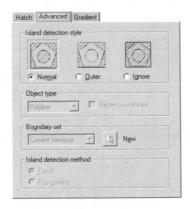

### Island Detection Style
- **Normal** hatches the alternate boundaries (default).
- **Outer** hatches only the outermost boundary.
- **Ignore** hatches everything within the outermost boundary.

*Other options shown in tab are unavailable.*

## Gradient tab
*New to AutoCAD 2004.*

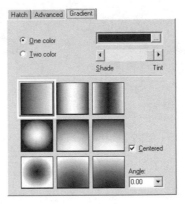

*See **BHatch** command for options.*

## -HATCHEDIT Command

**Command:** -hatchedit

**Select associative hatch object:** *(Select one object.)*

**Enter hatch option [Disassociate/Style/Properties] <Properties>:** *(Enter an option.)*

### COMMAND LINE OPTIONS

**Select** selects a single associative hatch pattern.

**Disassociate** removes associativity from hatch pattern.

Style options

**Enter hatching style [Ignore/Outer/Normal] <Normal>:** *(Enter an option.)*

**Normal** hatches alternate boundaries (default).

**Outer** hatches only outermost boundary.

**Ignore** hatches everything within outermost boundary.

Properties options

*See **Hatch** command.*

**Pattern name** specifies the name of a valid hatch pattern.

**?** lists the available hatch pattern names.

**Solid** replaces the hatch pattern with solid fill.

**User defined** creates a new on-the-fly hatch pattern.

**Scale** changes the hatch pattern scale.

**Angle** changes the hatch pattern angle.

### RELATED COMMANDS

**BHatch** applies associative hatch patterns.

**Explode** explodes a hatch pattern block into lines.

### RELATED SYSTEM VARIABLES

**HpAng** specifies the current hatch pattern angle.

**HpDouble** specifies doubled hatch pattern.

**HpName** specifies the current hatch pattern name.

**HpScale** specifies the current hatch pattern scale.

**HpSpace** specifies the current hatch pattern spacing.

**SnapBase** controls the origin of the hatch pattern.

**SnapAng** controls the angle of the hatch pattern.

### TIPS

- **HatchEdit** only works with associative hatch objects.

- To select a solid fill, click the outer edge of the hatch pattern, or use a crossing window selection on top of the solid fill.

- Even though the **Hatch Edit** dialog box looks identical to the **BHatch** dialog box, many options are not available in the **Hatch Edit** dialog box.

# 'Help

**V. 1.0**   Lists information for using AutoCAD's commands.

| Command | Alias | Ctrl+ | F-key | Alt+ | Menu Bar | Tablet |
|---------|-------|-------|-------|------|----------|--------|
| 'help | '? | ... | F1 | HH | Help | Y7 |
| | | | | | ↳Help | |

**Command:** help

*Displays window:*

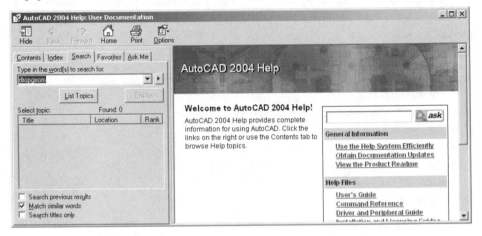

## RELATED COMMANDS
*All.*

## RELATED FILE
*acad.chm* is AutoCAD's primary help file.

## TIP
• Since **'Help**, **'?**, and F1 are transparent commands, you can use them during another command to get help with the command's options.

 # Hide

**V. 2.1**   Removes hidden lines from 3D drawings.

| Command | Alias | Ctrl+ | F-key | Alt+ | Menu Bar | Tablet |
|---------|-------|-------|-------|------|----------|--------|
| hide | hi | ... | ... | VH | View | M2 |
| | | | | | ↳Hide | |

**Command:** hide
**Regenerating drawing.**
**Removing hidden lines: 25**

*3D object without (left) and with (right) hidden lines removed.*

## COMMAND LINE OPTIONS
*None.*

## RELATED COMMANDS
**HLSettings** specifies settings for lines displayed by the Hide command.

**MView** removes hidden lines during plots of paper space drawings.

**Plot** removes hidden lines during plots of 3D drawings.

**Regen** returns the view to wireframe.

**Render** performs realistic renderings of 3D models.

**ShadeMode** performs quick renderings and quick hides of 3D models.

**VPoint** selects the 3D viewpoint.

**3dOrbit** removes hidden lines of perspective in 3D views.

## RELATED SYSTEM VARIABLES
**HaloGap** specifies the size of gap where a hidden object intersects with a visible object.

**HidePrecision** determines the calculated accuracy of hidden-line removal.

**HideText** determines whether text is involved in the hidden-line calculations.

**ObscuredColor** specifies the color of hidden lines.

**ObscuredLType** specifies the linetype of hidden lines.

**IntersectionDisplay** determines whether polylines are displayed at the intersection of faces.

**TIPS**

- This command considers the following objects as opaque: circles, 2D solids, traces, wide 2D polylines, 3D faces, polygon meshes, any objects with thickness, as well as regions and 3D solids.

- For a faster hide, use the **ShadeMode** command's **Hide** option.

- Use **MSlide** (or **SaveImg**) to save the hidden-line view as *.sld* (or *.tif*, *.tga*, or *.gif*) files. View the saved images with **VSlide** (or **Replay**).

- Freezing layers speeds up the hide process, because **Hide** ignores objects on frozen layers.

- **Hide** does considers the visibility of text and attributes, depending on the setting of the **HideText** system variable.

- The figure below shows (left to right) a 3D model, normal hidden line removal, and hidden lines turned on (see **HlSettings** command):

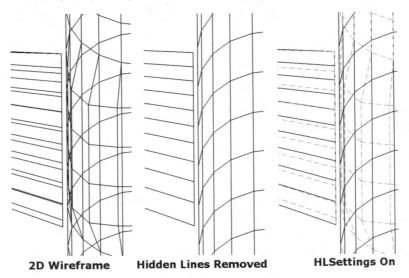

| 2D Wireframe | Hidden Lines Removed | HLSettings On |

- To create a hidden-line view when plotting in paper space, select the **HidePlot** option of the **MView** command.

## Removed Command

**HpConfig** was removed from AutoCAD 2000; it was replaced by the **Plot** command.

# HLSettings

**2004**    Controls the display of hidden (obscured) lines.

| Command | Alias | Ctrl+ | F-key | Alt+ | Menu Bar | Tablet |
|---------|-------|-------|-------|------|----------|--------|
| hlsettings | ... | ... | ... | ... | ... | ... |

**Command:** hlsettings

*Displays dialog box:*

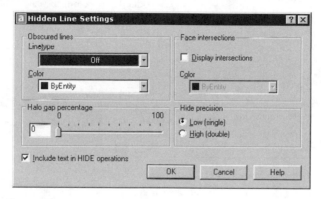

## DIALOG BOX OPTIONS

Obscured Lines options

**Linetype** selects linetype by which to display hidden lines; select Off to display no hidden lines.

**Color** selects the color in which to display hidden lines.

Face Intersection options

**Display intersections:**

   ☑ Displays polylines at the intersection of faces.

   ☐ Does not display face intersections.

**Color** selects the color of the face intersections.

Additional options

**Halo Gap Percentage** specifies the distance to shorten a *haloed* line. ,

**Hide Precision options:**

   • **Low (single)** calculates the display of hidden lines with single-precision; uses less memory.

   • **High (double)** calculates the display of hidden lines with double-precision; more accurate.

**Include text in HIDE operation:**

   ☑ Text is hidden by objects, and hides other objects.

   ☐ Text is ignored: it is not hidden, and does not hide other objects.

## RELATED COMMANDS

**Hide** performs hidden-line removal.

**ShadeMode** performs quick hides of 3D models.

## RELATED SYSTEM VARIABLES

**HaloGap** specifies the size of gap where a hidden object intersects with a visible object.

**HidePrecision** determines the calculated accuracy of hidden-line removal.

**HideText** determines whether text is involved in the hidden-line calculations.

**IntersectionColor** specifies the color of the polylines displayed at the intersection of faces.

**IntersectionDisplay** determines whether polylines are displayed at the intersection of faces.

**ObscuredColor** specifies the color of hidden lines:

| ObscuredColor | Meaning |
| --- | --- |
| 0 | Hidden lines are not displayed. |
| 1 | Hidden lines displayed in red. |
| 2 | Yellow. |
| 3 | Green. |
| 4 | Cyan. |
| 5 | Blue. |
| 6 | Magenta. |
| 7 | Black (white). |
| 8 | Dark gray. |
| 9 | Light gray. |
| 10 - 255 | Other colors. |

**ObscuredLtype** specifies the linetype of hidden lines:

| ObscuredLtype | Meaning |
| --- | --- |
| 0 | Hidden lines are not displayed. |
| 1 | Solid. |
| 2 | Dashed (traditional Hidden). |
| 3 | Dotted. |
| 4 | Short Dash. |
| 5 | Medium Dash. |
| 6 | Long Dash. |
| 7 | Double Short Dash. |
| 8 | Double Medium Dash. |
| 9 | Double Long Dash. |
| 10 | Medium Long Dash. |
| 11 | Sparse Dash. |

## TIPS

• AutoCAD sometimes calls hidden lines "obscured lines."

• A *haloed* line is a hidden line that is shortened at the point where it intersects another object. The value of the gap is the percentage of the current unit value; it is independent of zoom level.

• An *intersection polyline* a polyline displayed at the intersection of 3D faces.

• This command affects the display of the **Hide** and **ShadEdge** commands; it does not work with the **3dOrbit** command's **Hide** option.

• To disable the display of hidden-lines, set the **Linetype** to **Off** (**0**).

# Hyperlink

**2000**   Attaches hyperlinks (URLs) to objects in drawings.

| Commands | Alias | Ctrl+ | F-key | Alt+ | Menu Bar | Tablet |
|---|---|---|---|---|---|---|
| hyperlink | ... | K | ... | IH | Insert | ... |
| | | | | | ⮡Hyperlink | |
| -hyperlink | | | | | | |

**Command:** hyperlink
**Select objects:** *(Select one or more objects.)*
**Select objects:** *(Press **Enter** to end object selection.)*
  *Displays dialog box:*

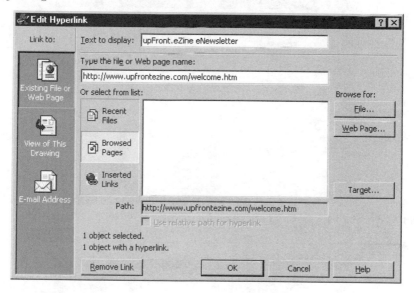

## COMMAND LINE OPTION

**Select object** selects the objects to which to attach the hyperlink.

## DIALOG BOX OPTIONS

### Existing File or Web Page page

**Text to display** describes the hyperlink; the description is displayed by the tooltip; when blank, the tooltip displays the URL.

**Type the file or Web page name** specifies the hyperlink (URL) to associate with the selected objects; the hyperlink may be any file on your computer, on any computer you can access on your local network, or on the Internet.

**Or select from list:**

- Recent files
- Browsed pages
- Inserted links

**File** opens the Browse the Internet - Select Hyperlink dialog box.

**Web Page** starts up a simple Web browser.

**Path** displays the full path and filename to the hyperlink; only the filename appears when Use relative path for hyperlink is checked.

**Target** specifies a location in the file, such as a target in an HTML file, a named view in AutoCAD, or a page in a spreadsheet document. Displays Select Place in Document dialog box.

**Use relative path for hyperlink** toggles use of the path for relative hyperlinks in the drawing; when "" (null), the drawing paths stored in AcadPrefix are used.

**Remove link** removes the hyperlink from the object; this button appears only if you select an object that already has a hyperlink.

### View of This Drawing page

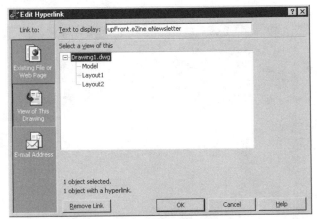

**Select a view of this** selects a layout, named view, or named plot.

### Email Address page

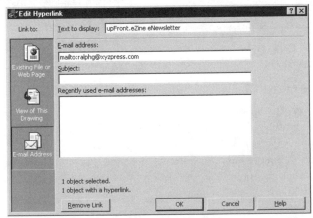

**Email address** specifies the email address; the *mailto:* prefix is added automatically.

**Subject** specifies the text that will be added to the Subject line.

**Recently used e-mail addresses** lists the email addresses you may have recently entered.

Select Place in Document option

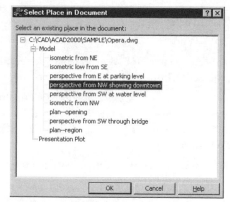

**Select an existing place in the document** selects a named location.

## -HYPERLINK Command
**Command:** -hyperlink
**Enter an option [Remove/Insert] <Insert>:** *(Enter an option.)*
**Enter hyperlink insert option [Area/Object] <Object>:** *(Enter an option.)*
**Select objects:** *(Select one or more objects.)*
**Select objects:** *(Press **Enter** to end object selection.)*
**Enter hyperlink <current drawing>:** *(Enter hyperlink address.)*
**Enter named location <none>:** *(Optional: Enter a bookmark.)*
**Enter description <none>:** *(Optional: Enter a description.)*

### COMMAND LINE OPTIONS
**Remove** removes a hyperlink from selected objects or areas.

**Insert** adds a hyperlink to selected objects or areas.

**Select objects** selects the object to which the hyperlink will be added.

**Enter hyperlink** specifies the filename or hyperlink address.

**Enter named location** *(optional)* specifies a location within the file or hyperlink.

**Enter description** *(optional)* specifies a description of the hyperlink.

### RELATED COMMANDS
**HyperlinkOptions** toggles the display of the hyperlink cursor, shortcut menu, and tooltip.

**HyperlinkOpen** opens a hyperlink (URL) via the command line.

**HyperlinkBack** returns to the previous URL.

**HyperlinkFwd** moves forward to the next URL; works only when the HyperlinkBack command was used, otherwise AutoCAD complains, "** No hyperlink to navigate to **".

**HyperlinkStop** stops the display of the current hyperlink.

**GoToUrl** displays a specific Web page.

**PasteAsHyperlink** pastes a hyperlink to the selected objects.

### RELATED SYSTEM VARIABLE

**HyperlinkBase** specifies the path for relative hyperlinks in the drawing; when "" (null), the drawing paths stored in **AcadPrefix** are used.

### TIPS

- If the drawing has never been saved, AutoCAD is unable to determine the default relative folder. For this reason, the **Hyperlink** command prompts you to save the drawing.

- By using hyperlinks, you can create a *project document* consisting of drawings, contacts (word processing documents), project timelines, cost estimates (spreadsheets pages), and architectural renderings. To do so, create a "title page" of an AutoCAD drawing with hyperlinks to the other documents.

- To edit a hyperlink with **Hyperlink**, select object(s), make editing changes, and click **OK**.

- To remove a hyperlink with **Hyperlink**, select the object, and then click **Remove Link**.

- An object can have just one hyperlink attached to it; more than one object, however, can share the same hyperlink.

- The alternate term for "hyperlink" is *URL*, which is short for "uniform resource locator," the universal file naming system used by the Internet.

# 'HyperlinkOpen, Back, Fwd, Stop

<u>2000</u> Controls the display of hyperlinked pages (*undocumented commands*).

| Commands | Alias | Ctrl+ | F-key | Alt+ | Menu Bar | Tablet |
|---|---|---|---|---|---|---|
| 'hyperlinkopen | ... | ... | ... | ... | ... | ... |
| 'hyperlinkback | | | | | | |
| 'hyperlinkfwd | | | | | | |
| 'hyperlinkstop | | | | | | |

**Command:** hyperlinkopen
**Enter hyperlink <current drawing>:** *(Enter a hyperlink option.)*
**Enter named location <none>:** *(Optional: Enter a bookmark.)*
*Displays the specified Web page or file, if possible.*

**Command:** hyperlinkback
*Returns to the previous hyperlinked page.*

**Command:** hyperlinkstop
*Stops displaying the Web page.*

**Command:** hyperlinkfwd
*Hyperlinks to the next page; can be used only after the **HyperLinkBack** command.*

## COMMAND LINE OPTIONS
**Enter hyperlink** enters a URL (uniform resource locator) or a filename.
**Enter named location** enters a named view or other valid target.

## RELATED COMMANDS
**Hyperlink** attaches a hyperlink to objects.
**HyperlinkOptions** specifies the options for hyperlinks.
**GoToUrl** displays a specific Web page.

## RELATED TOOLBAR ICONS

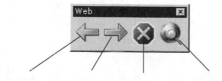

*Go Back    Go Forward    Stop Navigation    Browse the Web*

**Go Back** goes to the previous hyperlink; executes the HyperlinkBack command.
**Go Forward** goes to the next hyperlink; executes the HyperlinkFwd command.
**Stop Navigation** stops loading the current hyperlink file; executes the HyperlinkStop command.
**Browse the Web** displays the Web browser; executes the Browser command.

# 'HyperlinkOptions

<u>2000</u> Toggles the display of the hyperlink cursor, shortcut menu, and tooltip *(the longest command name in AutoCAD).*

| Command | Alias | Ctrl+ | F-key | Alt+ | Menu Bar | Tablet |
|---|---|---|---|---|---|---|
| 'hyperlinkoptions | ... | ... | ... | ... | ... | |

**Command:** hyperlinkoptions
**Display hyperlink cursor and shortcut menu? [Yes/No] <Yes>:** *(Type* **Y** *or* **N.***)*
**Display hyperlink tooltip? [Yes/No] <Yes>:** *(Type* **Y** *or* **N.***)*

*Hyperlink cursor and tooltip (left); hyperlink shortcut menu (right).*

## COMMAND LINE OPTIONS

**Display hyperlink cursor and shortcut menu** toggles the display of the hyperlink cursor and shortcut menu.

**Display hyperlink tooltip** toggles the display of the hyperlink tooltip.

## CURSOR MENU OPTIONS

*Select an object containing a hyperlink; then right-click to display the cursor menu.*

*Hyperlink options:*

**Open "url"** launches the appropriate applications and loads the file referenced by the URL.

**Copy Hyperlink** copies hyperlink data to the Clipboard; use the PasteAsHyperlink command to paste the hyperlink to selected objects.

**Add to Favorites** adds the hyperlink to the favorites list.

**Edit Hyperlink** displays the Edit Hyperlink dialog box; see the Hyperlink command.

## RELATED COMMANDS

**Hyperlink** attaches a hyperlink to objects.

**Options** determines options for most other aspects of AutoCAD.

## RELATED SYSTEM VARIABLE

**HyperlinkBase** specifies the path for relative hyperlinks in the drawing; when "" (null), the drawing paths stored in **AcadPrefix** are used.

# 'Id

<u>V. 1.0</u>    Identifies the 3D coordinates of picked points (*short for IDentify*).

| Command | Alias | Ctrl+ | F-key | Alt+ | Menu Bar | Tablet |
|---------|-------|-------|-------|------|----------|--------|
| 'id | ... | ... | ... | TYI | Tools | U9 |
| | | | | | ⵂInquiry | |
| | | | | | ⵂId Point | |

**Command:** id
**Specify point:** *(Pick a point.)*

*Sample output:*
**X = 1278.0018    Y = 1541.5993    Z = 0.0000**

## COMMAND LINE OPTION
**Specify point** picks a point.

## RELATED COMMANDS
**List** lists information about a picked object.

**Point** draws a point.

## RELATED SYSTEM VARIABLE
**LastPoint** contains the 3D coordinates of the last picked point.

## TIPS
- The **Id** command stores the picked point in the **LastPoint** system variable. Access that value by entering @ at the next prompt for a point value.

- Invoke the **Id** command to set the value of the **LastPoint** system variable, which can be used as relative coordinates in another command.

- The z-coordinate displayed by **Id** is the current elevation setting; if you use **Id** with an object snap, then the z-coordinate is the object-snapped value.

# Image

**Rel.14**  Controls the attachment of raster images.

| Commands | Aliases Ctrl+ | F-key | Alt+ | Menu Bar | Tablet |
|----------|---------------|-------|------|----------|--------|
| image | im    ... | ... | IM | Insert | T3 |
|  |  |  |  | ⇘Image Manager |  |
| -image | -im |  |  |  |  |

**Command:** image

*Displays dialog box:*

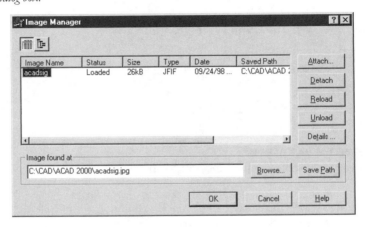

## DIALOG BOX OPTIONS

**Attach** displays the Attach Image File dialog box; see the ImageAttach command.

**Detach** erases the image from the drawing.

**Reload** reloads the image file into the drawing.

**Unload** removes the image from memory without erasing the image.

**Save Path** describes the drive and subdirectory location of the file.

**Browse** searches for the image file; displays the Attach Image File dialog box.

**Details** describes the technical details of the images; displays dialog box.

**Image File Details** dialog box

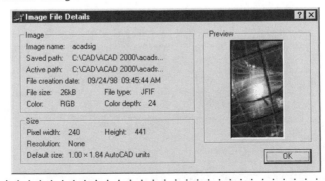

# -IMAGE Command

**Command:** -image

**Enter image option [?/Detach/Path/Reload/Unload/Attach] <Attach>:** *(Enter an option.)*

## COMMAND LINE OPTIONS

**?** lists currently-attached image files.

**Detach** erases the image from the drawing.

**Path** lists the names of images in the drawing.

**Reload** reloads the image file into the drawing.

**Unload** removes the image from memory without erasing the image.

**Attach** displays the Attach Image File dialog box; see the ImageAttach command.

## RELATED SYSTEM VARIABLE

**ImageHlt** toggles whether the entire image is highlighted.

## RELATED COMMANDS

**ImageAdjust** controls the brightness, contrast, and fading of the image.

**ImageAttach** attaches an image in the current drawing.

**ImageClip** creates a clipping boundary on an image.

**ImageFrame** toggles the display of the image's frame.

**ImageQuality** toggles the display between draft and high-quality mode.

**Transparency** changes the transparency of the image.

**Xref** attaches *.dwg* drawings as externally-referenced files.

## TIPS

- The **Image** command handles raster images of these color depths:

| Depth | Colors |
|-------|--------|
| **Bitonal** | Black and white (monochrome). |
| **8-bit gray** | 256 shades of gray. |
| **8-bit color** | 256 colors. |
| **24-bit color** | 16.7 million colors. |

- AutoCAD can display one or more images in any viewport.

- There is no theoretical limit to the number and size of images.

# ImageAdjust

**Rel.14** Controls brightness, contrast, and fading of attached raster images.

| Commands | Alias | Ctrl+ | F-key | Alt+ | Menu Bar | Tablet |
|---|---|---|---|---|---|---|
| imageadjust | iad | ... | ... | MOIA | Modify | X20 |
| | | | | | ⤷Object | |
| | | | | | ⤷Image | |
| | | | | | ⤷Adjust | |

-imageadjust

**Command:** imageadjust
**Select image(s):** *(Select one or more image objects.)*
**Select image(s):** *(Press **Enter** to end object selection.)*
*Displays dialog box:*

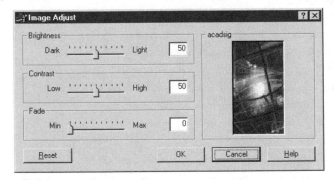

## DIALOG BOX OPTIONS

Brightness options
    **Dark** reduces the brightness of the image when values are closer to 0.

    **Light** increases the brightness of the image when values are closer to 100.

Contrast options
    **Low** reduces the image contrast when values are closer to 0.

    **High** increases the image contrast when values are closer to 100.

Fade options
    **Min** reduces the image fading when values are closer to 0.

    **Max** increases the image fading when values are closer to 100.

    **Reset** resets the image to its original parameters; default values are:

| Parameter | Original Setting |
|---|---|
| **Brightness** | 50 |
| Contrast | 50 |
| Fade | 0 |

## -IMAGEADJUST Command

**Command:** -imageadjust

**Select image(s):** *(Select one or more image objects.)*

**Select image(s):** *(Press **Enter** to end object selection.)*

**Enter image option [Contrast/Fade/Brightness] <Brightness>:** *(Enter an option.)*

### COMMAND LINE OPTIONS

Contrast option

**Enter contrast value (0-100) <50>:** *(Enter a value.)*

**Enter contract value** adjusts the contrast between 0% contrast and 100% contrast (*default = 50*).

Fade option

**Enter fade value (0-100) <0>:** *(Enter a value.)*

**Enter fade option** adjusts the fading between 0% faded and 100% faded (*default = 0*).

Brightness option

**Enter brightness value (0-100) <50>:** *(Enter a value.)*

**Enter brightness value** adjusts the brightness between 0% bright and 100% bright (*default = 50*).

### RELATED SYSTEM VARIABLES

*None.*

### RELATED COMMANDS

**Image** controls the loading of raster image files in the drawing.

**ImageAttach** attaches an image in the current drawing.

**ImageClip** creates a clipping boundary on an image.

**ImageFrame** toggles display of the image's frame.

**ImageQuality** toggles display between draft and high-quality mode.

**Transparency** changes the transparency of the image.

**TIPS**

* **Brightness** ranges from 0 (*left*) to 50 (*center*) and 100 (*right*).

* **Contrast** ranges from 0 (*left*) to 50 (*center*) and 100 (*right*).

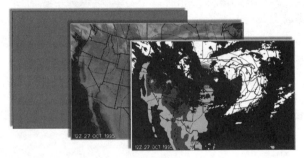

* **Fade** ranges from 0 (*left*) to 50 (*center*) and 100 (*right*).

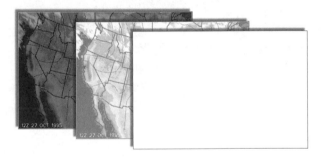

# ImageAttach

**Rel.14**  Selects raster files to attach to drawings.

| Command | Alias | Ctrl+ | F-key | Alt+ | Menu Bar | Tablet |
|---------|-------|-------|-------|------|----------|--------|
| imageattach | iat | ... | ... | II | Insert | ... |
| | | | | | ↳Raster Image | |

**Command:** imageattach

*Displays file dialog box:*

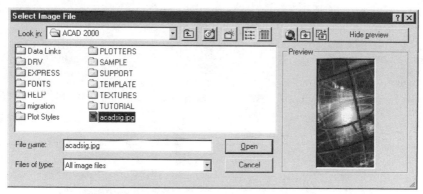

*Selecting an image, and then click* **Open.**

*Displays dialog box:*

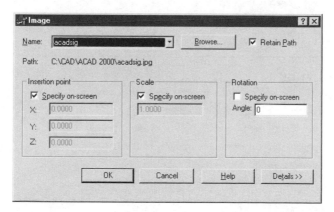

## DIALOG BOX OPTIONS

**Name** selects name from a list of previously-attached image names.

**Browse** selects file; displays **Select Image File** dialog box.

**Retain Path** saves the path to the image file.

Insertion Point options

**Specify On-Screen** specifies the insertion point of the image in the drawing, after the dialog box is dismissed.

**X, Y, Z** specifies the x, y, z coordinates of the lower-left corner of the image.

Scale options

**Specify on-screen** specifies the scale of the image (relative to the lower-left corner) in the drawing after the dialog box is dismissed.

**Scale** specifies the scale of the image; a positive value enlarges the image, while negative values reduce the image.

Rotation options

**Specify on-screen** specifies the rotation angle of the image about the lower-left corner in the drawing, after the dialog box is dismissed.

**Angle** specifies the angle to rotate the image; positive angles rotate the image counterclockwise.

*Button*

**Details** expands the dialog box to display information about the image:

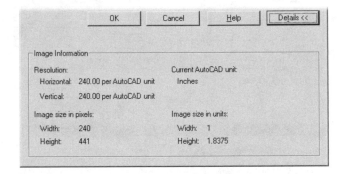

**RELATED COMMANDS**

**Image** controls the loading of raster image files in the drawing.

**ImageAttach** attaches an image in the current drawing.

**ImageClip** creates a clipping boundary on an image.

**ImageFrame** toggles the display of the image's frame.

**ImageQuality** adjusts the quality of the image.

**Transparency** changes the transparency of the image.

**RELATED SYSTEM VARIABLE**

**InsUnits** specifies the drawing units for the inserted image.

**TIPS**

- For a command-line version of the **ImageAttach** command, use the **-Image** command's **Attach** option.

- This dialog box no longer selects units from the **Current AutoCAD Unit** list box; as of AutoCAD 2000, use the **InsUnits** system variable.

# ImageClip

__Rel.14__   Clips raster images.

| Command | Alias | Ctrl+ | F-key | Alt+ | Menu Bar | Tablet |
|---------|-------|-------|-------|------|----------|--------|
| imageclip | icl | ... | ... | MCI | Modify | X22 |
| | | | | | ⤷Clip | |
| | | | | | ⤷Image | |

**Command:** imageclip
**Select image to clip:** *(Select one image object.)*
**Enter image clipping option [ON/OFF/Delete/New boundary] <New>:** *(Enter an option.)*

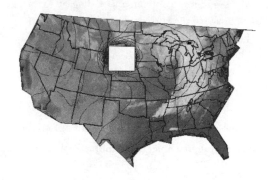

## COMMAND LINE OPTIONS

**Select image** selects one image to clip.

**ON** turns on a previous clipping boundary.

**OFF** turns off the clipping boundary.

**Delete** erases the clipping boundary.

New Boundary options
**Enter clipping type [Polygonal/Rectangular] <Rectangular>:** *(Enter an option.)*

**Polygonal** creates a polygonal clipping path.

**Rectangular** creates a rectangular clipping boundary.

Polygonal options
**Specify first point:** *(Pick a point.)*
**Specify next point or [Undo]:** *(Pick a point, or type **U**.)*
**Specify next point or [Undo]:** *(Pick a point, or type **U**.)*
**Specify next point or [Close/Undo]:** *(Pick a point, or enter an option.)*
**Specify next point or [Close/Undo]:** *(Enter **C** to close the polygon.)*

**Specify first point** specifies the start of the first segment of the polygonal clipping path.

**Specify next point** specifies the next vertex.

**Undo** undoes the last vertex.

**Close** closes the polygon clipping path.

Rectangular options
**Specify first corner point:** *(Pick a point.)*
**Specify opposite corner point:** *(Pick a point.)*

Specify first corner point specifies one corner of the rectangular clip.

Specify opposite corner point specifies the second corner.

*When you select an image with a clipped boundary, AutoCAD prompts:*
**Delete old boundary? [No/Yes] <Yes>:** *(Type **N** or **Y**.)*

Delete old boundary?

Yes removes previously-applied clipping path.

No exits the command.

## RELATED COMMANDS

**Image** controls the loading of raster image files in the drawing.

**ImageAdjust** controls the brightness, contrast, and fading of the image.

**ImageAttach** attaches an image in the current drawing.

**ImageFrame** toggles the display of the image's frame.

**ImageQuality** toggles the display between draft and high-quality mode.

**Transparency** changes the transparency of the image.

**XrefClip** clips a DWG drawing attached as an external reference file.

## TIPS

• You can use object snap modes on the image's frame, but not on the image itself.

• To clip a hole in the image, create the hole, then double back on the same path:

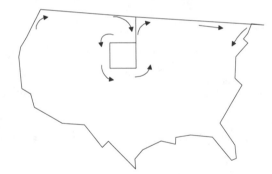

• For a rounded clipping path, apply the **PEdit** command.

# ImageFrame

<u>Rel.14</u> Toggles the display of the frame around raster images.

| Command | Alias | Ctrl+ | F-key | Alt+ | Menu Bar | Tablet |
|---|---|---|---|---|---|---|
| imageframe | ... | ... | ... | MOIF | Modify | ... |
| | | | | | ↳Object | |
| | | | | | ↳Image | |
| | | | | | ↳Frame | |

**Command:** imageframe
**Enter image frame setting [ON/OFF] <ON>:** *(Type* **ON** *or* **OFF**.*)*

*Image frame turned on (left) and turned off (right).*

## COMMAND LINE OPTIONS

**ON** turns on display of image frame (default).

**OFF** turns off display of image frame.

## RELATED SYSTEM VARIABLE

**ImageHlt** toggles whether the entire image is highlighted:

| ImageHlt | Meaning |
|---|---|
| 0 | Highlights image frame. |
| 1 | Highlights entire image (default). |

## RELATED COMMANDS

**Image** controls the loading of raster image files in the drawing.

**ImageAdjust** controls the brightness, contrast, and fading of the image.

**ImageAttach** attaches an image in the current drawing.

**ImageClip** creates a clipping boundary on an image.

**ImageQuality** toggles the display between draft and high-quality mode.

**Transparency** changes the transparency of the image.

## TIPS

- *Warning!* When **ImageFrame** is turned off, you cannot select the image.

- The frame is turned on (or off) in all viewports.

- The image frame is plotted.

# ImageQuality

**Rel.14**  Toggles the quality of raster images.

| Command | Alias | Ctrl+ | F-key | Alt+ | Menu Bar | Tablet |
|---|---|---|---|---|---|---|
| imagequality | ... | ... | ... | MOIQ | Modify | ... |
| | | | | | ↳Object | |
| | | | | | ↳Image | |
| | | | | | ↳Quality | |

**Command:** imagequality
**Enter image quality setting [High/Draft] <High>:** *(Type **H** or **D**.)*

## COMMAND LINE OPTIONS
**High** displays image at its highest quality.
**Draft** displays image at a lower quality.

## RELATED COMMANDS
**Image** controls the loading of raster image files in the drawing.

**ImageAdjust** controls the brightness, contrast, and fading of the image.

**ImageAttach** attaches an image in the current drawing.

**ImageClip** creates a clipping boundary on an image.

**ImageFrame** toggles the display of the image's frame.

**Transparency** changes the transparency of the image.

## TIPS
• High quality displays the image more slowly; draft quality displays the image more quickly.

• I find that draft quality looks better (crisper) than high quality (blurred).

# Import

**Rel.13**  Imports vector and raster files into drawings.

| Command | Alias | Ctrl+ | F-key | Alt+ | Menu Bar | Tablet |
|---------|-------|-------|-------|------|----------|--------|
| import | imp | ... | ... | ... | ... | T2 |

**Command:** import

*Displays dialog box:*

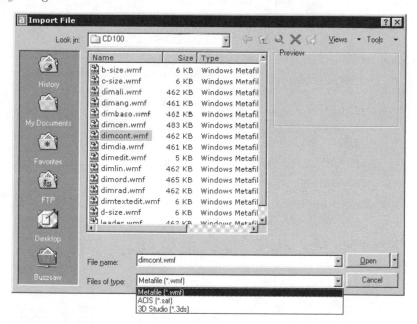

## DIALOG BOX OPTIONS

**Look in** selects the folder (*subdirectory*) and drive from which to import the file.

**File name** specifies the name of the file, or accepts the default.

**File of type** selects the file format in which to import the file.

**Open** imports the file.

**Cancel** dismisses the dialog box, and returns to AutoCAD.

## RELATED COMMANDS

**AppLoad** loads AutoLISP, VBA, and ObjectARX routines.

**DxbIn** imports a DXB file.

**Export** exports the drawing in several vector and raster formats.

**Load** imports SHX shape objects.

**Insert** places another drawing in the current drawing as a block.

**InsertObj** places an OLE object in the drawing via the Clipboard.

**LsLib** imports landscape objects.

**MatLib** imports rendering material definitions.

**MenuLoad** loads menu files into AutoCAD.

**Open** opens AutoCAD (any version) *.dwg* and *.dxf* files.

**PasteClip** pastes objects from the Clipboard.

**PasteSpec** pastes or links object from the Clipboard.

**Replay** displays renderings in TIFF, Targa, or GIF formats.

**VSlide** displays *.sld* slide files.

**XBind** imports named objects from another *.dwg* file.

**XRef** displays another *.dwg* file in the current drawing.

## TIPS

- The **Import** command acts as a "shell" command; it launches other AutoCAD commands that perform the actual import function. Other options may be available with the actual command, such as insertion point and scale.

| Format | Meaning |
|---|---|
| **Metafile** | Windows metafile WMF; executes **WmfIn**. |
| **ACIS** | ASCII SAT; executes **AcisIn**. |
| **3D Studio** | 3D Studio 3DS format, executes **3dsIn**. |

- To import *.dxf* files, use the **Open** command.

- AutoCAD no longer imports PostScript and EPS files.

## Removed Commands

**INetCfg** was removed from AutoCAD 2000; it was replaced by Windows' Internet configuration.

**INetHelp** was removed from AutoCAD 2000; it was replaced by AutoCAD's standard online help.

# Insert

**V. 1.0**   Inserts previously-defined blocks into drawings.

| Commands | Aliases | Ctrl+ | F-key | Alt+ | Menu Bar | Tablet |
|----------|---------|-------|-------|------|----------|--------|
| insert | i | ... | ... | IB | Insert | T5 |
| | inserturl | | | | ⇘Block | |
| -insert | -i | | | | | |

**Command:** insert

*Displays dialog box:*

## DIALOG BOX OPTIONS

**Name** selects from a list of previously-inserted blocks.

**Browse** displays file dialog box to select the block in one of the following formats:

- **Drawing (*.dwg)** AutoCAD drawing file.
- **DXF (*.dxf)** drawing interchange file.

### Insertion point options

**Specify On-Screen** specifies the insertion point in the drawing, after closing dialog box.

**X, Y, Z** specifies the x, y, z coordinates of the lower-left corner of the block.

### Scale options

**Specify on-screen** specifies the scale of the block (relative to the lower-left corner) in the drawing, after the dialog box is dismissed.

**X, Y, Z** specifies the x,y,z scale of the block; positive values enlarge the block, while negative values reduce the block.

**Uniform Scale** forces the y and z scale factors to be the same as the x scale factor.

### Rotation options

**Specify on-screen** specifies the rotation angle of the block (about the lower-left corner) in the drawing after the dialog box is dismissed.

**Angle** specifies the angle to rotate the block; positive angles rotate the block counterclockwise.

### Additional option

**Explode** explodes the block upon insertion.

. . . . . . . . . . . . . . . . . . . . . . . . . . . . . . . . . . . . . . . . . . . . .

## -INSERT Command

**Enter block name or [?]:** *(Enter name, or ?.)*

**Specify insertion point or [Scale/X/Y/Z/Rotate/PScale/PX/PY/PZ/PRotate]:** *(Pick a point, or enter an option.)*

**Enter X scale factor, specify opposite corner, or [Corner/XYZ] <1>:** *(Enter a scale factor, or enter an option.)*

**Enter Y scale factor <use X scale factor>:** *(Enter a scale factor, or press **Enter**.)*

**Specify rotation angle <0>:** *(Enter a rotation angle.)*

### COMMAND LINE OPTIONS

**Block name** specifies the name of the block to be inserted.

**?** lists the names of blocks stored in the drawing.

**Specify insertion point** specifies the lower-left corner of the block's insertion point.

**P** supplies a predefined block name, scale, and rotation values.

**X scale factor** indicates the x scale factor.

**Corner** indicates the x and y scale factors by pointing on the screen.

**XYZ** displays the x, y, and z scale submenu.

### INPUT OPTIONS

In response to the 'Block Name' prompt, you can enter:

| Option | Meaning |
|---|---|
| ~ | Display a dialog box of drawings stored on disk: |
| | **Block name:** ~ |
| * | Insert block exploded: |
| | **Block name:** *filename |
| = | Redefine existing block with a new block: |
| | **Block name:** oldname=newname |

In response to the 'Insertion point' prompt, you can enter:

| Option | Meaning |
|---|---|
| **Scale** | Specify x, y, and z-scale factors. |
| **PScale** | Preset x, y, and z-scale factors. |
| **XScale** | Specify x scale factor. |
| **PxScale** | Preset x scale factor. |
| **YScale** | Specify y scale factor. |
| **PyScale** | Preset y scale factor. |
| **ZScale** | Specify z scale factor. |
| **PzScale** | Preset z scale factor. |
| **Rotate** | Specify rotation angle. |
| **PRotate** | Preset rotation angle. |

## RELATED COMMANDS

**Block** creates a block out of a group of objects.

**Explode** reduces inserted blocks to their constituent objects.

**MInsert** inserts a block as a blocked rectangular array.

**Rename** renames blocks.

**WBlock** writes blocks to disk.

**XRef** displays drawings stored on disk in the drawing.

## RELATED SYSTEM VARIABLES

**ExplMode** toggles whether non-uniformly scaled blocks can be exploded:

| ExplMode | Meaning |
| --- | --- |
| 0 | Cannot explode; AutoCAD Release 12 compatible. |
| 1 | Can be exploded (default). |

**InsBase** specifies the name of the most-recently inserted block.

**InsUnits** specifies the drawing units for the inserted block.

## TIPS

- You can insert any other AutoCAD drawing into the current drawing.

- A *preset* scale factor or rotation means the dragged image is shown at that scale, but you can enter a new scale when inserting.

- Drawings are normally inserted as a block; prefix the filename with an * (*asterisk*) to insert the drawing as separate objects.

- Redefine all blocks of the same name in the current drawing by adding the = (*equal*) suffix after its name at the 'Block name' prompt.

- Insert a mirrored block by supplying a negative x- or y-scale factor, such as:

  **X scale factor:** -1

- AutoCAD converts a negative z scale factor into its absolute value, which always makes it positive.

- As of AutoCAD Release 13, you can explode a mirrored block and a block inserted with different scale factors when the system variable **ExplMode** is turned on.

- The ability of the **Insert** command to import drawings in XML format was removed from AutoCAD 2004.

# InsertObj

**Rel.13**  Places OLE objects as a linked or embedded objects (*short for INSERT OBJect*).

| Command | Alias | Ctrl+ | F-key | Alt+ | Menu Bar | Tablet |
|---------|-------|-------|-------|------|----------|--------|
| insertobj | io | ... | ... | IO | Insert | T1 |
| | | | | | ⬦OLE Object | |

**Command:** insertobj

*Displays dialog box:*

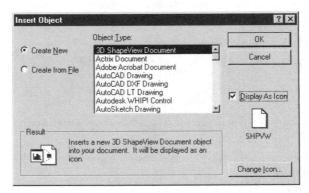

## DIALOG BOX OPTIONS

**Create New** creates a new OLE object in another application, then embeds the object in the current drawing.

**Create from File** selects a file to embed or link in the current drawing.

**Object Type** selects an object type from the list; the related application automatically launches if you select the Create New option.

**Display As Icon** displays the object as an icon, rather than as itself.

## RELATED COMMANDS

**OleLinks** controls the OLE links.

**PasteSpec** places an object from the Clipboard in the drawing as a linked object.

## RELATED WINDOWS COMMANDS

**Edit | Copy** copies an object to the Clipboard into another Windows application.

**File | Update** updates an OLE object from another application.

. . . . . . . . . . . . . . . . . . . . . . . . . . . . . . . . . . . . . . . . . . .

## Removed Command

**InsertUrl** was removed from AutoCAD 2000; it was replaced by the **Insert** command's **Browse | Search the Web** option.

. . . . . . . . . . . . . . . . . . . . . . . . . . . . . . . . . . . . . . . . . . .

# Interfere

Rel.11 Determines the interference of two or more 3D solid objects; creates a 3D solid body of the volumes in common.

| Command | Alias | Ctrl+ | F-key | Alt+ | Menu Bar | Tablet |
|---------|-------|-------|-------|------|----------|--------|
| interfere | inf | ... | ... | DII | Draw | ... |
| | | | | | ⌖Solids | |
| | | | | | ⌖Interference | |

**Command:** interfere
**Select first set of solids:**
**Select objects:** *(Select one or more solid objects.)*
**Select objects:** *(Press **Enter** to end object selection.)*
**Select second set of solids:**
**Select objects:** *(Select one or more solid objects.)*
**Select objects:** *(Press **Enter** to end object selection.)*
**Comparing 1 solid against 1 solid.**
**Interfering solids (first set)  : 1**
**                 (second set) : 1**
**Interfering pairs            : 1**
**Create interference solids? [Yes/No] <N>:** *(Enter **Y** or **N**.)*

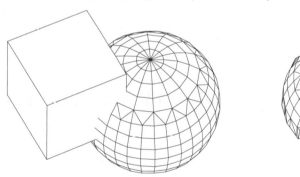

*A pair of interfering solids (left) and the interference (right).*

## COMMAND LINE OPTIONS

**Select objects** checks all solids in a single selection set for interference with one another.

**Create interference solids** creates a solid representing the volume of interference.

## RELATED COMMANDS

**Intersect** creates a new volume from the intersection of two volumes.

**Section** creates a 2D region from a 3D solid.

**Slice** slices a 3D solid with a plane.

## TIP

• When three or more solids interfere, AutoCAD asks "Highlight pairs of interfering solids?"

# Intersect

**Rel.11** Creates 3D solids of 2D regions through the Boolean intersection of two or more solids or regions.

| Command | Alias | Ctrl+ | F-key | Alt+ | Menu Bar | Tablet |
|---------|-------|-------|-------|------|----------|--------|
| intersect | in | ... | ... | MNI | Modify | X17 |
| | | | | | ⮡Solids Editing | |
| | | | | | ⮡Intersect | |

**Command:** intersect
**Select objects:** *(Select one or more solid objects.)*
**Select objects:** *(Press Enter to end object selection.)*

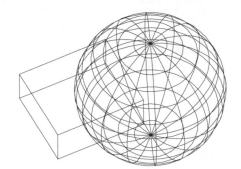

*Two intersecting solids (left) and the resulting intersection (right).*

## COMMAND LINE OPTION

**Select objects** selects two or more objects to intersect.

## RELATED COMMANDS

**Interfere** creates a new volume from the interference of two or more volumes.

**Subtract** subtracts one 3D solid from another.

**Union** joins 3D solids into a single body.

## TIPS

- You can use this command on 2D regions and 3D solids.

- The **Interference** and **Intersect** command may seem similar. Here is the difference between the two:

    **Intersect** *erases* all of the 3D solid parts that do not intersect.

    **Interfere** *creates a new object* from the intersection; it does not erase the original objects.

# 'Isoplane

**V. 2.0** Changes the crosshair orientation and grid pattern among the three isometric drawing planes.

| Command | Alias | Ctrl+ | F-key | Alt+ | Menu Bar | Tablet |
|---------|-------|-------|-------|------|----------|--------|
| 'isoplane | ... | E | F5 | ... | ... | ... |

**Command:** isoplane
**Enter isometric plane setting [Left/Top/Right] <Top>:** *(Enter an option.)*

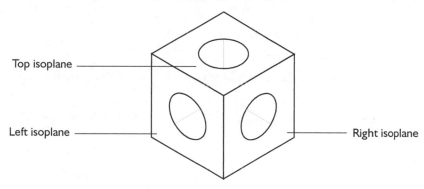

Top isoplane

Left isoplane

Right isoplane

## COMMAND LINE OPTIONS

**Left** switches to the left isometric plane.

**Top** switches to the top isometric plane.

**Right** switches to the right isometric plane.

**Toggle** switches to the next isometric plane in the following order: left, top, right.

## RELATED COMMANDS

**Options** displays a dialog box for setting isometric mode and planes.

**Ellipse** draws isocircles.

**Snap** turns on isometric drawing mode.

## RELATED SYSTEM VARIABLE

**SnapIsoPair** contains the current isometric plane.

**GridMode** current grid visibility:

| GridMode | Meaning |
|----------|---------|
| 0 | Grid is off (default). |
| 1 | Grid is on. |

**GridUnit** specifies the current grid x,y spacing.

**LimMin** holds the x,y coordinates of the lower-left corner of the grid display.

**LimMax** holds the x,y coordinates of the upper-right corner of the grid display.

**SnapStyl** displays a normal or isometric grid:

| SnapStyl | Meaning |
|----------|---------|
| 0 | Normal (default). |
| 1 | Isometric grid. |

# Creating Isometric Dimensions

AutoCAD's dimensions must be modified for isometric drawings, so that the dimension text looks "correct" in isometric mode. This involves two steps — (1) creating isometric text styles and (2) changing dimension variables — repeated three times, once for each isoplane.

## Step 1

How to create isometric text styles:

1. From the menu bar, select **Format | Text Style**.

2. When the **Text Style** dialog box appears, click **New**.

3. Enter **isotop** for the name of the new text style, which is used for text in the top isoplane.

4. Click **OK**.

5. When the **Text Style** dialog box reappears, select *Simplex.Shx* from **Font Name**.

6. Change the **Oblique Angle** to **-30**.

7. Click **Apply**.

8. Create text styles for the other two isoplanes:

| Style Name | Font Name | Oblique Angle |
|------------|-------------|---------------|
| **IsoTop** | *simplex.shx* | -30 |
| **IsoRight** | *simplex.shx* | 30 |
| **IsoLeft** | *simpelx.shx* | 30 |

Enter these values into the **Text Style** dialog box, and click **Apply;** then **Close**.

## Step 2

How to create isometric dimension styles:

1. Create the dimension styles for the three isoplanes by selecting **Format | Dimension Styles** from the menu bar. These dimension variables must be changed:

2. Create a new dimension style:

> Click **New**.
> Enter **IsoLeft** in the **New Style Name** field.
> Click **Continue**.

3. Force dimension text to align with the dimension line:

> Select the **Text** tab.
> Select **Aligned with dimension line** in the **Text Alignment** section.

4. Specify text style for dimension text:

> Select **ISOLEFT** from the **Text Style** list box.
> Click **OK**.

**5.** One of the three needed dimension styles has been created. Create dimstyles for all isoplanes using these parameters:

| Dimstyle Name | Text Style |
|---|---|
| **Isotop** | IsoTop |
| **Isoright** | IsoRight |
| **Isoleft** | IsoLeft |

**6.** Click **Close** to exit the **Dimension Style Manager** dialog box.

## Step 3

To place linear dimensions in an isometric drawing, you must use the **DimAligned** command, because it aligns the dimension along the isometric axes: place all dimensions in one isoplane, and then switch to the next isoplane.

**1.** Press **F5** to switch to the appropriate isoplane, such as **Top**.

**2.** Use the **DimStyle** command to select the associated dimension style, such as **IsoTop**.

**3.** Place the dimension with the **DimAligned** command; it is helpful to use **INTersection** object snaps.

**4.** Use the **DimEdit** command's **Oblique** option to skew the dimension by 30 or -30 degrees, as follows:

| IsoPlane | DimStyle | Oblique Angle |
|---|---|---|
| **Top** | IsoTop | 30 |
| **Left** | IsoLeft | 30 |
| **Right** | IsoRight | -30 |

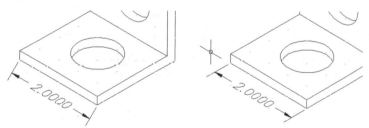

*Aligned dimension text before (left) and after (right) applying **DimEdit's Oblique** option.*

**5.** To place a leader, use the **Standard** dimstyle and **Standard** text style.

# JpgOut

2004 Exports drawings in JPEG format.

| Command | Alias | Ctrl+ | F-key | Alt+ | Menu Bar | Tablet |
|---------|-------|-------|-------|------|----------|--------|
| jpgout | ... | ... | ... | ... | ... | ... |

**Command:** jpgout

*Displays **Create Raster File** dialog box. Specify a filename, and then click **Save**.*

**Select objects or <all objects and viewports>:** *(Select objects, or press **Enter** to select all objects and viewports.)*

## COMMAND LINE OPTIONS

**Select objects** selects specific objects.

**All objects and viewports** selects all objects and all viewports, whether in model space or in layout mode.

## RELATED COMMANDS

**BmpOut** exports drawings in BMP (bitmap) format.

**Image** places raster images in the drawing.

**PngOut** exports drawings in PNG (portable network graphics) format.

**TifOut** exports drawings in TIFF (tagged image file format) format.

## TIPS

- This command provides no options for specifying the level of compression.

- The rendering effects of the **ShadEdge** command are preserved, but not of the **Render** command.

- JPEG files are often used by digital cameras and Web pages.

- JPEG is short for "joint photographic expert group."

- The drawback to saving drawings in JPEG format is that the image is less clear than in other formats; the advantage is that JPEG files are highly compressed.

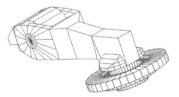 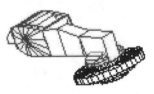

*Zoomed-in original image in AutoCAD (left), and enlarged JPEG image with artifacts (right).*

 # JustifyText

**2001** Changes the justification of text.

| Command | Alias | Ctrl+ | F-key | Alt+ | Menu Bar | Tablet |
|---------|-------|-------|-------|------|----------|--------|
| justifytext | ... | ... | ... | MOTJ | Modify | ... |
| | | | | | ⮑ Object | |
| | | | | | ⮑ Text | |
| | | | | | ⮑ Justify | |

**Command:** justifytext
**Select objects:** *(Select one or more text objects.)*
**Select objects:** *(Press **Enter** to end object selection.)*
**Enter a justification option**
**[Left/Align/Fit/Center/Middle/Right/TL/TC/TR/ML/MC/MR/BL/BC/BR]**
**<Left>:** *(Enter an option.)*

## COMMAND LINE OPTIONS

**Select objects** selects one or more text objects in the drawing.

**Align** aligns the text between two points with adjusted text height.

**Fit** fits the text between two points with fixed text height.

**Center** centers the text along the baseline.

**Middle** centers the text horizontally and vertically.

**Right** right-justifies the text.

**TL** justifies to top-left.

**TC** justifies to top-center.

**TR** justifies to top-right.

**ML** justifies to middle-left.

**MC** justifies to middle-center.

**MR** justifies to middle-right.

**BL** justifies to bottom-left.

**BC** justifies to bottom-center.

**BR** justifies to bottom-right.

## RELATED COMMANDS

**Text** places text in the drawing.

**DdEdit** edits text.

**ScaleText** changes the size of text.

**Style** defines text styles.

## TIPS

• This command works with text, mtext, leader text, and attribute text.

• When the justification is changed, the text does not move.

 **'Layer**

**V. 1.0** Controls the creation and visibility of layers.

| Commands | Aliases | Ctrl+ | F-key | Alt+ | Menu Bar | Tablet |
|----------|---------|-------|-------|------|----------|--------|
| 'layer | la | ... | ... | OL | Format | U5 |
| | ddlmodes | | | | ⬐Layer | |
| -layer | -la | | | | | |

**Command:** layer

*Displays dialog box:*

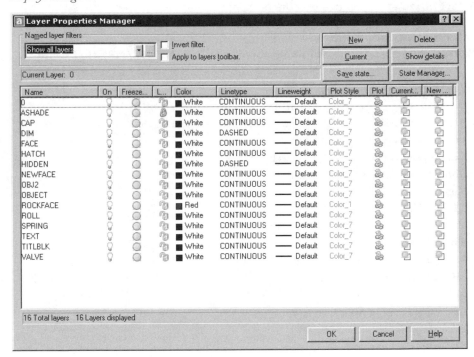

**DIALOG BOX OPTIONS**

*Name Layer Filter options*

**Named layer filters:**

- **Show all layers** displays all layers defined in the current drawing.
- **Show all used layers** displays layers with content.
- **Show all xref dependent layers** displays layers in externally-referenced drawings.

... displays Named Layer Filters dialog box.

**Invert filter** inverts the display of layer names; for example, when Show all used layers is selected, the Invert filter option causes all layers with no content to be displayed.

**Apply to Object Properties toolbar** applies the filter to the names of the layers displayed by the Object Properties toolbar.

**New** creates a new layer.

**Current** sets the selected layer as the current layer.

**Save state** displays the Layer States Manager dialog box; see below.

**Delete** purges the selected layer; some layers cannot be deleted, as described by the warning dialog box:

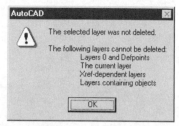

**Show Details** displays the Details portion of the **Layer Properties Manager** dialog box.

**Restore state** displays **Layer States Manager** dialog box.

## Details options

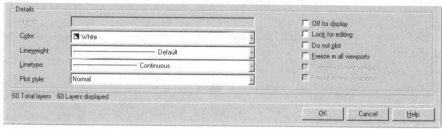

**Name** names the selected layer; if the name cannot be edited, the field is grayed out.

**Color** selects the color; click **Other** to display the **Select Color** dialog box.

**Lineweight** selects the lineweight.

**Linetype** selects the linetype.

**Plot style** selects the plot style.

**Off for display** turns off the display of the layer.

**Lock for editing** locks this layer; can be seen but not edited.

**Do not plot** prevents the layer from being plotted.

**Freeze in all viewports** freezes the layer; the current layer cannot be frozen.

**Freeze in active viewport** freezes the layer when TileMode = 0 (off); grayed out in model mode.

**Freeze in new viewport** freezes the layer when a new viewport is created, and when TileMode = 0; grayed out in model mode.

**Named Layer Filters** dialog box

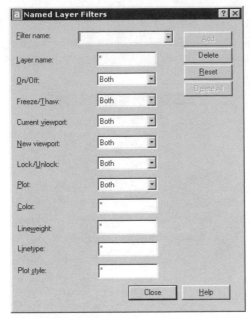

**Filter name** names the filter.

**Layer name** specifies the names of layers to filter; see wildcard metacharacters below.

**On/Off** selects the layers that are on, off, or both.

**Freeze/Thaw** selects the layers that are frozen, thawed, or both.

**Active viewport** selects the layers that are frozen, thawed, or both, in the current viewport.

**New viewport** selects the layers that are frozen, thawed, or both, in the new viewport.

**Lock/Unlock** selects the layers that are locked, unlocked, or both.

**Plot** selects the layers that plot, do not plot, or both.

**Colors** selects the layers of a specific color.

**Lineweight** selects the layers with a specific lineweight.

**Linetype** selects the layers with a specific linetype.

**Plot style** selects the layers with a specific plot style.

*Buttons*

**Add** adds the name to the list of filters; note that these names are also available in the Layer Properties Manager dialog box's Named layer filters list.

**Delete** removes a name.

**Reset** resets all filters to their default values.

**Delete All** deletes all filter names.

## WILDCARD METACHARACTERS

| Char | Meaning |
|------|---------|
| * | Matches any one or more characters. |
| ? | Matches any single character. |
| @ | Matches any alphabetic character (A - Z). |
| # | Matches any numeric digit (0 - 9). |
| . | Matches any non-alphanumeric character (!@#$%^&*, etc.) |
| ~ | Matches anything but the pattern of characters. |
| [ ] | Matches any one enclosed character. |
| [~] | Matches any character not enclosed. |
| [-] | Matches a range of characters. |
| ` | Allows literal use of metacharacters. |

**Save Layer States** dialog box

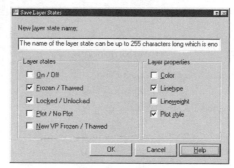

**New layer state name** specifies the name of the layer state.

**Layer states** specifies the layer states that will be saved.

**Layer properties** specifies the layer properties that will be saved.

**Layer States Manager** dialog box

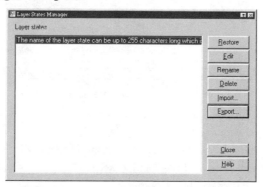

**Layer states** lists the names of layer states that can be restored.

*Buttons*

**Restore** restores the listed layer states.

**Edit** displays the Edit Layer States dialog box, identical to the Save Layer States dialog box.

**Rename** renames the selected layer state.

**Delete** deletes the selected layer state.

**Import** imports an *.las* (layer state) file; layer states can be shared among drawings.

**Export** exports the selected layer state to an *.las* file.

## SHORTCUT MENU OPTIONS

*Right-click any layer name in the **Layer Properties Manager** dialog box:*

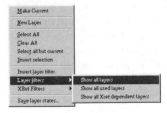

**Make Current** makes the selected layer current.

**New Layer** creates a new layer with the default name of Layer1.

**Select All** selects all layers.

**Clear All** deselects all layers.

**Select all but current** selects all layers, except the current layer; this allows you easily to freeze all layers except the current layer, which cannot be frozen.

**Invert selection** inverts the selection of layer names.

**Invert layer filter** inverts the layer names displayed by the current filter.

**Xref Filters** displays names of xref'ed drawings to display only their layers.

**Save Layer States** displays Save Layer States dialog box.

Layer Filters options

**Show all layer** displays all layers defined in the current drawing.

**Show all used layers** displays any layer with content.

**Show all xref dependent layers** displays layers in externally-referenced drawings.

## LIST OF LAYERS

*Click a header to sort alphabetically (A-Z); click a second time for reverse-alphabetical sort (Z-A):*

| Name | On | Freeze in all VP | Lock | Color | Linetype | Lineweight | Plot Style | Plot | Current VP Freeze | New VP Freeze |
|------|----|--------------|------|-------|----------|-----------|-----------|------|-----------------|--------------|
| 0 | | | | ■ White | CONTINUOUS | —— Default | Color_7 | | | |
| ASHADE | | | | □ Yellow | DASHED | ■■ 0.35 mm | Color_2 | | | |

**Name** lists the names of layers in the current drawing.

**On** toggles the layer between on and off.

**Freeze in all VP** toggles the layer between thawed and frozen in all viewports.

**Lock** toggles the layer between unlocked and locked.

**Color** specifies the color for all objects on the layer.

**Linetype** specifies the linetype for all objects on the layer.

**Lineweight** specifies the lineweight for all objects on the layer.

**Plot Style** specifies the plot style for all objects on the layer.

**Plot** toggles the layer between plot and no-plot.

**Current VP Freeze** specifies whether the layer is frozen in the current viewport.

**NewVP Freeze** specifies whether the layer is frozen in the new viewports.

## -LAYER Command

**Command:** -layer

**Current layer: "0"**

**Enter an option [?/Make/Set/New/ON/OFF/Color/Ltype/LWeight/Plot/ PStyle/Freeze/Thaw/LOck/Unlock/stAte]:** *(Enter an option.)*

### COMMAND LINE OPTIONS

**Color** indicates the color for all objects drawn on the layer.

**Freeze** disables the display of the layer.

**LOck** locks the layer.

**Ltype** indicates the linetype for all objects drawn on the layer.

**LWeight** specifies the lineweight.

**Make** creates a new layer, and makes it the current layer.

**New** creates a new layer.

**OFF** turns off the layer.

**ON** turns on the layer.

**Plot** determines whether the layer is plotted.

**PStyle** specifies the plotstyle; option available only when plotstyles are attached to the drawing.

**Set** makes the layer the current layer.

**stAte** sets and save layer sates.

**Thaw** unfreezes the layer.

**Unlock** unlocks the layer.

**?** lists the names of layers in the drawing.

### stAte options

**Enter an option [?/Save/Restore/Edit/Name/Delete/Import/EXport]:** *(Enter an option).*

**?** lists the names of layer states in the drawing.

**Save** saves the layer state and properties by name; properties include on, frozen, lock, plot, newvpfreeze, color, linetype, lineweight, and plotstyle.

**Restore** restores a named state.

**Edit** changes the settings of named state.

**Name** renames named states.

**Delete** erases named states from the drawing.

**Import** opens layer state *.las* files.

**Export** saves a selected named state to a *.las* file.

## RELATED COMMANDS

**LayerP** returns to the previous layer.

**Change** moves objects to different layers via the command line.

**Properties** moves objects to different layers via dialog box.

**LayTrans** translates layer names.

**Purge** removes unused layers from the drawing.

**Rename** renames layers.

**VpLayer** controls the visibility of layers in paper space viewports.

## RELATED SYSTEM VARIABLE

**CLayer** contains the name of the current layer.

## RELATED FILE

*__.las__* are layer state files, which use the DXF format.

## TIPS

* A *frozen* layer cannot be seen or edited.

* A *locked* layer can be seen, but not edited.

* Layer **Defpoints** is a non-plotting layer.

* To create more than one new layer at a time, use commas to separate layer names in **-Layer**.

* For new layers to take on properties of an existing layer, select the layer before clicking **New**.

* If layer names appear to be missing, they have been filtered from the list.

# LayerP

2002 Undoes changes made to layer settings *(short for LAYER Previous).*

| Command | Alias | Ctrl+ | F-key | Alt+ | Menu Bar | Tablet |
|---------|-------|-------|-------|------|----------|--------|
| layerp  | ...   | ...   | ...   | ...  | ...      | ...    |

**Command:** layerp
**Restored previous layer status**

## COMMAND LINE OPTIONS
*None.*

## RELATED COMMANDS
**LayerPMode** toggles layer previous mode on and off.

**Layer** creates and set layers and modes.

## RELATED SYSTEM VARIABLE
**CLayer** specifies the name of the current layer.

## TIPS
- This command acts like an "undo" command for changes made to layers only, such as changes to the layer's color or lineweight.

- **LayerPMode** command must be turned on for the **LayerP** command to work.

- The **LayerP** command does not undo the renaming of layers, the deleting or purging layers, and the creating new layers.

# LayerPMode

**2002** Turns on and off layer-previous mode *(short for LAYER Previous MODE).*

| Command | Alias | Ctrl+ | F-key | Alt+ | Menu Bar | Tablet |
|---|---|---|---|---|---|---|
| layerpmode | ... | ... | ... | ... | ... | ... |

**Command:** layerpmode
**Enter LAYERP mode [ON/OFF] <ON>:** *(Type **ON** or **OFF**.)*

## COMMAND LINE OPTIONS
**On** turns on layer previous mode.

**Off** turns off layer previous mode.

## RELATED COMMANDS
**LayerP** returns the drawing to the previous layer.

**Layer** creates and sets layers and modes.

## RELATED SYSTEM VARIABLES
*None.*

## TIP
* When layer previous mode is on, AutoCAD tracks changes to layers.

# Layout

2000 Creates and deletes paper space layouts on the command line.

| Commands | Alias | Ctrl+ | F-key | Alt+ | Menu Bar | Tablet |
|----------|-------|-------|-------|------|----------|--------|
| layout | lo | ... | ... | IL | Insert | ... |
| | | | | | ⓈLayout | |
| -layout | | | | | | |

**Command:** layout
**Enter layout option [Copy/Delete/New/Template/Rename/SAveas/Set/?]**
**<set>:** *(Enter an option.)*

## COMMAND LINE OPTIONS

**Copy** copies a layout to create a new layout.

**Delete** deletes a layout; the Model tab cannot be deleted.

**New** creates a new layout tab, automatically generating the name for the layout (default = Layout1), which you may override.

**Template** displays the Select File dialog box, which allows you to select a DWG drawing or DWT template file to use as a template for a new layout. If the file has layouts, it displays the **Insert Layout(s)** dialog box.

**Rename** renames a layout.

**SAveas** saves the layouts in a drawing template (DWT) file. The last current layout is used as the default for the layout to save.

**Set** makes a layout current.

**?** lists the layouts in the drawing in a format similar to the following:

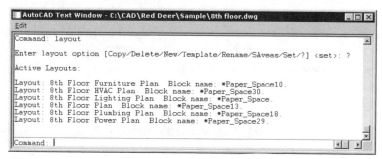

## SHORTCUT MENU

*Right-click any layout tab:*

**New layout** creates a new layout with the default name of **LAYOUT**.

**From template** displays the Select File and Insert Layout dialog boxes.

· · · · · · · · · · · · · · · · · · · · · · · · · · · · · · · · · · · · · · · · · · ·

**Delete** deletes the selected layout; displays a warning dialog box:

**Rename** Displays the Rename Layout dialog box:

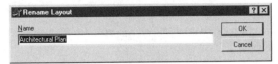

**Move or Copy** displays the Move or Copy dialog box.

**Select All Layouts** selects all layouts.

**Page Setup** displays the Page Setup dialog box; see the PageSetup command.

**Plot** displays the Plot dialog box; see the Plot command.

## Insert Layout(s) dialog box

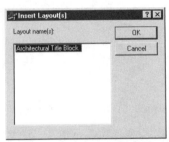

**Layout names(s)** lists the names of layouts found in the selected drawing; you may select more than one layout at a time by holding down the CTRL key.

**OK** adds the selected layouts to the current drawing.

**Cancel** dismisses the dialog box and cancels the command.

## Move or Copy dialog box

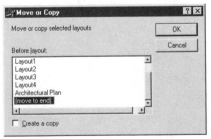

**Before layout** selects a layout to appear before the current layout.

**Move to end** moves the current layout to the end of layouts.

**Create a copy** makes a copy of the layout.

## -LAYOUT Command

**Enter layout option [Copy/Delete/New/Template/Rename/SAveas/Set/?] <set>:** *(Enter an option.)*

**Copy** copies existing layouts.

**Delete** erases layout from the drawing; model tab cannot be deleted.

**New** creates new layouts.

**Template** creates new layouts based on layouts found in template (*.dwt*), drawing (*.dwg*), and DXF (*.dxf*) files; displays the Insert Layouts dialog box.

**Rename** renames layouts.

**SAveas** saves layouts as template (*.dwt*) files; symbol tables and blocks are not saved.

**Set** sets a layout as current.

**?** lists layout names defined in the drawing.

### RELATED COMMAND

**LayoutWizard** creates and deletes paper space layouts via wizard.

### TIPS

- "Layout" is the new name for paper space as of AutoCAD 2000.

- A layout name can be up to 255 characters long; the first 31 characters are displayed in the tab.

- To switch between layouts, click the tab located below the drawing:

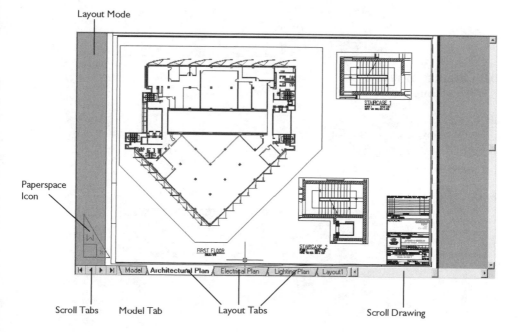

- The **Model** tab cannot be deleted, renamed, moved, or copied.

- A drawing can contain up to 255 layouts.

# LayoutWizard

<u>**2000**</u>   Creates and deletes paper space layouts via a wizard.

| Command | Alias | Ctrl+ | F-key | Alt+ | Menu Bar | Tablet |
|---------|-------|-------|-------|------|----------|--------|
| layoutwizard | ... | ... | ... | ILW | **Insert** | ... |
| | | | | | ⇘**Layout** | |
| | | | | |   ⇘**Layout Wizard** | |
| | | | | TZC | **Tool** | |
| | | | | | ⇘**Wizards** | |
| | | | | |   ⇘**Create Layout** | |

**Command:** layoutwizard

*Displays dialog box::*

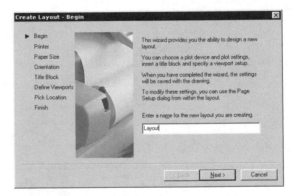

### DIALOG BOX OPTIONS

**Enter a name** specifies the name for the layout.

*Buttons*

**Back** displays the previous dialog box.

**Next** displays the next dialog box.

**Cancel** cancels the command.

**Begin** page

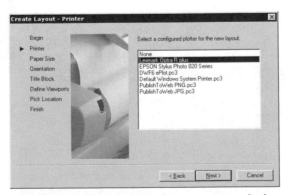

**Select a configured plotter** selects a printer or plotter to output the layout.

**Paper Size** page

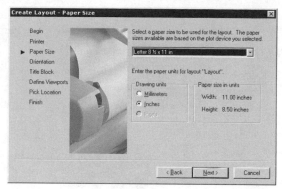

**Select a paper size...** selects a size of paper supported by the output device.

**Enter the paper units**

- **Millimeters** measures paper size in metric units.
- **Inches** measures paper size in Imperial units.
- **Pixels** measures paper size in dots per inch.

**Orientation** page

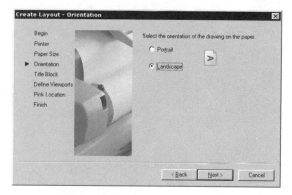

**Select the orientation**

- **Portrait** plots the drawing vertically.
- **Landscape** plots the drawing horizontally.

**Title Block** page

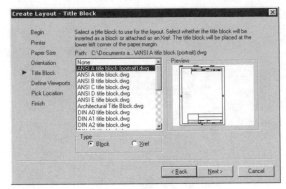

**Select a title block...** specifies a title border for the drawing as a block or an xref.

**Define Viewports** page

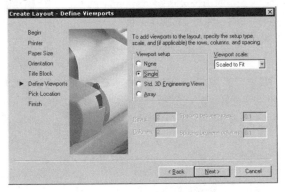

**Viewport Setup**
- **None** creates no viewport.
- **Single** creates a single paper space viewport.
- **Std. 3D Engineering Views** creates top, front, side, and isometric views.
- **Array** creates a rectangular array of viewports.

**Viewport scale**
- **Scaled to Fit** fits the model to the viewport.
- *mm:nn* specifies a scale factor, ranging from 1:1 to 1/128":1'0".

**Rows** specifies the number of rows for arrayed viewports.

**Columns** specifies the number of columns for arrayed viewports.

**Spacing between rows** specifies the vertical distance between viewports.

**Spacing between columns** specifies the horizontal distance between viewports.

**Pick Location** page

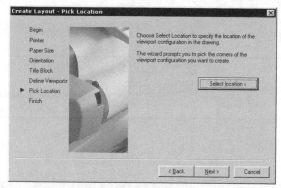

**Select location** specifies the corners of a rectangle holding the viewports. AutoCAD prompts:

**Regenerating layout.**
**Specify first corner:** *(Pick a point.)*
**Specify opposite corner:** *(Pick a point.)*
**Regenerating model.**

*AutoCAD returns to the* **Create Layout** *dialog box.*

**Finish** page

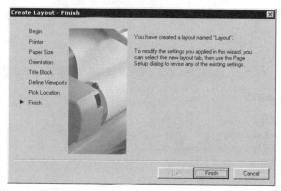

**Finish** exits the dialog box and creates the layout.

## RELATED COMMANDS

**Layout** creates a layout on the command line.

## RELATED SYSTEM VARIABLES

**CTab** contains the name of the current tab.

# LayTrans

**2002** Translates layer names *(short for LAYer TRANSlation).*

| Command | Alias | Ctrl+ | F-key | Alt+ | Menu Bar | Tablet |
|---------|-------|-------|-------|------|----------|--------|
| laytrans | ... | ... | ... | TSL | Tools | ... |
| | | | | | ↳CAD Standards | |
| | | | | | ↳Layer Translator | |

**Command:** laytrans

*Displays dialog box:*

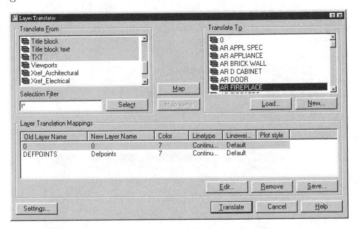

## DIALOG BOX OPTIONS

*Translate From options*

**Translate From** lists the names of layers in the current drawing; icons indicate whether the layer is being used (is being referenced):

▧ TXT            • **Green** icon means the layer contains at least one object.

▧ Unused layer  • **White** icon means the layer contains no objects, and can be purged.

**Selection Filter** specifies a subset of layer names; see Wildcard Metacharacters in the Layer command.

**Select** highlights the layer names that match the selection filter.

**Map** maps the selected layer(s) in the Translate From column to the selected layer in the Translate To column.

**Map same** maps layers automatically with the same name.

*Translate To options*

**Translate To** lists layer names in the drawing opened with the Load button.

**Load** accesses the layer names in another drawing via the Select Drawing File dialog box.

**New** creates a new layer via the New Layer dialog box.

## Layer Translation Mappings options

**Edit** edits the linetype, color, lineweight, and plot style settings via the Edit Layer dialog box; identical to the New Layer dialog box.

**Remove** removes the selected layer from the list.

**Save** saves the matching table to a DWS (drawing standard) file.

**Settings** specifies translation options via the Settings dialog box.

**Translate** changes the names of layers, as specified by the Layer Translation Mappings list.

## New Layer dialog box

*Identical to the **Edit Layer** dialog box.*

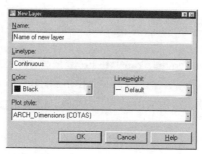

**Name** specifies the name of the layer, up to 255 characters long.

**Linetype** selects a linetype from those available in the drawing.

**Color** selects a color; or, select Other for the Select Color dialog box.

**Lineweight** selects a lineweight.

**Plot style** selects a plot style from those available in the drawing; this option is not available if plot styles have not been enabled in the drawing.

## Settings dialog box

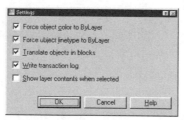

**Force object color to Bylayer** forces every translated layer to take on color Bylayer.

**Force object linetype to Bylayer** forces every translated layer to take on linetype Bylayer.

**Translate objects in blocks** forces objects in blocks to take on new layer assignments.

**Write transaction log** writes the results of the translation to a LOG file, using the same filename as the drawing. When command is complete, AutoCAD reports:

Writing transaction log to *filename*.log.

**Show layer contents when selected** lists the names of selected layers only in the Translate From list.

**RELATED COMMANDS**

**Standards** creates the standards for checking drawings.

**CheckStandards** checks the current drawing against a list of standards.

**Layer** creates and sets layers and modes.

**RELATED FILES**

*\*.dws* drawing standard file; saved in DWG format.

*\*.log* log file recording layer translation; saved in ASCII format.

**RELATED SYSTEM VARIABLES**

*None.*

**TIPS**

*   You can purge unused layers (those prefixed by a white icon) within the **Layer Translator** dialog box:

    1. Right click any layer name in the **Translate From** list.

    2. Select **Purge Layers**. The layers are removed from the drawing.

*   You can load layers from more than one drawing file; duplicate layer names are ignored.

# Leader

Rel.13 Draws leader lines with one or more lines of text.

| Command | Alias | Ctrl+ | F-key | Alt+ | Menu Bar | Tablet |
|---------|-------|-------|-------|------|----------|--------|
| leader | lead | ... | ... | ... | ... | R7 |

**Command:** leader
**Specify leader start point:** *(Pick a starting point.)*
**Specify next point:** *(Pick the shoulder point.)*
**Specify next point or [Annotation/Format/Undo] <Annotation>:** *(Pick another point, or enter an option.)*
**Specify next point or [Annotation/Format/Undo] <Annotation>:** *(Press **Enter** to specify the text.)*
**Enter first line of annotation text or <options>:** *(Enter text.)*
**Enter next line of annotation text:** *(Press **Enter**.)*

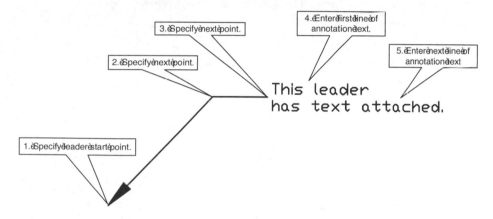

## COMMAND LINE OPTIONS

**Specify start point** specifies the location of the arrowhead.

**Specify next point** positions the leader line's vertex.

**Undo** undoes the leader line to the previous vertex.

Format options
**Enter leader format option [Spline/STraight/Arrow/None] <Exit>:** *(Enter an option.)*

**Spline** draws the leader line as a NURBS (short for non uniform rational Bezier spline) curve.

**STraight** draws a straight leader line (default).

**Arrow** draws the leader with an arrowhead (default).

**None** draws the leader with no arrowhead.

Annotation options
**Enter first line of annotation text or <options>:** *(Press* **Enter**.*)*
**Enter an annotation option [Tolerance/Copy/Block/None/Mtext] <Mtext>:**
*(Enter an option.)*

**Enter first line of annotation text** specifies the leader text.

**Tolerance** places one or more tolerance symbols; see the Tolerance command.

**Copy** copies text from another part of the drawing.

**Block** places a block; see the -Insert command.

**None** specifies no annotation.

**MText** displays the Multiline Text Editor dialog box; see the MText command.

## RELATED DIM VARIABLES

**DimAsz** specifies the size of the arrowhead and the hookline.

**DimBlk** specifies the type of arrowhead.

**DimClrd** specifies the color of the leader line and the arrowhead.

**DimGap** specifies the gap between hookline and annotation (gap between box and text).

**DimScale** specifies the overall scale of the leader.

## TIPS

• Autodesk recommends using the **QLeader** command, which has more options.

• This command can draw several types of leader:

• The text in a leader is an mtext (multiline text) object.

• Use the \P metacharacter to create line breaks in leader text.

# Lengthen

**Rel.13**   Lengthens and shortens open objects by several methods.

| Command | Alias | Ctrl+ | F-key | Alt+ | Menu Bar | Tablet |
|---------|-------|-------|-------|------|----------|--------|
| lengthen | len | ... | ... | MG | Modify | W14 |
| | | | | | ⤷Lengthen | |

**Command:** lengthen
**Select an object or [DElta/Percent/Total/DYnamic]:** *(Select an open object.)*
**Current length:** *n.nnnn*

## COMMAND LINE OPTIONS

**Select an object** displays length and included angle; does not change the object.

DElta option
**Enter delta length or [Angle] <0.0000>:** *(Enter a value, or type **A**.)*
**Specify second point:** *(Pick a point.)*
**Select an object to change or [Undo]:** *(Select an open object, or type **U**.)*

**Enter delta length** changes the length by an incremental amount.

Percent option
**Enter percentage length <100.0000>:** *(Enter a value.)*
**Select an object to change or [Undo]:** *(Select an open object, or type **U**.)*

**Enter percent length** changes the length by a percentage of the original length.

**Undo** undoes the most-recent lengthening operation.

Total option
**Specify total length or [Angle] <1.0000)>:** *(Enter a value, or type **A**.)*
**Select an object to change or [Undo]:** *(Select an open object, or type **U**.)*

**Specify total length** changes the length by an absolute value.

**Angle** changes the angle by an absolute value.

**Undo** undoes the most-recent lengthening operation.

DYnamic option
**Select an object to change or [Undo]:** *(Select an open object, or type **U**.)*
**Specify new end point:** *(Pick a point.)*

**Specify new end point** changes the length dynamically by dragging.

**Undo** undoes the most-recent lengthening operation.

## TIPS

- **Lengthen** command only works with open objects, such as lines, arcs, and polylines; it does not work with closed objects, such as circles, polygons, and regions.

- **DElta** option changes the length or angle using the following measurements: (1) the distance from endpoint of the selected object to the pick point; or (2) the incremental length measured from the endpoint of the angle.

# Light

**Rel.12**  Places several types of lights for use by **Render**.

| Command | Alias | Ctrl+ | F-key | Alt+ | Menu Bar | Tablet |
|---------|-------|-------|-------|------|----------|--------|
| light | ... | ... | ... | VEL | View | O1 |
| | | | | | ↳Render | |
| | | | | | ↳Light | |

**Command:** light

*Displays dialog box::*

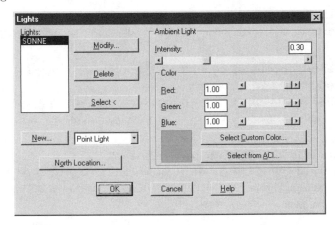

## DIALOG BOX OPTIONS

**Lights** lists the currently-defined lights in the drawing.

**Modify** modifies an existing light in the drawing; displays Modify Light dialog box.

**Delete** deletes the selected light.

**Select** selects a light from the drawing.

**New** creates a new point, spot, or direct light; displays New Light dialog box.

**North Location** selects the direction for North; displays North Location dialog box.

Ambient Light option

**Ambient Light Intensity** adjusts the intensity of ambient light from 0 (dark) to 1.0 (bright).

Color options

**Red** adjusts the level of red from 0 (black) to 1.0 (full red).

**Green** adjusts the level of green from 0 (black) to 1.0 (full green).

**Blue** adjusts the level of blue from 0 (black) to 1.0 (full blue).

**Select Custom Color** displays Windows' Color dialog box.

**Select from ACI** displays AutoCAD's Select Color dialog box.

**North Location** dialog box

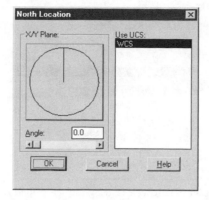

XY Plane options
   **Angle** selects the angle from icon; enter a number, or drag the slider bar.
   **Use UCS** selects a named UCS (user-defined coordinate system).

**New Point Light** dialog box
   *Point lights must be positioned, and they radiate light in all directions.*

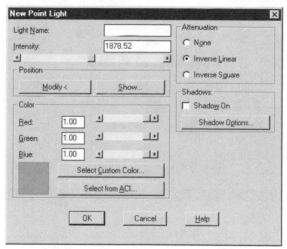

   **Light Name** names the light; maximum = 8 characters, no spaces.
   **Intensity** specifies the intensity of the light, from 0 (turned off) to 31.33.

Attenuation options
   **None** specifies a light that does not diminish in intensity with distance.
   **Inverse Linear** specifies a light whose intensity decreases with distance.
   **Inverse Square** specifies a light whose intensity decreases with the square of the distance.

Position options
   **Modify** changes the location of the light.
   **Show** displays Show dialog box.

Shadows options

**Shadows On** turns on shadow casting.

**Shadow Options** displays Shadow Options dialog box.

### Shadow Options dialog box

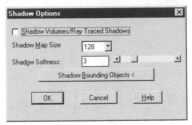

**Shadow Volumes/Raytrace Shadows** creates volumetric shadows; raytracer creates ray-traced shadows; disables shadow map.

**Shadow Map Size** specifies the size of one side of the shadow map; ranges from 64 to 4096 pixels; larger values give more accurate shadows.

**Shadow Softness** specifies the number of pixels at the shadow's edge blended with the underlying image; ranges from 1 to 10.

**Shadow Bounding Objects** selects objects to clip the shadow maps.

**Color** selects a color for the light.

### Show Light Position dialog box

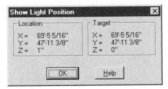

### New Distant Light dialog box

*Distant lights must be positioned, and they radiate light in parallel rays in specified directions.*

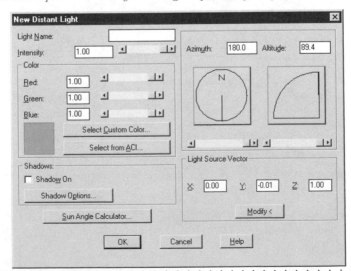

**Name** names the light; maximum = 8 characters.

**Intensity** specifies the intensity of the light; 0 is off.

**Color** specifies the color of the light.

**Shadows** creates shadows.

**Azimuth** sets light's position between -180 and 180 degrees.

**Altitude** sets an angle for the light between 0 and 90 degrees.

Light Source Vector options

**X** specifies the vector x coordinate range from -1.0 to 1.0.

**Y** specifies the vector y coordinate range from -1.0 to 1.0.

**Z** specifies the vector z coordinate range from -1.0 to 1.0.

**Modify** changes the position of the light.

**Sun Angle Calculator** displays **Sun Angle Calculator** dialog box.

## Sun Angle Calculator dialog box

*This calculator eliminates the need to specify the azimuth, altitude, and light source vectors for a distant light.*

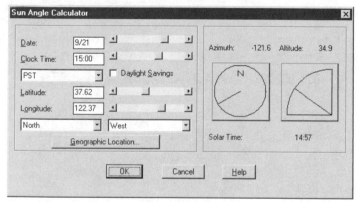

**Date** displays today's date or any date of the year.

**Clock Time** displays the current time or any time of day.

**Latitude** displays the latitude on earth.

**Longitude** displays the longitude.

**Geographic Locator** displays dialog box.

**Geographic Locator** dialog box

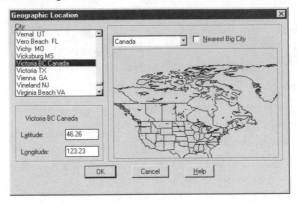

**City** selects the name of a city.

**Latitude** displays the latitude of the city.

**Longitude** displays the longitude of the city.

**Nearest Big City** selects a city from its list closest to your pick point.

**New Spotlight** dialog box
*Spotlights radiate a cone of light, from the light to a spot centered on the target position.*

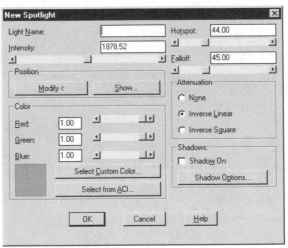

**Light Name** names the light; maximum = 8 characters, no spaces.

**Intensity** specifies the intensity of the light, from 0 (*turned off*) to 31.33.

Attenuation options
**None** specifies that the light's intensity does not diminish with distance.

**Inverse Linear** specifies that the light's intensity decreases with distance.

**Inverse Square** specifies that the light's intensity decreases with the square of the distance.

Position options
**Modify** changes the location of the light.

**Show** displays the Show Light Position dialog box.

Shadows options

**Shadows On** turns on shadow casting.

**Shadow Options** displays the Shadows Options dialog box.

**Color** specifies the color of the light.

## RELATED COMMANDS

**Render** renders the drawing.

**Scene** specifies the lights and view to use in rendering.

## RELATED FILES

*In \autocad 2004\support subdirectory:*

**direct.dwg** is the direct light block

**overhead.dwg** is the overhead drawing block.

**sh_spot.dwg** is the spotlight drawing block.

*Direct (left), Overhead (center), and Spotlight (right).*

## TIPS

- This command works in model space only.

- When the drawing has no lights defined, AutoCAD assumes ambient light.

- While it is not necessary to define any lights to use the **Render** command, a light must be included in a **Scene** definition for the **Render** command to make use of the light.

- In a spotlight, the light beam travels from the *light location* (light block) to the *light target*.

- Ambient light ensures every object in the scene has illumination; ambient light is an omnipresent light source.

- Set ambient light to 0 to turn off for night scenes.

- Place one distant light to simulate the Sun; distant lights have parallel light beams with constant intensity.

- Place several point lights as light bulbs (*lamps*); a point light beams light in all directions, with inverse linear, inverse square, or constant intensity.

- Spotlights beam light in a cone.

## DEFINITIONS

*Constant light* — attenuation is 0; default intensity is 1.0.

*Inverse linear light* — light strength decreases to ½-strength two units of distance away, and ¼-strength four units away; default intensity is ½ extents distance.

*Inverse square light* — light strength decreases to ¼-strength two units away, and $1/_8$-strength four units away; default intensity is ½ the square of the extents distance.

*Extents distance* — distance from minimum lower-left coordinate to the maximum upper-right.

*RGB color* — three primary colors — red, green, blue — shaded from black to white.

*HLS color* — changes colors by hue (color), lightness, and saturation (less gray).

*Hotspot* — brightest cone of light; beam angle ranges from 0 to 160 degrees (default: 45 degrees).

*Falloff* — angle of the full light cone; field angle ranges 0 to 160 degrees (default: 45 degrees).

# 'Limits

**V. 1.0**   Defines the 2D limits in the WCS for the grid markings and the **Zoom All** command; optionally prevents drawing outside of limits.

| Command | Alias | Ctrl+ | F-key | Alt+ | Menu Bar | Tablet |
|---------|-------|-------|-------|------|----------|--------|
| 'limits | ... | ... | ... | OA | Format<br>⮡Drawing Limits | V2 |

**Command:** limits

*In model space:*

**Reset Model space limits:**

*In paper space:*

**Reset Paper space limits:**

*In either model or paper space:*

**Specify lower left corner or [ON/OFF] <0.0000,0.0000>:** *(Pick a point, or type* **ON** *or* **OFF**.*)*

**Specify upper right corner <12.0000,9.0000>:** *(Pick a point.)*

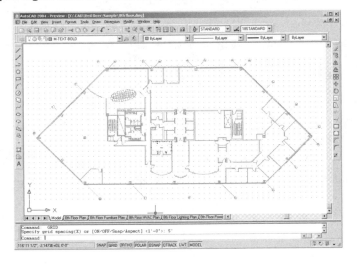

## COMMAND LINE OPTIONS

**OFF** turns off limits checking.

**ON** turns on limits checking.

ENTER retains limits values.

## RELATED COMMANDS

**Grid** displays grid dots, which are bounded by limits.

**Zoom** displays the drawing's extents or limits with the All option.

## RELATED SYSTEM VARIABLES

**LimCheck** toggles the limit's drawing check.

**LimMin** specifies the lower-right 2D coordinates of current limits.

**LimMax** specifies the upper-left 2D coordinates of current limits.

# Line

**V. 1.0** Draws straight 2D and 3D lines.

| Command | Alias | Ctrl+ | F-key | Alt+ | Menu Bar | Tablet |
|---------|-------|-------|-------|------|----------|--------|
| line | l | ... | ... | DL | Draw ⤷Line | J10 |

**Command:** line
**Specify first point:** *(Pick a starting point.)*
**Specify next point or [Undo]:** *(Pick another point, or type **U**.)*
**Specify next point or [Undo]:** *(Pick another point, or type **U**.)*
**Specify next point or [Close/Undo]:** *(Pick another point, or enter an option.)*
**Specify next point or [Close/Undo]:** *(Press **Enter** to end the command.)*

### Single Segment Line:

### Multi Segment Line:

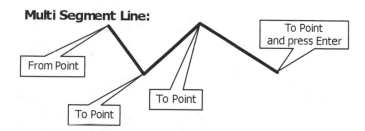

### Closed, MultiSegment Line:

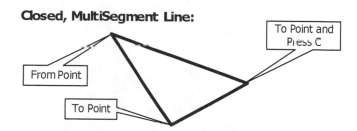

### COMMAND LINE OPTIONS

**Close** closes the line from the current point to the starting point.

**Undo** undoes the last line segment drawn.

ENTER continues the line from the last endpoint at the 'From point' prompt; terminates the **Line** command at the 'To point' prompt.

## RELATED COMMANDS

**MLine** draws up to 16 parallel lines.

**PLine** draws polylines and polyline arcs.

**Trace** draws lines with width.

**Ray** creates a semi-infinite construction line.

**XLine** creates an infinite construction line.

## RELATED SYSTEM VARIABLES

**Elevation** specifies the distance above (or below) the x,y plane a line is drawn.

**Lastpoint** specifies the last-entered coordinate triple (x,y,z-coordinate).

**Thickness** determines the thickness of the line.

## TIPS

- To draw a 2D line, enter x,y coordinate pairs; the z coordinate takes on the value of the **Elevation** system variable.

- To draw a 3D line, enter x,y,z coordinate triples.

- When system variable **Thickness** is not zero, the line has thickness, which makes it a plane perpendicular to the current UCS.

# 'Linetype

V. 2.0 Loads linetype definitions into the drawing, creates new linetypes, and sets
the working linetype.

| Commands | Aliases Ctrl+ | F-key | Alt+ | Menu Bar | Tablet |
|---|---|---|---|---|---|
| 'linetype | lt    ... | ... | ON | Format | U3 |
| | ltype | | | ⬦Linetype | |
| | ddltype | | | | |
| -linetype | -lt | | | | |
| | -ltype | | | | |

**Command:** linetype

*Displays dialog box::*

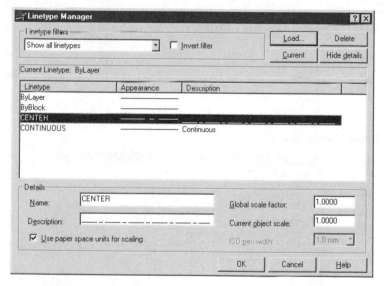

## DIALOG BOX OPTIONS

**Linetype filters** displays the following groups of linetypes:

- **Show all linetypes** displays all linetypes defined in the current drawing.
- **Show all used linetypes** displays all linetypes being used.
- **All xref dependent linetypes** displays linetypes in externally referenced drawings.

**Invert filter** inverts the display of layer names; for example, when Show all used linetypes is
selected, the Invert filter option displays all linetypes not used in the drawing.

*Buttons*

**Loads** displays the Load or Reload Linetypes dialog box.

**Current** sets the selected layer as the current layer.

**Delete** purges the selected linetypes; some linetypes cannot be deleted, as described by the warning dialog box:

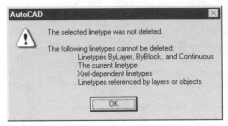

**Show/Hide Details** toggles the display of the Details portion of the Linetype Properties Manager dialog box.

## Details options

**Name** names the selected linetype.

**Description** displays the description associated with the linetype.

**Use paper space units for scaling** specifies that paper space linetype scaling is used, even in model space.

**Global scale factor** specifies the scale factor for all linetypes in the drawing.

**Current object scale** specifies the individual object scale factor for all subsequently-drawn linetypes, multiplied by the global scale factor.

**ISO pen width** applies standard scale factors to ISO (international standards) linetypes.

## Load or Reload Linetypes dialog box

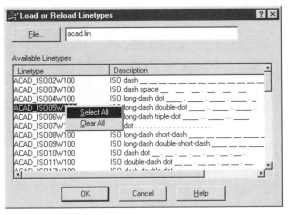

**File** names the LIN linetype definition file.

## SHORTCUT MENU

*Right-click any linetype name in the **Linetype Manager** dialog box:*

**Select All** selects all linetypes.

**Clear All** selects no linetypes.

# -LINETYPE Command

**Command:** -linetype

**Enter an option [?/Create/Load/Set]:** *(Enter an option.)*

## COMMAND LINE OPTIONS

**Create** creates new user-defined linetypes; see the Quick Start Tutorial.

**Load** loads linetypes from linetype definition (*.lin*) files.

**Set** sets the working linetype.

**?** lists the linetypes loaded into the drawing.

## RELATED COMMANDS

**Change** changes objects to a new linetype; changes linetype scale.

**ChProp** changes objects to a new linetype.

**LtScale** sets the scale of the linetype.

**Rename** changes the name of the linetype.

## RELATED SYSTEM VARIABLES

**CeLtype** specifies the current linetype setting.

**LtScale** specifies the current linetype scale.

**PsLtScale** specifies the linetype scale relative to paper scale.

**PlineGen** controls how linetypes are generated for polylines.

## TIPS

*   The only linetypes defined initially in a new AutoCAD drawing are:

    **Continuous** draws unbroken lines.

    **Bylayer** specifies linetype by the layer setting.

    **Byblock** specifies linetypes by the block definition.

*   Linetypes must be loaded from LIN definition files before being used in a drawing.

*   When loading one or more linetypes, it is faster to load all linetypes, then use the **Purge** command to remove the linetype definitions that have not been used in the drawing.

*   As of AutoCAD Release 13, objects can have independent linetype scales.

## RELATED FILE

- The following standard linetypes are in \autocad 2004\support\acad.lin:

| | |
|---|---|
| ACAD_ISO02W100 | ISO dash __ __ __ __ __ __ __ __ __ __ __ __ |
| ACAD_ISO03W100 | ISO dash space __   __   __   __   __   __ |
| ACAD_ISO04W100 | ISO long-dash dot ____ . ____ . ____ . ____ . _ |
| ACAD_ISO05W100 | ISO long-dash double-dot ____ .. ____ .. ____ . |
| ACAD_ISO06W100 | ISO long-dash triple-dot ____ ... ____ ... ____ |
| ACAD_ISO07W100 | ISO dot . . . . . . . . . . . . . . . . |
| ACAD_ISO08W100 | ISO long-dash short-dash ____ __ ____ __ ____ _ |
| ACAD_ISO09W100 | ISO long-dash double-short-dash ____ __ __ ____ |
| ACAD_ISO10W100 | ISO dash dot __ . __ . __ . __ . __ . |
| ACAD_ISO11W100 | ISO double-dash dot __ __ . __ __ . __ __ . __ |
| ACAD_ISO12W100 | ISO dash double-dot __ .. __ .. __ .. __ .. |
| ACAD_ISO13W100 | ISO double-dash double-dot __ __ .. __ __ .. |
| ACAD_ISO14W100 | ISO dash triple-dot __ ... __ ... __ ... |
| ACAD_ISO15W100 | ISO double-dash triple-dot __ __ ... __ __ . |
| BATTING | Batting SSSSSSSSSSSSSSSSSSSSSSSSSSSSSSSSSSSSSSSSSSS |
| BORDER | Border __ __ . __ __ . __ __ . __ __ . |
| BORDER2 | Border (.5x) __.__.__.__.__.__.__.__. |
| BORDERX2 | Border (2x) ____ ____ . ____ ____ . ____ |
| CENTER | Center ____ _ ____ _ ____ _ ____ _ ____ |
| CENTER2 | Center (.5x) ____ _ ____ _ ____ _ ____ _ ____ |
| CENTERX2 | Center (2x) _____ __ _____ __ _____ |
| DASHDOT | Dash dot __ . __ . __ . __ . __ . |
| DASHDOT2 | Dash dot (.5x) _._._._._._._._._._._. |
| DASHDOTX2 | Dash dot (2x) ____ . ____ . ____ . ____ |
| DASHED | Dashed __ __ __ __ __ __ __ __ __ __ __ |
| DASHED2 | Dashed (.5x) _ _ _ _ _ _ _ _ _ _ _ _ _ _ |
| DASHEDX2 | Dashed (2x) ____ ____ ____ ____ ____ ____ |
| DIVIDE | Divide ____ .. ____ .. ____ .. ____ .. ____ |
| DIVIDE2 | Divide (.5x) __.._.__.._.__.._.__.._ |
| DIVIDEX2 | Divide (2x) _____ .. _____ .. _ |
| DOT | Dot . . . . . . . . . . . . . . . . . |
| DOT2 | Dot (.5x) . . . . . . . . . . . . . . . . . . . . . . . . . |
| DOTX2 | Dot (2x) . . . . . . . . . . . . |
| FENCELINE1 | Fenceline circle ----O-----O----O-----O----O--- |
| FENCELINE2 | Fenceline square ----[]-----[]----[]-----[]---- |
| GAS_LINE | Gas line ----GAS----GAS----GAS----GAS----GAS--- |
| HIDDEN | Hidden __ __ __ __ __ __ __ __ __ __ __ __ __ |
| HIDDEN2 | Hidden (.5x) _ _ _ _ _ _ _ _ _ _ _ _ _ _ _ |
| HIDDENX2 | Hidden (2x) ____ ____ ____ ____ ____ ____ |
| HOT_WATER_SUPPLY | Hot water supply ---- HW ---- HW ---- HW ---- |
| PHANTOM | Phantom ____ __ __ ____ __ __ ____ |
| PHANTOM2 | Phantom (.5x) ___ _ _ ___ _ _ ___ _ _ |
| PHANTOMX2 | Phantom (2x) _____ ___ ___ _ |
| TRACKS | Tracks -I-I-I-I-I-I-I-I-I-I-I-I-I-I-I-I-I |
| ZIGZAG | Zig zag /\/\/\/\/\/\/\/\/\/\/\/\/\/\/\/ |

# Creating Custom Linetypes

You can create custom linetypes on-the-fly. This method does not work for complex linetypes that include shapes or text:

**Step 1**

Enter the **-Linetype** command and use the **Create** option:

> **Command:** -linetype
> **Current line type: "ByLayer"**
> **Enter an option [?/Create/Load/Set]:** c

**Step 2**

Name the linetype in three steps:

First, specify the linetype name:

> **Enter name of linetype to create:** *(Enter up to 31 characters.)*

Second, specify the LIN filename. When you select *acad.lin*, AutoCAD appends your new linetype description to *acad.lin*; when you enter a new filename, AutoCAD creates a new *.lin* file. Third, describe the linetype:

> **Descriptive text:** *(Enter up to 47 characters.)*

**Step 3**

Define the linetype pattern by using five codes:

Positive number for dashes; for example, **0.5** is a dash 0.5 units long.

Negative number for gaps; for example, **-0.25** is a gap 0.25 units long.

Zero for dots: **0** is a single dot.

An **A** forces the linetype to align between two endpoints; linetypes always start and stop with a dash.

Commas ( **,** ) separate values.

*Example:*
> *DASHDOT,__ . __ . __ . __ . __ . __ . __ . __ .
> A,.5,-.25,0,-.25 *(Press **Enter**.)*

**Step 4**

Press ENTER to end the linetype definition.

**Step 5**

Use the **-Linetype** command's **Load** option to load the pattern into the drawing.

> **Enter linetype(s) to load:** *(Enter name.)*

**Step 6**

Use the **Set** option to set the linetype.

> **New object linetype (or ?) <>:** *(Enter name.)*

Alternatively, use the **Properties** command to change objects to the new linetype.

# List

**V. 1.0** Lists information about selected objects in the drawing.

| Command | Alias | Ctrl+ | F-key | Alt+ | Menu Bar | Tablet |
|---------|-------|-------|-------|------|----------|--------|
| list | li | ... | ... | TYL | Tools | U8 |
| | ls | | | | ⟡Inquiry | |
| | | | | | ⟡List | |

**Command:** list
**Select objects:** *(Select one or more objects.)*
**Select objects:** *(Press **Enter** to end object selection.)*
   *Sample output:*

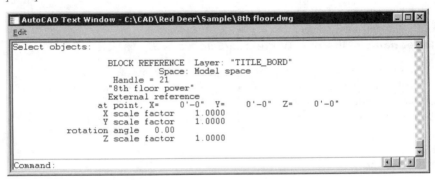

## COMMAND LINE OPTIONS

ENTER continues the display.

ESC cancels the display.

F2 returns to graphics screen.

## RELATED COMMANDS

**Area** calculates the area and perimeter of selected objects.

**DbList** lists information about *all* objects in the drawing.

**MassProp** calculates the properties of 2D regions and 3D solids.

## TIPS

• The **List** command lists the following information only under certain conditions:

| Information | Condition |
|-------------|-----------|
| **Color** | When not set BYLAYER. |
| **Linetype** | When not set BYLAYER. |
| **Thickness** | When not 0. |
| **Elevation** | When z coordinate is not 0. |
| **Extrusion direction** | When z axis differs from current UCS. |

• Object handles are described by hexadecimal numbers.

## Removed Command

**ListURL** was removed from AutoCAD 2000; it was replaced by **-Hyperlink**.

# Load

**V. 1.0** Loads SHX-format shape files into drawings.

| Command | Alias | Ctrl+ | F-key | Alt+ | Menu Bar | Tablet |
|---------|-------|-------|-------|------|----------|--------|
| load | ... | ... | ... | ... | ... | ... |

**Command:** load

*Displays **Load Shape File** dialog box. Select a .shx file, and the click **Open**.*

## COMMAND LINE Options

*None.*

## RELATED AUTOCAD COMMAND

**Shape** inserts shapes into the current drawing.

## RELATED FILES

***.shp** are source code for shape files.

***.shx** are compiled shape files.

*In \autocad 2004\support folder:*

**gdt.shx** and **gdt.shp** are geometric tolerance shapes used by the Tolerance command.

**ltypeshp.shx** and **ltypeshp.shp** are linetype shapes used by the Linetype command.

## TIPS

- Shapes are more efficient than blocks, but are harder to create.

- The **Load** command cannot load .*shx* files meant for fonts. AutoCAD complains, "gdt.shx is a normal text font file, not a shape file."

# LogFileOff

Rel.13 Closes the *acad.log* command logging file.

| Command | Alias | Ctrl+ | F-key | Alt+ | Menu Bar | Tablet |
|---------|-------|-------|-------|------|----------|--------|
| logfileoff | ... | ... | ... | ... | ... | ... |

**Command:** logfileoff

## COMMAND LINE OPTIONS
*None.*

## RELATED AUTOCAD COMMAND
**LogFileOn** turns on the recording of 'Command' prompt text to file *acad.log*.

## RELATED SYSTEM VARIABLE
**LogFileMode** text window written to log file:

| LogFileMode | Meaning |
|-------------|---------|
| 0 | Text not written to file (default). |
| 1 | Text written to file. |

## TIPS
- AutoCAD places a dashed line at the end of each log file session.

- The CTRL+Q shortcut now quits AutoCAD, insted of toggling the log file.

# LogFileOn

Rel.13 Opens *acad.log* file and records 'Command' prompt text to the file.

| Command | Alias | Ctrl+ | F-key | Alt+ | Menu Bar | Tablet |
|---------|-------|-------|-------|------|----------|--------|
| logfileon | ... | ... | ... | ... | ... | ... |

**Command:** logfileon

## COMMAND LINE OPTIONS
*None.*

## RELATED AUTOCAD COMMANDS
**CopyHist** copies all command text from the Text window to the Clipboard.

**LogFileOff** turns off recording 'Command' prompt text to file *acad.log*.

## RELATED SYSTEM VARIABLES
**LogFileMode** determines whether text is written to the log file:

| LogFileMode | Meaning |
|-------------|---------|
| 0 | Text not written to file (default). |
| 1 | Text written to file. |

**LogFileName** is the name of the log file (default = *drawingname.log*).

## RELATED FILE
**\*.log** is the log file.

## TIPS

- If log file recording is left on, it resumes when AutoCAD is next loaded.

- AutoCAD places a dashed line at the end of each log file session.

- You can give the log file a different name with the **Preferences** command's **Files** tab or with system variable **LogFileName**.

- *Historical note:* In some early versions of AutoCAD, CTRL+Q meant "quick screen print," which outputted the current screen display to the printer. CTRL+Q reappeared in AutoCAD Release 14 to record command text to a file. As of AutoCAD 2004, the CTRL+Q shortcut changed meaning to quit AutoCAD, insted of toggling the log file — curious, given that there already is a keyboard shortcut, ALT+F4, that quits AutoCAD.

 # LsEdit

**Rel.14**    Edits the properties of landscape objects *(short for LandScape EDIT)*.

| Command | Alias | Ctrl+ | F-key | Alt+ | Menu Bar | Tablet |
|---------|-------|-------|-------|------|----------|--------|
| lsedit | ... | ... | ... | VEE | View | ... |
| | | | | | ⮡ Render | |
| | | | | | ⮡ Landscape Edit | |

**Command:** lsedit
**Select a landscape object:** *(Select a single landscape object.)*

*Displays dialog box:*

## DIALOG BOX OPTIONS

**Height** changes height of the object, by entering a new value or moving the slider bar.

**Position** moves the object to another position in the drawing.

Geometry options

**Single Face** renders faster, but is less realistic.

**Crossing Face** produces more realistic ray-traced shadows.

**View Aligned** forces object always to face the camera.

## RELATED COMMANDS

**LsLib** lets you add and remove raster images from the *render.lli* file.

**LsNew** places a landscape object in the drawing.

**Render** renders the landscape object.

## TIP

• Landscape objects are rendered only when using the **Render** command's photoreal or photo ray trace options.

 # LsLib

**Rel.14**  Maintains libraries of landscape objects (*short for LandScape LIBrary*).

| Command | Alias | Ctrl+ | F-key | Alt+ | Menu Bar | Tablet |
|---------|-------|-------|-------|------|----------|--------|
| lsedit | ... | ... | ... | VEC | View | ... |
| | | | | | ⟂Render | |
| | | | | | ⟂Landscape Library | |

**Command:** lslib
**Select a landscape object:** *(Select a single landscape object.)*

*Displays dialog box:*

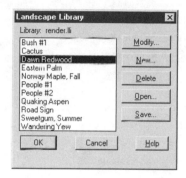

## DIALOG BOX OPTIONS

**Library** indicates current LLI landscape library filename; selects a landscape object.

*Buttons*

**Modify** changes the properties of landscape objects; displays Landscape Library Edit dialog box.

**New** assigns default values to landscape objects; displays Landscape Library New dialog box.

**Delete** removes landscape objects from the library.

**Open** opens landscape library files; displays the Open Landscape Library dialog box.

**Save** saves landscape objects to *.lli* files; displays dialog box.

## Landscape Library Edit dialog box

### Default Geometry options

**Single Face** renders faster, but is less realistic.

**Crossing Face** produces more realistic ray-traced shadows.

**View Aligned** forces the object always to face the camera.

**Preview** previews the landscape image.

**Name** names the landscape object.

**Image File** specifies the type of raster file, *.bmp*, *.png*, *.gif*, *.jpg*, *.pcx*, *.tga*, or *.tif*.

**Opacity Map File** names the raster file that provides opacity.

**Find File** finds the file; displays the Find Image File dialog box.

## Landscape Library New dialog box

*Options are identical to those found in the **Landscape Library Edit** dialog box.*

## RELATED COMMANDS

**LsEdit** edits the properties of a landscape object.

**LsNew** places a landscape object in the drawing.

**MatLib** provides a library of surface textures.

# LsNew

**Rel.14**   Places landscape objects in drawings *(short for LandScape NEW).*

| Command | Alias | Ctrl+ | F-key | Alt+ | Menu Bar | Tablet |
|---------|-------|-------|-------|------|----------|--------|
| lsnew | ... | ... | ... | VEN | View | ... |
| | | | | | ⮥Render | |
| | | | | | ⮥Landscape New | |

**Command:** lsnew

*Displays dialog box:*

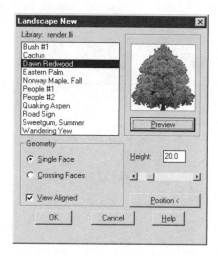

## DIALOG BOX OPTIONS

**Preview** views the raster image.

**Height** changes the height of the object by entering a new value or moving the slider bar.

**Position** moves the object to another position in the drawing.

Geometry options

**Single Face** renders faster, but is less realistic.

**Crossing Face** produces more realistic ray-traced shadows.

**View Aligned** forces the object always to face the camera.

## RELATED COMMANDS

**LsLib** adds and removes raster images from the *render.lli* file.

**LsEdit** edits the properties of a landscape object.

**Render** renders the landscape object.

## TIPS

* A *landscape object* is defined as a **Plant** object in the AutoCAD database.

* Turn on **View Aligned** when you want the landscape object — such as a tree — always to face the camera.

- Turn off **View Aligned** to fix the orientation of the landscape object, such as a store front.

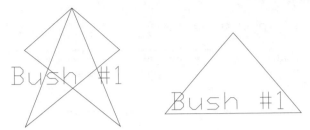

*A landscape object with **crossing faces** (left) and **single face** (right).*

- The grips at the base, top, and corners of landscape objects have special meaning:

| Grip | Meaning |
| --- | --- |
| **Top** | Changes the object's height. |
| **Bottom corner** | Rotates (if not view aligned) and scales the object. |
| **Base** | Moves the object. |

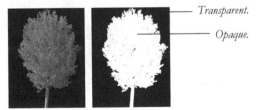

- An *opacity map* determines which part of a raster image is opaque and which is transparent. The opacity map should be a bi-color (*black and white*) image file.

*Transparent.*

*Opaque.*

- The landscape object does not appear until you use the **Render** command:

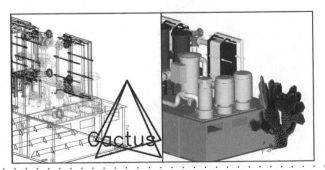

# 'LtScale

**V. 2.0**  Sets the global scale factor of linetypes (*short for Line Type SCALE*).

| Command | Alias | Ctrl+ | F-key | Alt+ | Menu Bar | Tablet |
|---------|-------|-------|-------|------|----------|--------|
| 'ltscale | lts | ... | ... | ... | ... | ... |

**Command:** ltscale
**Enter new linetype scale factor <1.0000>:** *(Enter a scale factor.)*
**Regenerating drawing.**

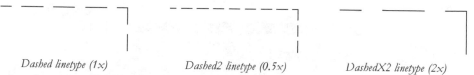

Dashed linetype (1x)　　　Dashed2 linetype (0.5x)　　　DashedX2 linetype (2x)

## COMMAND LINE OPTION

**Enter new linetype scale factor** changes the global scale factor of all linetypes in the drawing.

## RELATED COMMANDS

**ChProp** changes the linetype scale of one or more objects.

**Properties** changes the linetype scale of objects.

**Linetype** loads, creates, and sets the working linetype.

## RELATED SYSTEM VARIABLES

**LtScale** contains the current linetype scale factor.

**PlineGen** controls how linetypes are generated for polylines.

**PsLtScale** specifies that the linetype scale is relative to paper space.

## TIPS

- If the linetype scale is too large, the linetype appears solid.

- If the linetype scale is too small, the linetype appears as a solid line that redraws very slowly.

- In addition to setting the scale with the **LtScale** command, the *acad.lin* file contains each linetype in three scales: normal, half-size, and double-size.

- You can change the linetype scaling of individual objects, which is then multiplied by the global scale factor specified by the **LtScale** command.

## Removed Command

**MakePreview** was removed from AutoCAD Release 14.

# 'LWeight

2000 Sets the current lineweight (*display width*) of objects.

| Commands | Aliases | Ctrl+ | Status Bar | Alt+ | Menu Bar | Tablet |
|----------|---------|-------|------------|------|----------|--------|
| 'lweight | lw | ... | LWT | OW | Format | W14 |
| | lineweight | | | | ⤷Lineweight | |
| -lweight | | | | | | |

**Command:** lweight

*Displays dialog box:*

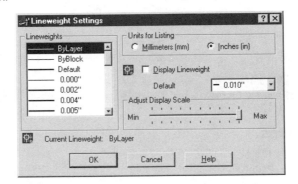

## DIALOG BOX OPTIONS

**Lineweights** lists lineweight values.

**Units for Listing** specifies the units of lineweights:

- **Millimeters (mm)** specifies lineweight values in millimeters.
- **Inches (in)** specifies lineweight values in inches.

**Display Lineweight** toggles the display of lineweights; when checked, lineweights are displayed.

**Default** specifies the default lineweight for layers (default = 0.01" or 0.25 mm).

**Adjust Display Scale** controls the scale of lineweights in the Model tab, which displays lineweights in pixels.

## SHORTCUT MENU OPTIONS

*Right-click **LWT** on status bar to display shortcut menu:*

**On** turns on lineweight display.

**Off** turns off lineweight display.

**Settings** displays **Lineweight Settings** dialog box.

## -LWEIGHT Command

**Command:** -lweight

**Enter default lineweight for new objects or [?]:** *(Enter a value, or type ?.)*

### COMMAND LINE OPTIONS

**Enter default lineweight** specifies the current lineweight; valid values include Bylayer, Byblock, and Default.

**?** lists the valid values for lineweights:

```
ByLayer ByBlock Default
     0.000" 0.002"  0.004"   0.005" 0.006"  0.007"
     0.008" 0.010"  0.012"   0.014" 0.016"  0.020"
     0.021" 0.024"  0.028"   0.031" 0.035"  0.039"
     0.042" 0.047"  0.055"   0.062" 0.079"  0.083"
```

### RELATED SYSTEM VARIABLES

**LwDefault** specifies the default linewidth; default = 0.01" or 0.25 mm.

**LwDisplay** toggles the display of lineweights in the drawing.

**LwUnits** determines whether the lineweight is measured in inches or millimeters.

### TIPS

* To create custom lineweights for plotting, use the **Plot Style Table Editor**.

* A lineweight of 0 plots the lines at the thinnest width of which the plotter is capable, usually one pixel or one dot wide.

# MassProp

**Rel.11**  Reports the mass properties of 3D solid models, bodies, and 2D regions
(*short for MASS PROPerties*).

| Command | Alias | Ctrl+ | F-key | Alt+ | Menu Bar | Tablet |
|---------|-------|-------|-------|------|----------|--------|
| massprop | ... | ... | ... | TYM | Tools | U7 |
| | | | | | ⌐Inquiry | |
| | | | | | ⌐Mass Properties | |

**Command:** massprop
**Select objects:** *(Select one or more solid model objects.)*
**Select objects:** *(Press **Enter**.)*

*Example output of a solid sphere:*

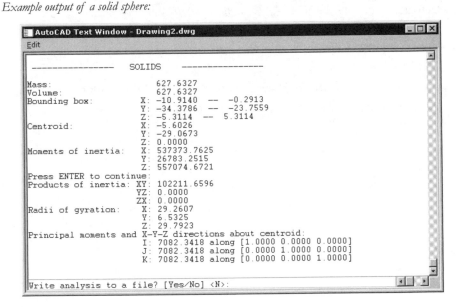

```
-------------------  SOLIDS  -------------------

Mass:                  627.6327
Volume:                627.6327
Bounding box:      X: -10.9140  --  -0.2913
                   Y: -34.3786  --  -23.7559
                   Z: -5.3114   --   5.3114
Centroid:          X: -5.6026
                   Y: -29.0673
                   Z: 0.0000
Moments of inertia: X: 537373.7625
                    Y: 26783.2515
                    Z: 557074.6721
Press ENTER to continue:
Products of inertia: XY: 102211.6596
                     YZ: 0.0000
                     ZX: 0.0000
Radii of gyration:  X: 29.2607
                    Y: 6.5325
                    Z: 29.7923
Principal moments and X-Y-Z directions about centroid:
                    I: 7082.3418 along [1.0000 0.0000 0.0000]
                    J: 7082.3418 along [0.0000 1.0000 0.0000]
                    K: 7082.3418 along [0.0000 0.0000 1.0000]
```

`Write analysis to a file? [Yes/No] <N>:`

## COMMAND LINE OPTIONS

**Select objects**  selects the solid model objects — 2D regions, 3D solids, and bodies — to
analyze.

**Write to a file:**

**Yes** writes mass property reports to *.mpr* files.

**No** doesn't write reports to file.

## RELATED COMMAND

**Area** calculates the area and perimeter of non-solid objects.

## RELATED FILE

***.mpr*** is the file to which MassProp writes its results (mass properties report).

- This command can be used with 2D regions as well as 3D solids; it cannot be used with 3D surface models or 2D non-region objects.

- As of Release 13, AutoCAD's solid modeling no longer allows you to apply a material density to a solid model. All solids and bodies have a density of 1.

- AutoCAD only analyzes regions coplanar (laying in the same plane) to the first region selected.

## DEFINITIONS

*Area* — total surface area of the selected 3D solids, bodies, or 2D regions.

*Bounding Box* — the lower-right and upper-left coordinates of a rectangle enclosing the 2D region; the x,y,z coordinate triple of a 3D box enclosing the 3D solid or body.

*Centroid* — the x,y,z coordinates of the center of the 2D region; the center of mass for 3D solids and bodies.

*Mass* — equal to the volume, because density − 1; not calculated for regions.

*Moment of Inertia*

     — for 2D regions = **Area** * **Radius**$^2$

     — for 3D bodies = **Mass** * **Radius**$^2$

*Perimeter* — Total length of inside and outside loops of 2D regions; not calculated for 3D solids and bodies.

*Product of Inertia*

     — for 2D regions = **Mass** * **Distance** (of centroid to y,z axis) * **Distance** (of centroid to x,z axis).

     — for 3D bodies = **Mass** * **Distance** (of centroid to y,z axis) * **Distance** (of centroid to x,z axis)

*Radius of Gyration* — for 2D regions and 3D solids = (**MomentOfInertia** / **Mass**)$^{1/2}$

*Volume* — 3D space occupied by a 3D solid or body; not calculated for regions.

# 'MatchProp

<u>Rel.14</u>  Matches the properties between selected objects *(short for MATCH PROPerties)*.

| Command | Alias | Ctrl+ | F-key | Alt+ | Menu Bar | Tablet |
|---------|-------|-------|-------|------|----------|--------|
| 'matchprop | ma | ... | ... | MM | Modify | Y14 |
| | painter | | | | ⬦Match Properties | |

**Command:** matchprop
**Select source object:** *(Select a single object.)*
**Current active settings: Color Layer Ltype Ltscale Lineweight Thickness PlotStyle Text Dim Hatch Polyline Viewport**
**Select destination object(s) or [Settings]:** *(Pick one or more objects, or type **S**.)*
**Select destination object(s) or [Settings]:** *(Press **Enter** to exit command.)*

## COMMAND LINE OPTIONS

**Select source object** gets property settings from the source object.

**Select destination object(s)** passes property settings to the destination objects.

**Settings** displays dialog box:

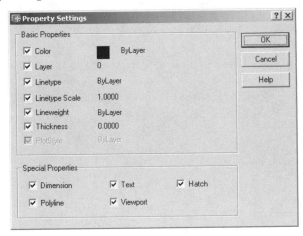

## DIALOG BOX OPTIONS

Basic Properties options

**Color** specifies the color for destination objects; not available when OLE objects are selected.

**Layer** specifies the layer name for destination objects; not available when OLE objects are selected.

**Linetype** specifies the linetype for destination objects; not available when attributes, hatch patterns, mtext, OLE objects, points, or viewports are selected.

**Linetype Scale** specifies the linetype scale for destination objects; not available when attributes, hatch patterns, mtext, OLE objects, points, or viewports are selected.

**Lineweight** specifies the lineweight for destination objects.

**Thickness** specifies the thickness for destination objects; available only for objects that can have thickness: arcs, attributes, circles, lines, mtext, points, 2D polylines, regions, text, and traces.

**Plot Style** specifies the plot style; not available when PStylePolicy = 1 (color-dependent plot style mode) or when OLE objects are selected.

## Special Properties options

**Dimension** copies the dimension style of dimension, leader, and tolerance objects.

**Text** copies the text style of text and mtext objects.

**Hatch** copies the hatch pattern of hatched objects.

**Polyline** copies the width and linetype generation of polylines; curve fit, elevation, and variable width properties are not copied.

**Viewport** copies all properties of viewport objects, except clipping, UCS-per-viewport, and freeze-thaw settings.

## RELATED SYSTEM VARIABLE

**PStylePolicy** determines whether the **PlotStyle** option is available.

## RELATED COMMAND

**Properties** changes most aspects of one selected object.

## TIP

• In other Windows applications, this command is known as **Format Painter**.

# MatLib

**Rel.13** Imports and exports material-look definitions for use by the RMat command (*short for MATerial LIBrary*).

| Command | Alias | Ctrl+ | F-key | Alt+ | Menu Bar | Tablet |
|---------|-------|-------|-------|------|----------|--------|
| matlib  | ...   | ...   | ...   | VEY  | View     | Q1     |
|         |       |       |       |      | ⌐Render  |        |
|         |       |       |       |      | ⌐Materials Library | |

**Command:** matlib

*Displays dialog box:*

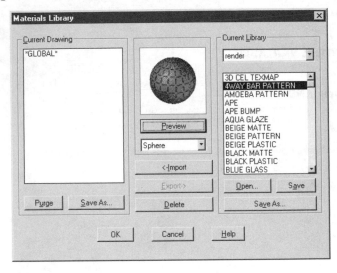

## DIALOG BOX OPTIONS

**Import** brings the selected material definition into the drawing; when there is a conflict, displays Reconcile Imported Material Names dialog box.

**Preview** previews the selected material mapped to sphere and box objects.

**Export** adds material definitions to *.mli* library files; if there is a conflict, displays the Reconcile Exported Material Names dialog box, which is identical to Reconcile Imported Material Names dialog box.

**Purge** deletes unattached material definitions from the Materials list.

**Save** saves to *.mli* files.

**Delete** deletes selected material definitions from the Materials or Library lists.

**Open** loads material definitions from *.mli* files; displays file dialog box.

## Reconcile Imported Material Names dialog box

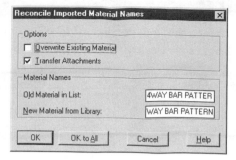

### Options options

**Overwrite Existing Material** overwrites existing material definition with selected material definition.

**Transfer Attachments** keeps objects attached to material definition.

### Material Names options

**Old Material in List** allows you to edit the name of the material.

**New Material from Library** allows you to edit the name of the material.

### RELATED COMMAND

**RMat** attaches a material definition to objects, colors, and layers.

### RELATED FILE

*render.mli* is the material library that contains the material definitions.

### TIPS

- To use materials in rendering, you must take these steps:
    1. Use **MatLib** to load and purge material definitions.
    2. Use **RMat** attach the definitions to objects.
    3. In the **Render** dialog box, turn on the **Apply Materials** option.
- A *material* defines the look of a rendered object: coloring, reflection or shine, roughness, and ambient reflection.
- Materials only appear with **Render**; they do no appear with the **Shade** command.
- By default, a drawing contains a single material definition, called *GLOBAL*, with the default parameters for color, reflection, roughness, and ambience.

*Materials applied to spheres.*

- Materials do not define the density of 3D solids and bodies.

# Measure

**V. 2.5**   Divides lines, arcs, circles, and polylines into equidistant segments, placing points or blocks at each segment.

| Command | Alias | Ctrl+ | F-key | Alt+ | Menu Bar | Tablet |
|---------|-------|-------|-------|------|----------|--------|
| measure | me | ... | ... | DOM | Draw | V12 |
| | | | | | ⬫Point | |
| | | | | | ⬫Measure | |

**Command:** measure
**Select object to measure:** *(Pick a single object.)*
**Specify length of segment or [Block]:** *(Enter a value, or type* **B**.*)*

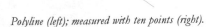

*Polyline (left); measured with ten points (right).*

## COMMAND LINE OPTIONS

**Select object** selects a single object for measurement.

**Specify length of segment** indicates the distance between markers.

Block options
**Enter name of block to insert:** *(Enter name.)*
**Align block with object? [Yes/No] <Y>:** *(Enter* **Y** *or* **N**.*)*
**Specify length of segment:** *(Enter a value.)*

**Enter name of block** indicates the name of the block to use as a marker; the block must already exist in the drawing.

**Align block with object?** aligns the block's x axis with the object.

## RELATED COMMANDS

**Block** creates blocks that can be used with the Measure command.

**Divide** divides an object into a number of segments.

## RELATED SYSTEM VARIABLES

**PdMode** controls the shape of a point.

**PdSize** controls the size of a point.

## TIPS

- You must define the block before it can be used with this command.

- The **Measure** command does not place a point or block at the beginning of the measured object.

## Removed Command

MeetNow was removed from AutoCAD 2004.

# Menu

<u>V. 1.0</u>  Loads MNC, MNS, and MNU menu files.

| Command | Alias | Ctrl+ | F-key | Alt+ | Menu Bar | Tablet |
|---------|-------|-------|-------|------|----------|--------|
| menu | ... | ... | ... | ... | ... | ... |

**Command:** menu

*Displays the **Select Menu File** dialog box. Select a .mnc, .mns, or .mnu file, and then click **Open**.*

## COMMAND LINE OPTIONS

*None.*

## RELATED COMMANDS

**MenuLoad** loads a partial menu file.

**Tablet** configures digitizing tablet for use with overlay menus.

## RELATED SYSTEM VARIABLES

**MenuName** specifies the name of the currently-loaded menu file.

**MenuEcho** suppresses menu echoing.

**ScreenBoxes** specifies the number of menu lines displayed on the side menu.

## RELATED FILES

***.mnc** compiled menu file; stored in binary format.

***.mnc** source menu file; stored in ASCII format.

***.mnu** menu template file; stored in ASCII format.

## TIPS

- AutoCAD automatically compiles .mns and .mns files into .mnc files for faster loading.

- The .mnu file defines the function of the screen menu, menu bar, cursor menu, icon menus, digitizing tablet menus, pointing device buttons, toolbars, help strings, and the AUX: device.

- To access the menu source code, use the **Tools | Customize | Edit Custom Files | Current Menu**. AutoCAD displays the current .mns file in Notepad.

# MenuLoad

Rel.13 Loads a part of a menu file.

| Command | Alias | Ctrl+ | F-key | Alt+ | Menu Bar | Tablet |
|---------|-------|-------|-------|------|----------|--------|
| menuload | ... | ... | ... | TC | Tools | Y9 |
| | | | | | ⮑Customize Menus | |

**Command:** menuload

*Displays tabbed dialog box::*

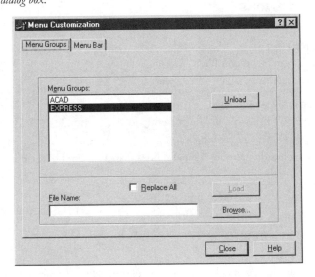

## DIALOG BOX OPTIONS

**Menu Groups** tab

**Menu Groups** lists the names of loaded menu groups and files.

**Unload** unloads selected menu group.

**Replace All** replaces all currently-loaded menus with the newly-loaded menu.

**Load** loads the selected menu group into AutoCAD.

**File Name** displays the name of the menu file.

**Browse** displays the Select Menu File dialog box.

**Close** closes the dialog box.

**Help** provides context-sensitive help.

**Menu Bar** tab

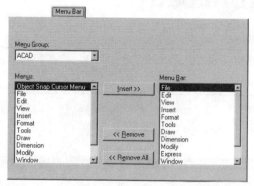

**Menu Group** selects a menu group or file.

**Menus** provides the names of the menu items in the selected menu group.

**Insert** inserts a menu item on the menu bar, immediately above the selected item.

**Remove** removes a menu item from the menu bar.

**Remove All** removes all menu items from the menu bar.

**Menu Bar** provides the names of menu items on the menu bar.

## RELATED COMMANDS

**Menu** loads a full menu file.

**MenuUnload** unloads part of the menu file.

**Tablet** configures the digitizing tablet for use with overlay menus.

## RELATED SYSTEM VARIABLES

**MenuName** specifies the name of the currently-loaded menu file.

**MenuEcho** suppresses menu echoing.

**ScreenBoxes** specifies the number of menu lines displayed on the side menu.

## RELATED FILES

*\*.mnc* is the compiled menu file; stored in binary format.

*\*.mns* is the source menu file; stored in ASCII format.

*\*.mnu* is the menu template file; stored in ASCII format.

## TIP

- The **MenuLoad** command allows you to add *partial* menus to the menu bar, without replacing the entire menu structure.

# MenuUnLoad

**Rel.13** Unloads a partial menu file.

| Command | Alias | Ctrl+ | F-key | Alt+ | Menu Bar | Tablet |
|---|---|---|---|---|---|---|
| **menuload** | ... | ... | ... | **TC** | **Tools**<br>**↳Customize Menus** | **Y9** |

**Command:** menuunload

*Displays tabbed dialog box:*

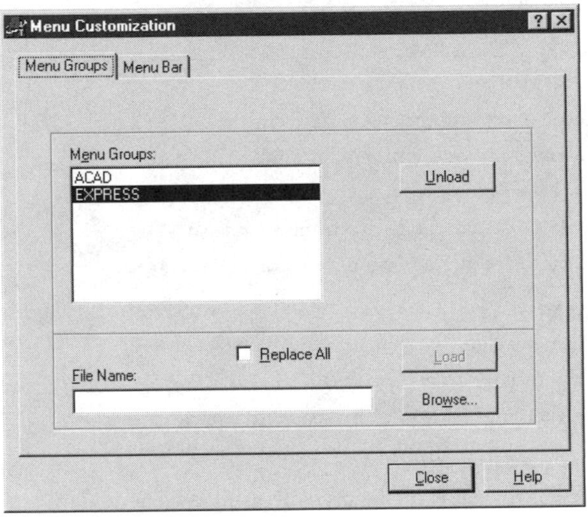


## DIALOG BOX OPTIONS

*See* ***MenuLoad*** *command.*

## TIP

- The **MenuUnload** command allows you to remove partial menu files, such as *db_con.mnu* and *accov.mns*.

# MInsert

**V. 2.5**    Inserts an array of blocks as a single block (*short for Multiple INSERT*).

| Command | Alias | Ctrl+ | F-key | Alt+ | Menu Bar | Tablet |
|---------|-------|-------|-------|------|----------|--------|
| minsert | ... | ... | ... | ... | ... | ... |

**Command:** minsert
**Enter block name or [?]:** *(Enter a name, or type **?**.)*
**Specify insertion point or [Scale/X/Y/Z/Rotate/PScale/PX/PY/PZ/PRotate]:** *(Pick a point, or enter an option.)*
**Enter X scale factor, specify opposite corner, or [Corner/XYZ] <1>:** *(Enter a value, pick a point, or enter an option.)*
**Enter Y scale factor <use X scale factor>:** *(Enter a value, or press **Enter**.)*
**Specify rotation angle <0>:** *(Enter a value, or press **Enter**.)*
**Enter number of rows (---) <1>:** *(Enter a value.)*
**Enter number of columns (|||) <1>:** *(Enter a value.)*
**Enter distance between rows or specify unit cell (---):** *(Enter a value.)*
**Specify distance between columns (|||):** *(Enter a value.)*

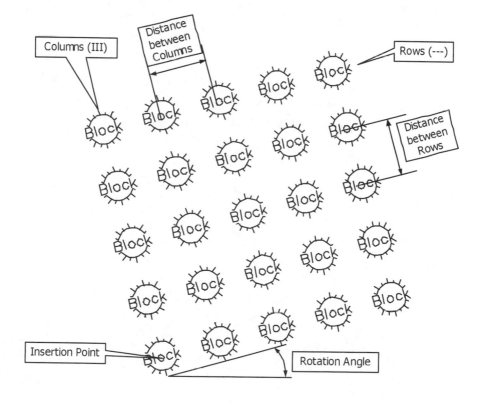

## COMMAND LINE OPTIONS

**Enter block name** indicates the name of the block to be inserted; the block must already exist in the drawing.

**?** lists the names of blocks stored in the drawing.

**Specify insertion point** specifies the x,y coordinates of the first block.

**P** supplies predefined scale and rotation values.

**X scale factor** indicates the x scale factor.

**Specify opposite corner** specifies a second point that indicates the x,y-scale factor.

**Corner** indicates the x and y scale factors by picking two points on the screen.

**XYZ** specifies x, y, and z scaling.

**Specify rotation angle** specifies the angle for the array.

**Number of rows** specifies the number of horizontal rows.

**Number of columns** specifies the number of vertical columns.

**Distance between rows** specifies the distance between rows.

**Specify unit cell** shows the cell distance by picking two points on the screen.

**Distance between columns** specifies the distance between columns.

## RELATED COMMANDS

**3dArray** creates 3D rectangular and polar arrays.

**Array** creates 2D rectangular and polar arrays.

**Block** creates a block.

## TIPS

- The array placed by the **MInsert** command is a single block.

- You *cannot* explode the block created by the **MInsert** command.

- You may redefine the block created by the **MInsert** command.

# Mirror

**V. 2.0**  Creates a mirror copy of a group of objects in 2D space.

| Command | Alias | Ctrl+ | F-key | Alt+ | Menu Bar | Tablet |
|---------|-------|-------|-------|------|----------|--------|
| mirror | mi | ... | ... | MI | Modify | V16 |
|  |  |  |  |  | ⤷Mirror |  |

**Command:** mirror
**Select objects:** *(Pick one or more objects.)*
**Select objects:** *(Press **Enter** to end object selection.)*
**Specify first point of mirror line:** *(Pick a point.)*
**Specify second point of mirror line:** *(Pick another point.)*
**Delete source objects? [Yes/No] <N>:** *(Type **Y** or **N**.)*

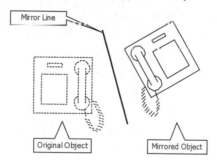

Mirror Line

Original Object

Mirrored Object

## COMMAND LINE OPTIONS

**Select objects** selects the objects to mirror.

**First point** specifies the starting point of the mirror line.

**Second point** specifies the end point of the mirror line.

**Delete source objects** deletes selected objects.

## RELATED COMMANDS

**Array** mirrors object around a circle.

**Copy** creates a non-mirrored copy of a group of objects.

**Mirror3d** mirrors objects in 3D-space.

## RELATED SYSTEM VARIABLE

**MirrText** determines whether text is mirrored by the Mirror command.

## TIPS

* The **Mirror** command is excellent for cutting your drawing work in half for symmetrical objects. For double-symmetrical objects, use **Mirror** twice.

* Although you can mirror a viewport in paper space, this does not mirror the model space objects inside the viewport.

* Turn on **Ortho** mode to ensure that the mirror is perfectly horizontal or vertical.

* The mirror line becomes a mirror plane in 3D; it is perpendicular to the x,y plane of the UCS containing the mirror line.

# Mirror3d

**Rel.11** Mirrors objects about a plane in 3D space.

| Command | Alias | Ctrl+ | F-key | Alt+ | Menu Bar | Tablet |
|---------|-------|-------|-------|------|----------|--------|
| mirror3d | ... | ... | ... | M3M | Modify | W21 |
| | | | | | ⮡3D Operation | |
| | | | | | ⮡Mirror 3D | |

**Command:** mirror3d
**Select objects:** *(Pick one or more objects.)*
**Select objects:** *(Press **Enter** to end object selection.)*
**Specify first point of mirror plane (3 points) or**
**[Object/Last/Zaxis/View/XY/YZ/ZX/3points] <3points>:** *(Pick a point, or enter an option.)*
**Delete old objects? <N>:** *(Type **Y** or **N**.)*

## COMMAND LINE OPTIONS

**Select objects** selects the objects to be mirrored in space.

**Specify first point** specifies the first point of the mirror plane.

**Object** selects a circle, arc or 2D polyline segment as the mirror plane.

**Last** selects the last-picked mirror plane.

**View** specifies that the current view plane is the mirror plane.

**XY** specifies that the x,y plane is the mirror plane.

**YZ** specifies that the y,z plane is the mirror plane.

**ZX** specifies that the z,x plane is the mirror plane.

**Zaxis** defines the mirror plane by a point on the plane and the normal to the plane, i.e., the z axis.

**3points** defines three points on the mirror plane.

## RELATED COMMANDS

**Align** translates and rotates objects in 2D planes and 3D space.

**Mirror** mirrors objects in 2D space.

**Rotate3d** rotates objects in 3D space.

## RELATED SYSTEM VARIABLE

**MirrText** determines whether text is mirrored by the **Mirror** command:

| MirrText | Meaning |
|----------|---------|
| 0 | Text is not mirrored about the horizontal axis (default). |
| 1 | Text is mirrored. |

# MlEdit

<u>Rel.13</u>  Edits multiline vertices (*short for MultiLine EDITor*).

| Command | Alias | Ctrl+ | F-key | Alt+ | Menu Bar | Tablet |
|---------|-------|-------|-------|------|----------|--------|
| mledit | ... | ... | ... | ... | Modify | Y19 |
| | | | | | ↳Multiline | |
| -mledit | | | | | | |

**Command:** mledit

*Displays dialog box:*

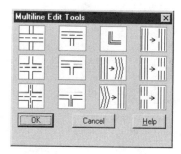

## DIALOG BOX OPTIONS

**Closed Cross** closes the intersection of two multilines.

**Open Cross** opens the intersection of two multilines.

**Merged Cross** merges a pair of multilines: opens exterior lines; closes interior lines.

**Closed Tee** closes T-intersections.

**Open Tee** opens T-intersections.

**Merged Tee** merges T-intersection by opening exterior lines and closing interior lines.

**Corner Joint** creates corner joints with pairs of intersecting multilines.

**Add Vertex** adds vertcies (*joints*) to multiline segments.

**Delete Vertex** removes vertices from multiline segments.

**Cut Single** places gaps in a single line of multilines.

**Cut All** places gaps in all lines of multilines.

**Weld All** removes gaps from multilines.

# -MLEDIT Command

**Command:** -mledit

**Enter mline editing option [CC/OC/MC/CT/OT/MT/CJ/AV/DV/CS/CA/WA]:**
*(Enter an option.)*

## COMMAND LINE OPTIONS

**AV** adds vertices.

**DV** deletes vertices.

**CC** closes crossings.

**OC** opens crossings.

**MC** merges crossings.

**CT** closes tees.

**OT** opens tees.

**MT** merges tees.

**CJ** creates corner joints.

**CS** cuts a single line.

**CA** cuts all lines.

**WA** welds all lines.

**U** undoes the most-recent multiline edit.

## RELATED COMMANDS

**MLine** draws up to 16 parallel lines.

**MlStyle** defines the properties of a multiline.

## RELATED SYSTEM VARIABLES

**CMlJust** specifies the current multiline justification mode:

| CMlJust | Meaning |
|---------|-----------------|
| 0 | Top (default). |
| 1 | Middle. |
| 2 | Bottom. |

**CMlScale** specifies the current multiline scale factor (default = 1.0).

**CMlStyle** specifies the current multiline style name (default = " ").

## RELATED FILE

*\*.mln* is the multiline style definition file.

## TIPS

- Use the **Cut All** option to open up a gap before placing door and window symbols in a multiline wall.

- Use the **Weld All** option to close up a gap after removing the door or window symbol in a multiline.

- Use the **Stretch** command to move a door or window symbol in a multiline wall.

- When you open a gap in a multiline, AutoCAD does not cap the sides of the gap. You may need to add the endcaps with the **Line** command.

# MLine

<u>Rel.13</u>  Draws up to 16 parallel lines (*short for Multiple LINE*).

| Command | Alias | Ctrl+ | F-key | Alt+ | Menu Bar | Tablet |
|---------|-------|-------|-------|------|----------|--------|
| mline | ml | ... | ... | DM | Draw<br>⤷Multiline | M10 |

**Command:** mline
**Current settings: Justification = Top, Scale = 1.00, Style = STANDARD**
**Specify start point or [Justification/Scale/STyle]:** *(Pick a point, or enter an option.)*
**Specify next point:** *(Pick a point.)*
**Specify next point or [Undo]:** *(Pick a point, or type U.)*
**Specify next point or [Close/Undo]:** *(Pick a point, or else enter an option.)*

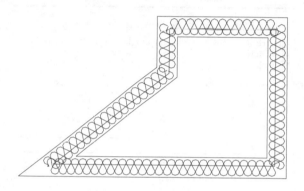

## COMMAND LINE OPTIONS

**Specify start point** indicates the start of the multiline.

**Specify next point** indicates the next vertex.

**Undo** removes the most recently-added segment.

**Close** closes the multiline to its start point.

Justification options
**Enter justification type [Top/Zero/Bottom] <top>:** *(Enter an option.)*

**Top** draws top line of the multiline at the cursor; remainder of multiline is "below" the cursor.

**Zero** draws the center (*zero offset point*) of the multiline at the cursor.

**Bottom** draws the bottom of the multiline at the cursor; remainder of the multiline is "above" the cursor.

Scale option
**Enter mline scale <1.00>:** *(Enter a value.)*

**Enter mline scale** specifies the scale of the width of the multiline; see Tips for examples.

STyle options
**Enter mline style name or [?]:** *(Enter style name, or type **?**.)*

**Enter mline style name** specifies the name of the multiline style.

**?** lists the names of the multiline styles defined in drawing.

## RELATED COMMANDS

**MlEdit** edits multilines.

**MlProp** defines the properties of a multiline.

## RELATED SYSTEM VARIABLES

**CMlJust** specifies the current multiline justification:

| CMlJust | Meaning |
|---------|-----------------|
| 0 | Top (default). |
| 1 | Middle. |
| 2 | Bottom. |

**CMlScale** specifies the current multiline scale factor (default = 1.0).

**CMlStyle** specifies the current multiline style name (default = "").

## RELATED FILE

*****.mln** is the multiline style definition file.

## TIPS

- Examples of scale factors:

| Scale | Meaning |
|-------|------------------------------------------|
| 1.0 | Default scale factor. |
| 2.0 | Draws multiline twice as wide |
| 0.5 | Draws multiline half as wide. |
| -1.0 | Flips multiline. |
| 0 | Collapses multiline to a single line. |

- Multiline styles are stored in *.mln* files in DXF-like format.

# MlStyle

<u>Rel.13</u>   Defines the characteristics of multilines (*short for MultiLine STYLE*).

| Command | Alias | Ctrl+ | F-key | Alt+ | Menu Bar | Tablet |
|---------|-------|-------|-------|------|----------|--------|
| mlstyle | ... | ... | ... | OM | Format | V5 |
| | | | | | ⌐Multiline Style | |

**Command:** mlstyle

*Displays dialog box:*

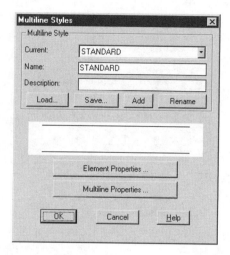

## DIALOG BOX OPTIONS

Multiline Style options

**Current** lists the currently-loaded multiline style names (default = STANDARD).

**Name** gives a new multiline style a name, or renames an existing style.

**Description** describes the multiline style, with up to 255 characters.

**Load** loads styles from the multiline library file *acad.mln* or another *.mln* file; displays the Load Multiline Styles dialog box.

**Save** saves a multiline style or renames a style; displays dialog box.

**Add** adds the multiline style from the Name box to the Current list.

**Remove** removes the multiline style from the Current list.

Additional options

**Element Properties** specifies properties of multiline elements; displays the Element Properties dialog box.

**Multiline Properties** specifies additional properties for multilines; displays the Multiline Properties dialog box.

**Load Multiline Styles** dialog box

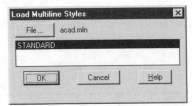

**File** selects an *.mln* multiline definition file.

**Element Properties** dialog box

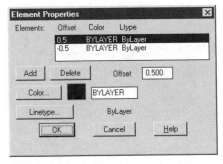

**Add** adds an element (line).

**Delete** deletes an element.

**Offset** specifies the distance from origin to element.

**Color** specifies the element color; displays Select Color dialog box.

**Linetype** specifies the element linetype; displays Select Linetype dialog box.

**Multiline Properties** dialog box

**Display Joints** toggles the display of joints (miters) at vertices; affects all multiline segments.

Caps options
    **Line** draws a straight line start and/or end cap.

    **Outer Arc** draws an arc to cap the outermost pair of lines.

    **Inner Arcs** draws an arc to cap all inner pairs of lines.

    **Angle** specifies the angle for straight line caps.

Fill options
    **On** specifies the fill color.

    **Color** displays the Select Color dialog box.

## RELATED COMMANDS

**MlEdit** edits multilines.

**MLine** draws up to 16 parallel lines.

## RELATED SYSTEM VARIABLES

**CMlJust** specifies the current multiline justification:

| CMlJust | Meaning |
|---------|-----------------|
| 0 | Top (default). |
| 1 | Middle. |
| 2 | Bottom. |

**CMlScale** specifies the current multiline scale factor (default = 1.0).

**CMlStyle** specifies the current multiline style name (default = " ").

## RELATED FILE

*acad.mln* is the multiline style definition file.

## TIPS

- Use the **MlEdit** command to create (or close up) gaps to place door and window symbols in multiline walls.

- The multiline scale factor has the following effect on the look of a multiline:

| Scale | Meaning |
|-------|------------------------------------------------|
| 1.0 | The default scale factor. |
| 0.5 | Draws multiline half as wide. |
| 2.0 | Draws multiline twice as wide, not twice as long. |
| -1.0 | Flips multiline about its origin. |
| 0.0 | Collapses multiline to a single line. |

- The *.mln* file describes multiline styles in a DXF-like format.

- You cannot change the element or multiline properties once the drawing contains a multiline using the style.

# Model

<u>**2000**</u>  Switches to model tab.

| Command | Alias | Ctrl+ | F-key | Alt+ | Menu Bar | Tablet |
|---------|-------|-------|-------|------|----------|--------|
| model | ... | ... | ... | ... | ... | ... |

**Command:** model

*Switches to the model tab.*

## COMMAND LINE OPTIONS

*None.*

## RELATED COMMANDS

**Layout** creates layouts.

**MSpace** switches to model space.

## RELATED SYSTEM VARIABLE

**Tilemode** switches between model tab and layout tab.

## TIPS

• This command automatically sets **TileMode** to 1.

• As an alternative to this command, you can select the **Model** tab:

• The **Model** tab replaces the **TILE** button on the status bar of AutoCAD Release 13 and 14.

 # Move

**V. 1.0** Moves a group of objects to a new location.

| Command | Alias | Ctrl+ | F-key | Alt+ | Menu Bar | Tablet |
|---------|-------|-------|-------|------|----------|--------|
| move | m | ... | ... | MV | Modify<br>↳Move | V19 |

**Command:** move

**Select objects:** *(Select one or more objects.)*

**Select objects:** *(Press* **Enter** *to end object selection.)*

**Specify base point or displacement:** *(Pick a point.)*

**Specify second point of displacement or <use first point as displacement>:** *(Pick a point.)*

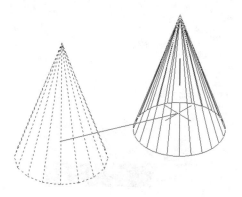

## COMMAND LINE OPTIONS

**Select objects** selects the objects to copy.

**Specify base point** indicates the starting point for the move.

**Displacement** specifies relative x,y,z displacement when you press ENTER at the next prompt.

**Specify second point of displacement** indicates the distance to move.

## RELATED COMMANDS

**Copy** copies of the selected objects.

**MlEdit** moves the vertices of a multiline.

**PEdit** moves the vertices of a polyline.

 # MRedo

**2004**   Reverses the effect of the Undo command (*short for Multiple REDO*).

| Command | Alias | Ctrl+ | F-key | Alt+ | Menu Bar | Tablet |
|---------|-------|-------|-------|------|----------|--------|
| mslide | ... | ... | ... | ... | ... | ... |

**Command:** mredo
**Enter number of actions or [All/Last]:** *(Enter an option.)*

## COMMAND LINE OPTIONS

**Enter number of actions** redoes the specifies number of steps.

**All** redoes all commands undone.

**Last** redoes the last command.

## RELATED COMMANDS

**Redo** redoes a single undo.

**U** undoes a single command.

**Undo** undoes one or more commands.

## TIPS

- The **MRedo** button on the toolbar lists the redoable actions:

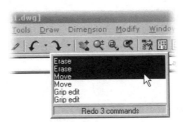

- This command allows you to undo several undoes, but does not allow you to skip over actions.

# MSlide

Ver.2.0 Saves the current viewport as SLD slide files on disk (*short for Make SLIDE*).

| Command | Alias | Ctrl+ | F-key | Alt+ | Menu Bar | Tablet |
|---------|-------|-------|-------|------|----------|--------|
| mslide | ... | ... | ... | ... | ... | ... |

**Command:** mslide

*Displays* **Create Slide File** *dialog box. Specify a file name, and then click* **Save**.

## COMMAND LINE OPTIONS
*None.*

## RELATED COMMANDS
**Save** saves the current drawing as a DWG-format drawing file.

**SaveImg** saves the current view as a TIFF, Targa, or GIF-format raster file.

**VSlide** displays an SLD-format slide file in AutoCAD.

## RELATED AUTODESK PROGRAM
**slidelib.exe** compiles a group of slides into an SLB-format slide library file.

## TIPS
• You view the slide with the **VSlide** command.

• Slides were a predecessor to viewing raster and vector images inside AutoCAD.

• Slide files are used to create the images in palette dialog boxes.

M Commands / 365

# MSpace

**Rel.11** Switches the drawing from paper space to model space (*short for Model SPACE*).

| Command | Alias | Ctrl+ | F-key | Alt+ | Menu Bar | Tablet |
|---------|-------|-------|-------|------|----------|--------|
| mspace | ms | ... | ... | ... | ... | L4 |

**Command:** mspace

*In model space, AutoCAD complains, "** Command not allowed in Model Tab **."*

*In paper space, AutoCAD switches to model space in layout mode, and highlights a viewport:*

**COMMAND LINE OPTIONS**

*None.*

**RELATED COMMANDS**

**PSpace** switches from model space to paper space.

**Model** switches from layout mode to model mode.

**Layout** switches from model mode to layout mode.

**RELATED SYSTEM VARIABLES**

**MaxActVp** specifies the maximum number of viewports with visible objects; default=64.

**TileMode** specifies the current setting of tiled viewports.

**TIPS**

- To switch quickly between paper space and model space, click the **MODEL** and **PAPER** buttons on the status bar:

| 4.0679, 0.4813, 0.0000 | SNAP | GRID | ORTHO | POLAR | OSNAP | OTRACK | LWT | MODEL |
| 4.0679, 0.4813, 0.0000 | SNAP | GRID | ORTHO | POLAR | OSNAP | OTRACK | LWT | PAPER |

*Click PAPER to switch from paper space to model space.*

- AutoCAD clears the selection set when moving between paper space and model space.

# MTEdit

**Rel.13** Edits mtext objects *(short for Multiline Text EDITor; undocumented command).*

| Command | Alias | Ctrl+ | F-key | Alt+ | Menu Bar | Tablet |
|---------|-------|-------|-------|------|----------|--------|
| mtedit | ... | ... | ... | ... | ... | ... |

**Command:** mtedit
**Select an MTEXT object:** *(Pick an mtext object.)*
  *Displays* **Mutiline Text Editor** *dialog box; see* **MText** *command.*

## COMMAND LINE OPTION
**Select an MTEXT object** selects one paragraph text object for editing.

## RELATED COMMANDS
**DdEdit** displays the text editor appropriate for the text object.

**Properties** changes the properties of an mtext object.

**MtProp** specifies the properties of an mtext object.

## RELATED SYSTEM VARIABLE
**MTextEd** specifies the name of the external text editor to place and edit multiline text.

## TIP
• This command displays the same dialog box as the **DdEdit** command when an mtext object is selected.

**MText**

<u>Rel.13</u> Creates multiline, or paragraph, text objects that fit the width defined by the boundary box (*short for Multline TEXT*).

| Command | Alias | Ctrl+ | F-key | Alt+ | Menu Bar | Tablet |
|---------|-------|-------|-------|------|----------|--------|
| mtext | t | ... | ... | DXM | Draw | J8 |
| | mt | | | | ⬙Text | |
| | | | | | ⬙Multiline Text | |
| -mtext | -t | | | | | |

**Command:** mtext
**Current text style: "Standard"   Text height: 0.20**
**Specify first corner:** *(Pick a point.)*
*AutoCAD displays the mtext bounding box:*

**Specify opposite corner or [Height/Justify/Line spacing/Rotation/Style/Width]:** *(Pick another point, or enter an option.)*
*Displays toolbar.*

**TOOLBAR OPTIONS**

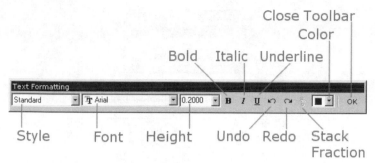

Close Toolbar
Color
Bold  Italic  Underline

Style  Font  Height  Undo  Redo  Stack
Fraction

**Style** selects a predefined text style; see Style command.
**Font** selects a TrueType (TTF) or AutoCAD (SHX) font name (default=TXT).
**Height** specifies the height of the text in units (default = 0.2 units).

**Bold** boldfaces the text, if allowed by the font.
**Italic** italicizes the text, if allowed by the font.
**Underline** underlines the text.
**Undo** undoes the last action.
**Redo** undoes the last undo.
**Stack Fraction** stacks a pair of characters separated by slash.
**Color** selects color for text; click Other Color to display Select Color dialog box.
**OK** closes the toolbar, and exits the MText command.

**TAB BAR OPTIONS**

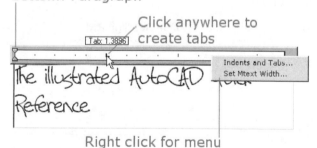

Drag Indent Markers
Top: First line
Bottom: Paragraph

Click anywhere to
create tabs

Right click for menu

**Indents and Tabs** displays Indents and Tabs dialog box.
**Set MText Width** displays Set MText Width dialog box.

## SET MTEXT WIDTH DIALOG BOX

**Width** changes the width of the mtext bounding box; as an alternative, you can change the boundary box's size by dragging its right and bottom borders.

## SHORTCUT MENU

*Right-click the text to display this menu:*

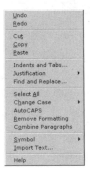

**Undo** undoes the last action.

**Redo** undoes the last undo.

**Cut** removes the selected text, and places it in the Clipboard.

**Copy** copies the selected text, and places it in the Clipboard.

**Paste** inserts text stored in the Clipboard.

**Indents and Tabs** displays the Indents and Tabs dialog box.

**Justification** selects a justification mode:

| Mode | Meaning |
| --- | --- |
| **TL** | Top left (default). |
| **TC** | Top center. |
| **TR** | Top right. |
| **ML** | Middle left. |
| **MC** | Middle center. |
| **MR** | Middle right. |
| **BL** | Bottom left. |
| **BC** | Bottom center. |
| **BR** | Bottom right. |

**Find and Replace** displays the Replace dialog box.

**Select All** selects all the text in the bounding box.

**Change Case** changes the case of the text:

- **UPPERCASE** changes selected characters to uppercase.
- **lowercase** changes selected characters to lowercase.

**AutoCAPS** places text as uppercase as it is typed.

**Remove Formatting** removes bold, italic, and underlinned formatting from selected text.

**Combine Paragraphs** combines selected text into a single paragraph.

**Symbol** inserts symbols in the text:

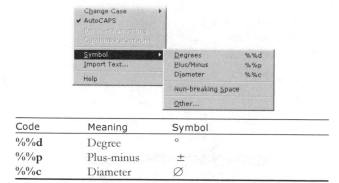

| Code | Meaning | Symbol |
|------|---------|--------|
| %%d | Degree | ° |
| %%p | Plus-minus | ± |
| %%c | Diameter | ∅ |

Click **Other** for **Character Map** dialog box.

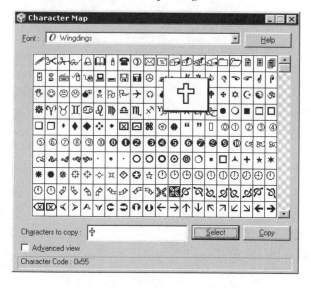

**Import Text** imports text from ASCII and RTF files; displays Open dialog box. *Caution!* The maximum size of text file is limited to 32KB.

## INDENTS AND TABS DIALOG BOX

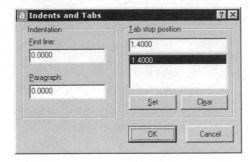

Indentation options
> **First line** specifies the indent distance for the first line of a paragraph; enter a negative number to create a hanging indent.
>
> **Paragraph** specifies the indent distance for the entire paragraph.

Tab Stop Position options
> **Set** adds the tab position.
>
> **Clear** removes the selected tab position.

## REPLACE DIALOG BOX

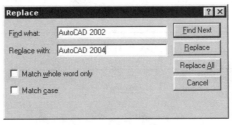

**Find what** specifies the text to search for. This dialog box searches only text within the bounding box; to search for text in the entire drawing, use the Find command.

**Replace with** specifies the text to replace with; leave blank to search only.

**Match whole word only**

> ☑ matches the entire word(s).
>
> ☐ matches parts of the word(s).

**Match case**

> ☑ matches the case of the words.
>
> ☐ ignores the word case.

**Find Next** finds the next occurance of the word(s).

**Replace** replaces the found occurance.

**Replace All** replaces all occurances.

**Cancel** dismisses the dialog box.

## -MTEXT Command

**Command:** -mtext
**Current text style: STANDARD. Text height: 0.2000**
**Specify first corner:** *(Pick a point.)*
**Specify opposite corner or [Height/Justify/Rotation/Style/Width]:** *(Pick another point.)*
**MText:** *(Enter text.)*
**MText:** *(Press **Enter** to end the command.)*

### COMMAND LINE OPTIONS

**Height** specifies the height of UPPERCASE text *(default = 0.2 units)*.

**Justify** specifies a justification mode.

**Rotation** specifies the rotation angle of the boundary box.

**Style** selects the text style for multiline text (default = STANDARD).

**Width** sets the width of the boundary box; a width of 0 eliminates the boundary box.

### RELATED COMMANDS

**Properties** changes all aspects of mtext.

**MtProp** changes properties of multiline text.

**MtEdit** edits mtext.

**PasteSpec** pastes formatted text from the Clipboard into the drawing.

**Style** creates a named text style from a font file.

### RELATED SYSTEM VARIABLE

**MTextEd** names the external text editor for placing and editing multiline text.

### TIPS

- Use the **MTextEd** system variable to define a different text editor.

- The **Import Text** option is limited to ASCII (unformatted) and RTF (rich text format) text files no more than 32KB in size.

- To import Word documents, copy the text to the Clipboard, and then press CTRL+V in the MText editor. Most, but not all, formatting is retained.

- To import formatted text, copy text from the word processor to the Clipboard, then use AutoCAD's **PasteSpec** command.

- To link text in the drawing with a word processor, use the **InsertObj** command. When the word processor updates, the linked text is updated in the drawing.

# MtProp

**Rel.13** Changes the properties of multiline text *(short for Multiline Text PROPerties; undocumented command)*.

| Command | Alias | Ctrl+ | F-key | Alt+ | Menu Bar | Tablet |
|---------|-------|-------|-------|------|----------|--------|
| mtprop | ... | ... | ... | ... | ... | ... |

**Command:** mtprop
**Select an MText object:** *(Pick an mtext object.)*
  *Displays **Multiline Text Editor** dialog box; see the **MText** command.*

## COMMAND LINE OPTIONS
  *See the **MText** command.*

## RELATED COMMANDS
  **DdEdit** edits multiline text.
  **MText** places multiline text.
  **Style** creates a named text style from a font file.

# Multiple

<u>V. 2.5</u>  Automatically repeats commands that do not repeat on their own.

| Command | Alias | Ctrl+ | F-key | Alt+ | Menu Bar | Tablet |
|---------|-------|-------|-------|------|----------|--------|
| multiple | ... | ... | ... | ... | ... | ... |

**Command:** multiple
**Enter command name to repeat:** *(Enter command name.)*

*This command can be used as a command modifier:*

**Command:** multiple circle
**3P/2P/TTR/<Center point>:** *(Pick a point, or enter an option.)*
**Diameter/<Radius>:** *(Enter an option.)*
**circle 3P3P/2P/TTR/<Center point>:** *(Pick a point, or enter an option.)*
**Diameter/<Radius>:** *(Enter an option.)*
**circle 3P3P/2P/TTR/<Center point>:** *(Press **Esc** to end command.)*

## COMMAND LINE OPTIONS

**Enter command name to repeat** specifies the name of the command to repeat.

ESC stops the command from automatically repeating itself.

## COMMAND INPUT OPTIONS

SPACEBAR repeats the previous command.

CLICK repeats a command by clicking on any blank spot of the tablet menu.

## RELATED COMMANDS

**Redo** undoes an undo.

**U** undoes the previous command; undoes one multiple command at a time.

## RELATED COMMAND MODIFIERS

**'** *(apostrophe)* allows the use of some commands within another command.

**.** *(period)* forces the use of an undefined command.

**-** *(dash)* forces the display of prompts on the command line for some commands.

**+** *(plus)* prompts for the tab number of tabbed dialog box.

**_** *(underscore)* uses the English command in an international version of AutoCAD.

**(** *(open parenthesis)* executes an AutoLISP function on the command line.

**$(** *(dollar and parenthesis)* executes a Diesel function on the command line.

## TIPS

• Use the **Multiple** command to repeat commands that do not repeat on their own.

• **Multiple** repeats the command name only; it does not repeat command options.

# MView

Rel.11 Creates and manipulates overlapping viewports *(short for Make VIEWports).*

| Command | Alias | Ctrl+ | F-key | Alt+ | Menu Bar | Tablet |
|---------|-------|-------|-------|------|----------|--------|
| mview   | mv    | R     | ...   | ...  | ...      | M4     |

**Command:** mview

*In model space, AutoCAD complains, "** Command not allowed in Model Tab **."*

*In paper space, AutoCAD prompts:*

**Specify corner of viewport or**
**[ON/OFF/Fit/Shadeplot/Lock/Object/Polygonal/Restore/2/3/4]<Fit>:** *(Pick a point, or enter an option.)*
**Specify opposite corner:** *(Pick a point.)*
**Regenerating drawing.**

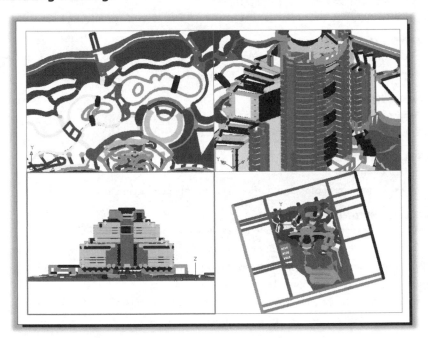

## COMMAND LINE OPTIONS

**Specify corner of viewport** indicates the first point of a single viewport (default).

**Fit** creates a single viewport that fits the screen.

**Shadeplot** creates a hidden-line or shaded view during plotting and printing.

**Lock** locks the selected viewport.

**Object** converts a circle, closed polyline, ellipse, spline, or region into a viewport.

**OFF** turns off a viewport.

**ON** turns on a viewport.

**Polygonal** creates a multisided viewport of straight lines and arcs.

**Restore** restores a saved viewport configuration.

2 options
**Enter viewport arrangement [Horizontal/Vertical] <Vertical>:** *(Enter an option.)*
**Specify first corner or [Fit] <Fit>:** *(Pick a point, or enter an option.)*

**Horizontal** stacks two viewports.

**Vertical** places two viewports side-by-side (default).

3 options
**[Horizontal/Vertical/Above/Below/Left/Right]<Right>:** *(Enter an option.)*
**Specify first corner or [Fit] <Fit>:** *(Pick a point, or enter an option.)*

**Horizontal** stacks the three viewports.

**Vertical** places three side-by-side viewports.

**Above** places two viewports above the third.

**Below** places two viewports below the third.

**Left** places two viewports to the left of the third.

**Right** places two viewports to the right of the third *(default)*.

4 options
**Specify first corner or [Fit] <Fit>:** *(Pick a point, or enter an option.)*

**Fit** creates four identical viewports that fit the viewport.

**First Point** indicates the area of the four viewports (default).

Shadeplot options
**Shade plot? [As displayed/Wireframe/Hidden/Rendered] <As displayed>:**
*(Enter an option.)*

**As displayed** plots the drawing as displayed.

**Wireframe** plots the drawing as a wireframe.

**Hidden** plots the drawing with hidden lines removed.

**Rendered** plots the drawing rendered.

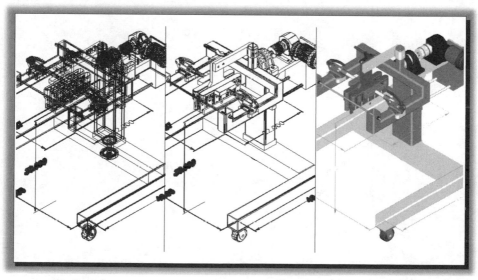

Wireframe          Hidden-line          Rendered

## RELATED COMMANDS

**Layout** creates new layouts.

**MSpace** switches to model space.

**PSpace** switches to paper space before creating viewports.

**RedrawAll** redraws all viewports.

**RegenAll** regenerates all viewports.

**VpLayer** controls the visibility of layers in each viewport.

**VPorts** creates tiled viewports in model space.

**Zoom** zooms a viewport relative to paper space via the XP option.

## RELATED SYSTEM VARIABLES

**CvPort** specifies the number of the current viewport.

**MaxActVp** controls the maximum number of visible viewports:

| MaxActVP | Meaning |
|----------|---------|
| 1 | Minimum. |
| 64 | Default. |
| 32767 | Maximum. |

**TileMode** controls the availability of overlapping viewports.

## TIPS

- Although the system variable **MaxActVp** limits the number of simultaneously-visible viewports, the **Plot** command plots all viewports.

- **TileMode** must be set to zero to switch to paper space and use the **MSpace** command.

- **Snap**, **Grid**, **Hide**, **Shade**, and so on can be set separately in each viewport.

- The preset viewports created by the **MView** command have these shapes:

Fit option

**Fit**     Creates a single viewport.

2 options

**Horizontal**     Creates one viewport over another viewport.

**Vertical**     Creates one viewport beside another (default).

3 options

**Horizontal**     Creates three viewports over each other.

**Vertical**     Creates three viewports side-by-side.

**Above**     Creates one viewport over of two viewports.

**Below**     Creates one viewport below two viewports.

**Left**     Creates one viewport left of two viewports.

**Right**     Creates one viewport right of two viewports (default).

4 option

**4**     Splits the current viewport into four viewports.

- Press CTRL+R to switch between viewports.

# MvSetup

**Rel. 11** Quickly sets up a drawing, complete with a predrawn border. Optionally sets up multiple viewports, sets the scale, and aligns views in each viewport (*short for Model View SETUP*).

| Command | Alias | Ctrl+ | F-key | Alt+ | Menu Bar | Tablet |
|---------|-------|-------|-------|------|----------|--------|
| mvsetup | mvs | ... | ... | ... | ... | ... |

**Command:** mvsetup

*When in model space:*

**Enable paper space? [No/Yes] <Y>:** *(Type* **Y** *or* **N**.*)*

*Command prompts in model tab (not paper space):*

**Enter units type [Scientific/Decimal/Engineering/Architectural/Metric]:** *(Enter an option.)*

**Enter the scale factor:** *(Enter a value.)*

**Enter the paper width:** *(Enter a value.)*

**Enter the paper height:** *(Enter a value.)*

*Command prompts in layout mode (paper space):*

**Enter an option [Align/Create/Scale viewports/Options/Title block/Undo]:** *(Enter an option.)*

## COMMAND LINE OPTIONS

Align options

*Pans the view to align a base point with another viewport.*

**Enter an option [Angled/Horizontal/Vertical alignment/Rotate view/Undo]:** *(Enter an option.)*

**Angled** specifies the distance and angle from a base point to a second point.

**Horizontal** aligns views horizontally with a base point in another viewport.

**Vertical alignment** aligns views vertically with a base point in another viewport.

**Rotate view** rotates the view about a base point.

**Undo** undoes the last action.

Create options

**Enter option [Delete objects/Create viewports/Undo] <Create>:** *(Enter an option.)*

**Delete objects** erases existing viewports.

**Create viewports** creates viewports in these configurations:

| Layout | Meaning |
|--------|---------|
| 0 | No layout. |
| 1 | Single viewport. |
| 2 | Standard engineering layout. |
| 3 | Array viewports along x and y axes. |

**Undo** undoes the last action.

. . . . . . . . . . . . . . . . . . . . . . . . . . . . . . . . . . . . . . . . . . . . . . . . . . . . . . . . . . .

Scale Viewports options

**Select the viewports to scale...**

**Select objects:** *(Pick a viewport.)*

**Select objects:** *(Press* **Enter** *to end object selection.)*

**Set the ratio of paper space units to model space units...**

**Enter the number of paper space units <1.0>:** *(Enter a value.)*

**Enter the number of model space units <1.0>:** *(Enter a value.)*

**Select objects** selects one or more viewports.

**Enter the number of paper space units** scales the objects in the viewport with respect to drawing objects.

**Enter the number of model space units** scales the objects in the viewport with respect to drawing objects.

Options options

**Enter an option [Layer/LImits/Units/Xref] <exit>:** *(Enter an option.)*

**Layer** specifies the layer name for the title block.

**Limits** specifies whether to reset limits after title block insertion.

**Units** specifies inch or millimeter paper units.

**Xref** specifies whether title is inserted as a block or as an external reference.

Title Block options

**Enter title block option [Delete objects/Origin/Undo/Insert] <Insert>:** *(Enter an option.)*

**Delete objects** erases an existing title block from the drawing.

**Origin** relocates the origin.

**Undo** undoes the last action.

**Insert** displays the available title blocks.

**RELATED SYSTEM VARIABLE**

**TileMode** specifies the current setting of TileMode.

**RELATED FILES**

*mvsetup.dfs* is the MvSetup default settings file.

*acadiso.dwg* is a template drawing with ISO (international standards) defaults.

*Plus all .dwt template drawings.*

**RELATED COMMANDS**

**LayoutWizard** sets up the viewports via a "wizard."

## TIPS

- When option **2 (Std. Engineering)** is selected at the **Create** option, the following views are created (counterclockwise from upper left):

  Top view.

  Isometric view.

  Front view.

  Right view.

- To create the title block, **MvSetup** searches the path specified by the **AcadPrefix** variable. If the appropriate drawing cannot be found, **MvSetup** creates the default border.

- **MvSetup** makes use the following predefined title blocks:

  | | |
  |---|---|
  | 0: | None |
  | 1: | ISO A4 Size(mm) |
  | 2: | ISO A3 Size(mm) |
  | 3: | ISO A2 Size(mm) |
  | 4: | ISO A1 Size(mm) |
  | 5: | ISO A0 Size(mm) |
  | 6: | ANSI-V Size(in) |
  | 7: | ANSI-A Size(in) |
  | 8: | ANSI-B Size(in) |
  | 9: | ANSI-C Size(in) |
  | 10: | ANSI-D Size(in) |
  | 11: | ANSI-E Size(in) |
  | 12: | Arch/Engineering (24 x 36in) |
  | 13: | Generic D size Sheet (24 x 36in) |

- The metric A0 size is similar to the imperial E-size, while the metric A4 size is similar to A-size.

- You can add your own title block with the **Add** option. Before doing so, create the title block as an AutoCAD drawing.

- This command provides the following preset scales (scale factor shown in parentheses):

| Architectural Scales | Scientific Scales | Decimal Scales | Engineering Scales | Metric Scales |
|---|---|---|---|---|
| **(480) 1/40"=1'** | **(4.0) 4 TIMES** | **(4.0) 4 TIMES** | **(120) 1"=10'** | **(5000) 1:5000** |
| **(240) 1/20"=1'** | **(2.0) 2 TIMES** | **(2.0) 2 TIMES** | **(240) 1"=20'** | **(2000) 1:2000** |
| **(192) 1/16"=1'** | (1.0) FULL | (1.0) FULL | **(360) 1"=30'** | **(1000) 1:1000** |
| **(96) 1/8"=1'** | **(0.5) HALF** | **(0.5) HALF** | **(480) 1"=40'** | **(500) 1:500** |
| **(48) 1/4"=1'** | **(0.25) QUARTER** | **(0.25) QUARTER** | **(600) 1"=50'** | **(200) 1:200** |
| **(24) 1/2"=1'** | | | **(720) 1"=60'** | **(100) 1:100** |
| **(16) 3/4"=1'** | | | **(960) 1"=80'** | **(75) 1:75** |
| **(12) 1"=1'** | | | **(1200) 1"=100'** | **(50) 1:50** |
| **(4) 3"=1'** | | | | **(20) 1:20** |
| **(2) 6"=1'** | | | | **(10) 1:10** |
| (1) FULL | | | | **(5) 1:5** |
| | | | | (1) FULL |

# Using MvSetup

**MvSetup** has many options, but does not present them in a logical fashion. To set up a drawing with **MvSetup**, follow these basic steps:

**Step 1**

Start the **MvSetup** command:

> **Command:** mvsetup
> **Enter an option [Align/Create/Scale viewports/Options/Title block/ Undo]:** *(Type* **C.***)*

**Step 2**

Select options:

> **Enter an option [Layer/LImits/Units/Xref] <exit>:** *(Type* **L.***)*

Decide on the layer for the title block with the **Layer** option. Specify the paper space units with the **Units** option.

**Step 3**

Place title block:

> **Enter title block option [Delete objects/Origin/Undo/Insert] <Insert>:** *(Type* **I.***)*

Place the title block with the **Title block** option's **Insert** option.

**Step 4**

Create viewports:

> **Enter option [Delete objects/Create viewports/Undo] <Create>:** *(Type* **C.***)*

Set up the viewports with the **Create** option's **Create viewports** option. For standard drawings, select option **#2, Std. Engineering**.

**Step 5**

Scale the viewports. Make the object the same size in all four viewports with the **Scale viewports** option. When you are prompted to 'Select objects', select the four *viewports*, not the objects in the viewports.

**Step 6**

Align the views in each viewport with the **Align** option.

- You can interrupt the **MvSetup** command at any time with the ESC key, then and resume the command to complete the setup.

- Save your work when done!

# New

<u>Rel.12</u>   Starts new drawings from scratch, from template drawings, or through step-by-step drawing setup "wizards."

| Command | Alias | Ctrl+ | F-key | Alt+ | Menu Bar | Tablet |
|---------|-------|-------|-------|------|----------|--------|
| new | ... | N | ... | FN | File | T24 |
| | | | | | ⁖New | |

**Command:** new

*Display depends on **Startup** option in **General Options** section of the **System** tab (**Options** dialog box):*

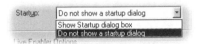

*Startup:*

- *Do not show a startup dialog option displays the Select Template dialog box.*
- *Show Startup dialog option displays the Startup or Create New Drawing dialog boxes.*

## DIALOG BOX OPTIONS

 **Open a Drawing** view

*Not found in the **Create New Drawing** variation of this dialog box.*

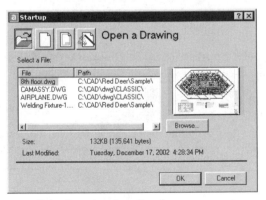

**Select a File** selects one of the four drawings listed.

**Browse** displays the Select File dialog box; see the Open command.

**Start from Scratch** view

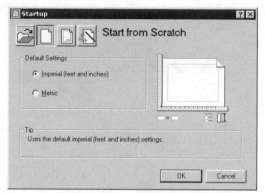

**English** creates a new drawing based on the *acad.dwt* (English units) template file.

**Metric** creates a new drawing based on the *acadiso.dwt* (metric units) template file.

**Use a Template** view

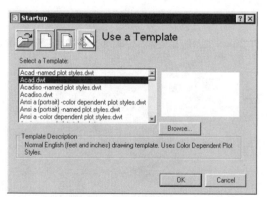

**Select a Template** creates a new drawing based on the selected *.dwt* template file.

**Browse** displays the Select a Template File dialog box.

**Use a Wizard** view

Select a Wizard

**Select a Wizard**

- **Advanced Setup** sets up a new drawing in several steps.
- **Quick Setup** sets up a new drawing in two steps.

. . . . . . . . . . . . . . . . . . . . . . . . . . . . . . . . . . . . . . . . . . . . . .

## Quick Setup wizard

**Units** page

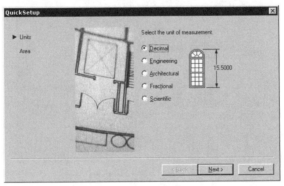

**Decimal** displays units in decimal (or "metric") notation (default): 123.5000.
**Engineering** displays units in feet and decimal inches: 10'-3.5000".
**Architectural** displays units in feet, inches, and fractional inches: 10' 3-1/2".
**Fractional** displays units in inches and fractions: 123 1/2.
**Scientific** displays units in scientific notation: 1.235E+02.

*Buttons*
 **Cancel** cancels the wizard, and returns to the previous drawing.
 **Back** moves back one step.
 **Next** moves forward one step.

**Area** page

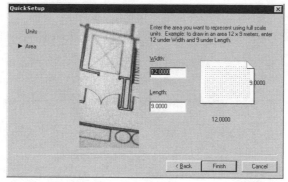

**Width** specifies the width of the drawing in real-world (not scaled) units; default = 12 units.
**Length** specifies the length or depth of the drawing in real-world units; default = 9 units.

## Advanced Setup wizard

**Units** page

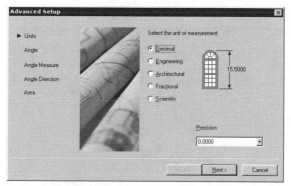

**Decimal** displays units in decimal (or "metric") notation (default): 123.5000.
**Engineering** displays units in feet and decimal inches: 10'-3.5000".
**Architectural** displays units in feet, inches, and fractional inches: 10' 3-1/2".
**Fractional** displays units in inches and fractions: 123 1/2.
**Scientific** displays units in scientific notation: 1.235E+02.
**Precision** selects the precision of display up to 8 decimal places or 1/256.

**Angle** page

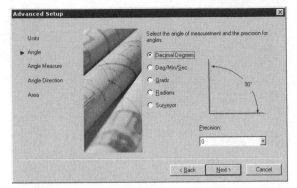

**Decimal Degrees** displays decimal degrees (default): 22.5000.

**Deg/Min/Sec** displays degrees, minutes, and seconds: 22 30.

**Grads** displays grads: 25g.

**Radians** displays radians: 25r.

**Surveyor** displays surveyor units: N 25d0'0" E.

**Precision** selects a precision ranging up to 8 decimal places.

**Angle Measure** page

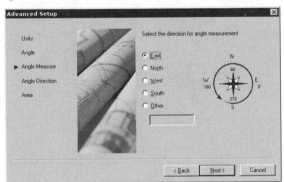

**East** specifies that zero degrees points East (default).

**North** specifies that zero degrees points North.

**West** specifies that zero degrees points West.

**South** specifies that zero degrees points South.

**Other** specifies any of the 360 degrees as zero degrees.

## Angle Direction page

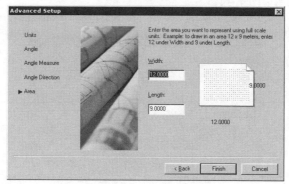

**Counter-Clockwise** measures positive angles counterclockwise from 0 degrees (default).

**Clockwise** measures positive angles clockwise from 0 degrees.

## Area page

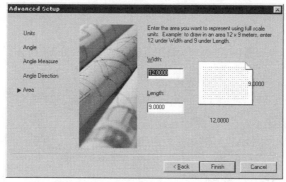

**Width** specifies the width of the drawing in real-world (not scaled) units; default = 12 units.

**Length** specifies the length or depth of the drawing in real-world units; default = 9 units.

## Command Line Switches

*Switches used by **Target** field on **Shortcut** tab in the AutoCAD desktop icon's **Properties** dialog box:*

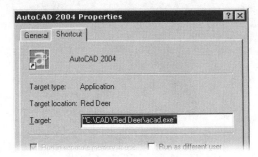

**/b** runs a script file after AutoCAD starts; uses the following format:

**acad.exe "\acad 2004\drawing.dwg" /b "file name.scr"**

**/c** specifies the path for alternative hardware configuration file; default = *acad2004.cfg*.

**/nologo** suppresses the display of the AutoCAD logo screen.

**/p** specifies a user-defined profile to customize AutoCAD's user interface.

**/r** restores the default pointing device.

**/s** specifies additional support folders; maximum is 15 folders, with each folder name separated by a semicolon.

**/t** specifies the *.dwt* template drawing to use.

**/v** specifies the named view to display upon startup of AutoCAD.

### RELATED COMMANDS

**QNew** starts a new drawing based on a predetermined template file.

**SaveAs** saves the drawing in *.dwg* or *.dwt* formats; creates template files.

### RELATED SYSTEM VARIABLES

**DbMod** indicates whether the drawing has changed since being loaded.

**DwgPrefix** indicates the path to the drawing.

**DwgName** indicates the name of the current drawing.

**FileDia** displays prompts at the 'Command' prompt.

### RELATED FILES

*wizard.ini* holds the names and descriptions of template files.

*\*.dwt* are template files stored in *.dwg* format.

### TIPS

• Until you give the drawing a name, AutoCAD names it *drawing1.dwg*.

• The default template drawing is *acad.dwg*.

• Edit and save *.dwt* template drawings to change the defaults for new drawings.

• When you press CTRL+N, AutoCAD's behavior differs from Microsoft Office programs: AutoCAD displays the **Startup** dialog box; Office programs display a new document that takes on the properties of the current document.

• The **Today** window was removed with AutoCAD 2004.

# Offset

**V. 2.5**  Draws parallel lines, arcs, circles and polylines; repeats automatically until cancelled.

| Command | Alias | Ctrl+ | F-key | Alt+ | Menu Bar | Tablet |
|---------|-------|-------|-------|------|----------|--------|
| offset | o | ... | ... | MS | Modify | V17 |
| | | | | | ⌁Offset | |

**Command:** offset
**Specify offset distance or [Through] \<Through\>:** *(Enter a number, or type **T**.)*
**Select object to offset or \<exit\>:** *(Select an object.)*
**Specify point on side to offset:** *(Pick a point.)*
**Select object to offset or \<exit\>:** *(Select another object, or press **Enter** to end the command.)*

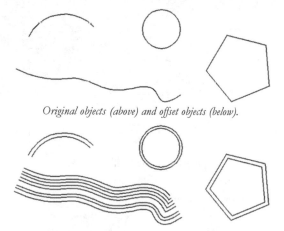

*Original objects (above) and offset objects (below).*

## COMMAND LINE OPTIONS

**Offset distance** specifies the perpendicular distance to offset.

**Through** indicates the offset distance.

ESC exits the command.

Through options
**Select object to offset or \<exit\>:** *(Select an object.)*
**Specify through point:** *(Pick a point.)*

**Select object to offset** selects the object to be offset.

**Specify through point** specifies the offset distance.

## RELATED COMMANDS

**Copy** creates one or more copies of a group of objects.

**MLine** draws up to 16 parallel lines.

## RELATED SYSTEM VARIABLE

**OffsetDist** specifies the current offset distance.

# OleLinks

**Rel.13** Changes, updates, and cancels OLE links between the drawing and other Windows applications (*short for Object Linking and Embedding LINKS*).

| Command | Alias | Ctrl+ | F-key | Alt+ | Menu Bar | Tablet |
|---------|-------|-------|-------|------|----------|--------|
| olelinks | ... | ... | ... | EO | Edit | ... |
| | | | | | ⬐OLE Links | |

**Command:** olelinks

*When no OLE links are in the drawing, the command does nothing.*

*When at least one OLE object is in the drawing, displays dialog box:*

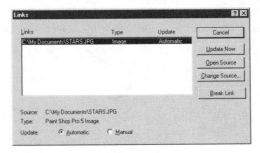

## DIALOG BOX OPTIONS

**Links** displays a list of linked objects: source filename, type of file, and update mode — automatic or manual.

**Update** selects either automatic or manual updates.

**Update Now** updates selected links.

**Open Source** starts the source application program.

**Break Link** cancels the OLE link; keeps the object in place.

**Change Source** displays the Change Source dialog box:

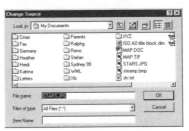

## RELATED COMMANDS

**InsertObj** places an OLE object in the drawing.

**PasteSpec** pastes objects from the Clipboard as linked objects in the drawing.

## RELATED WINDOWS COMMANDS

**Edit | Copy** copies objects from the source application to the Clipboard.

**File | Update** updates the linked object in the source application.

. . . . . . . . . . . . . . . . . . . . . . . . . . . . . . . . . . . . . . . . . . . . . .

# OleScale

<u>2000</u>  Modifies the properties of OLE objects.

| Command | Alias | Ctrl+ | F-key | Alt+ | Shortcut Menu | Tablet |
|---------|-------|-------|-------|------|---------------|--------|
| olescale | ... | ... | ... | ... | ... | ... |

**Command:** olescale

*Displays dialog box:*

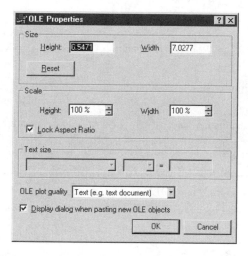

## DIALOG BOX OPTIONS

Size options

**Height** changes the height of the OLE object; displays the current height.

**Width** changes the width of the OLE object; displays the current width.

**Reset** resets the OLE object to its original size when first inserted into the drawing.

Scale options

**Height** changes the height of the OLE object by a percentage of the original height.

**Width** changes the width of the OLE object by a percentage of the original width.

**Lock Aspect Ratio** changes the Width size and ratio to match the Height, and vice versa.

Text Size options

**Font** displays the fonts used by the OLE object, if any.

**Point Size** displays the text height in point sizes; limited to the point sizes available for the selected font, if any (1 point = $^1/_{72}$ inch).

**Text Height** specifies the text height in drawing units.

**OLE Plot Quality** determines the quality of the pasted object when plotted:

| Plot Quality | Meaning |
| --- | --- |
| **Line Art** | Text is plotted as text; no colors or shading are preserved; some graphical images are not plotted, while others are plotted as monochrome images (black and white only, no shades of gray). |
| **Text** | All text formatting is preserved; text is plotted as graphics, which plots less cleanly than the **Line Art** setting; graphics are plotted less cleanly and at reduced colors than **Graphics** and **Photograph** settings. |
| **Graphics** | Graphics are plotted at a reduced number of colors (fewer shades of gray or "posterization"); all text formatting is preserved; text is plotted as graphics, but more cleanly than **Text** and **Photograph** settings. |
| **Photograph** | Graphics are plotted at reduced resolution and colors; all text formatting is preserved; text is plotted as graphics. |
| **High Quality** | Graphics are plotted at full resolution and colors. |
| **Photograph** | All text formatting preserved; text plotted more eanly than with **Text** and **Photograph** settings. |

### Display dialog when pasting new OLE object:

☑ displays the OLE Properties dialog box automatically when inserting an OLE object.

☐ does not display the dialog box.

## RELATED SYTSTEM VARIABLES

**OleHide** toggles the display of OLE objects in the drawing and in plots.

**OleQuality** specifies the quality of display and plotting of embedded OLE objects.

**OleStartup** loads the source application of an embedded OLE object for plots.

## RELATED COMMANDS

**InsertObj** inserts an OLE object into the drawing.

**OleLinks** modifies the link between the object and its source.

**PasteSpec** allows you to paste an object with a link.

## TIPS

• The **U** and **Undo** commands do not reverse the effect of changes made by the **OLE Properties** dialog box. Instead, right-click the OLE object and select **Undo** from the shortcut menu.

• Change **OleStartup** to 1 to load the OLE source application, which may help improve the plot quality of OLE objects.

• The **OLE Plot Quality** list box determines the quality of the pasted object when plotted. I recommend the **Line Art** setting for text, unless the text contains shading and other graphical effects.

# Oops

<u>V. 1.0</u>  Restores the last-erased group of objects; restores objects removed by the **Block** and **-Block** commands.

| Command | Alias | Ctrl+ | F-key | Alt+ | Menu Bar | Tablet |
|---------|-------|-------|-------|------|----------|--------|
| oops | ... | ... | ... | ... | ... | ... |

**Command:** oops

## COMMAND LINE OPTIONS
*None.*

## RELATED COMMANDS
**Block**: use **Oops** after the **Block** command to return erased objects.

**Erase**: use **Oops** after the **Erase** command to return erased objects.

**U** undoes the most recent command.

## TIPS
* **Oops** only restores the most-recently erased object; use the **Undo** command to restore earlier objects.

* Use **Oops** to bring back objects after turning them into a block with the **Block** and **WBlock** commands.

 # Open

**Rel.12**  Loads one or more drawings and DXF files into AutoCAD.

| Command | Alias | Ctrl+ | F-key | Alt+ | Menu Bar | Tablet |
|---------|-------|-------|-------|------|----------|--------|
| open | openurl | O | ... | FO | File | T25 |
| | | | | | ↳Open | |

**Command:** open

*Displays dialog box:*

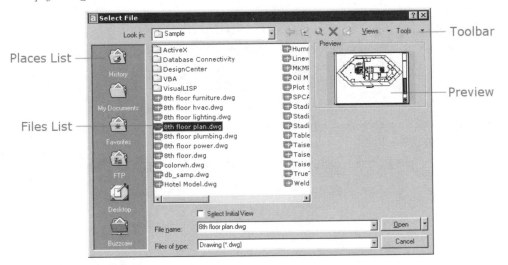

## DIALOG BOX OPTIONS

**Look in** selects the network drive, hard drive, or folder (subdirectory).

**Preview** displays preview image of AutoCAD drawings.

**Select initial view** selects a named view from a dialog box, if the drawing has saved views; after drawing is opened, AutoCAD displays the Select Initial View dialog box listing named views :

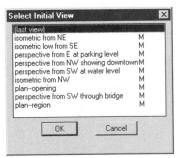

**M** indicates a view created in model space.

**P** indicates a view created in paper space (layout mode).

**File name** specifies the name of the drawing.

**Files of type** specifies the type of file:

- **Drawing (*.dwg)** AutoCAD drawing file.
- **Standard (*.dws)** drawing standards file; see the Standards command.
- **DXF (*.dxf)** drawing interchange file; see the DxfIn command.
- **Drawing Template File (*.dwt)** template drawing file; see the New command.

**Open** opens the selected drawing file(s); to open more than one drawing at a time:

- In the files list, hold down the SHIFT key to select a continuous range of files:

- Hold down the CTRL key to select two or more non-continuous files:

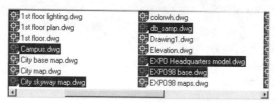

**Cancel** dismisses the dialog box without opening a file.

Open button options

**Open** opens the drawing.

**Open as read-only** loads the drawing, but you cannot save changes to the drawing except under another file name. AutoCAD displays "Read Only" on the title bar:

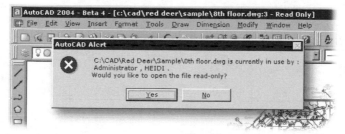

**Partial Open** loads selected layers or named views; displays Partial Open dialog box; not available for *.dxf* and template files.

**Partial Open Read-Only** partially loads the drawing in read-only mode.

. . . . . . . . . . . . . . . . . . . . . . . . . . . . . . . . . . . . . . . . . . . . . . . . . . . . . . . .

**Partial Open** dialog box

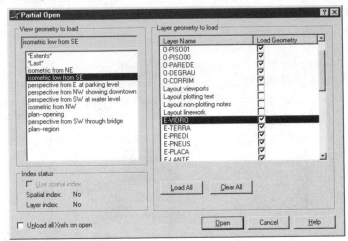

**View geometry to load** selects the model space views to load; paper space views are not available for partial loading.

## Layer Geometry to Load options

**Load Geometry** selects the layers to load.

**Load All** loads all layers.

**Clear All** deselects all layer names.

## Index Status options

*Available only when the drawing was saved with spatial indices.*

**Use spatial index** determines whether to use the spatial index for loading, if available.

**Spatial Index** indicates whether the drawing contains the spatial index.

**Layer Index** indicates whether the drawing contains the layer index.

## Additional options

**Unload all xrefs on open** loads externally-referenced drawings when opening the drawing.

**Open** opens the drawing, and then partially loads the geometry.

## TOOLBAR ICONS

**Back** returns to the previous folder (keyboard shortcut **ALT+1**).

**Up** moves up one level to the next folder or drive (**ALT+2**).

**Search the Web** displays the Browse the Web window (**ALT+3**); see Browser command.

**Delete** removes the selected file(s); does not delete folders or drives (**DEL**).

**Create New Folder** creates new folders (**ALT+5**).

**Views** provides display options:

- **List** displays the file and folder names only.
- **Details** displays file and folder names, type, size, and date.
- **Thumbnails** displays thumbnail images of *.dwg* files.
- **Preview** toggles display of the preview window.

**Tools** provides file-oriented tools:

- **Find** displays the Find dialog box for searching files.
- **Locate** searches for the file along AutoCAD's search paths.
- **Add/Modify FTP Locations** displays a dialog box for storing the logon names and passwords for FTP (file transfer protocol) sites.
- **Add Current Folder to Places** adds the selected folders to the places sidebar.
- **Add to Favorites** adds the selected files and folders to the Favorites list.

## PLACES LIST

**History** displays files opened by AutoCAD during the last four weeks.

**My Documents** displays files and folders in the *my documents* folder.

**Favorites** displays files and folders in the *favorites* folder.

**FTP** displays the **FTP Locations** list.

**Desktop** displays the contents of the \\*desktop* folder.

**Buzzsaw.com** goes to the www.buzzsaw.com Web site.

The **Point A** and **RedSpark** items were removed from AutoCAD 2004.

## SHORTCUT MENUS

### Places List menu
*Right-click icons in the Places List:*

**Remove** removes a folder from the list.

**Add Current Folder** adds the selected folder to the list; you can also drag a folder from the file list into the places list.

**Add** displays the Add Places Item dialog box.

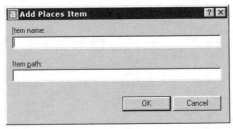

**Properties** displays the Places Item Properties dialog box, which is identical to the Add Places Item dialog box, and is available only for items you add.

**Restore Standard Folders** restores the folders shown above.

### File List menu
*Right-click the file list without selecting a file or folder:*

**View** switches between filename views: large icons, small icons, list, thumbnails, and details.

**Arrange Icons** arranges icons by name, type, size, and date.

**Line Up Icons** places icons in an orderly pattern.

**Refresh** updates the folder listing.

**Paste** pastes a file from the Clipboard.

**Paste Shortcut** pastes a file from the Clipboard as a shortcut.

**Undo Copy** undoes the copy-paste operation; available only after a copy or paste.

**New** creates a new folder (subdirectory) or shortcut.

**Properties** displays the Properties dialog box of the folder selected by the Look in list.

*Right-click a file or folder name:*

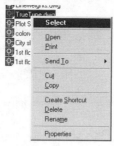

**Select** opens the drawing in AutoCAD.

**Open** opens the drawing in AutoCAD.

**Print** does not work.

**Send To** copies the file to another drive; may not work with some software.

**Cut** cuts the file to the Clipboard.

**Copy** copies the file to the Clipboard.

**Create Shortcut** creates a shortcut icon for the selected file.

**Delete** erases the file; displays a warning dialog box.

**Rename** renames the file; displays a warning dialog box if you change the extension.

**Properties** displays the Properties dialog box in read-only mode; use the DwgProps command within AutoCAD to change the settings.

### RELATED SYSTEM VARIABLES

**DbMod** indicates whether the drawing has been modified.

**DwgCheck** checks if the drawing was last edited by AutoCAD.

**DwgName** contains the drawing's filename.

**DwgPrefix** contains the drive and folder of the drawing.

**DwgTitled** indicates whether the drawing has a name other than *drawing1.dwg*.

**FullOpen** indicates whether the drawing is fully or partially opened.

### RELATED COMMANDS

**FileOpen** opens drawings without the dialog box.

**SaveAs** saves drawings with new names.

**PartiaLoad** loads additional portions of partially-opened drawings.

### TIPS

- A drawing can be opened by dragging its filename from Explorer into AutoCAD. When the dragged into an open drawing, it is inserted as a block; when dragged to AutoCAD's titlebar, it is opened as a drawing.

- DXF and template files cannot be partially opened.

- After a drawing is partially opened, use **PartiaLoad** to load additional parts of the drawing.

- When a partially-opened drawing contains a bound xref, only the portion of the xref defined by the selected view is bound to the partially-open drawing.

### Removed Command

**OpenUrl** was removed from AutoCAD 2000; it was replaced by **Open**'s FTP option.

# Options

2000 Sets system and user preferences.

| Commands | Aliases | Ctrl+ | F-key | Alt+ | Menu Bar | Tablet |
|----------|---------|-------|-------|------|----------|--------|
| options | op | ... | ... | TN | Tools | Y10 |
| | gr | | | | ⮑Options | |
| | preferences | | | | | |
| | ddgrips | | | | | |
| | ddselect | | | | | |

**+options**

**Command:** options

*Displays dialog box.*

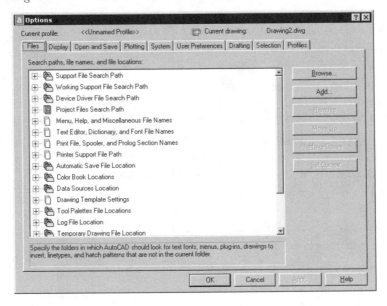

## DIALOG BOX OPTIONS

**OK** applies the changes, and closes the dialog box.

**Cancel** cancels the changes, and closes the dialog box.

**Apply** applies the changes, and keeps the dialog box open.

### Files tab

**Search paths, file names, and file locations** specifies the folders and support files used by AutoCAD.

**Browse** displays the Browse for Folder dialog box for selecting folders, and the Select a File dialog box for selecting files.

**Add** adds an item below the selected path or file name.

**Remove** removes the selected item without warning; click CANCEL to undo the removal.

**Move Up** moves the selected item above (or before) the preceding item; applies to search paths only.

**Move Down** moves the selected item below the following item; applies to search paths only.

**Set Current** makes current the selected project names and spelling dictionaries only.

### Display tab

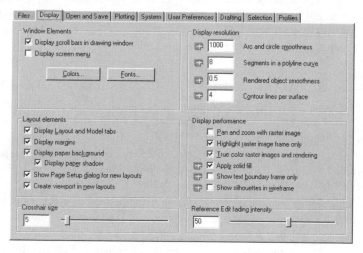

### Window Elements options

**Display scroll bars in drawing window** toggles the presence of the horizontal and vertical scroll bars; default = on.

**Display screen menu** toggles the presence of the screen menu; default = off.

**Text lines in command line window** specifies the number of lines of text in the docked command line window; range is 1 to 100 lines; default = 3.

**Colors** displays the Color Options dialog box to select colors for the AutoCAD graphics and text windows.

**Fonts** displays the Command Line Window Font dialog box to select the font for text on the command line.

### Layout Elements options

**Display Layout and Model tabs** toggles the presence of the Model and Layout tabs.

**Display margins** toggles the display of dashed margin lines in layout modes.

**Display paper background** toggles the presence of the page in layout modes.

**Display paper shadow** toggles the presence of the drop shadow under the page in layout modes.

**Show Page Setup dialog for new layouts** specifies whether the Page Setup dialog box is displayed when you create a new layout. Use this dialog box to set options related to paper and plot settings.

**Create viewport in new layouts** toggles the automatic creation of a single viewport for new layouts.

**Crosshair size** specifies the size of the crosshair cursor; range is 1% to 100% of the viewport (stored in system variable CursorSize); default = 5%.

## Display Resolution options

**Arc and circle smoothness** controls the displayed smoothness of circles, arcs, and other curves; range is 1 to 20000 (ViewRes); default = 100.

**Segments in a polyline curve** specifies the number of line segments used to display polyline curves; range is -32767 to 32767 (SplineSegs); default = 8.

**Rendered object smoothness** controls the displayed smoothness of shaded and rendered curveds; range is 0.01 to 10 (FaceTRes) default = 0.5.

**Contour lines per surface** specifies the number of contour lines on solid 3D objects; range is 0 to 2047 (IsoLines); default = 4.

## Display Performance options

**Pan and zoom with raster image** toggles the display of raster images during realtime pan and zoom (RtDisplay); default = off.

**Highlight raster image frame only** highlights only the frame, and not the entire raster image, when on (ImageHlt); default = off.

**True color raster images and rendering** toggles between displaying raster images and renderings at True Color — 24-bit color or 16.7 million colors — and at the highest number of colors available with your computer; default = off.

**Apply solid fill** toggles the display of solid fills in multilines, traces, solids, solid fills, and wide polylines; this option does not come into effect until you click OK, and then use the Regen command (FillMode); default = on.

**Show text boundary frame only** toggles the display of rectangles in place of text this option does not come into effect until you click OK, and then use the Regen command (QTextMode); default = off.

**Show silhouettes in wireframe** toggles the display of silhouette curves for 3D solid objects; when off, isolines are drawn when hidden-line removal is applied to the 3D object (DispSilh); default = off.

**Reference fading intensity** specifies the amount of fading during in-place reference editing; range is 0% to 90% (XFadeCtl); default = 50%.

## Open and Save tab

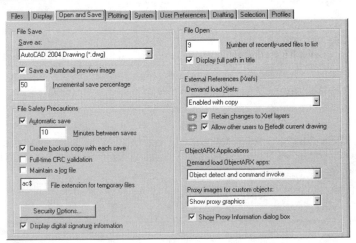

**File Save options**

**Save as** specifies the default file format used by the **Save** and **SaveAs** commands; default = "AutoCAD 2004 Drawing (*.dwg)."

**Save a thumbnail preview image** saves a thumbnail preview image with drawings (RasterPreview); default = on.

**Incremental save percentage** indicates the percentage of wasted space allowed in a drawing file before a full save is performed; range is 0% to 100% (ISavePercent); default = 50.

**File Safety Precautions options**

**Automatic save** automatically saves the drawing at prescribed time intervals (SaveFile and SaveFilePath); default = on.

**Minutes between saves** specifies the duration between automatic saves (SaveTime); default = 10 minutes.

**Create backup copy with each save** creates backup copies when drawings are saved; (ISavBak); default = on.

**Full-time CRC validation** performs cyclic redundancy check (*CRC*) error-checking each time an object is read into the drawing.

**Maintain a log file** saves the Text window text to a log file (LogFileMode); default = off.

**File extension for temporary files** specifies the filename extension for temporary files created by AutoCAD (NodeName); default = .ac$.

**Security Options** displays the Security Options dialog box; see the SecurityOptions command.

**Display digital signature information** displays digital signature information when opening files with valid digital signatures (SigWarn); default = on.

**File Open options**

**Number of recently-used files to list** specifies the number of recently-opened filenames to list in the Files menu; default = 4; maximum = 9.

**Display full path in title** displays the drawing file's path in AutoCAD's titlebar.

**External References (Xrefs) options**

**Demand load Xrefs** specifies the style of demand loading of externally-referenced drawings a.k.a. xrefs (XLoadCtl):

- **Disabled** turns off demand loading.
- **Enabled** turns on demand loading to improve performance, but the drawing cannot be edited by another user; default.
- **Enabled with copy** turns on demand loading; loads a copy of the drawing so that another user can edit the original.

**Retain changes to Xref layers** saves changes to properties for xref-dependent layers (VisRetain); default = on.

**Allow other users to Refedit current drawing** allows another user to edit the current drawing when referenced by another drawing (XEdit); default = on.

**ObjectARX Applications options**

**Demand load ObjectARX apps** demand-loads an ObjectARx application when the drawing contains proxy objects (DemandLoad):

- **Disable load on demand** turns off demand loading.

- **Custom object detect** demand-loads the application when the drawing contains proxy objects.
- **Command invoke** demand-loads the application when a command of the application is invoked.
- **Object detect and command invoke** demand-loads the application when the drawing contains proxy objects, or when one of the application's commands is invoked; default.

**Proxy images for custom objects** specifies how proxy objects are displayed:
- **Do not show proxy graphics** does not display proxy objects.
- **Show proxy graphics** displays proxy objects.
- **Show proxy bounding box** displays a rectangle instead of the proxy object.

**Show Proxy Information dialog box** displays a warning dialog box when a drawing contains proxy objects; (ProxyNotice); default = on.

### Plotting tab

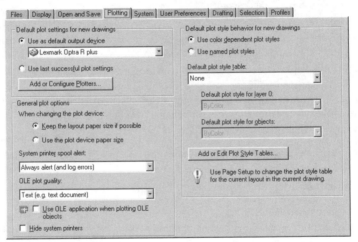

### Default Plot Settings for New Drawings options

**Use as default output device** selects the default output device.

**Use last successful plot settings** reuses the plot settings from the last successful plot.

**Add or configure plotters** displays Autodesk Plotter Manager window; see the PlotterManager command.

### General Plot Options options

**When changing the plot device:**
- **Keep the layout paper size if possible** uses the paper size specified by the Page Setup dialog box's Layout Settings tab, provided the output device can handle the paper size (PaperUpdate); default = on.
- **Use the plot device paper size** uses the paper size specified by the PC3 plotter configuration file (PaperUpdate); default = off.

**System printer spool alert** displays an alert when a spooled drawing has a conflict:
- **Always alert (and log errors)** displays alert, and logs the error message.

- **Alert first time only (and log errors)** displays the alert once, but logs all error messages.
- **Never alert (and log first error)** does not display an alert, but logs the first error message.
- **Never alert (do not log errors)** neither displays an alert, nor logs any error messages.

**OLE plot quality** determines the quality of OLE objects when plotted (OleQuality); default = text; see the OleScale command.

**Use OLE application when plotting OLE objects** launches the application that created the OLE object when plotting a drawing with an OLE object (OleStarup); default = off.

**Hide system printers** hides the names of Windows system printers not specific to CAD.

## Default Plot Style Behavior for New Drawings options

**Use color dependent plot styles** uses color-dependent plot styles, which use the AutoCAD color index ranging from 1 to 255 (PStylePolicy); default = 1.

**Use named plot styles** uses the plot property settings specified by the plot style definition (PStylePolicy); default = 0.

**Default plot style table** names the default plot style table to attach to new drawings; default = None.

**Default plot style for layer 0** specifies the default plot style for layer 0 (DefLPlStyle); default = ByColor.

**Default plot style for objects** specifies the name of the default plot style assigned to new objects (DefLPlStyle); default = ByColor.

**Add or Edit Plot Style Tables** displays the Autodesk Plot Style Table Manager window; see the StylesManager command.

## System tab

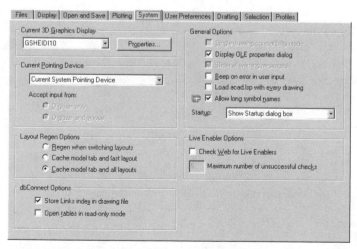

## Current 3D Graphics Display options

**Current 3D Graphics Display** selects the 3D graphics display driver (default = GSHEIDI10).

**Properties** displays the 3D Graphics System Configuration dialog box.

## Current Pointing Device options

**Current Pointing Device** selects the pointing device driver:

- **Current System Pointing Device** selects the pointing device used by Windows.
- **Wintab Compatible Digitizer** selects a Wintab-compatible digitizer driver.

**Accept input from:**

- **Digitizer only** reads input from the digitizer, and ignores the mouse.
- **Digitizer and mouse** reads input from the digitizer and the mouse.

## Layout Regen options

**Regen when switching layouts** regenerates the drawing each time layouts are switched.

**Cache model tab and last layout** saves the display list of the model tab and last layout accessed.

**Cache model tab and all layouts** saves the display list of the model tab and all layouts.

## dbConnect Options options

**Store links index in drawing file** stores the database index in the drawing file; default = on.

**Open tables in read-only mode** opens database tables in read-only mode; default = off.

## General Options options

**Single-drawing compatibility mode** forces the Single-drawing Interface (SDI), which limits AutoCAD to opening a single drawing at a time; this may be required for compatibility with some third-party applications (SDI); default = off.

**Display OLE Properties dialog** displays the OLE Properties dialog box after an OLE object is inserted in the drawing; see the OleScale command; default = on.

**Show all warning messages** displays all dialog boxes with the Don't Display This Warning Again option; default = on.

**Beep on error in user input** beeps the computer when AutoCAD detects a user error.

**Load acad.lsp with every drawing** loads the *acad.lsp* file with every drawing (AcadLspAsDoc); default = off).

**Allow long symbol names** allows symbol names — layers, dimension styles, blocks, linetypes, text styles, layouts, UCS names, views, and viewport configurations — to be up to 255 characters long, and to include letters, numbers, blank spaces, and most punctuation marks; when off, names are limited to 31 characters, and spaces may not be used (ExtNames); default = on.

**Show Startup dialog:**

- **Show Startup dialog box** displays Startup dialog box when AutoCAD launches or Create New Drawing dialog box when AutoCAD is already running.
- **Do not show a startup dialog** displays the Select Template File dialog box.

## Live Enabler Options options

**Check Web for Live Enablers** checks whenever an Internet connection is present.

**Maximum number of unsuccessful checks** Default = 5.

## User Preferences tab

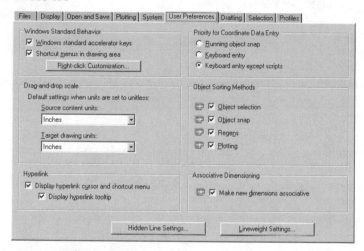

### Windows Standard Behavior options

**Windows standard accelerator keys:**

☑ uses Windows' keyboard accelerators (default).

☐ uses DOS-based AutoCAD keyboard accelerators; the differences are:

| Accelerator | On | Off |
|---|---|---|
| CTRL+C | CopyClip | Cancel command. |
| CTRL+O | Open | Ortho toggle. |
| CTRL+V | Paste | Viewports switch. |

**Shortcut menus in drawing area:**

☑ right-clicks in the drawing area displays a shortcut menu.

☐ right-clicks is equivalent to pressing the ENTER key (ShortCutMenu); default = on.

**Right-click customization** displays the Right-Click Customization dialog box (ShortCutMenu).

### Drag and Drop Scale options (formerly AutoCAD DesignCenter)

**Source content units** specifies the default units when an object is inserted into the drawing from AutoCAD DesignCenter; Unspecified-Unitless means the object is not scaled when inserted (InsUnitsDefTarget); default = inches or mm.

**Target drawing units** specifies the default units when "insert units" are not specified by the InsUnits system variable (InsUnitsDefTarget); default = inches or mm.

### Hyperlink options

**Display hyperlink cursor and shortcut menu** displays the hyperlink cursor — looks like chain links and the planet earth — when the cursor passes over an object containing a hyperlink; displays the Hyperlink option when right-clicking an object containing a hyperlink; see the HyperlinkOptions command; default = on.

**Display hyperlink tooltip** displays a tooltip when the cursor pauses over an object containing a hyperlink; default = on.

Priority for Coordinate Data Entry options
- **Running object snap** means that osnap overrides coordinates entered at the keyboard (OSnapCoord); default = off.
- **Keyboard entry** means that coordinates entered at the keyboard override osnaps (OSnapCoord); default = off.
- **Keyboard entry except scripts** means taht coordinates entered at the keyboard override running object   snaps, except when coordinates are provided by a script (OSnapCoord); default = on.

Object Sorting Methods options

**Object selection** sorts objects in the drawing for selection from first to last created (SortEnts); default = off.

**Object snap** sorts objects in the drawing during object snap for selection from first to last created  (SortEnts); default = off.

**Regens** sorts objects in the drawing from first to last created for display during commands that cause a regeneration, such as Regen (SortEnts); default = off.

**Plotting** sorts objects in the drawing from first to last created for plotted output to file or printer (SortEnts); default = on.

Associative Dimensioning options
**Associate new dimension with objects**

☑ means that dimensions are associated with objects.

☐ means that dimensions are associated with defpoints.

**Hidden Line Settings** displays the Hidden Line Settings dialog box; see HLSettings command.

**Lineweight Settings** displays the Lineweight Settings dialog box; see the LWeight command.

**Drafting** tab

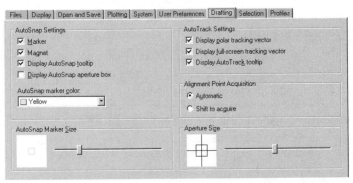

AutoSnap Settings options
**Marker** displays the AutoSnap icon (AutoSnap); default = on.

**Magnet** turns on the AutoSnap magnet (AutoSnap); default = on.

**Display AutoSnap tooltip** displays the AutoSnap tooltip (AutoSnap); default = on.

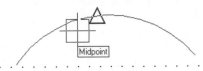

**Display AutoSnap aperture box** displays the AutoSnap aperture box (ApBox); default = off.

**AutoSnap marker color** specifies the color of the AutoSnap icons; choose from seven colors; default = yellow.

**AutoSnap marker size** sets the size for the AutoSnap icon; range is 1 to 20 pixels.

## AutoTrack Settings options

**Display polar tracking vector** displays the Polar Tracking vectors at specific angles (TrackPath); default = on.

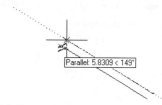

*Polar tracking vector (dotted line) and AutoTrack tootip.*

**Display full-screen tracking vector** displays the tracking vectors (TrackPath); default = on.

*Full-screen tracking vector and AutoTrack tooltip.*

**Display AutoTrack tooltip** displays the AutoTrack tooltip (AutoSnap); default = on.

## Alignment Point Acquisition options

- **Automatic** displays tracking vectors automatically when aperture moves over an object snap.
- **Shift to acquire** displays tracking vectors when pressing **SHIFT** and moving the aperture over an object snap.

**Aperture Size** sets the size for the aperture; range is 1 to 50 pixels (Aperture); default = 10.

## Selection tab

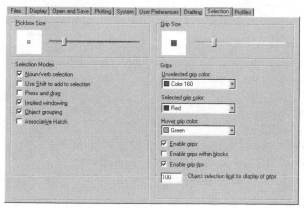

**Pickbox Size** specifies the size of the pickbox; range is 1 to 20 pixels (PickBox); default = 3.

**Grip size** specifies the size of grips; range is from 1 to 20 pixels (GripSize); default = 5.

**Selection Modes options** (replaces DdSelect command)

**Noun/verb selection** allows you to select an object before executing an editing command (PickFirst); default = on.

**Use Shift to add to selection** allows you to press SHIFT to add or remove objects from the selection set (PickAdd); default = off.

**Press and drag** allows you to create the selection window by dragging (PickDrag); default = off.

**Implied windowing** creates a selection window when you pick a point in the drawing that does not pick an object (PickAuto); default = on.

**Object grouping** selects the entire group when an object in the group is selected (PickStyle); default = off.

**Associative hatch** selects boundary objects, along with the associative hatch patterns (PickStyle); default = off.

**Grips options** (replaces the DdGrips command)

**Unselected grip color** specifies color of unselected — cold – grips (GripColor); default = blue.

**Selected grip color** specifies the color of selected — hot – grip (GripHot); default = red.

**Hover grip color** specifies the color of the grip when the cursor hovers over it; default = green.

**Enable grips** displays grips on selected objects (Grips); default = on.

**Enable grips within blocks** displays all grips for every object in the selected block; when off, a single grip at the block's insertion point is displayed (GripBlock); default = off.

**Enable grip tips** toggles the display of grip tips on custom objects (GripTips); default = on.

**Object selection limit for display of grips** limits the number of selected objects that display grips (GripObjLimit); default = 100.

**Profiles** tab

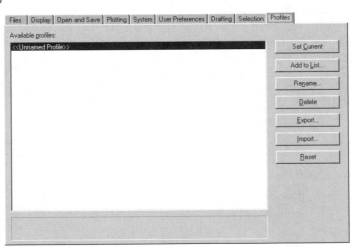

**Available Profiles** lists available profiles, which customize the AutoCAD user interface.

**Set Current** sets the selected profile as the current profile.

**Add to List** displays the Add Profile dialog box; allows you to enter a name and description for the new profile.

**Rename** displays the Change Profile dialog box; allows you to change the name and the description of the selected profile.

**Delete** erases the selected profile; the current profile cannot be erased.

**Export** exports profiles as *.arg* files.

**Import** imports *.arg* profiles into AutoCAD.

**Reset** resets the values of the selected profile to AutoCAD's default settings.

. . . . . . . . . . . . . . . . . . . . . . . . . . . . . . . . . . . . . . . . . . . . . . . . . .

## +OPTIONS COMMAND
**Command:** +options
**Tab index <0>:** *(Enter a digit between 0 and 8.)*

### COMMAND LINE OPTION
**Tab index** specifies the tab to display:

| Tab Index | Meaning |
|-----------|---------|
| 0 | Files tab. |
| 1 | Display tab. |
| 2 | Open and Save tab. |
| 3 | Plotting tab. |
| 4 | System tab. |
| 5 | User Preferences tab. |
| 6 | Drafting tab. |
| 7 | Selection tab. |
| 8 | Profiles tab. |

## TIPS

• *Grips* are small squares that appear on an object when the object is selected at the 'Command' prompt. In other Windows applications, grips are known as *handles*.

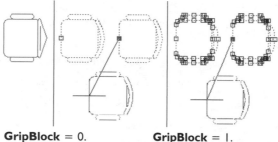

**GripBlock = 0.**          **GripBlock = 1.**

• When an object is first selected, the grips are blue, and are called *unselected* or *cold* grips.

• When a grip is selected, it turns into a solid red square; this is called a *hot* grip.

• Press ESC to turn off unselected grips; press ESC twice to turn off hot grips.

• A larger pickbox makes it easier to select objects, but also easier to accidentally select unintended objects.

• Use **Object Sort Method** if the drawing requires objects to be processed in the order they appear in the drawing, such as for NC (numerically controlled) applications.

• **Plotting** and **PostScript Output** are turned on, by default; setting more sort methods increases processing time.

• The first time you use the **DrawOrder** command, it turns on all object sort method options.

. . . . . . . . . . . . . . . . . . . . . . . . . . . . . . . . . . . . . . . . . . . . . . . . . .

# 'Ortho

**V. 1.0**  Constrains drawing and editing commands to the vertical and horizontal directions only (*short for ORTHOgraphic*).

| Command | Alias | Ctrl+ | F-key | Alt+ | Status Bar | Tablet |
|---------|-------|-------|-------|------|------------|--------|
| 'ortho | ... | L | F8 | ... | ORTHO | ... |

**Command:** ortho
**Enter mode [ON/OFF] <OFF>:** *(Enter ON or OFF.)*

## COMMAND LINE OPTIONS
**OFF** turns off ortho mode.

**ON** turns on ortho mode.

## STATUS BAR OPTIONS
*Ortho mode is toggled on and off by clicking ORTHO on the status bar:*

| 10.1046, 0.9275 , 0.0000 | | SNAP | GRID | ORTHO | POLAR | OSNAP | OTRACK | LWT | MODEL |

| 7.3372, 0.2001 , 0.0000 | | SNAP | GRID | ORTHO | POLAR | OSNAP | OTRACK | LWT | MODEL |

## RELATED COMMANDS
**DSettings** toggles ortho mode via a dialog box.

**Snap** rotates the ortho angle.

## RELATED SYSTEM VARIABLES
**OrthoMode** stores the current ortho modes.

**SnapAng** specifies the rotation angle of the ortho cursor.

## TIPS
- Use ortho mode when you want to constrain your drawing and editing to right angles.

- Rotate the angle of ortho with the **Snap** command's **Rotate** option

- In isoplane mode, ortho mode constrains the cursor to the current isoplane.

- AutoCAD ignores ortho mode when you enter coordinates by keyboard, and in perspective mode; ortho is also ignored by object snap modes.

- Ortho is not necessarily horizontal or vertical; its orientation is determined by the current UCS and snap alignment.

# '-OSnap

**V. 2.0**  Sets and turns on and off object snap modes at the command line (*short for Object SNAP*).

| Command | Alias | Ctrl+ | F-key | Alt+ | Status Bar | Tablet |
|---------|-------|-------|-------|------|-----------|--------|
| '-osnap | -os | ... | F3 | ... | OSNAP | T15 - U22 |

*Note that the **OSnap** command displays the **Drafting Settings** dialog box; see the **DSettings** command.*

**Command:** -osnap
**Current osnap modes: Ext**
**Enter list of object snap modes:** *(Enter one or more modes separated by commas.)*

## COMMAND LINE OPTIONS

*You only need enter the first three letters as the abbreviation for each option:*

**APParent** snaps to the the intersection of two objects that don't physically cross, but appear to intersect on the screen, or would intersect if extended.

**CENter** snaps to the center point of arcs and circles.

**ENDpoint** snaps to the endpoint of lines, polylines, traces, and arcs.

**EXTension** snaps to the extension path of objects.

**FROm** extends from a point by a given distance.

**INSertion** snaps to the insertion point of blocks, shapes, and text.

**INTersection** snaps to the intersection of two objects, or to a self-crossing object, or to objects that would intersect if extended.

**MIDpoint** snaps to the middle point of lines and arcs.

**NEArest** snaps to the object nearest to the crosshair cursor.

**NODe** snaps to a point object.

**NONe** turns off all object snap modes temporarily.

**OFF** turns off all object snap modes.

**PARallel** snaps to a parallel offset.

**PERpendicular** snaps perpendicularly to objects.

**QUAdrant** snaps to the quadrant points of circles and arcs.

**QUIck** snaps to the first object found in the database.

**TANgent** snaps to the tangent of arcs and circles.

## STATUS BAR OPTIONS

*Right-click **OSNAP** on the status bar:*

**On** turns on previously-set object snap modes.

**Off** turns off all running object snaps.

**Settings** displays the Object Snap tab of the Drafting Settings dialog box.

## SHORTCUT MENU

*Hold down the CTRL key, and then right-click anywhere in the drawing:*

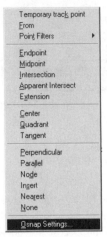

**Temporary track point** invokes Tracking mode; see the Tracking command.

**Point filters** invokes point filter modes:

| Point Filter | Meaning |
|---|---|
| .x | Prompts for y,z-coordinate. |
| .y | Prompts for x,z-coordinate. |
| .z | Prompts for x,y-coordinate. |
| .xy | Prompts for z-coordinate. |
| .xz | Prompts for y-coordinate. |
| .yz | Prompts for x-coordinate. |

**Osnap Settings** displays the Object Snap tab of the Drafting Settings dialog box; see DSettings command.

## RELATED SYSTEM VARIABLES

**Aperture** specifies the size of the object snap aperture in pixels.

**OsMode** stores the current object snap mode(s):

| OsMode | Meaning |
|---|---|
| 0 | NONe (default). |
| 1 | ENDpoint. |
| 2 | MIDpoint. |
| 4 | CENter. |
| 8 | NODe. |
| 16 | QUAdrant. |
| 32 | INTersection. |
| 64 | INSertion. |
| 128 | PERpendicular. |
| 256 | TANgent. |
| 512 | NEArest. |
| 1024 | QUIck. |
| 2048 | APParent intersection. |
| 4096 | EXTension. |
| 8192 | PARallel. |

**AutoSnap** controls the display of AutoSnap (default = 63):

| AutoSnap | Meaning |
|---|---|
| 0 | Turns off marker, SnapTip, and magnet. |
| 1 | Turns on the marker. |
| 2 | Turns on the SnapTip. |
| 4 | Turns on the magnet. |
| 8 | Turns on polar tracking. |
| 16 | Turns on object snap tracking. |
| 32 | Turns on polar and object snap tracking tooltips. |

**OsnapCoord** overrides object snaps when entering coordinates at 'Command' prompt.

| OsnapCoord | Meaning |
|---|---|
| 0 | Object snap overrides keyboard. |
| 1 | Keyboard overrides object snap settings. |
| 2 | Keyboard overrides object snap settings, except during a script. |

## TIPS

- The **Aperture** command controls the drawing area AutoCAD searches through.

- If AutoCAD finds no snap matching the current modes, then the pick point is selected.

- The **APPint** and **INT** object snap modes should not be used together.

- To turn on more than one object snap at a time, use a comma to separate mode names:

  **Enter list of object snap modes:** int,end,qua

- The location of all object snaps:

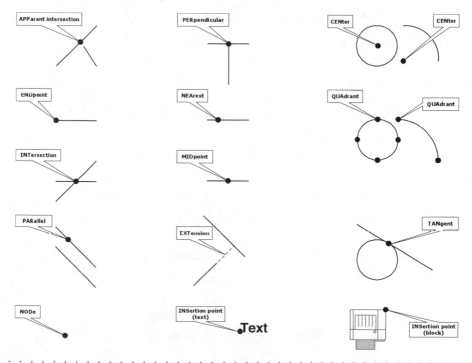

# PageSetup

**2000** Sets up model and layout views in preparation for plotting drawings.

| Command | Alias | Ctrl+ | F-key | Alt+ | Menu Bar | Tablet |
|---------|-------|-------|-------|------|----------|--------|
| pagesetup | ... | ... | ... | FG | File | V25 |
| | | | | | ⤷Page Setup | |

**Command:** pagesetup

*Displays dialog box:*

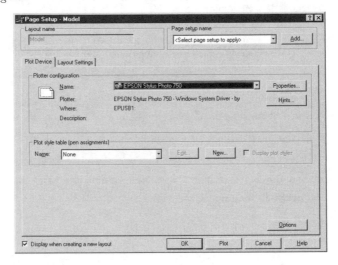

## DIALOG BOX OPTIONS

*See **Plot** command.*

## RELATED COMMANDS

**Layout** creates new layouts.

**Plot** plots the drawing based on the settings of the PageSetup command.

## RELATED SYSTEM VARIABLE

**CTab** contains the name of the current model or layout tab; default = "Model."

## TIPS

• The **Page Setup** dialog box displays automatically when new layouts are created.

• Right-click the **Model** or **Layout** tabs to display a shortcut menu; select **Page Setup**.

# 'Pan

**V. 1.0**  Moves the view in the current viewport to different positions.

| Commands | Aliases Ctrl+ | F-key | Alt+ | Menu Bar | Tablet |
|----------|---------------|-------|------|----------|--------|
| 'pan | p        ... | ... | VPT | View | N11- |
|  | rtpan |  |  | ⇘Pan | P11 |
|  |  |  |  | ⇘Realtime |  |
| -pan | -p |  | VPP | View |  |
|  |  |  |  | ⇘Pan |  |
|  |  |  |  | ⇘Point |  |

**Command:** pan
**Press Esc or Enter to exit, or right-click to display shortcut menu.** *(Move cursor to pan, and then press* **Esc** *to exit command.)*

*Enters real-time panning mode, and displays hand cursor:*

*Drag the hand cursor to pan the drawing in the viewport.*
*Press* ENTER *or* ESC *to return to the 'Command' prompt.*

**SHORTCUT MENU OPTIONS**
*During real-time pan mode, right-click the drawing to display shortcut menu:*

**Exit** exits real-time pan mode; returns to the 'Command' prompt.

**Pan** switches to real-time pan mode, when in real-time zoom mode.
**Zoom** switches to real-time zoom mode; see the Zoom command.
**3D Orbit** switches to 3D orbit mode; see the 3dOrbit command.

**Zoom Window** prompts you to "Press pick button and drag to specify zoom window."
**Zoom Original** returns to the view when you first started the Pan command.
**Zoom Extents** displays the entire drawing.

## -PAN Command

**Command:** -pan
**Displacement:** *(Pick a point, or enter x,y-coordinates.)*
**Second point:** *(Pick another point, or enter x,y-coordinates.)*

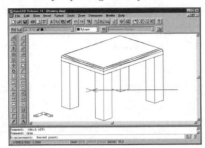

*Before panning to the left:*

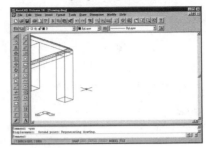

*After panning to the left:*

### COMMAND LINE OPTIONS

ENTER exits real-time panning mode.

ESC exits real-time panning mode.

**Displacement** specifies the distance and direction to pan the view.

**Second point** pans to this point.

### RELATED COMMANDS

**DsViewer** displays the Aerial View window, which pans in an independent window.

**RegenAuto** determines how regenerations are handled.

**View** saves and restores named views.

**Zoom** pans with the Dynamic option.

**3dOrbit** pans during perspective mode.

### RELATED SYSTEM VARIABLES

**MButtonPan** determines the action of a mouse's third button or wheel.

**ViewCtr** specifies the x,y-coordinate of the view's center.

**ViewDir** specifies the view direction relative to UCS.

**ViewRes** turns off real-time pan off when FastZoom is disabled.

**ViewSize** specifies the height of view.

**TIPS**

- You pan in each viewport independently.

- You can use the **'Pan** command transparently to start drawing an object in one area of the drawing, pan over, and then continue working in another area of the drawing.

- In the **Aerial View** window, change the **Static** button to **Dynamic** to perform real-time panning; the drawing pans as quickly as you move the mouse.

- You cannot use transparent pan under the following conditions:

     Paper space.
     Perspective mode.
     **VPoint** command.
     **DView** command.
     Another **Pan** command.
     **View** command
     **Zoom** command.
     **3dOrbit** command.

- When the drawing no longer moves during real-time panning, you have reached the panning limit; AutoCAD changes the hand icon to show the limit:

- As an alternative to the **Pan** command, you can use the horizontal and vertical scroll bars to pan the drawing.

- When **MButtonPan** is set to 1, you can pan the drawing by holding down the wheelbutton and moving the mouse.

# PartiaLoad

Loads additional views and layers of a partially-loaded drawing; this command works only with drawings that have been partially loaded.

| Commands | Alias | Ctrl+ | F-key | Alt+ | Menu Bar | Tablet |
|----------|-------|-------|-------|------|----------|--------|
| partiaload | ... | ... | ... | FR | File | |
| | | | | | ⤷Partial Load | |

| -partiaload |
|-------------|

**Command:** partiaload

*If drawing was not been partially opened (via the **Open** command's **Partial Open** option), AutoCAD reports:*

**Command not allowed unless the drawing has been partially opened.**

*Otherwise, displays dialog box:*

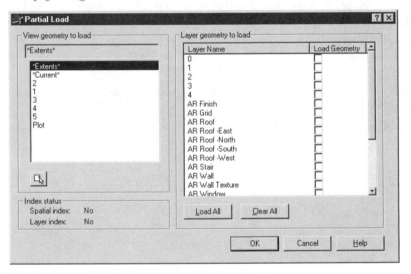

## DIALOG BOX OPTIONS

**View geometry to load** selects the model space views to load; paper space views are not available for partial loading.

**Pick a Window** dismisses the dialog box temporarily to allow you specify an area that becomes the view to load; prompts you:

**Specify first corner:** *(Pick a point.)*
**Specify opposite corner:** *(Pick a point.)*

When the Partial Load dialog box returns, *New View* is listed by View Geometry to Load list.

## Layer Geometry to Load options

**Load Geometry** selects the layers to load.

**Load All** loads all layers.

**Clear All** deselects all layer names.

**422** / The Illustrated AutoCAD 2004 Quick Reference

# -PARTIALOAD Command

**Command:** -partiaload

**Specify first corner or [View]:** *(Pick a point, or type* **V**.*)*

**Enter view to load or [?] <\*Extents\*>:** *(Enter a view name, or type* **?**.*)*

**Enter layers to load or [?] <none>:** *(Enter a layer name, or type* **?**.*)*

## COMMAND LINE OPTIONS

**Specify first corner** specifies a corner to create a new view.

**Specify opposite corner** specifies the second corner; causes geometry in the new view to be loaded into the drawing.

*View options:*

**Enter view to load** loads  into the drawing the geometry found in the view.

**?** list the names of views in the drawing:

**Enter view name(s) to list <\*>:** *(Enter a view name, or type* **\***.*)*

*Sample display:*

```
Saved model space views:
View name"1"
"Plot"
```

**\*Extents\*** loads all geometry into the drawing.

**Enter layers to load** loads geometry on the layers in the drawing.

**?** list the names of layers in the drawing:

**Enter layers to list <\*>:** *(Press* **Enter**.*)*

*Sample display:*

```
Layer names:
"0"
"3"
```

**None** loads no layers.

## RELATED COMMANDS

**Open** partially opens a drawing via a dialog box.

**PartialOpen** partially opens a drawing via the command line.

## RELATED SYSTEM VARIABLE

**FullOpen** indicates whether the drawing is fully or partially opened.

# -PartialOpen

**2000** Opens a drawing and loads selected layers and views.

| Command | Alias | F-key | Alt+ | Menu Bar | Tablet |
|---|---|---|---|---|---|
| -partialopen | partialopen | ... | ... | ... | ... |

**Command:** -partialopen
**Enter name of drawing to open <filename.dwg>:** *(Enter a file name.)*
**Enter view to load or [?]<\*Extents\*>:** *(Enter a view name, or type **?**.)*
**Enter layers to load or [?]<none>:** *(Enter a layer name, or type **?**.)*
**Unload all Xrefs on open? [Yes/No] <N>:** *(Type **Y** or **N**.)*

## COMMAND LINE OPTIONS

**Enter name of drawing to open** specifies the name of the DWG file.

**Enter view to load** loads the geometry found in the view into the drawing.

**?** list the names of the views in the drawing.

**\*Extents\*** loads all geometry into the drawing.

**Enter layers to load** loads geometry on the layers into the drawing.

**?** list the names of layers in the drawing.

**None** loads no layers.

**Unload all Xrefs on open:**

   **Yes** does not load externally-referenced drawings.

   **No** loads all externally-referenced drawings.

## RELATED COMMANDS

**Open** displays the Select Drawing dialog box; includes the Partial Open option.

**PartiaLoad** loads additional views or layers of a partially-opened drawing.

## RELATED SYSTEM VARIABLES

**FullOpen** indicates whether the drawing is fully or partially opened.

## TIPS

- **PartialOpen** is an alias for the **-PartialOpen** command.

- As an alternative, you may use the the **Partial Open** option of the **Open** command's **Select File** dialog box.

# PasteAsHyperlink

<u>2000</u>  Pastes object as hyperlinks in drawings; works only if the Clipboard contains appropriate data (*undocumented command*).

| Command | Alias | Ctrl+ | F-key | Alt+ | Menu Bar | Tablet |
|---|---|---|---|---|---|---|
| pasteashyperlink | ... | ... | ... | EH | Edit | ... |
| | | | | | ⬉Paste as Hyperlink | |

**Command:** pasteashyperlink
**Select objects:** *(Select one or more objects.)*
**Select objects:** *(Press* **Enter**.*)*

## COMMAND LINE OPTION
**Select objects** selects the objects to which the hyperlink will be pasted.

## RELATED COMMANDS
**CopyClip** copies a hyperlink to the Clipboard.

**Hyperlink** adds hyperlinks to selected objects.

**SelectURL** highlights all objects with hyperlinks.

## RELATED SYSTEM VARIABLES
*None.*

# PasteBlock

<u>2000</u>  Pastes objects as blocks in drawings; works only when the Clipboard contains AutoCAD block objects.

| Command | Alias | Ctrl+ | F-key | Alt+ | Menu Bar | Tablet |
|---------|-------|-------|-------|------|----------|--------|
| pasteblock | ... | Shift+V ... | | EK | Edit | ... |
| | | | | | ⮑Paste as Block | |

**Command:** pasteblock
**Specify insertion point:** *(Pick a point.)*

## COMMAND LINE OPTION
**Specify insertion point** specifies the position for the block.

## RELATED COMMANDS
**Insert** inserts blocks in the drawing.

**CopyClip** copies a block to the Clipboard.

## RELATED SYSTEM VARIABLES
*None.*

## TIPS
- This command does not work when the Clipboard contains data that cannot be pasted as a block.

- This command pastes any AutoCAD object as a block, generating a generic block name similar to "A$C65D94228."

- If the Clipboard contains a block, the block is nested by this command.

- As of AutoCAD 2004, you can also use the CTRL+SHIFT+V shortcut.

# PasteClip

<u>Rel.13</u>  Places an object from the Clipboard in the drawing (*short for PASTE CLIPboard*).

| Command | Alias | Ctrl+ | F-key | Alt+ | Menu Bar | Tablet |
|---------|-------|-------|-------|------|----------|--------|
| pasteclip | ... | V | ... | EP | Edit<br>🖑Paste | U13 |

**Command:** pasteclip

*When the Clipboard contains an AutoCAD object, the following prompt is displayed:*

**Specify insertion point:** *(Pick a point.)*

*When the Clipboard contains a non-AutoCAD object, it is pasted into the upper-left corner of the current viewport.*

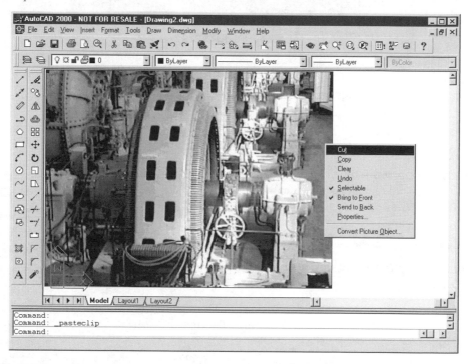

## COMMAND LINE OPTION

**Specify insertion point** specifies the position for the AutoCAD object.

## SHORTCUT MENU OPTIONS

*Right-click pasted objects to display shortcut menu. Not all options appear with all pasted objects.*

**Cut** cuts the object from drawing to the Clipboard.

**Copy** copies the object to the Clipboard.

**Clear** erases the pasted object.

**Undo** undoes the last action.

**Selectable** toggles the selectability of the object; the handles disappear.

**Bring to Front** displays the object as topmost in the drawing.

**Send to Back** displays the objects as lowermost in the drawing.

**Properties** displays the OLE Properties dialog box; see the OleScale command.

**Convert** converts (or, in some cases, does not convert) the object to another format; displays dialog box:

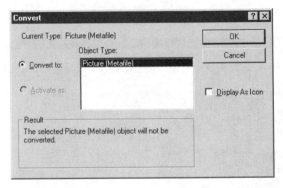

## RELATED COMMANDS

**CopyClip** copies drawing to the Clipboard.

**Insert** inserts an AutoCAD drawing in the drawing.

**InsertObj** inserts an OLE object in the drawing.

**PasteSpec** places the Clipboard object as pasted or linked object.

## RELATED SYSTEM VARIABLE

**OleHide** toggles the display of the OLE object (1 = off).

## TIPS

* The **PasteClip** command places all objects in the upper-left corner of the current viewport, unless they are AutoCAD objects.

* Graphical objects are placed in the drawing as OLE objects.

* Text is usually — but not always — placed in the drawing as an Mtext object.

* Use the **PasteSpec** command to paste the object as an AutoCAD block.

. . . . . . . . . . . . . . . . . . . . . . . . . . . . . . . . . . . . . . . . . . . . . . .

## Removed Command

**PcxIn** was removed from AutoCAD Release 14. Use the **ImageAttach** command instead.

. . . . . . . . . . . . . . . . . . . . . . . . . . . . . . . . . . . . . . . . . . . . . . .

# PasteOrig

<u>2000</u> Pastes a block from the Clipboard into drawings at the block's original insertion point (*short for PASTE at ORIGin*).

| Command | Alias | Ctrl+ | F-key | Alt+ | Menu Bar | Tablet |
|---------|-------|-------|-------|------|----------|--------|
| pasteorig | ... | ... | ... | ED | Edit | ... |
| | | | | | ⬚ Paste to Original Coordinates | |

**Command:** pasteorig

## COMMAND LINE OPTIONS
*None.*

## RELATED COMMANDS
**CopyClip** copies a drawing to the Clipboard.

**Insert** inserts an AutoCAD drawing in the drawing.

**PasteBlock** pastes AutoCAD objects as a block at a user specified insertion point.

## RELATED SYSTEM VARIABLES
*None.*

## TIPS
- Use this command to copy objects from one drawing to another.

- This command cannot be used to paste objects into the drawing from which they originate.

# 'PasteSpec

__Rel.13__ Pastes Clipboard objects in drawings as embedded, linked, pasted, or converted objects (*short for PASTE SPECial*).

| Command | Alias | Ctrl+ | F-key | Alt+ | Menu Bar | Tablet |
|---------|-------|-------|-------|------|----------|--------|
| 'pastespec | pa | ... | ... | ES | Edit | ... |
| | | | | | ⌐Paste Special | |

**Command:** pastespec

*Displays dialog box:*

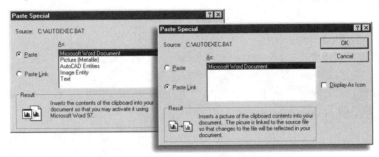

*Paste (left) and Paste Link (right).*

## DIALOG BOX OPTIONS

**Paste** pastes the object as an embedded object.

**Paste Link** pastes the object as a linked object.

**Display as Icon** displays the object as an icon from the originating application.

**Change Icon** allows you to select the icon; displays dialog box:

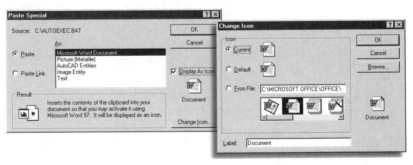

## RELATED COMMANDS

**CopyClip** copies the drawing to the Clipboard.

**InsertObj** inserts an OLE object in the drawing.

**OleLinks** edits the OLE link data.

**PasteClip** places the Clipboard object as a pasted object.

## RELATED SYSTEM VARIABLE

**OleHide** toggles the display of OLE objects (1 = off).

# PcInWizard

<u>2000</u>  Converts PCP and PC2 plot configuration files to PC3 format.

| Command | Alias | Ctrl+ | F-key | Alt+ | Menu Bar | Tablet |
|---|---|---|---|---|---|---|
| pcinwizard | ... | ... | ... | TZI | Tools | ... |
| | | | | | ⬑Wizard | |
| | | | | | ⬑Import R14 Plot Settings | |

**Command:** pcinwizard

*Displays dialog box.*

## DIALOG BOX OPTIONS

**Introduction** page

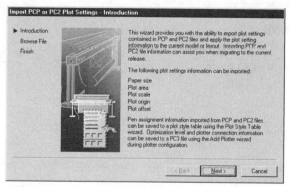

**Back** returns to the previous step.

**Next** continues to the next step.

**Cancel** cancels the wizard.

**Browse File** page

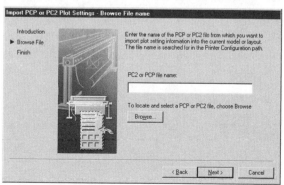

**PCP or PC2 filename** specifies *.pcp* or *pc2* name.

**Browse** displays the **Import** dialog box.

**Finish** page

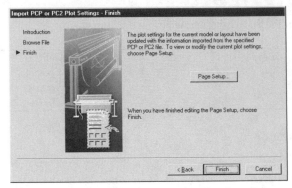

**Page Setup** displays the Page Setup dialog box; see the PageSetup command.

**Finish** completes the importation process.

## RELATED COMMAND

**Plot** uses PC3 files to control the plotter configuration.

## RELATED SYSTEM VARIABLES

*None.*

## TIPS

- PCP is short for "plotter configuration parameters"; PCP files are used by AutoCAD Release 13.

- PC2 files are used by AutoCAD Release 14.

- PC3 files are used by AutoCAD 2000, 20 and , and 2004.002.

- This command imports PCP and PC2 files, and applies them to the current layout or model tab.

- You would only use this command if you created PCP or PC2 files with earlier versions of AutoCAD.

- AutoCAD 2004 imports the following information from PCP and PC2 files: paper size, plot area, plot scale, plot origin, and plot offset.

- To import color-pen mapping, use the **Plot Style Table** wizard, run by the *StyShWiz.Exe* program.

- To import the optimization level and plotter connection, use the **Add-a-Plotter** wizard, run by the *AddPlWiz.Exe* program.

# PEdit

**V. 2.1** Edits 2D polylines, 3D polylines, or 3D meshes — depending on the selected object (*short for Polyline EDIT*).

| Command | Alias | Ctrl+ | F-key | Alt+ | Menu Bar | Tablet |
|---------|-------|-------|-------|------|----------|--------|
| pedit | pe | ... | ... | MP | Modify | Y17 |
| | | | | | ⮑Polyline | |

**Command:** pedit

*Options vary, depending on whether a 2D polyline, 3D polyline, or polymesh is picked.*

. . . . . . . . . . . . . . . . . . . . . . . . . . . . . . . . . . . . . . . . . . . . . . . . . . . .

## 2D Polyline
**Select polyline or [Multiple]:** *(Pick a 2D polyline, or type **M**.)*
**Enter an option [Open/Join/Width/Edit vertex/Fit/Spline/Decurve/Ltype gen/Undo]:** *(Enter an option.)*

Original polyline

Pedit Close

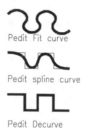
Pedit Fit curve

Pedit Width

Pedit spline curve

Pedit Edit vertex

Pedit Decurve

## COMMAND LINE OPTIONS
**Multiple** selects more than one polyline to edit.

**Close** closes an open polyline by joining the two endpoints with a single segment.

**Decurve** reverses the effects of a Fit or Spline operation.

**Edit vertex** options are listed below.

**Fit** fits a curve to the tangent points of each vertex.

**Ltype gen** specifies the linetype generation style.

**Join** adds other polylines to the current polyline.

**Open** opens a closed polyline by removing the last segment.

**Spline** fits a splined curve along the polyline.

**Undo** undoes the most-recent PEdit operation.

**Width** changes the width of the entire polyline.

**eXit** exits the PEdit command.

Edit Vertex options
**Enter a vertex editing option**
**[Next/Previous/Break/Insert/Move/Regen/Straighten/Tangent/Width/eXit]**
**<N>:** *(Enter an option.)*

**Break** removes a segment or breaks the polyline at a vertex:

    **Next** moves the x-marker to the next vertex.

    **Previous** moves the x-marker to the previous vertex.

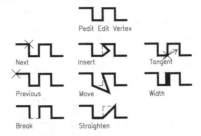

    **Go** performs the break.

    **eXit** exits the **Break** sub-submenu.

**Insert** inserts another vertex.

**Move** relocates a vertex.

**Next** moves the x-marker to the next vertex.

**Previous** moves the x-marker to the previous vertex.

**Regen** regenerates the screen to show the effect of the PEdit command.

**Straighten** draws a straight segment between two vertices:

    **Next** moves the x-marker to the next vertex.

    **Previous** moves the x-marker to the previous vertex.

    **Go** performs the straightening.

    **eXit** exits the Straighten sub-submenu.

**Tangent** shows the tangent to current vertex.

**Width** changes the width of a segment.

**eXit** exits the Edit vertex submenu.

## 3D Polyline Operations

**Command:** pedit

**Select polyline or [Multiple]:** *(Pick a 3D polyline, or type **M**.)*

**Enter an option [Open/Edit vertex/Spline curve/Decurve/Undo]:** *(Enter an option.)*

### COMMAND LINE OPTIONS

**Multiple** selects more than one 3D polyline to edit.

**Close** closes an open polyline.

**Decurve** reverses the effects of a Fit-curve or Spline-curve operation.

**Open** removes the last segment of a closed polyline.

**Spline curve** fits a splined curve along the polyline.

**Undo** undoes the most-recent **PEdit** operation.

**eXit** exits the **PEdit** command.

Edit Vertex options

### Enter a vertex editing option

**[Next/Previous/Break/Insert/Move/Regen/Straighten/eXit] <N>:** *(Enter an option.)*

**Break** removes a segment or breaks the polyline at a vertex.

> **Next** moves the x marker to the next vertex.
>
> **Previous** moves the x-marker to the previous vertex.
>
> **Go** performs the break.
>
> **eXit** exits the Break sub-submenu.

**Insert** inserts another vertex.

**Move** relocates a vertex.

**Next** moves the x-marker to the next vertex.

**Previous** moves the x-marker to the previous vertex.

**Regen** regenerates the screen to show the effect of PEdit options.

**Straighten** draws a straight segment between two vertices:

> **Next** moves the x-marker to the next vertex.
>
> **Previous** moves the x-marker to the previous vertex.
>
> **Go** performs the straightening.
>
> **eXit** exits the Straighten sub-submenu.

**eXit** exits the Edit-vertex submenu.

## 3D Mesh Operations
**Select polyline or [Multiple]:** *(Pick a 3D mesh, or type* **M***.)*
**Enter an option [Edit vertex/Smooth surface/Desmooth/Mclose/Nclose/ Undo]:** *(Enter an option.)*

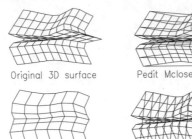

Original 3D surface   Pedit Mclose

Pedit Smooth surface   Pedit Nclose

### COMMAND LINE OPTIONS

**Multiple** selects more than one 3D mesh to edit.

**Desmooth** reverses the effect of the **Smooth** surface options.

**Mclose** closes the mesh in the m-direction.

**Mopen** opens the mesh in the m-direction.

**Nclose** closes the mesh in the n-direction.

**Nopen** opens the mesh in the n-direction.

**Smooth surface** smooths the mesh with a Bezier-spline.

**Undo** undoes the most recent **PEdit** operation.

**eXit** exits the **PEdit** command.

Edit Vertex options
**Current vertex (0,0).**
**Enter an option [Next/Previous/Left/Right/Up/Down/Move/REgen/eXit] <N>:** *(Enter an option.)*

**Down** moves the x-marker down the mesh by one vertex.

**Left** moves the x-marker along the mesh by one vertex left.

**Move** relocates the vertex to a new position.

**Next** moves the x-marker along the mesh to the next vertex.

**Previous** moves the x-marker along the mesh to the previous vertex.

**REgen** regenerates the drawing to show the effects of the PEdit command.

**Right** moves the x-marker along the mesh by one vertex right.

**Up** moves the x-marker up the mesh by one vertex.

**eXit** exits the Edit-vertex submenu.

## RELATED COMMANDS

**Break** breaks a 2D polyline at any position.

**Chamfer** chamfers all vertices of a 2D polyline.

**Convert** converts older polylines to the new lwpolyline format.

**EdgeSurf** draws a 3D mesh.

**Fillet** fillets all vertices of a 2D polyline.

**PLine** draws a 2D polyline.

**RevSurf** draws a 3D surface of revolution mesh.

**RuleSurf** draws a 3D ruled surface mesh.

**TabSurf** draws a 3D tabulated surface mesh.

**3D** draws a 3D surface objects.

**3dPoly** draws a 3D polyline.

## RELATED SYSTEM VARIABLES

**SplFrame** determines the visibility of a polyline spline frame:

| SplFrame | Meaning |
|----------|---------|
| 0 | Do not display control frame (default). |
| 1 | Display control frame. |

**SplineSegs** specifies the number of lines used to draw a splined polyline (default = 8).

**SplineType** determines the Bezier-spline smoothing for 2D and 3D polylines:

| SplineType | Meaning |
|-----------|---------|
| 5 | Quadratic Bezier spline. |
| 6 | Cubic Bezier spline. |

**SurfType** determines the smoothing using the Smooth option:

| SurfType | Meaning |
|----------|---------|
| 5 | Quadratic Bezier spline. |
| 6 | Cubic Bezier spline. |
| 7 | Bezier surface. |

## TIP

• During vertex editing, mouse button #2 moves the x-marker to the next or previous vertex, whichever was used last.

# PFace

**Rel.11**  Draws multisided 3D meshes; meant for use by AutoLISP and ARx programs (*short for Poly FACE*).

| Command | Alias | Ctrl+ | F-key | Alt+ | Menu Bar | Tablet |
|---------|-------|-------|-------|------|----------|--------|
| pface | ... | ... | ... | ... | ... | ... |

**Command:** pface
**Specify location for vertex 1:** *(Pick a point.)*
**Specify location for vertex 2 or <define faces>:** *(Pick a point, or press **Enter**.)*
**Face 1, vertex 1:**
**Enter a vertex number or [Color/Layer]:** *(Type a number, or enter an option.)*
**Face 1, vertex 2:**
**Enter a vertex number or [Color/Layer] <next face>:** *(Type a number, or enter an option.)*
   *...and so on.*

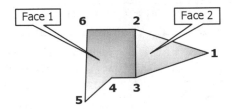

## COMMAND LINE OPTIONS

**Vertex** defines the location of a vertex.

**Face** defines the faces, based on vertices.

**Color** gives the face a different color.

**Layer** places the face on a different layer.

## RELATED COMMANDS

**3dFace** draws three- and four-sided 3D meshes.

**3dMesh** draws a 3D mesh with polyfaces.

## RELATED SYSTEM VARIABLE

**PFaceVMax** specifies the maximum number of vertices per polyface (default = 4).

**SplFrame** controls the display of invisible faces (default = 0, not displayed).

## TIPS

- **3dFace** creates 3- and 4-sided meshes, while **PFace** creates meshes of an arbitrary number of sides, and allows you to control the layer and the color of each face.

- The maximum number of vertices in the m- and n-direction is 256 when entered from the keyboard, and 32767 vertices when entered from a DXF file or created by programming.

- Make an edge invisible by entering a negative number for the beginning vertex of the edge.

- **PFace** is meant for programmers; to draw 3D surface objects, use these commands instead: **3d**, **3dMesh**, **RevSurf**, **RuleSurf**, **EdgeSurf**, or **TabSurf**.

# Plan

**Rel.10** Displays the plan view of the WCS or the UCS.

| Command | Alias | Ctrl+ | F-key | Alt+ | Menu Bar | Tablet |
|---------|-------|-------|-------|------|----------|--------|
| plan | ... | ... | ... | V3P | View | N3 |
| | | | | | ⌦3D Views | |
| | | | | | ⌦Plan View | |

**Command:** plan
**Enter an option [Current ucs/Ucs/World] <Current>:** *(Enter an option.)*

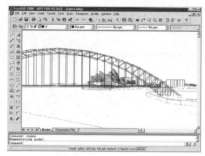

*Example of a 3D view.*

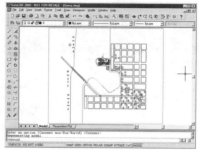

*After using the **Plan World** command.*

## COMMAND LINE OPTIONS

**Current UCS** shows the plan view of the current UCS.

**Ucs** shows the plan view of the named UCS.

**World** shows the plan view of the WCS.

## RELATED COMMANDS

**UCS** creates new UCS views.

**VPoint** changes the viewpoint of 3D drawings.

## RELATED SYSTEM VARIABLES

**UcsFollow** displays automatically the plan view for UCS or WCS.

**ViewDir** contains the x,y,z coordinates of the current view direction.

## TIPS

- Entering **VPoint 0,0,0** is an alternative to the **Plan** command.

- The **Plan** command turns off perspective mode and clipping planes.

- **Plan** does not work in paper space; AutoCAD complains, "** Command only valid in Model space **."

- The **Plan** command is an excellent method for turning off perspective mode.

# PLine

**V. 1.4** Draws complex 2D lines made of straight and curved sections of constant and variable width; treated as a single object (*short for Poly LINE*).

| Command | Alias | Ctrl+ | F-key | Alt+ | Menu Bar | Tablet |
|---------|-------|-------|-------|------|----------|--------|
| pline | pl | ... | ... | DP | Draw | N10 |
| | | | | | ⇘Polyline | |

**Command:** pline
**Specify start point:** *(Pick a point.)*
**Current line-width is 0.0000**
**Specify next point or [Arc/Halfwidth/Length/Undo/Width]:** *(Pick a point, or enter an option.)*
**Specify next point or [Arc/Close/Halfwidth/Length/Undo/Width]:** *(Pick a point, or enter an option.)*

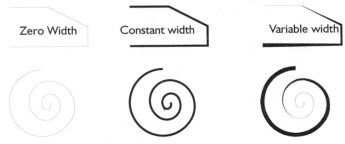

*Examples of polylines created with the **PLine** command.*

## COMMAND LINE OPTIONS

**Specify start point** indicates the start of the polyline.

**Close** closes the polyline with a line segment.

**Halfwidth** indicates the half-width of the polyline.

**Length** draws a polyline tangent to the last segment.

**Undo** erases the last-drawn segment.

**Width** indicates the width of the polyline.

**Endpoint of line** indicates the endpoint of the polyline.

Arc options
**Specify endpoint of arc or**
**[Angle/CEnter/CLose/Direction/Halfwidth/Line/Radius/Second pt/Undo/Width]:** *(Enter an option.)*

**Endpoint of arc** indicates the arc's endpoint.

**Angle** indicates the arc's included angle.

**CEnter** indicates the arc's center point.

**CLose** uses an arc to close a polyline.

**Direction** indicates the arc's starting direction.

**Halfwidth** indicates the arc's halfwidth.

**Line** switches back to the menu for drawing lines.

**Radius** indicates the arc's radius.

**Second pt** draws a three-point arc.

**Undo** erases the last drawn arc segment.

**Width** indicates the width of the arc.

## RELATED COMMANDS

**Boundary** draws a polyline boundary.

**Donut** draws solid-filled circles as polyline arcs.

**Ellipse** draws ellipses as polyline arcs when PEllipse = 1.

**Explode** reduces a polyline to lines and arcs with zero width.

**Fillet** fillets polyline vertices with a radius.

**PEdit** edits the polyline's vertices, widths, and smoothness.

**Polygon** draws polygons as polylines of up to 1024 sides.

**Rectang** draws a rectangle out of a polyline.

**Sketch** draws polyline sketches, when SkPoly = 1.

**Xplode** explodes a group of polylines into line and arcs of zero width.

**3dPoly** draws 3D polylines.

## RELATED SYSTEM VARIABLES

**PLineGen** specifies the style of linetype generation:

| PlineGen | Meaning |
| --- | --- |
| 0 | Vertex to vertex (default). |
| 1 | End to end (compatible with Release 12). |

**PLineType** controls the conversion of old (pre-R14) polylines and the creation of lwpolyline objects by the PLine command:

| PLineType | Meaning |
| --- | --- |
| 0 | Old polylines not converted; old-format polylines created. |
| 1 | Not converted; lwpolylines created. |
| 2 | Polylines in pre-R14 drawings converted on open; **PLine** command creates lwpolyline objects (default). |

**PLineWid** sets the current width of polyline (default = 0.0).

## TIPS

- **Boundary** uses a polyline to outline a region automatically; use the **List** command to find its area.

- If you cannot see a linetype on a polyline, change system variable **PLineGen** to 1; this regenerates the linetype from one end of the polyline to the other.

- If the angle between a joined polyline and polyarc is less than 28 degrees, the transition is chamfered; at greater than 28 degrees, the transition is not chamfered.

- Use the object snap modes **INTersection** or **ENDpoint** to snap to the vertices of a polyline.

· · · · · · · · · · · · · · · · · · · · · · · · · · · · · · · · · · · · ·

# Plot

**V.1.0**   Creates hardcopies of drawings on vector, raster, and PostScript plotters; also plots to files on disk.

| Commands | Alias | Ctrl+ | F-key | Alt+ | Menu Bar | Tablet |
|----------|-------|-------|-------|------|----------|--------|
| plot | print | P | ... | FP | File ↳ **Plot** | W25 |

-plot

**Command:** plot

*Displays dialog box:*

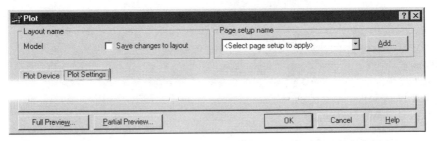

## DIALOG BOX OPTIONS

**Save changes to layout** toggles the saving of the settings in this dialog box.

**Page setup name** specifies the name of the page setup.

**Add** displays the User Defined Page Setup dialog box.

**OK** begins the plot.

**Cancel** dismisses the dialog box without plotting the drawing.

**Full Preview** displays a detailed preview of the drawing; see the Preview command.

**Partial Preview** displays an overview preview:

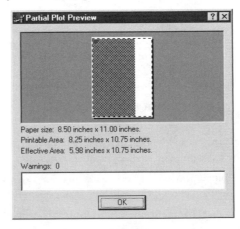

**Plot Device** tab

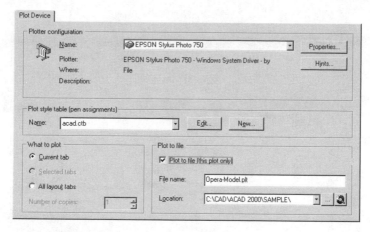

**Plotter Configuration options**

**Name** selects a system printer, or a named plotter configuration.

**Properties** displays the Plotter Configuration Editor dialog box; see PlotterManager command.

**Hints** displays an AutoCAD Help dialog box containing specific information about plotting with AutoCAD plotter configurations, or general information about plotting with the Windows system printer.

**Plot Style Table (Pen Assignments) options**

**Name** selects a plot style table assigned to the current tab; see the PlotStyle command.

**Edit** displays the Plot Style Table Editor dialog box.

**New** displays the Add Plot Style Table wizard.

**What to Plot options**

- **Current tab** plots the current model or layout tab; see the Layout command.
- **Selected tabs** plots all selected tabs; hold down CTRL key to select more than one tab.
- **All Layout Tabs** plots all layout tabs.

**Number of copies** specifies the number of copies to be plotted.

**Plot to File options**

**Plot to file** sends the plots to a file on disk.

**File name** specifies the file name for the plot (default = *filename-tab.plt*).

**Location** specifies the folder in which to place the plot file.

**...** displays a standard Browse for Folder dialog box.

**Browse** displays the Browse the Web window; see the Browser command.

## Plot Settings tab

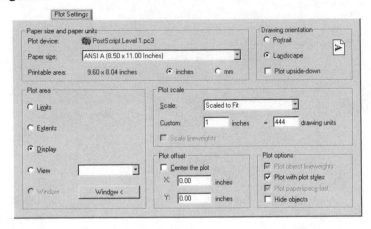

Paper Size and Paper Units options
   **Paper size** displays paper sizes supported by the selected output device.

   **Printable area** selects inches or mm as the plotting units; raster images show the size in pixels.

Drawing Orientation options
   • **Portrait** plots with the long edge at the side of the page.
   • **Landscape** plots with the long edge at the top of the page.
   **Plot upside-down** plots the drawing upside down, rotated by 180 degrees.

Plot Area options
   • **Layout** plots all parts of the drawing within the margins of the specified paper size; the origin is calculated from 0,0 in the layout.
   • **Limits** plots the drawing in the model tab within a rectangle defined by the Limits command.
   • **Extents** plots the entire drawing.
   • **Display** plots all parts of the drawing within the current viewport.
   • **View** plots a view saved with the View command.
   • **Window** plots all parts of the drawing within a picked rectangle.

Plot Scale options
   **Scale** specifies the plotting scale; select Scaled to Fit to plot the entire drawing to fit the page.
   **Custom** specifies a user-defined scale.
   **Scale lineweights** scales lineweights proportionately to the plot scale.

Plot Offset options
   **Center the plot** centers the plot on the paper.
   **X** specifies the plot origin offset from x = 0.
   **Y** specifies the plot origin offset from y = 0.

Plot Options options
   **Plot with lineweights** plots objects with lineweights.
   **Plot with plot styles** plots objects using the plot styles.
   **Plot paperspace last** plots the model space objects first, followed by paper space objects.
   **Hide objects** removes hidden lines before plotting.

## -PLOT Command

**Command:** -plot

**Detailed plot configuration? [Yes/No] <No>:** *(Type Y or N.)*

Brief Plot Configuration options

**Enter a layout name or [?] <Model>:** *(Enter a layout name, or type ?.)*
**Enter a page setup name <>:** *(Enter a page setup name, or press Enter for none.)*
**Enter an output device name or [?] <default printer>:** *(Enter a printer name, or type ? for a list of printers.)*
**Write the plot to a file [Yes/No] <N>:** *(Type Y or N.)*
**Save changes to page setup [Yes/No]? <N>** *(Type Y or N.)*
**Proceed with plot [Yes/No] <Y>:** *(Type Y or N.)*

Detailed Plot Configuration options

**Enter a layout name or [?] <Model>:** *(Enter a layout name, or type ?.)*
**Enter an output device name or [?] <default printer>:** *(Enter a printer name, or type ? for a list of printers.)*
**Enter paper size or [?] <Letter 8 ½ x 11 in>:** *(Enter a paper size, or type ?.)*
**Enter paper units [Inches/Millimeters] <Inches>:** *(Type I or M.)*
**Enter drawing orientation [Portrait/Landscape] <Landscape>:** *(Type P or L.)*
**Plot upside down? [Yes/No] <No>:** *(Type Y or N.)*
**Enter plot area [Display/Extents/Limits/View/Window] <Display>:** *(Enter an option.)*
**Enter plot scale (Plotted Inches=Drawing Units) or [Fit] <Fit>:** *(Enter a scale factor, or type F.)*
**Enter plot offset (x,y) or [Center] <0.00,0.00>:** *(Enter an offset, or type C.)*
**Plot with plot styles? [Yes/No] <Yes>:** *(Type Y or N.)*
**Enter plot style table name or [?] (enter . for none) <>:** *(Enter a name, or enter an option.)*
**Plot with lineweights? [Yes/No] <Yes>:** *(Type Y or N.)*
**Enter shade plot setting [As displayed/Wireframe/Hidden/Rendered] <As displayed>:** *(Enter an option.)*
**Write the plot to a file [Yes/No] <N>:** *(Type Y or N.)*
**Save changes to page setup? Or set shade plot quality? [Yes/No/Quality] <N>:** *(Enter an option.)*
**Proceed with plot [Yes/No] <Y>:** *(Type Y or N.)*

### COMMAND OPTIONS

**Enter a layout name** specifies the name of a layout; enter Model for model view or ? for a list of layout names.

**Enter an output device name** specifies the name of a printer, or enter ? for a list of print devices.

**Enter paper size** specifies the name of a paper size, or enter ? for a list of paper size supported by the print device.

**Enter paper units** specifies Inches (imperial) or Millimeters (metric) units.

**Enter drawing orientation** specifies Portrait (vertical) or Landscape (horizontal) orientation.

**Plot upside down:**

Yes plots the drawing upside down.

No plots the drawing normally.

**Enter plot area** specifies that the current Display, Extents of the drawing, Limits defined by the Limits command, View name, or Window area be plotted.

**Enter plot scale** specifies the scale using the Plotted Inches=Drawing Units format; or enter F to fit the drawing to the page

**Enter plot offset** specifies the distance between the lower left corner of the paper and the drawing; or enter C to center the drawing on the page.

**Plot with plot styles:**

Yes prompts you to specify the name of a plot style.

No ignores plot styles, and uses color-dependent styles.

**Plot with lineweights:**

Yes uses lineweight settings to draw thicker lines.

No ignores lineweights.

**Enter shade plot setting** specifies whether the plot should be rendered As displayed, Wireframe, Hidden lines removed, or Rendered.

**Write the plot to a file:**

Yes sends the plot to a file.

No sends the plot to the plot device.

**Save changes to page setup:**

Yes saves the settings you specified during the previous set of questions.

No does not save the settings.

**Quality** specifies the quality of the plot.

**Proceed with plot:**

Yes plots the drawing.

No cancels the plot.

Quality options

**Enter shade plot quality [Draft/Preview/Normal/pResentation/Maximum/ Custom] <Normal>:** *(Select an option.)*

**Draft, Preview, Normal, pResentation, and Maximum** apply preset dpi (dots per inch) settings to the plot. The higher the dpi, the better the quality, but the slower the plot.

**Custom** allows you to specify the dpi setting; default = 150dpi.

## RELATED COMMANDS

**PageSetup** selects one or more plotter devices.

**Preview** goes directly to the plot preview screen.

**Publish** plots multi-page drawings.

**PublishToWeb** plots drawings to HTML format.

**PsOut** saves drawings in *.eps* format.

## RELATED SYSTEM VARIABLES

**PlotId** (*Obsolete*) specifies the currently-selected plotter number.

**Plotter** (*Obsolete*) specifies the currently-selected plotter name.

**TextFill** toggles the filling of TrueType fonts.

**TextQlty** specifies the quality of TrueType fonts.

## RELATED FILES

**\*.pc3** holds plotter configuration parameter files.

**\*.plt** holds plot files created with the Plot command.

## TIPS

- As of AutoCAD Release 12, the **Plot** command replaces the **PrPlot** command.

- As of AutoCAD Release 13, the freeplot feature, which starts AutoCAD to plot with the **-p** parameter (without using up a network license) is no longer available.

- AutoCAD cannot plot perspective view to scale, only to fit.

- To plot drawing sets, use the **Publish** command.

# 'PlotStamp

<u>2000i</u>  Adds information about drawings to hardcopy plots.

| Commands | Alias | Ctrl+ | F-key | Alt+ | Menu Bar | Tablet |
|----------|-------|-------|-------|------|----------|--------|
| 'plotstamp | ... | ... | ... | FPS | File | ... |
| | | | | | ⮬Plot | |
| | | | | | ⮬Settings | |

-plotstamp

**Command:** plotstamp

*Displays dialog box:*

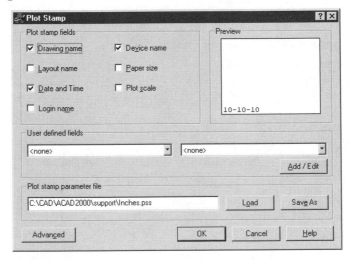

## DIALOG BOX OPTIONS

Plot Stamp Fields options

**Drawing name** adds the path and name of the drawing.

**Layout name** adds the layout name.

**Date and time** adds the date (short format) and time of the plot.

**Login name** adds the Windows login name, as stored in the LogInName system variable.

**Device name** adds the plotting device's name.

**Paper size** adds the paper size, as currently configured.

**Plot scale** adds the plot scale.

User Defined Fields options

**Add/Edit** displays the User Defined Fields dialog box.

Plot Stamp Parameter File options

**Load/Save** displays the Plotstamp Parameter File Name dialog box for opening (or saving) *.pss* files.

**Advanced** displays the Advanced Options dialog box.

**Advanced** dialog box

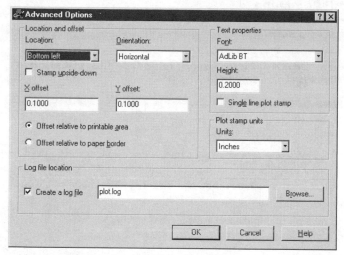

## Location and Offset options

**Location** specifies the location of the plot stamp: Top Left, Bottom Left (default), Bottom Right, or Top Right — relative to the orientation of the drawing on the plotted page.

**Stamp upside down:**

    **Yes** draws the plot stamp upside down.

    **No** draws the plot stamp rightside up.

**Orientation** rotates the plot stamp relative to the page: Horizontal or Vertical.

**X offset** specifies the distance from the corner of the printable area or page; default = 0.1.

**Y offset** specifies the distance from the corner of the printable area or page; default = 0.1

• **Offset relative to printable area.**

• **Offset relative to paper border.**

## Text Properties options

**Font** selects the font for the plot stamp text.

**Height** specifies the height of the plot stamp text.

**Single line plot stamp:**

    **Yes** plotstamp text is placed on a single line.

    **No** plotstamp test is placed on two lines.

## Plot Stamp Units options

**Units** selects either inches, millimeters, or pixels.

## Log File Location options

**Create a log file** saves the plot stamp text to a text file.

**Browse** displays the Log File Name dialog box.

## -PLOTSTAMP Command
**Command: -plotstamp**
**Current plot stamp settings:**

*Displays the current setting of nearly 20 plot stamp settings.*

**Enter an option [On/OFF/Fields/User Fields/Log file/LOCation/Text properties/UNits]:** *(Enter an option.)*

### COMMAND LINE OPTIONS

**On** turns on plot stamping.

**OFF** turns off plot stamping.

**Fields** specifies plot stamp data: drawing name, layout name, date and time, login name, plot device name, paper size, plot scale, comment, write to log file, log file path, location, orientation, offset, offset relative to, units, font, text height, and stamp on single line.

**User fields** specifies two user-defined fields.

**Log file** writes the plotstamp data to a file instead of the drawing.

**LOCation** specifies the location and orientation of the plotstamp.

**Text properties** specifies the font name and height.

**UNits** specifies the units of measurement: inches, millimeters, or pixels.

### RELATED COMMAND

**Plot** plots the drawing.

### RELATED FILES

***.log** is the plotstamp log file; stored in ASCII text format.

***.pss** holds plotstamp parameter file; stored in binary format.

### TIPS

- When the options of the **Plot Stamp** dialog box are grayed out, or when the **-Plotstamp** command reports "Current plot stamp file or directory is read only," this means that the Inches.Pss or Mm.Pss file in the \Support folder is read-only. To change, in Explorer: (1) right-click the file; (2) select **Properties**; (3) uncheck **Read-only**; and (4) click **OK**.

- You can access the **Plot Stamp** dialog box via the **Plot** dialog box's **Plot Device** tab:

- *Caution!* Too large of an offset value positions the plotstamp text beyond the plotter's printable area, which causes the text to be cut off. To prevent this, use the **Offset relative to printable area** option.

# PlotStyle

2000 Selects and assigns plot styles to objects.

| Commands | Alias | Ctrl+ | F-key | Alt+ | Menu Bar | Tablet |
|----------|-------|-------|-------|------|----------|--------|
| plotstyle | ... | ... | ... | OY | Format | ... |
| | | | | | ⌐Plot Style | |
| -plotstyle | | | | | | |

**Command:** plotstyle

*When no objects are selected, displays the* **Current Plot Style** *dialog box.*

*When one or more objects are selected, displays the* **Select Plot Style** *dialog box.*

## DIALOG BOX OPTIONS

**Plot styles** lists the available plot styles.

**Active plot style table** lists the available plot style tables.

**Editor** displays the Plot Style Table Editor dialog box; see the StylesManager command.

**OK** accepts the changes, and closes the dialog box.

**Cancel** cancels the changes, and closes the dialog box.

## Select Plot Style dialog box

*Displayed when one or more objects are selected:*

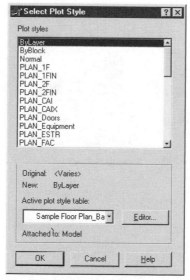

**Plot styles** lists the plot styles available in the drawing.

**Active plot style table** selects the plot style table to attach to the current drawing.

**Editor** displays the Plot Style Table Editor dialog box; see the StylesManager command.

**Current Plot Style** dialog box

*Displayed when no objects are selected:*

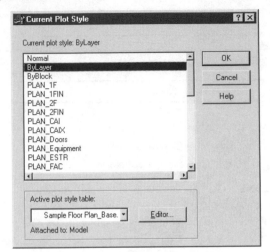

**Plot styles** lists the plot styles available in the drawing.

**Active plot style table** selects the plot style table to attach to the current drawing.

**Editor** displays the Plot Style Table Editor dialog box; see the StylesManager command.

## -PLOTSTYLE Command

**Command:** -plotstyle
**Current plot style is "Default"**
**Enter an option [?/Current] :** *(Type ? or C.)*
**Set current plot style :** *(Enter a name.)*
**Current plot style is "*plotstylename*"**
**Enter an option [?/Current] :** *(Press Enter.)*

### COMMAND LINE OPTIONS

**Current** prompts you to change plot styles.

**Set current plot style** specifies the name of the plot style.

ENTER exits the command.

**?** displays plot style names; sample output:

```
Plot Styles:
------------------
ByLayer
ByBlock
Normal
ARCH_Dimensions for stairway
Current plot style is "PLAN_ESTR"
```

### RELATED COMMANDS

**Plot** plots drawings with plot styles.

**PageSetup** attaches plot style tables to layouts.

**StylesManager** modifies plot style tables.

**Layer** specifies plot styles for layers.

### RELATED SYSTEM VARIABLES

**CPlotStyle** specifies the plot style for new objects; defined values include "ByLayer," "ByBlock," "Normal," and "User Defined."

**DeflPlStyle** specifies the current plot style name.

### TIPS

- This command does not operate until a plot style table has been created for the drawing; the **PlotStyle** command displays this message box:

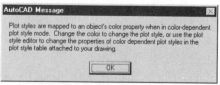

- A plot style can be assigned to any object and to any layer.

- A plot style can override the following plot settings: color, dithering, gray scale, pen assignment, screening, linetype, lineweight, end style, join style, and fill style.

- Plot styles are useful for plotting the same layout in different ways, such as emphasizing objects using different lineweights or colors in each plot.

- Plot style tables can be attached to the **Model** tab and layout tabs, and attach different plot style tables to layouts, to create different looks for plots.

## Quick Start Tutorial
# Applying Plotstyles

Before you can apply plot styles to a new drawing, you must turn on plot styles; notice that **Plot Style Control** in the **Object Properties** toolbar is grayed out.

Follow these steps to turn on plot style:

**Step 1**

From the menu bar, select **Tools | Options | Plotting**.

**Step 2**

In the **Options** dialog box, click **Use named plot styles** to turn on the feature.

**Step 3**

In the **Default** plot style list, select any plot style — *except* **Default R14 pen assignments.stb**, since it turns off plot styles.

**Step 4**

Click **OK** to dismiss the **Options** dialog box. Notice that the **Plot Style Control** in the **Object Properties** toolbar is now available for use.

# PlotterManager

**2000** Displays the Plotters window, the Add-A-Plotter wizard, and PC3 configuration Editor.

| Command | Alias | Ctrl+ | F-key | Alt+ | Menu Bar | Tablet |
|---------|-------|-------|-------|------|----------|--------|
| plottermanager | ... | ... | ... | FM | File | Y24 |
| | | | | | ⬐Plotter Manager | |

**Command:** plottermanager

*Displays window:*

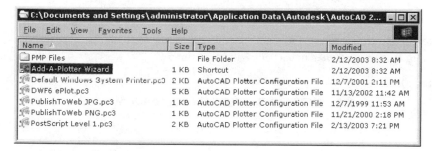

## WINDOW OPTIONS

**Add-a-Plotter Wizard** adds a plotter configuration; double-click to display the **Add Plotter** wizard.

**\*.pc3** specifies parameters for creating plotted output; double-click to display the **Plotter** Configuration Editor; see the StylesManager command.

## ADD PLOTTER WIZARD

*The following steps show how to create a PC3 plotter configuration for plotting drawings to an EPS (encapsulated PostScript) file; the steps are similar to creating PC3 files for other types of plotters:*

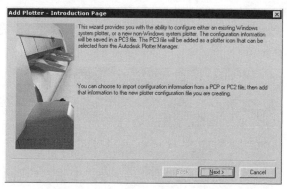

**Next** displays the next dialog box.
**Cancel** dismisses the dialog box.

**Begin** page

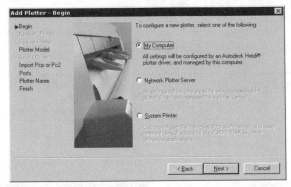

**My Computer** configures printers and plotters connected to your computer.

**Network Plotter Server** configures printers and plotters connected other computers on network.

**System Printer** configures the default Windows printer.

**Plotter Model** page

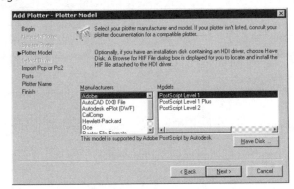

**Manufacturer** selects a brand name of plotter.

**Model** selects a specific model number.

**Have Disk** selects plotter drivers provided by the manufacturer; displays the Open dialog box.

**Import Pcp or Pc2** page

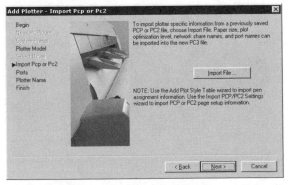

**Import File** imports *.pcp* and *.pc2* plotter configuration files from earlier versions of AutoCAD.

**Ports** page

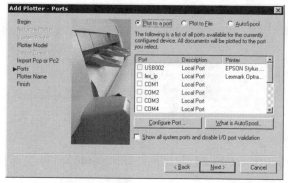

**Plot to a port** sends the plot to an output port.

**Plot to File** plots the drawing to a file with a user-definable filename; default filename is the same as the drawing name with the *.plt* extension; PostScript plot files are given the *.eps* extension.

**AutoSpool** plots the drawing to a file with a filename generated by AutoCAD, and then executes the command specified in the Option dialog box's Files tab. Enter the name of the program that should process the AutoSpool file in **Print Spool Executable**. You may include these DOS command-line arguments:

| Argument | Meaning |
| --- | --- |
| %s | Substitutes path, spool filename, and extension. |
| %d | Specifies the path, AutoCAD drawing name, and extension. |
| %c | Specifies the description for the device. |
| %m | Returns the plotter model. |
| %n | Specifies the plotter name. |
| %p | Specifies the plotter number. |
| %h | Returns the height of the plot area in plot units. |
| %w | Returns the width of the plot area in plot units. |
| %i | Specifies the first letter of the plot units. |
| %l | Specifies the login name (**LogInName** *system variable*). |
| %u | Specifies the user name. |
| %e | Specifies the equal sign (=). |
| %% | Specifies the percent sign (%). |

Port options

**Port** lists the virtual ports defined by the Windows operating system:

| Port | Meaning |
| --- | --- |
| **USB** | Universal serial bus. |
| **COM** | Serial port. |
| **LPT** | Parallel port. |
| **HDI** | Autodesk's Heidi Device Interface. |

**Description** describes the type of port:

| Description | Meaning |
| --- | --- |
| **Local Port** | Printer is connected to your computer. |
| **Network Port** | Printer is accessible through the network. |

**Printers** describes the brand name of the printer connected to the port.

**Configure Port** displays the Configure Port dialog box; allows you to specify parameters specific to the port, such as timeout and protocol.

**What is AutoSpool?** displays an explanatory dialog box:

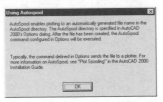

**Show all system ports and disable I/O port validation** prevents AutoCAD from checking whether the port is valid.

**Plotter Name** page

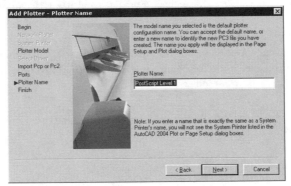

**Plotter name** Specifies a user-defined name for the plotter configuration; you may have many different configurations for a single plotter.

**Finish** page

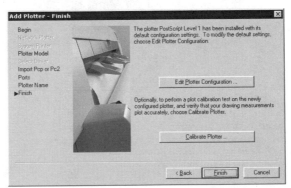

**Edit Plotter Configuration** displays the Edit Plotter Configuration dialog box.

**Calibrate Plotter** displays the Calibrate Plotter wizard.

## Plotter Configuration Editor dialog boxes

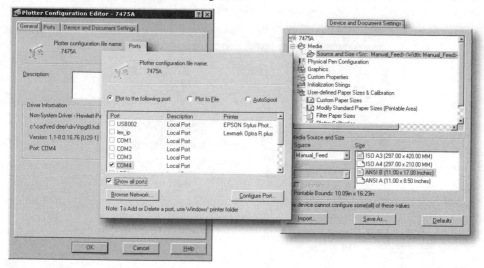

### General tab

**Description** allows you to provide a detailed description of the plotter configuration.

### Ports tab

**Plot to the following port** sends the plot to an output port.

**Plot to File** plots the drawing to a file with a user-definable filename; default filename is the same as the drawing name, with a PLT extension; PostScript plot files are given the EPS extension.

**AutoSpool** plots the drawing to a file with a filename generated by AutoCAD, and then executes the command specified in the Option dialog box's Files tab.

**Show all ports** lists all ports on the computer.

**Browse Network** displays the Browse for Printer dialog box; selects a printer on the network.

**Configure Port** displays the Configure Port dialog box; allows you to specify the parameters specific to the port, such as timeout and protocol.

### Device and Document Settings tab

**Media** specifies the paper source, paper size, type, and destination.

**Physical Pen Configuration** (for pen plotters only) specifies the physical pens in the pen plotter.

**Graphics** specifies settings for plotting vector and raster graphics and TrueType fonts.

**Custom Properties** specifies settings specific to the plotter, printer, or other output device.

**Initialization Strings** (for non-system plotters only) specifies the control codes for pre-initialization, post-initialization, and termination.

**Calibration Files and Paper Sizes** calibrates the plotter by specifying the *.pmp* file; adds and modifies custom paper sizes; see the Calibrate Plotter wizard.

**Import** imports *.pcp* and *.pc2* plotter configuration files from earlier versions of AutoCAD.

**Save As** saves the plotter configuration data to a *.pc3* file.

**Defaults** resets the plotter configuration settings to the previously-saved values.

## CALIBRATE PLOTTER WIZARD

*Before using this wizard, use AutoCAD to draw and plot a rectangle — say, 11 inches by 8 inches — then use this wizard to check the accuracy of the plotted drawing:*

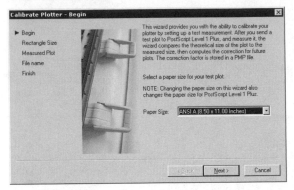

**Paper Size** selects the size of media or paper.

**Next** displays the next dialog box.

**Cancel** dismisses the dialog box.

### Rectangle Size page

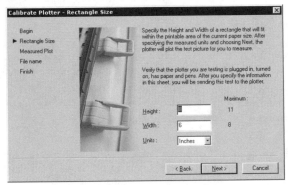

**Length** specifies the length of the calibration rectangle.

**Width** specifies the width of the calibration rectangle.

**Units** selects imperial inches or metric millimeters.

**Measured Plot** page

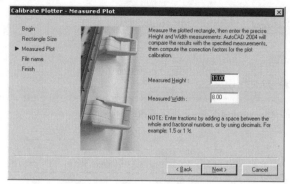

**Measured Length** specifies the length of the rectangle, which is plotted by AutoCAD, and then measured by you.

**Measured Width** specifies the width of the rectangle, which is plotted by AutoCAD, and then measured by you.

**File Name** page

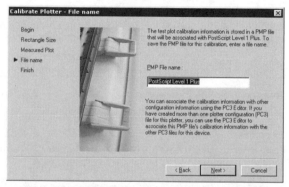

**PMP File name** specifies the name of the file in which to store the calibration data.

**Finish** page

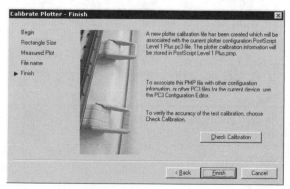

**Check Calibration** returns to the Calibrate Plotter - Rectangle Size dialog box.

**Finish** returns to the Add Plotter - Finish dialog box.

## RELATED COMMANDS

**Plot** plots drawings with plot styles.

**PageSetup** attaches plot style tables to layouts.

**StylesManager** modifies plot style tables.

**AddPlWiz.Exe** runs the Add Plotter wizard.

## RELATED SYSTEM VARIABLES

**PaperUpdate** toggles the display of a warning before AutoCAD plots a layout with a paper size different from the size specified by the plotter configuration file:

| PaperUpdate | Meaning |
|---|---|
| 0 | Displays warning dialog box (default). |
| 1 | Changes paper size to match the size specified by the plotter configuration file. |

**PlotId** (*Obsolete*) holds the current plotter configuration ID number.

**PlotRotMode** controls the orientation of plots:

| PlotRotMode | Meaning |
|---|---|
| 0 | Aligns rotation icon with media at the lower left for 0 degrees; calculates x and y origin offsets relative to lower-left corner. |
| 1 | Aligns the lower-left corner of plotting area with lower-left corner of the paper. |
| 2 | Same as 0, except x and y origin offsets relative to the rotated origin position. |

**Plotter** (*Obsolete*) holds the current plotter name.

## TIPS

• You can create and edit PC3 plotter configuration files without AutoCAD. From the Start button on the Windows toolbar, select **Settings | Control Panel | Autodesk Plotter Manager**.

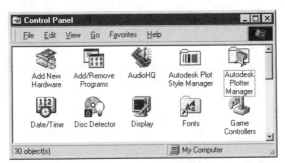

• The *dwf classic.pc3* file specifies parameters for creating *.dwf* files via the Release 14-compatible **DwfOut** command.

• The *dwf eplot.pc3* file specifies parameters for creating *.dwf* files via the AutoCAD 2004-preferred **Plot** command.

# PngOut

<u>2004</u>  Exports the current view in PNG raster format.

| Command | Alias | Ctrl+ | F-key | Alt+ | Menu Bar | Tablet |
|---------|-------|-------|-------|------|----------|--------|
| pngout  | ...   | ...   | ...   | ...  | ...      | ...    |

**Command:** pngout

*Displays* **Create Raster File** *dialog box. Specify a filename, and then click* **Save.**

**Select objects or <all objects and viewports>:** *(Select objects, or press* **Enter** *to select all objects and viewports.)*

## COMMAND LINE OPTIONS

**Select objects** selects specific objects.

**All objects and viewports** selects all objects and all viewports, whether in model space or in layout mode.

## RELATED COMMANDS

**BmpOut** exports drawings in BMP (bitmap) format.

**Image** places raster images in the drawing.

**JpgOut** exports drawings in JPEG (joint photographic expert group) format.

**TifOut** exports drawings in TIFF (tagged image file format) format.

## TIPS

- The rendering effects of the **ShadEdge** command are preserved, but not those of the **Render** command.

- PNG files are used as a royalty-free alternative to JPEG files.

- PNG is short for "portable network graphics."

# Point

**V. 1.0**  Draws a 3D point.

| Command | Alias | Ctrl+ | F-key | Alt+ | Menu Bar | Tablet |
|---------|-------|-------|-------|------|----------|--------|
| point | po | ... | ... | DO | Draw | O9 |
|  |  |  |  |  | ⸜Point |  |

**Command:** point
**Current point modes: PDMODE=0 PDSIZE=0.0000**
**Specify a point:** *(Pick a point.)*

## COMMAND LINE OPTION
**Point** positions a point, or enters a 2D or 3D coordinate.

## RELATED COMMANDS
**DdPType** displays a dialog box for selecting PsMode and PdSize.

**Regen** regenerates display to see the new point mode or size.

## RELATED SYSTEM VARIABLES
**PDMode** determines the appearance of a point:

| 0 | 1 | 2 | 3 | 4 |
|---|---|---|---|---|
| · | | + | ✕ | ' |

| 32 | 33 | 34 | 35 | 36 |
|----|----|----|----|----|
| ⊙ | ○ | ⊕ | ⊗ | ◔ |

| 64 | 65 | 66 | 67 | 68 |
|----|----|----|----|----|
| □ | □ | ⊞ | ⊠ | ⊓ |

| 96 | 97 | 98 | 99 | 100 |
|----|----|----|----|-----|
| ▢ | ▢ | ⊕ | ⊠ | ◫ |

**PDSize** determines the size of a point:

| PdSize | Meaning |
|--------|---------|
| 0 | 5% of height of the **ScreenSize** system variable (default). |
| 1 | No display. |
| -10 | *(Negative)* Ten percent of the viewport size. |
| 10 | *(Positive)* Ten pixels in size. |

## TIPS
- The size and shape of the point is determined by **PdSize** and **PdMode**; changing these changes the appearance and size of all points in the drawing with the next regeneration.

- Entering only x,y coordinates places the point at a z coordinate of the current elevation; setting **Thickness** to a value draws the point as a line in 3D space.

- Prefix coordinates with * ( such as *1,2,3) to place points in the WCS, rather than in the UCS.

- Use the object snap mode **NODe** to snap to a point.

# Polygon

**V. 2.5**  Draws 2D polygons of between three to 1024 sides.

| Command | Alias | Ctrl+ | F-key | Alt+ | Menu Bar | Tablet |
|---------|-------|-------|-------|------|----------|--------|
| polygon | pol | ... | ... | DY | Draw | P10 |
| | | | | | ↳Polygon | |

**Command:** polygon
**Enter number of sides <4>:** *(Enter a number.)*
**Specify center of polygon or [Edge]:** *(Pick a point, or type E.)*
**Enter an option [Inscribed in circle/Circumscribed about circle] <I>:** *(Type I or C.)*
**Specify radius of circle:** *(Enter a value, or pick two points.)*

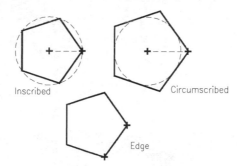

Inscribed    Circumscribed

Edge

## COMMAND LINE OPTIONS
**Center of polygon** indicates the center point of the polygon; then:
**C** (*Circumscribed*) fits the polygon outside a circle.
**I** (*Inscribed*) fits the polygon inside a circle.
**Edge** draws the polygon based on the length of one edge.

## RELATED COMMANDS
**PEdit** edits polylines, include polygons.
**Rectang** draws squares and rectangles.

## RELATED SYSTEM VARIABLE
**PolySides** specifiers the most-recently entered number of sides (default = 4).

## TIPS
* Polygons are drawn as a polyline; use **PEdit** to edit the polygon, such as its width.

* The pick point determines the polygon's first vertex; polygons are drawn counterclockwise.

* Use the system variable **PolySides** to preset the default number of polygon sides.

* Use the **Snap** command to place the polygon precisely; use **INTersection** or **ENDpoint** object snap modes to snap to the polygon's vertices.

## Removed Command
**Preferences** was removed from AutoCAD 2000; it was replaced by **Options**.

# Preview

<u>Rel.13</u> Displays plot preview; bypasses the **Plot** command.

| Command | Alias | Ctrl+ | F-key | Alt+ | Menu Bar | Tablet |
|---------|-------|-------|-------|------|----------|--------|
| preview | pre | ... | ... | FV | File | X24 |
| | | | | | ↳Plot Preview | |

**Command:** preview
**Press ESC or ENTER to exit, or right-click to display shortcut menu.**
*Displays preview screen:*

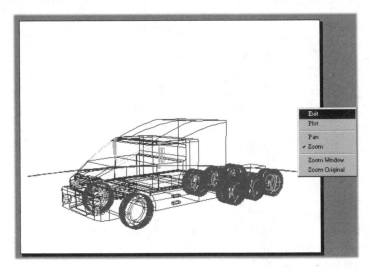

## COMMAND LINE OPTION
ESC returns to the drawing window.

## SHORTCUT MENU OPTIONS
Exit preview ——— Exit
Plot drawing ——— Plot
Pan view ——— Pan
Zoom view ——— ✓ Zoom
Zoom window ——— Zoom Window
Original view ——— Zoom Original

## RELATED COMMANDS
**Plot** plots the drawing.

**PageSetup** enables plot preview once a plotter is assigned to the layout.

## TIPS
• This command does not operate when no plotter is assigned; use the **PageSetup** command to assign a plotter to the model and layout tabs.

• Press ESC to exit preview mode.

# 'Properties

**2000** Opens the **Properties** window for modifying properties of selected objects.

| Command | Aliases | Ctrl+ | F-key | Alt+ | Menu Bar | Tablet |
|---------|---------|-------|-------|------|----------|--------|
| 'properties | ch | 1 | ... | TP | Tools | Y12-Y13 |
| | props | | | | ⌐Properties | |
| | ddchprop | | | MP | Modify | |
| | ddmodify | | | | ⌐Properties | |
| | mo | | | | | |

**Command:** properties

*When no objects are selected:*

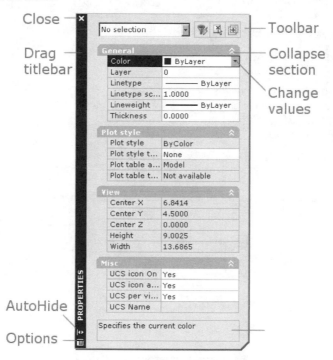

Close — Toolbar
Drag titlebar — Collapse section
Change values
AutoHide
Options

**TOOLBAR OPTIONS**

*Selection* lists the selected objects.

**Quick Select** displays the Quick Select dialog box; see the QSelect command.

**Select Objects** prompts at the 'Command:' prompt, "Select objects: ."

**Toggle Value of PickAdd Variable** controls additional selections:

**Off** adds objects to the selection set while the SHIFT key is held.

**On** adds objects to the selection set, unless SHIFT key is held.

## OPTIONS MENU

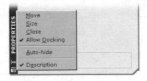

**Move** moves the window.

**Size** changes the size of the window.

**Close** closes the window.

**Allow Docking** toggles whether the window can be docked at the side of AutoCAD.

**Auto-hide** toggles whether the window reduces its size when the cursor is elsewhere.

**Description** toggles the display of the description box.

### RELATED COMMANDS

**ChProp** changes an object's color, layer, linetype, and thickness.

**Style** creates and changes text styles.

### RELATED SYSTEM VARIABLES

**CeColor** specifies the current color.

**CeLtScale** specifies the current linetype scale factor.

**CeLtype** specifies the current linetype.

**CLayer** specifies the current layer.

**Elevation** specifies the current elevation in the z-direction.

**Thickness** specifies the current thickness in the z-direction.

### TIPS

• Use the **Selection** list to count objects in the drawing. Use CTRL+A to select all objects, and then click the **Selection** list:

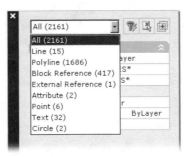

• Double-click the title bar to dock and undock the window within the AutoCAD window.

• The **Properties** window can be dragged larger and smaller by its edges.

• When an item is displayed by gray text, it cannot be modified.

• Bodies, 3D solids, and 2D regions cannot be edited beyond the items in the **General** section.

- When one or more objects are selected, the items displayed by the **Properties** window vary; here are three examples:

### Arc

### Mtext

### Linear Dimension

**'PropertiesClose**

**2000** Closes the **Properties** window.

| Command | Alias | Ctrl+ | F-key | Alt+ | Menu Bar | Tablet |
|---|---|---|---|---|---|---|
| 'propertiesclose | prclose | 1 | ... | TP | Tools | |
| | | | | | ↳ **Properties** | |

**Command:** propertiesclose

**COMMAND LINE OPTIONS**
*None.*

**RELATED COMMANDS**
> **Properties** displays the Properties window.
> **AdClose** closes the AutoCAD DesignCenter.
> **Close** closes the current drawing.
> **Exit** closes AutoCAD.

**RELATED SYSTEM VARIABLES**
*None.*

**TIP**
- As an alternative to entering the **PropertiesClose** command, you can click the **x** button on the **Properties** window title bar:

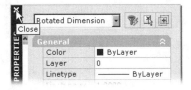

.  .  .  .  .  .  .  .  .  .  .  .  .  .  .  .  .  .  .  .  .  .  .  .  .  .  .  .  .  .  .  .  .  .  .  .  .  .  .  .  .  .  .  .  .  .  .  .  .  .  .  .  .  .  .  .  .  .  .  .  .

**Removed Command**
The **PsDrag** command was discontinued with AutoCAD 2000i. It has no replacement.

.  .  .  .  .  .  .  .  .  .  .  .  .  .  .  .  .  .  .  .  .  .  .  .  .  .  .  .  .  .  .  .  .  .  .  .  .  .  .  .  .  .  .  .  .  .  .  .  .  .  .  .  .  .  .  .  .  .  .  .  .

# PSetupIn

<u>**2000**</u>  Imports user-defined page setups into the current drawing layout (*short for Page SETUP IN*).

| Commands | Alias | Ctrl+ | F-key | Alt+ | Menu Bar | Tablet |
|----------|-------|-------|-------|------|----------|--------|
| psetupin | ... | ... | ... | ... | ... | ... |
| -psetupin | | | | | | |

**Command:** psetupin

*Displays the **Select File** dialog box. Select a .dwg, .dwt, or .dxf file, and then click Open.*
*AutoCAD displays dialog box:*

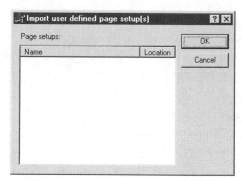

## DIALOG BOX OPTIONS

**Name** lists the names of page setups.

**Location** lists the location of the setups.

**OK** closes the dialog box, and loads the page setup.

. . . . . . . . . . . . . . . . . . . . . . . . . . . . . . . . . . . . . . . . . . . . . .

## -PSETUPIN Command

**Command:** -psetupin
**Enter filename:** *(Enter .dwg file name.)*
**Enter user defined page setup to import or [?]:** *(Enter a name, or type ?.)*

## COMMAND LINE OPTIONS

**Enter filename** enters the name of a drawing file.

**Enter user defined page setup to import** enters the name of a page setup.

**?** lists the names of page setups in the drawing.

## RELATED COMMANDS

**PageSetup** creates a new page setup configuration.

**Plot** plots the drawing.

## RELATED SYSTEM VARIABLE

**CTab** contains the the the name of the current model or layout tab in the drawing (default = "Model").

. . . . . . . . . . . . . . . . . . . . . . . . . . . . . . . . . . . . . . . . . . . . . .

# PsFill

<u>Rel.12</u>  Fills 2D polyline outlines with raster PostScript patterns (*short for PostScript FILL; undocumented command*).

| Command | Alias | Ctrl+ | F-key | Alt+ | Menu Bar | Tablet |
|---------|-------|-------|-------|------|----------|--------|
| psfill | ... | ... | ... | ... | ... | ... |

**Command:** psfill
**Select polyline:** *(Pick an outline.)*
**Enter PostScript fill pattern name (. = none) or [?] <.>:** *(Enter pattern name, or type **?**.)*

## COMMAND LINE OPTIONS

**Select polyline** selects the closed polyline to fill.

**PostScript pattern** specifies the name of the fill pattern.

**.** (dot) selects no fill pattern.

**?** lists the available fill patterns.

**\*** specifies that the pattern should not be outlined with a polyline.

## RELATED COMMANDS

**BHatch** fills an area with vector hatch patterns and solid colors.

**PsOut** exports drawings as PostScript files.

## RELATED SYSTEM VARIABLE

**PSQuality** specifies the display options for PostScript images:

| PsQuality | Meaning |
|-----------|---------|
| 75 | (*Positive*) Displays filled image at 75dpi (default). |
| 0 | Displays bounding box and filename; no image. |
| -75 | (*Negative*) Displays image outline at 75dpi; no fill. |

## TIPS

- These fill patterns are available:

| | |
|---|---|
| Grayscale | RGBcolor |
| AIlogo | Lineargray |
| Radialgray | Square |
| Waffle | Zigzag |
| Stars | Brick |
| Specks | |

- You cannot see fill patterns in AutoCAD drawings until they are exported with the **PsOut** command.

## Removed Command

**PsIn** command was removed with AutoCAD 2000i; there is no replacement.

# PsOut

Rel.12 Exports the current drawing as an encapsulated PostScript file (*undocumented command*).

| Command | Alias | Ctrl+ | F-key | Alt+ | Menu Bar | Tablet |
|---------|-------|-------|-------|------|----------|--------|
| psout | ... | ... | ... | FE | File | ... |
| | | | | ⮦EPS | ⮦Export | |
| | | | | | ⮦Encapsulated PS | |

**Command:** psout

*Displays the **Create PostScript File** dialog box.*

*Select **Tools | Options** to display dialog box:*

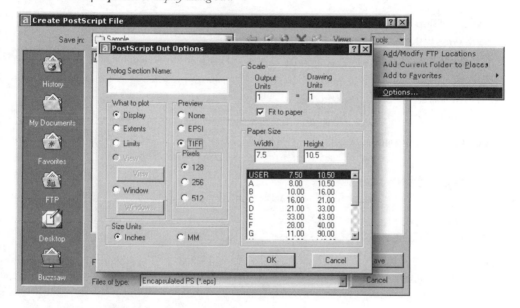

## DIALOG BOX OPTIONS

**Prolog Section Name** *(optional)* specifies the name of the prolog section, which is read from the *acad.psf* file and customizes the PostScript output.

What to Plot options

**Display** selects the current display in the current viewport.

**Extents** selects the drawing extents.

**Limits** selects the drawing limits.

**View** selects a named view.

**Window** picks two corners of a window.

### Preview options
**None** specifies no preview image (default).

**EPSI** specifies Macintosh preview image format.

**TIFF** specifies Tagged Image File Format.

### Pixels options
**128** specifies a preview image size of 128x128 pixels (default).

**256** specifies a preview image size of 256x256 pixels.

**512** specifies a preview image size of 512x512 pixels.

### Size Units options
**Inches** specifies the plot parameters in inches.

**MM** specifies the plot parameters in millimeters.

### Scale options
**Output Units** scales the output units.

**Drawing Units** specifies the drawing units.

**Fit to Paper** forces the drawing to fit paper size.

### Paper Size options
**Width** enters a width for the output size.

**Height** enters a height for the output size.

### RELATED COMMAND
**Plot** exports the drawing in a variety of formats, including raster EPS.

### RELATED SYSTEM VARIABLE
**PSProlog** specifies the PostScript prologue information.

### RELATED FILE
***.eps** extension of file produced by PsOut.

### TIPS
- The "screen preview image" is only used for screen display purposes, since graphics software generally cannot display PostScript graphic files.

- When you select the **Window** option, AutoCAD prompts you for the window corners *after* you finish selecting options.

- Although Autodesk recommends using the smallest screen preview image size (128x128), even the largest preview image (512x512) has a minimal effect on file size and screen display time.

- Some software programs, such as those from Microsoft, might reject an *.eps* file when the preview image is larger than 128x128.

- The screen preview image size has no effect on the quality of the PostScript output.

- If you're not sure which screen preview format to use, select TIFF.

- AutoCAD no longer imports PostScript files.

# PSpace

Rel.11 Switches from model space to paper space/layout mode (*short for Paper SPACE*).

| Command | Alias | Ctrl+ | F-key | Alt+ | Menu Bar | Tablet |
|---------|-------|-------|-------|------|----------|--------|
| pspace | ps | ... | ... | ... | ... | L5 |

**Command:** pspace

## COMMAND LINE OPTIONS
*None.*

## RELATED COMMANDS
**MSpace** switches from paper space to model space.

**MView** creates viewports in paper space.

**UcsIcon** toggles the display of the paper space icon.

**Zoom** scales paper space relative to model space with the XP option.

## RELATED SYSTEM VARIABLES
**MaxActVP** specifies the maximum number of viewports displaying an image.

**PsLtScale** specifies the linetype scale relative to paper space.

**TileMode** allows paper space when set to 0.

## TIPS
• Use paper space to layout multiple views of a single drawing.

• You can switch to paper space by double-clicking the word **MODEL** on the status bar; switch back to model space by double-clicking the word **PAPER**.

• When a drawing is in paper space, AutoCAD displays **PAPER** on the status line and the paper space icon:

• Paper space is known as "drawing composition" in other CAD packages.

# Publish

<u>2004</u>  Outputs multiple layout sheets from one or more drawings as a single, multi-page DWF file or hardcopy plot.

| Command | Alias | Ctrl+ | F-key | Alt+ | Menu Bar | Tablet |
|---------|-------|-------|-------|------|----------|--------|
| publish | ... | ... | ... | FH | File | ... |
| | | | | | ⌐Publish | |

**-publish**

**Command:** publish

*Displays dialog box:*

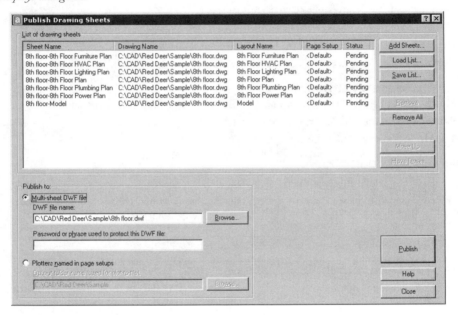

## DIALOG BOX OPTIONS

List of Drawing Sheets options

**Sheet Name** concantenates the drawing name and the layout name with a dash ( - ); edit sheet names with the Rename option.

**Drawing Name** specifies the path and file name of the *.dwg* file.

**Layout Name** specifies the name of the layout page.

**Page Setup** lists the named page setup for each layout.

**Status** displays a message as layouts are published.

Publish To options

**Multi-sheet DWF file** generates a multi-sheet *.dwf* file.

**DWF file name** specifies the path and name of the *.dwf* file.

**Browse** opens the Select DWF File dialog box.

**Password or phrase to protect this DWF file** specifies a password to protect the *.dwf* file.

**Plotters named in page setups** generates a hardcopy plot, or plots to file.

*Buttons*

**Add Sheets** displays the Select Drawings dialog box; when duplicate sheet names are encountered, AutoCAD prompts to change their names.

**Load List** displays the Load List of Sheets dialog box. Select a *.dsd* (drawing set descriptions) or *.bp3* (batch plot list) file, and then click Load.

Save List saves the list of sheets as a *.dsd* file.

**Remove** removes selected sheets from the list.

**Remove All** removes all sheets.

**Move Up** moves the selected sheet up the list.

**Move Down** moves the selected sheet down the list.

**Publish** generates the *.dwf* file or plot; displays Now Plotting dialog box:

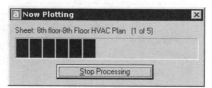

**SHORTCUT MENU**

*Right-click in the List of Drawing Sheets area:*

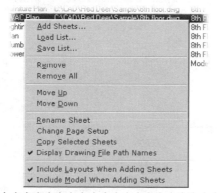

**Rename Sheet** allows you to rename the selected sheet.

**Change Page Setup** displays the Change Page Setup dialog box.

**Copy Selected Sheets** copies selected sheets, and adds the -Copy*n* suffix to the sheet name.

**Display Drawing File Path Names** toggles the display of paths with the file names.

**Include Layouts when Adding Sheets** includes all layouts.

**Include Model When Adding Sheets** determines whether the model layout is included with the list of sheets.

## PUBLISHING COMPLETE DIALOG BOX

*After processing is complete, displays dialog box:*

**Save Log File** saves the log file as a *.csv* (comma separated values) file, which can be opened as a spreadsheet.

**View DWF File** opens the published file in Autodesk Express Viewer, installed with AutoCAD 2004; this button is unavailable when Publish encounters errors during processing.

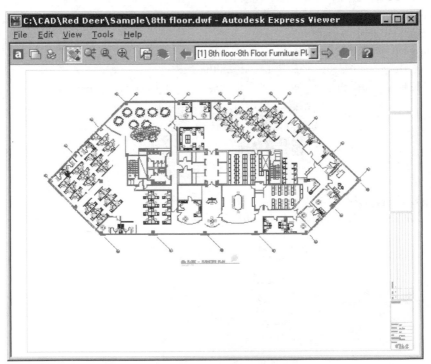

*The multi-sheet .dwf file displayed by Autodesk Express Viewer.*

## CHANGE PAGE SETUP DIALOG BOX

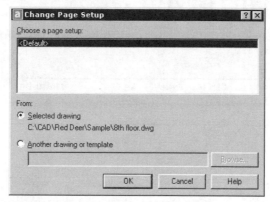

Chose a page setup selects a page setup from the list.

From:

- **Selected drawing** specifies the current drawing.
- **Another drawing or template** selects a different drawing.

**Browse** opens the Select Drawing dialog box.

## -PUBLISH Command
**Command:** -publish

*If drawing is not saved, displays **Drawing Modified - Save Changes** dialog box. Click **OK**.*

*Displays **Select List of Sheets** dialog box. Select a .dsd file, and then click **Select**.*

*Immediately plots the drawing set, and generates a log file.*

## RELATED COMMANDS
**Plot** outputs drawings as *.dwf* files.

**PublishToWeb** coverts drawings to Web pages.

## TIPS
- The Model tab is included only when the **Include Model When Adding Sheets** setting is turned on.

- DWF passwords are case sensitive. If the password is lost, it cannot be recovered, and the *.dwf* file cannot be opened. Erase the *.dwf* file, and create a new set.

- The **-Publish** command is good for generated sheets when a *.dsd* (drawing set description) file exists.

# PublishToWeb

**2000i** Exports drawings as DWF, JPG, and PNG images embedded in pre-formatted Web pages.

| Command | Alias | Ctrl+ | F-key | Alt+ | Menu Bar | Tablet |
|---------|-------|-------|-------|------|----------|--------|
| publishtoweb | ... | ... | ... | FW | File | X25 |
| | | | | | ↳Publish to Web | |

**Command:** publishtoweb

*Displays wizard.*

## WIZARD OPTIONS

**Begin** page

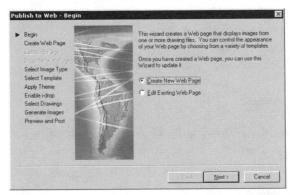

**Create New Web Page** guides you through the steps in creating new Web pages from drawings.

**Edit Existing Web Page** guides you through the steps in editing existing Web pages.

**Create Web Page** page

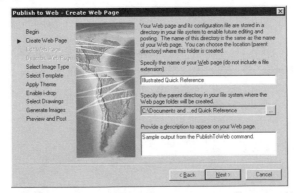

**Specify the name of your Web page** requires you to enter a file name. (AutoCAD uses the name for the files making up the Web page, which allows you later to edit the Web page.) The name also appears at the top of the Web page.

**Provide a description to appear on your Web page** specifies a description that appears below the name on the Web page.

**Select Image Type** page

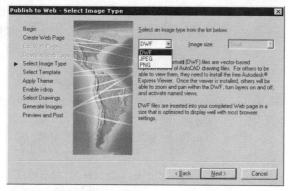

**Select an image type from the list below**

- **DWF** (drawing Web format) is a vector format that displays cleanly, and can be zoomed and panned; not all Web browsers can display DWF.
- **JPEG** (joint photographic experts group) is a raster format that all Web browsers display; may create *artifacts* (details that don't exist).
- **PNG** (portable network graphics) is a raster format that does not suffer the artifact problem; some older Web browsers do not display PNG.

**Image size** selects a size of raster image (available for JPEG and PNG only):

| Image Size | Resolution | Approximate PNG Filesize |
|---|---|---|
| Small | 789 x 610 | 60KB |
| Medium | 1009 x 780 | 90KB |
| Large | 1302 x 1006 | 130KB |
| Extra Large | 1576 x 1218 | 170KB |

**Select Template** page

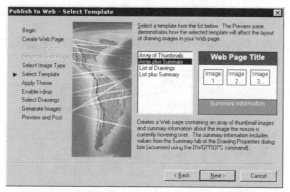

**Select a template from the list below** selects one of the pre-designed formats for the Web page.

**Apply Theme** page

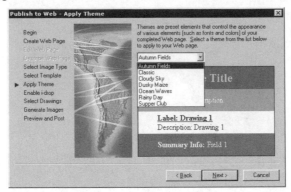

**Select a theme from the list below** selects one of the pre-designed themes (colors and fonts) for the Web page.

**Enable iDrop** page

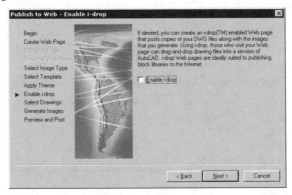

**Enable i-drop** adds i-drop capability to the Web page.

**Select Drawings** page

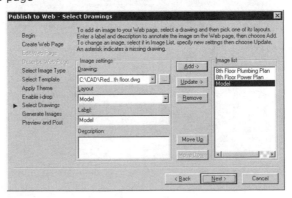

Image Settings options

   **Drawing** selects the drawing; the current drawing is the default.

**Layout** selects the name of a layout, or Model space.

**Label** specifies a name, such as the filename or a more descriptive name.

**Description** specifies a description that appears with the drawing on the Web page.

*Buttons*

**Add** adds the image setting to the image list.

**Update** changes the image setting in the image list.

**Remove** removes the image setting from the image list.

**Move up** moves the image setting up the image list.

**Move down** moves the image setting down the image list.

**Generate Images** page

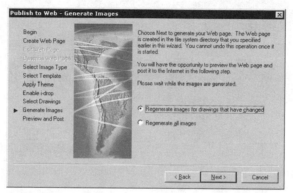

**Regenerate images for drawings that have changed** updates images for those drawings that have been edited.

**Regenerate all images** regenerates all images from all drawings; ensures all are up to date.

**Preview and Post** page

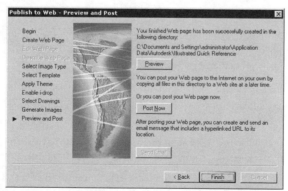

**Preview** launches the Web browser to preview the resulting Web page.

**Post Now** uploads the files (HTML, JPEG, PNG, DWF, and so on) to the Web site.

**Send Email** sends an email to alert others of the posted Web page.

*Examples of resulting Web pages:*

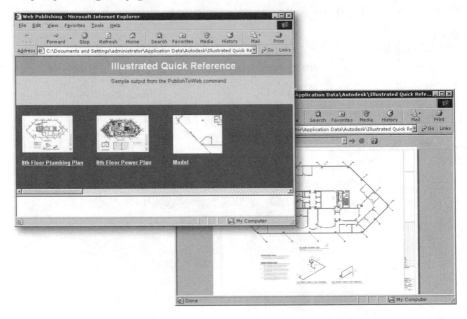

## RELATED COMMANDS

**Publish** exports drawings as multi-sheet *.dwf* files.

**Plot** exports drawings as *.dwf* files via the ePlot option.

**Hyperlink** places hyperlinks in drawings.

## RELATED FILES

*\*.ptw* are PublishToWeb parameter files, stored in tab-delimited ASCII file.

*\*.js* are JavaScript files.

*\*.jpg* are joint photographic experts group (raster image) files.

*\*.png* are portable network graphics (raster image) files.

*\*.dwf* are drawing Web format (vector image) files.

## TIPS

* Use the **Regenerate all images** option, unless you have an exceptionally slow computer or large number of drawings to process. The **Generate Images** step can take a long time.

* After you click **Preview** to view the Web page (and after AutoCAD launches the Web browser), click the **Back** button to make changes, if the result is not to your liking.

* The **Post Now** option works only if you have correctly set up the FTP (file transfer protocol) parameters. If so, you can have AutoCAD upload the HTML files directly to a Web site. If not, use a separate FTP program to upload the files from the *\windows\applications data\autodesk*.

* You can customize the themes and templates by editing the *acwebpublish.css* (themes) and *acwebpublish.xml* (templates) files.

# Purge

**V. 2.1** Removes unused, named objects from the drawing: block, dimension style, layer, linetype, plot style, shape, text style, application ID table, and multi-line style.

| Commands | Alias | Ctrl+ | F-key | Alt+ | Menu Bar | Tablet |
|----------|-------|-------|-------|------|----------|--------|
| purge | pu | ... | ... | FUP | File | X25 |
| | | | | | ⓑDrawing Utilities | |
| | | | | | ⓑPurge | |
| -purge | | | | | | |

**Command:** purge

*Displays dialog box:*

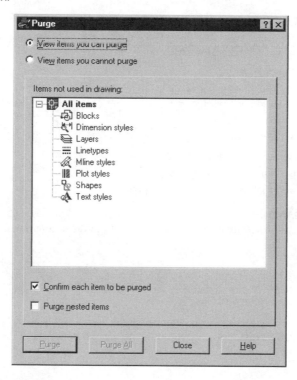

## DIALOG BOX OPTIONS

**View items you can purge** lists objects that can be purged from the drawing.

**View items you cannot purge** lists objects that cannot be purged from the drawing.

**Confirm each item to be purged** displays a confirmation dialog box for each object being purged.

**Purge nested items** purges nested objects, such as unused blocks within unused blocks.

## -PURGE Command
**Command:** -purge
**Enter type of unused objects to purge**
**[Blocks/Dimstyles/LAyers/LTypes/Plotstyles/SHapes/textSTyles/**
**Mlinestyles/All]:** *(Enter an option.)*

*Sample response:*
No unreferenced blocks found.
Purge layer DOORWINS? <N> **y**
Purge layer TEXT? <N> **y**
Purge linetype CENTER? <N> **y**
Purge linetype CENTER2? <N> **y**
No unreferenced text styles found.
No unreferenced shape files found.
No unreferenced dimension styles found.

### COMMAND LINE OPTIONS
**Blocks** purges named but unused (not inserted) blocks.

**Dimstyles** purges unused dimension styles.

**LAyers** purges unused layers.

**LTypes** purges unused linetypes.

**Plotstyle** purges unused plot styles.

**SHapes** purges unused shape files.

**STyles** purges unused text styles.

**APpids** purges unused application ID tables of AutoLISP applications.

**Mlinestyles** purges unused multiline styles.

**All** purges drawing of all named objects, if possible.

### RELATED COMMAND
**WBlock** writes the current drawing to disk with the * option, and removes spurious information from the drawing.

### TIPS
• As of AutoCAD Release 13, the **Purge** command can be used at any time; it no longer must be the first command after a drawing is loaded.

• It may be necessary to use the **Purge** command several times; follow each purge with the **Close** command, then open the drawing, and purge again. Repeat until the **Purge** command reports nothing to purge.

 # QDim

<u>**2000**</u>  Draws continuous, baseline, ordinate, radius, diameter, and staggered dimensions with just three picks *(short for Quick DIMensioning).*

| Command | Alias | Ctrl+ | F-key | Alt+ | Menu Bar | Tablet |
|---------|-------|-------|-------|------|----------|--------|
| qdim | ... | ... | ... | NQ | Dimension | W1 |
| | | | | | ⌖QDIM | |

**Command:** qdim
**Select geometry to dimension:** *(Select one or more objects.)*
**Select geometry to dimension:** *(Press **Enter** to end object selections.)*
**Specify dimension line position, or**
**[Continuous/Staggered/Baseline/Ordinate/Radius/Diameter/datumPoint/**
**Edit/seTtings] <Continuous>:** *(Enter an option.)*

## COMMAND LINE OPTIONS

**Select geometry to dimension** selects a single object to dimension.

**Specify dimension line position** specifies the location of the dimension.

**Continuous** draws continuous dimensions.

**Staggered** draws staggered dimensions.

**Baseline** draws baseline dimensions.

**Ordinate** draws ordinate dimensions relative to the UCS origin.

**Radius** draws radial dimensions; prompts 'Specify dimension line position:'.

**Diameter** draws diameter dimensions; prompts 'Specify dimension line position:'.

**datamPoint** sets a new datum point for ordinate and baseline dimensions; prompts 'Select new datum point:'

### Edit options
**Indicate dimension point to remove, or [Add/eXit] <eXit>:** *(Select a dimension, or enter an option.)*

**Indicate dimension to remove** selects the dimension to remove from the continuous dimension.

**Add** adds a dimension to the continuous dimension.

**eXit** returns to dimension drawing mode.

### seTtings options
**Associative dimension priority [Endpoint/Intersection] <Endpoint>:** *(Enter an option.)*

**Endpoint** applies associative dimensions to endpoints over intersections.

**Intersection** applies associative dimensions to intersections over endpoints.

## RELATED COMMANDS

**DimStyle** creates dimension styles, which specify the look of a dimension.

**Dim**xxx draws other kinds of dimensions.

**QLeader** draws leaders.

**RELATED SYSTEM VARIABLE**

**DimStyle** specifies the current dimension style.

**TIPS**

- Example of continuous dimensions:

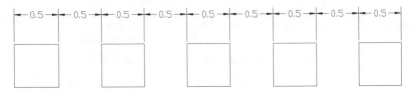

- Example of staggered dimensions:

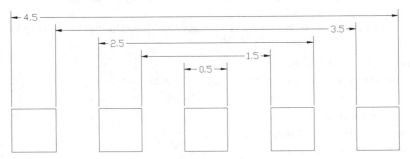

- Example of ordinate dimensions:

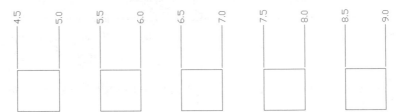

- Example of radial dimensions:

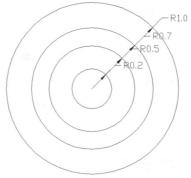

- As of AutoCAD 2004, dimensions created with **QDim** are fully associative.

# QLeader

**Rel.14** Draws leaders; dialog box specifies options for custom leaders and annotations (*short for Quick LEADER*).

| Command | Alias | Ctrl+ | F-key | Alt+ | Menu Bar | Tablet |
|---------|-------|-------|-------|------|----------|--------|
| qleader | le | ... | ... | NE | Dimension | W2 |
| | | | | | ↳ Leader | |

**Command:** qleader
**Specify first leader point, or [Settings] <Settings>:** *(Pick a point for the arrowhead, or type **S**.)*

*When S is entered, displays dialog box:*

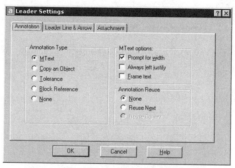

*Click **OK** to continue with the command; the prompts vary, depending on the options selected in the dialog box:*
**Specify next point:** *(Pick a point for the leader shoulder.)*
**Specify next point:** *(Pick a point, or press **Enter** for text options.)*
**Specify text width <0.0000>:** *(Enter a value.)*
**Enter first line of annotation text <Mtext>:** *(Enter text, or type **M**.)*
**Enter next line of annotation text:** *(Press **Enter** to end command.)*

## COMMAND LINE OPTIONS

**Specify first leader point** picks the location for the leader's arrowhead; press ENTER to display tabbed dialog box.

**Specify next point** picks the vertices of the leader; press ENTER to end the leader line.

**Specify text width** specifies the width of the bounding box for the leader text.

**Enter first line of annotation text** specifies the text for leader annotation; press ENTER twice to end.

**Enter next line of annotation text** specifies more text; press ENTER once to end.

## DIALOG BOX OPTIONS

### Annotation/Format tab

Annotation Type options

**MText** prompts you to enter text for the annotation.

**Copy an Object** attaches any object in the drawing as an annotation.

**Tolerance** prompts you to select tolerance symbols for the annotation.

**Block Reference** prompts you to select a block for the annotation.

**None** attaches no annotation.

MText Options options

**Prompt for width** displays the 'Specify text width' prompt.

**Always left justify** forces the text to be left-justified, even when the leader is drawn to the right.

**Frame text** places a rectangle around the text.

Annotation Reuse options

**None** does not retain annotation for next leader.

**Reuse Next** remembers the current annotation for the next leader.

**Reuse Current** uses the last annotation for the current leader.

### Leader Line & Arrow tab

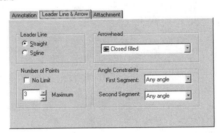

Leader Line options

**Straight** draws the leader with straight lines.

**Spline** draws the leader as a spline curve.

Number of Points options

**No limit** keeps prompting for leader vertex points until you press ENTER.

**Maximum** stops the command prompting for leader vertex points; default=3.

Arrowhead option

**Arrowhead** selects the type of arrowhead, including Closed filled (default), None, and User Arrow.

Angle Constraints options

**First Segment** selects from Any angle (user-specified), Horizontal (0 degrees), 90, 45, 30, or 15-degree leader line, first segment.

**Second Segment** selects from Any angle (user-specified), Horizontal, 90, 45, 30, or 15-degree leader line, second segment.

**Attachment** tab

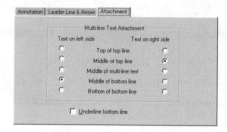

**Multiline Text Attachment options**

**Left Side** positions the annotation at one of several locations relative to the last leader segment, when the annotation is located to the left of the leader:

- Top of top line.
- Middle of top line.
- Middle of multi-line text.
- Middle of bottom line (default).
- Bottom of bottom line.

**Right Side** positions the annotation at one of different locations relative to the last leader segment, when the annotation is located to the right of the leader.

**Underline bottom line** underlines the last line of leader text.

### RELATED COMMANDS

**DdEdit** edits leader text; see the **MText** command.

**DimStyle** sets dimension variables, including leaders.

**Leader** draws leaders without dialog boxes.

### TIPS

- The **QLeader** command draws leaders, just like the **Leader** command in AutoCAD Release 13 and 14. The difference is that it brings up a triple-tab dialog box for setting the leader options.

- Some options have interesting possibilities, such as using any object in the drawing in place of the leader text.

# QNew

**2004** Starts new drawings based on a specified template drawing (*short for Quick NEW*).

| Command | Alias | Ctrl+ | F-key | Alt+ | Menu Bar | Tablet |
|---------|-------|-------|-------|------|----------|--------|
| qnew | ... | ... | ... | ... | ... | ... |

**Command:** qnew

*Display depends on **Startup** option in **General Options** section of the **System** tab (**Options** dialog box). See the **New** command.*

## RELATED COMMANDS

**New** starts a new drawing.

**SaveAs** saves the drawing in *.dwg* or *.dwt* formats; creates template files.

## RELATED SYSTEM VARIABLES

**DbMod** indicates whether the drawing has changed since being loaded.

**DwgPrefix** indicates the path to the drawing.

**DwgName** indicates the name of the current drawing.

**FileDia** displays prompts at the 'Command' prompt.

## RELATED FILES

*wizard.ini* holds the names and descriptions of template files.

***.dwt** are template files stored in *.dwg* format.

## TIPS

- The **New** and **QNew** commands operate in identical manner, except when you specify a template drawing in the **Option** command's **Files** tab:

- The icon executes the **QNew** command, while the menu pick executes the **New** command.

# QSave

**Rel.12**  Saves the current drawing without requesting a filename (*short for Quick SAVE*).

| Command | Alias | Ctrl+ | F-key | Alt+ | Menu Bar | Tablet |
|---------|-------|-------|-------|------|----------|--------|
| qsave | ... | S | ... | FS | File | U24- |
| | | | | | ⬦Save | U25 |

**Command:** qsave

*If the drawing has never been saved, displays the* **Drawing Save As** *dialog box.*

## COMMAND LINE OPTIONS
*None.*

## RELATED COMMANDS
**Quit** ends AutoCAD, with or without saving the drawing.

**Save** saves the drawing, after requesting the filename.

## RELATED SYSTEM VARIABLES
**DBMod** indicates whether the drawing has changed since it was loaded.

**DwgName** specifies the current drawing filename (default = *drawing1*).

**DwgTitled** specifies the status of drawing's filename:

| DwgTiled | Meaning |
|----------|---------|
| 0 | Drawing is named *drawing1* (default). |
| 1 | Drawing was given another name. |

## TIPS
• When the drawing is unnamed, the **QSave** command displays the **Save Drawing As** dialog box to request a file name; see the **SaveAs** command.

• When the drawing file, its folder, or drive (such as a CD-ROM drive) is marked read-only, use the **SaveAs** command to save the drawing by another file name, or to another folder or drive.

# QSelect

Creates selection sets of objects *(short for Quick SELECT)*.

**2000**

| Command | Alias | Ctrl+ | F-key | Alt+ | Menu Bar | Tablet |
|---------|-------|-------|-------|------|----------|--------|
| qselect | ... | ... | ... | TQ | Tools<br>⤷Quick Select | X9 |

**Command:** qselect

*Displays dialog box:*

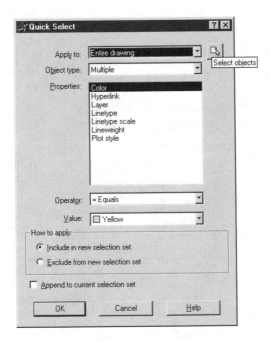

## DIALOG BOX OPTIONS

**Apply to** applies the selection criteria to the entire drawing or current selection set; click Select Objects to create a selection set.

**Select Objects** allows you to select objects; AutoCAD prompts: 'Select objects: '. Right-click or press ENTER to return to this dialog box.

**Object type** lists the objects in the selection set; allows you to narrow the selection criteria to specific types of objects (default = Multiple).

**Properties** lists the properties valid for the selected object types; when you select more than one object type, only the properties in common are listed.

**Operator** lists logical operators available for the selected property; operators include:

| Operator | Meaning |
| --- | --- |
| = Equals | Selects objects equal to the property. |
| <> Not Equal To | Selects objects different from the property. |
| > Greater Than | Selects objects greater than the property. |
| < Less Than | Selects objects less than the property. |
| * Wildcard Match | Selects objects with matching text. |

**Value** specifies the property value for the filter. If known values for the selected property are available, Value becomes a list from which you can choose a value. Otherwise, enter a value.

## How to Apply options

**Include in new selection set** creates a new selection set.

**Exclude from new selection set** inverts the selection set, excluding all objects that match the selection criteria.

## Additional options

**Append to current selection set**

☑ adds to the current selection set.

☐ replaces the current selection set.

## RELATED COMMANDS

**Select** selects objects on the command line.

**Filter** runs a more sophisticated version of the QSelect command.

## RELATED SYSTEM VARIABLES

*None.*

## TIPS

- This command works with the properties of proxy objects created by ObjectARX applications.

- You may select objects before entering the **QSelect** command, and then add or remove objects from the selection set with the **Quick Select** dialog box's options.

- Since this command is not transparent, you cannot use it within other commands; instead, use the **'Filter** command.

# 'QText

**V. 2.0**  Displays lines of text as rectangular boxes (*short for Quick TEXT*).

| Command | Alias | Ctrl+ | F-key | Alt+ | Menu Bar | Tablet |
|---------|-------|-------|-------|------|----------|--------|
| 'qtext  | ...   | ...   | ...   | ...  | ...      | ...    |

**Command:** qtext
**Enter mode [ON/OFF] <OFF>:** *(Enter* **ON** *or* **OFF**.*)*

# The Illustrated AutoCAD 2004 Quick Reference

*Normal text (at left) and quick text after regeneration (at right).*

## COMMAND LINE OPTIONS
**ON** turns on quick text after the next Regenall command.

**OFF** turns off quick text after the next Regenall command.

## RELATED COMMAND
**Regenall** regenerates the screen, which makes quick text take effect.

## RELATED SYSTEM VARIABLE
**QTextMode** holds the current state of quick text mode.

## TIPS
- A regeneration is required before AutoCAD displays text in quick outline form:

  **Command:** regenall
  **Regenerating model.**
- To reduce regen time, use qtext to turn lines of text into rectangles, which redraw faster.

- The length of a qtext box does not necessarily match the actual length of text.

- Turning on qtext affects text during plotting; qtext blocks are plotted as rectangles.

- To find invisible text (such as text made of spaces), turn on qtext, thaw all layers, zoom to extents, and use the **Regenall** command.

# Quit

**V. 1.0** Exits AutoCAD without saving changes made to the drawing after the most recent **QSave** or **SaveAs** command.

| Command | Alias | Ctrl+ | F-key | Alt+ | Menu Bar | Tablet |
|---------|-------|-------|-------|------|----------|--------|
| quit | exit | Q | ... | FX | File | Y25 |
| | | | F4 | | ⬐Exit | |

**Command:** quit

*Displays dialog box:*

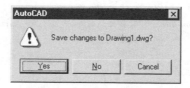

## DIALOG BOX OPTIONS

**Yes** saves changes before leaving AutoCAD.

**No** does not save changes.

**Cancel** does not quit AutoCAD; returns to drawing.

## RELATED COMMANDS

**Close** closes the current drawing.

**SaveAs** saves the drawing by another name and to another folder or drive.

## RELATED SYSTEM VARIABLE

**DBMod** indicates whether the drawing has changed since it was loaded.

## RELATED FILES

*\*.dwg* are AutoCAD drawing files.

*\*.bak* are backup files.

*\*.bkn* are additional backup files.

## TIPS

- You can change a drawing, yet preserve its original format:
    1. Use the **SaveAs** command to save the drawing by another name.
    2. Use the **Quit** command to preserve the drawing in its original state.
- Even if you accidently save over a drawing, you can recover the previous version:
    1. Use Windows Explorer to rename the drawing file.
    2. Use Windows Explorer to rename the *.bak* (backup) extension to *.dwg* — if you have AutoCAD set up to save backup files; see the **Options** command.
- You cannot save changes to a read-only drawing with the **Quit** command; use the **SaveAs** command instead.

# R14PenWizard

2000 Helps create color-dependent plot style tables (*undocumented command*).

| Command | Alias | Ctrl+ | F-key | Alt+ | Menu Bar | Tablet |
|---------|-------|-------|-------|------|----------|--------|
| r14penwizard | ... | ... | ... | TZD | Tools | ... |
| | | | | | ⌐Wizard | |
| | | | | |   ⌐Add Color-Dependent | |
| | | | | |     Plot Style Table | |

**Command:** r14penwizard

*Displays **Add Color-Dependent Plot Style Table** wizard.*

*You can also access this command through the Windows Control Panel's **Autodesk Plot Style Manager**. (This command makes AutoCAD 2004's **Plot** command compatible with versions of AutoCAD prior to 2000.)*

## DIALOG BOX OPTIONS

**Begin** page

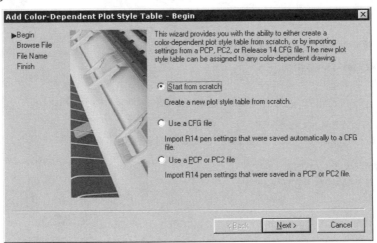

**Start from scratch** creates a new color-dependent plot style table file.

**Use a CFG file** converts the plotter pen settings stored in AutoCAD Release 14 *acad.cfg* files.

**Use a PCP or PC2 file** converts the plotter pen settings stored in the plotter configuration parameter *.pcp* and *.pc2* files of earlier versions of AutoCAD.

*Buttons*

**Back** goes back to the previous dialog box.

**Next** moves to the next dialog box.

**Cancel** exits the wizard.

**File Name** page

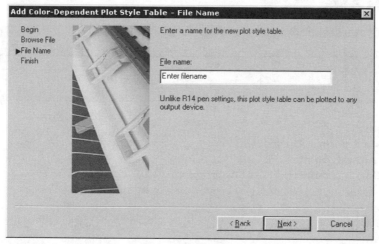

File name specifies name of the file in which to store the new color dependent plot style table.

**Finish** page

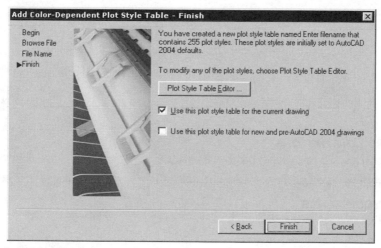

**Plot Style Table Editor** displays the Plot Style Table Editor dialog box; see StylesManager command.

**Use this plot style table for the current drawing** applies the plot style table to the current drawing.

**Use this plot style table for new and pre-AutoCAD 2004 drawings** applies the plot style table to all new drawings and drawings created by versions of AutoCAD prior to 2004.

## RELATED COMMANDS

**PcinWizard** imports *.pcp* and *.pc2* configuration plot files into the current layout.

**PlotterManager** accesses the Add Plotter wizard and Assign Plot Style wizard.

**Plot** plots the drawing.

**StylesManager** displays the Plot Style Table Editor dialog box.

 **Ray**

**Rel.13**   Draws semi-infinite construction lines.

| Command | Alias | Ctrl+ | F-key | Alt+ | Menu Bar | Tablet |
|---------|-------|-------|-------|------|----------|--------|
| ray | ... | ... | ... | DR | Draw | K10 |
|  |  |  |  |  | ⇘Ray |  |

**Command:** ray
**Specify start point:** *(Pick a point.)*
**Specify through point: :** *(Pick another point.)*
**Specify through point: :** *(Press **Enter** to end the command.)*

**COMMAND LINE OPTIONS**
   **Start point** specifies the starting point of the ray.

   **Through point** specifies the point through which the ray passes.

**RELATED COMMANDS**
   **Properties** modifies rays.

   **Line** draws finite lines.

   **XLine** creates infinite construction lines.

**TIPS**
- The *ray* object is semi-infinite in length.

- A ray is a "construction line"; it displays and plots, but does not affect the extents.

- A ray has all the properties of a line (including color, layer, and linetype), and can be used as a cutting edge for the **Trim** command.

**Removed Command**
**RConfig** — render configuration — was removed from Release 14. It is replaced by **Render**.

# Recover

<u>Rel.12</u>  Recovers damaged drawings without user intervention.

| Command | Alias | Ctrl+ | F-key | Alt+ | Menu Bar | Tablet |
|---------|-------|-------|-------|------|----------|--------|
| recover | ... | ... | ... | FUR | File | ... |
| | | | | | ⌐Drawing Utilities | |
| | | | | | ⌐Recover | |

**Command:** recover

*Displays the **Select** **File** dialog box. Select a .dwg file, and then click **Open**.*

*Sample output:*

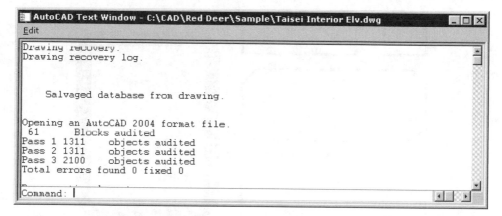

## COMMAND LINE OPTIONS
*None.*

## RELATED COMMAND
**Audit** checks a drawing for integrity.

## TIPS
- The **Recover** command does not ask permission to repair damaged parts of the drawing file; use the **Audit** command if you want control over the repair process.

- The **Quit** command discards changes made by the **Recover** command.

- If the **Recover** and **Audit** commands do not fix the problem, try using the **DxfOut** command, followed by the **DxfIn** command.

# Rectang

Draws squares and rectangles with a variety of options.

| Command | Aliases | Ctrl+ | F-key | Alt+ | Menu Bar | Tablet |
|---------|---------|-------|-------|------|----------|--------|
| rectang | rec | ... | ... | DG | Draw | Q10 |
|  | rectangle |  |  |  | ↳Rectangle |  |

**Command:** rectang
**Specify first corner point or [Chamfer/Elevation/Fillet/Thickness/Width]: :**
*(Pick a point, or enter an option)*
**Specify other corner point or [Dimensions]:** *(Pick another point.)*

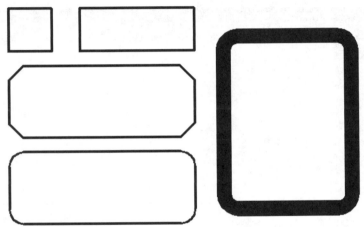

*Square and rectangles drawn with the **Rectangle** command's **Chamfer**, **Fillet**, and **Width** options.*

## COMMAND LINE OPTIONS

**Specify first corner point** picks the first corner of the rectangle.

**Specify other corner point** picks the opposite corner of the rectangle.

Chamfer options
**Specify first chamfer distance for rectangles <0.0000>: :** *(Enter a value.)*
**Specify second chamfer distance for rectangles <0.0000>:** *(Enter a value.)*

**First chamfer distance for rectangles** sets the first chamfer distance for all four corners.

**Second chamfer distance for rectangles** sets the second chamfer distance for all four corners.

Dimensions options
**Specify length for rectangles <0.0>:** *(Enter length.)*
**Specify width for rectangles <0.0>:** *(Enter width.)*
**Specify other corner point or [Dimensions]:** *(Pick a point.)*

**Specify length for rectangles** specifies the length along the x axis.

**Specify width for rectangles** specifies the length along the y axis.

**Specify other corner point** specifies the orientation of rectangle.

Elevation option
**Specify the elevation for rectangles <0.0000>:** *(Enter a value.)*

Elevation for rectangles sets the elevation (*height of the rectangle in the z direction*).

Fillet option
**Specify fillet radius for rectangles <1.0000>:** *(Enter a value.)*

Fillet radius for rectangles sets the fillet radius for all four corners of the rectangle.

Thickness option
**Specify thickness for rectangles <0.0000>:** *(Enter a value.)*

Thickness for rectangles sets the thickness of the rectangle's sides in the z direction.

Width option
**Specify line width for rectangles <0.0000>:** *(Enter a value.)*

Width for rectangles sets the width of all segments of the rectangle's four sides.

## RELATED COMMANDS

**Donut** draws solid-filled circles with a polyline.

**Ellipse** draws ellipsis with a polyline, when PEllipse = 1.

**PEdit** edits polylines, including rectangles.

**PLine** draws polylines and polyline arcs.

**Polygon** draws a polygon — 3 to 1024 sides — from a polyline.

## RELATED SYSTEM VARIABLES

*None.*

## TIPS

- Rectangles are drawn from polylines; use the **PEdit** command to change the rectangle, such as the width of the polyline.

- The values you set for the **Chamfer**, **Elevation**, **Fillet**, **Thickness**, and **Width** options become the default for the next execution of the **Rectangle** command.

- The pick point determines the location of the rectangle's first vertex; rectangles are drawn counterclockwise.

- Use the **Snap** command and object snap modes to place the rectangle precisely.

- Use object snap modes **ENDpoint** or **INTersection** to snap to the rectangle's vertices.

- This command ignores the settings in the **ChamferA**, **ChamferB**, **Elevation**, **FilletRad**, **PLineWid**, and **Thickness** system variables.

# Redefine

Rel. 9 Restores the meaning of AutoCAD commands after being disabled by the Undefine command.

| Command | Alias | Ctrl+ | F-key | Alt+ | Menu Bar | Tablet |
|---|---|---|---|---|---|---|
| redefine | ... | ... | ... | ... | ... | ... |

**Command:** redefine
**Enter command name:** *(Enter command name.)*

**COMMAND LINE OPTION**

**Enter command name** specifies the name of the AutoCAD command to redefine.

**RELATED COMMANDS**

*All commands* allows all AutoCAD commands to be redefined.

**Undefine** disables the meaning of an AutoCAD command.

**TIPS**

- Prefix any command with a . *(period)* to redefine the undefinition temporarily, as in:

  **Command:** .line

- Prefix any command with an _ *(underscore)* to make an English-language command work in any linguistic version of AutoCAD, as in:

  **Command:** _line

- You must undefine a command with the **Undefine** command before using the **Redefine** command.

# Redo

V. 2.5  Reverses the effect of the most-recent U and Undo commands.

| Command | Alias | Ctrl+ | F-key | Alt+ | Menu Bar | Tablet |
|---------|-------|-------|-------|------|----------|--------|
| redo | ... | Y | ... | ER | Edit ⮑Redo | U12 |

**Command:** redo

## COMMAND LINE OPTIONS
*None.*

## RELATED COMMANDS
**MRedo** redoes more than one undo.

**Oops** un-erases the most recently-erased objects.

**U** undoes the most recent AutoCAD command.

**Undo** undoes the most recent series of AutoCAD commands.

## TIPS
- The **Redo** command is limited to reversing a single undo, while the **Undo** and **U** commands undo operations all the way back to the beginning of the editing session.

- The **Redo** command must be used immediately following the **U** or **Undo** command.

# 'Redraw

**V. 1.0** Redraws the current viewport to clean up the screen.

| Command | Alias | Ctrl+ | F-key | Alt+ | Menu Bar | Tablet |
|---------|-------|-------|-------|------|----------|--------|
| 'redraw | r | ... | ... | ... | ... | ... |

**Command:** redraw

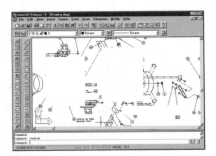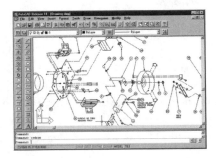

*Before redraw, portions of the drawing are "missing" (at left);*
*the drawing is clean after the redraw (at right).*

## COMMAND LINE OPTION

ESC stops the redraw.

## RELATED COMMANDS

**RedrawAll** redraws all viewports.

**Regen** regenerates the current viewport.

## RELATED SYSTEM VARIABLE

**SortEnts** controls the order of redrawing objects:

| SortEnts | Meaning |
|----------|---------|
| 0 | Sorts by the order in the drawing database. |
| 1 | Sorts for object selection. |
| 2 | Sorts for object snap. |
| 4 | Sorts for redraw. |
| 8 | Sorts for creating slides. |
| 16 | Sorts for regenerations. |
| 32 | Sorts for plotting. |
| 64 | Sorts for PostScript plotting. |

## TIPS

- Use the **Redraw** command to clean up the screen after a lot of editing; some commands automatically redraw the screen when they are done.

- **Redraw** does not affect objects on frozen layers.

- Use the **RedrawAll** command to redraw all viewports.

# 'RedrawAll

**Rel.10**  Redraws all viewports to clean up the screen.

| Command | Alias | Ctrl+ | F-key | Alt+ | Menu Bar | Tablet |
|---------|-------|-------|-------|------|----------|--------|
| 'redrawall | ra | ... | ... | VR | View | Q11- |
| | | | | | ⬆Redraw | R11 |

**Command:** redrawall

## COMMAND LINE OPTION
ESC cancels the redraw.

## RELATED COMMANDS
**Redraw** redraws only the current viewport.

**RegenAll** regenerates all viewports.

## RELATED SYSTEM VARIABLE
**SortEnts** controls the order of redrawing objects:

| SortEnts | Meaning |
|----------|---------|
| 0 | Sorts by order in the drawing database. |
| 1 | Sorts for object selection. |
| 2 | Sorts for object snap. |
| 4 | Sorts for redraw. |
| 8 | Sorts for creating slides. |
| 16 | Sorts for regeneration. |
| 32 | Sorts for plotting. |
| 64 | Sorts for PostScript plotting. |

## TIPS
• The **RedrawAll** command does not affect objects on frozen layers.

• Use the **Redraw** command to redraw a single viewport.

# RefClose

<u>**2000**</u>  Saves or discards changes made to reference objects edited in-place (*short for REFerence CLOSE*).

| Command | Alias | Ctrl+ | F-key | Alt+ | Menu Bar | Tablet |
|---------|-------|-------|-------|------|----------|--------|
| refclose | ... | ... | ... | MBD | Modify | ... |
| | | | | | ↳In-place Xref and Block Edit | |
| | | | | | ↳Discard Changes to Reference | |
| | | | | MBS | Modify | |
| | | | | | ↳In-place Xref and Block Edit | |
| | | | | | ↳Save Changes Back to Reference | |

**Command:** refclose
**Enter option [Save/Discard reference changes] <Save>:** *(Enter an option.)*

**COMMAND LINE OPTIONS**

**Save** saves the editing changes made to the block or externally-referenced file.

**Discard** discards the changes.

**RELATED COMMANDS**

**RefEdit** edits blocks and externally-referenced files attached to the current drawing.

**Insert** inserts a block in the drawing.

**XAttach** attaches an externally-referenced drawing.

**RELATED SYSTEM VARIABLE**

**RefEditName** specifies the filename of the referenced file being edited.

**TIPS**

• AutoCAD prompts you with a warning dialog box to ensure you really want to discard or save the changes made to the reference:

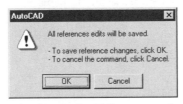

• You can use this command only after the **RefEdit** command, otherwise AutoCAD reports, "** Command not allowed unless a reference is checked out with RefEdit command **."

# RefEdit

<u>**2000**</u> Edits blocks and externally-referenced files attached to the current drawing
(*short for REFerence EDIT*).

| Commands | Alias | Ctrl+ | F-key | Alt+ | Menu Bar | Tablet |
|----------|-------|-------|-------|------|----------|--------|
| refedit | ... | ... | ... | MBE | Modify | ... |
| | | | | | ⮡ **In-place Xref and Block Edit** | |
| | | | | | ⮡ **Edit Block or Xref** | |
| -refedit | | | | | | |

**Command:** refedit
**Select reference:** *(Pick an externally-referenced drawing or block.)*
*Displays dialog box:*

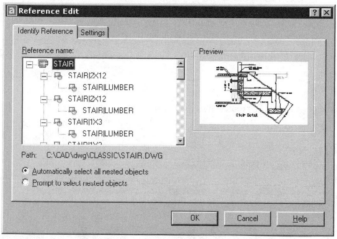

*Click* **OK** *to continue with the command:*
**Select nested objects:** *(Pick one or more objects.)*
**Use REFCLOSE or the Refedit toolbar to end reference editing session.**
*Displays RefEdit toolbar.*

## COMMAND LINE OPTIONS

**Select reference** selects an externally-referenced drawing or an inserted block for editing.

**Select nested objects** selects objects within the reference — this becomes the selection set
of objects that you may edit; you may select all nested objects with the All option (with the
exception of OLE objects and objects inserted with the MInsert command, which cannot be
refedited).

## DIALOG BOX OPTIONS

### Identify Reference tab

**Reference name** lists a tree of the selected reference object and its nested references; a
single reference can be edited at a time.

**Preview** displays a preview image of the selected reference.

- **Automatically select all nested objects** selects all nested objects.
- **Prompt to select nested objects** prompts you with "Select nested objects."

## Settings tab

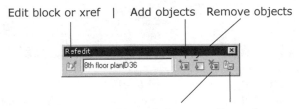

Identify Reference | Settings

☑ Create unique layer, style, and block names
☐ Display attribute definitions for editing
☑ Lock objects not in working set

**Create unique layer, style, block names**

    ☑ prefixes layer and symbol names of extracted objects with **$n$**.

    ☐ retains the names of layers and symbols, as in the reference.

**Display attribute definitions for editing** option is available only when an xref contains attributes:

    ☑ makes non-constant attributes invisible; attribute definitions can be edited.

    ☐ attribute definitions cannot be edited.

*Caution!* When edited attributes are saved back to the block reference, the attributes of the original reference are not changed; instead, the modified attribute definitions come into effect with the next insertion of the block.

**Lock objects not in working set** option is new to AutoCAD 2004:

    ☑ locks objects in a manner similar to locked layer.

    ☐ objects are not locked.

## TOOLBAR OPTIONS

*This toolbar appears automatically after you select nested objects to edit:*

Edit block or xref | Add objects  Remove objects

Refedit

8th floor plan|D36

Discard changes   Save changes

**Edit block or xref** executes the RefEdit command.

**Add objects to working set** executes the RefSet Add command.

**Remove objects from working set** executes the RefSet Remove command.

**Discard changes to reference** executes the RefClose Discard command.

**Save back changes to reference** executes the RefClose Save command.

## -REFEDIT Command

**Command:** -refedit
**Select reference:** *(Pick an externally-referenced drawing, or a block.)*
**Select nesting level [Ok/Next] <Next>:** *(Type O or N.)*
**Enter object selection method [All/Nested] <All>:** *(Type A or N.)*
**Display attribute definitions [Yes/No] <No>:** *(Type Y or N.)*
**Use REFCLOSE or the Refedit toolbar to end reference editing session.**

*Displays **Refedit** toolbar.*

### COMMAND LINE OPTIONS

**Select reference** selects an externally-referenced drawing or an inserted block for editing.

**Select nesting level** selects objects within the reference — this becomes the selection set of objects that you may edit; you may select all nested objects with All (with the exception of OLE objects and objects inserted with the MInsert command, which cannot be refedited).

**Enter object selection method:**

**All** selects all objects.

**Nested** selected nested objects.

**Display attribute definitions:**

**Yes** makes non-constant attributes invisible; attribute definitions can be edited.

**No** means attribute definitions cannot be edited.

### RELATED COMMANDS

**RefSet** adds and removes objects from a working set.

**RefClose** saves or discards editing changes to the reference.

### RELATED SYSTEM VARIABLES

**RefEditName** stores the name of the externally-referenced file or block being edited.

**XEdit** determines whether the current drawing can be edited while being referenced by another drawing.

**XFadeCtl** specifies the amount of fading for objects not being edited in place.

### TIPS

* OLE objects and objects inserted with the **MInsert** command cannot be refedited.

* AutoCAD identifies the "working set" as those objects that you have selected to edit in-place.

* Objects *not* selected to be edited have their layers locked.

 # RefSet

**2000** Adds and removes objects from working sets (*short for REFerence SET*).

| Command | Alias | Ctrl+ | F-key | Alt+ | Menu Bar | Tablet |
|---------|-------|-------|-------|------|----------|--------|
| refset | ... | ... | ... | MBA | **Modify** | ... |
| | | | | | ↳**In-place Xref and Block Edit** | |
| | | | | | ↳**Add Objects to Working Set** | |
| | | | | MBR | **Modify** | |
| | | | | | ↳**In-place Xref and Block Edit** | |
| | | | | | ↳**Remove Objects from Working Set** | |

**Command:** refset
**Transfer objects between the Refedit working set and host drawing...**
**Enter an option [Add/Remove] <Add>:** *(Type A or R.)*
**Select objects:** *(Pick one or more objects.)*

## COMMAND LINE OPTIONS

 **Add** prompts you to select objects to add to the working set.

 **Remove** prompts you to select objects to remove from the working set.

**Select objects** selects the objects to be added or removed.

## RELATED COMMANDS

**RefEdit** edit reference objects in place.

**RefClose** saves or discards editing changes to the reference.

## RELATED SYSTEM VARIABLES

**RefEditName** stores the name of the xref or block being edited.

**XEdit** determines whether the current drawing can be edited while being referenced by another drawing.

**XFadeCtl** specifies the amount of fading for objects not being edited in place.

## TIPS

- The purpose of this command is to add objects to — or remove them from — the "working set" of objects, while you are performing in-place editing of a block or an externally-referenced drawing.

- When you select an object that cannot be added or removed from the working set, AutoCAD prompts: "** *n* selected objects are on a locked layer."

# Regen

<u>**V. 1.0**</u>  Regenerates the current viewport to update the drawing.

| Command | Alias | Ctrl+ | F-key | Alt+ | Menu Bar | Tablet |
|---------|-------|-------|-------|------|----------|--------|
| regen | re | ... | ... | VG | View | J1 |
| | | | | | ⬐Regen | |

**Command:** regen
**Regenerating model.**

## COMMAND LINE OPTION

ESC cancels the regeneration.

## RELATED COMMANDS

**Redraw** cleans up the current viewport quickly.

**RegenAll** regenerates all viewports.

**RegenAuto** checks with you before doing most regenerations

**ViewRes** controls whether zooms and pans are performed at redraw speed.

## RELATED SYSTEM VARIABLES

**RegenMode** holds the current setting of automatic regeneration:

| RegenMode | Meaning |
|-----------|---------|
| 0 | Off. |
| 1 | On (default). |

**WhipArc** determines how circles and arcs are displayed:

| WhipArc | Meaning |
|---------|---------|
| 0 | Circles and arcs displayed as vectors. |
| 1 | Circles and arcs displayed as true circles and arcs. |

## TIPS

• Some commands automatically force a regeneration of the screen; other commands queue the regeneration.

• The **Regen** command reindexes the drawing database for better display and object selection performance.

• To save on regeneration time, freeze layers you are not working with, apply **QText** to turn text into rectangles, and place hatching on its own layer.

• Use the **RegenAll** command to regenerate all viewports.

# RegenAll

Rel.10 Regenerates all viewports.

| Command | Alias | Ctrl+ | F-key | Alt+ | Menu Bar | Tablet |
|---------|-------|-------|-------|------|----------|--------|
| regenall | rea | ... | ... | VA | View<br>⊱Regen All | K1 |

**Command:** regenall
**Regenerating model.**

## COMMAND LINE OPTION
ESC cancels the regeneration process.

## RELATED COMMANDS
**RedrawAll** redraws all viewports.

**Regen** regenerates the current viewport.

**RegenAuto** checks with you before performing most regenerations.

**ViewRes** controls whether zooms and pans are performed at redraw speed.

## RELATED SYSTEM VARIABLE
**RegenMode** holds the current setting of automatic regeneration:

| RegenMode | Meaning |
|-----------|---------|
| 0 | Off. |
| 1 | On (default). |

**WhipArc** determines how circles and arcs are displayed:

| WhipArc | Meaning |
|---------|---------|
| 0 | Circles and arcs displayed as vectors. |
| 1 | Circles and arcs displayed as true circles and arcs. |

## TIPS
- The **RegenAll** command does not regenerate objects on frozen layers.

- Use the **Regen** command to regenerate a single viewport.

# 'RegenAuto

<u>V. 1.2</u>    Prompts before performing regenerations, when turned off (*short for REGENeration AUTOmatic*).

| Command | Alias | Ctrl+ | F-key | Alt+ | Menu Bar | Tablet |
|---|---|---|---|---|---|---|
| 'regenauto | ... | ... | ... | ... | ... | ... |

**Command:** regenauto
**Enter mode [ON/OFF] <ON>:** *(Enter* **ON** *or* **OFF***.)*

## COMMAND LINE OPTIONS
**OFF** turns on "About to regen, proceed?" message.
**ON** turns off "About to regen, proceed?" message.

## RELATED COMMANDS
**Regen** forces a regeneration in the current viewport.
**RegenAll** forces a regeneration in all viewports.

## RELATED SYSTEM VARIABLES
**Expert** suppresses "About to regen, proceed?" message when value is greater than 0.
**RegenMode** specifies the current setting of automatic regeneration:

| RegenMode | Meaning |
|---|---|
| 0 | Off. |
| 1 | On (default). |

## TIPS
*   If a regeneration is caused by a transparent command, AutoCAD delays it and responds with the message, "Regen queued."
*   When off, results in the following prompt:
    **Command:** regen
    **About to regen, proceed? <Y>:** *(Press* **Enter***.)*
*   AutoCAD Release 12 reduces the number of regenerations by expanding the virtual screen from 16 bits to 32 bits.

# Region

**Rel.11**  Creates 2D regions from closed objects.

| Command | Alias | Ctrl+ | F-key | Alt+ | Menu Bar | Tablet |
|---------|-------|-------|-------|------|----------|--------|
| region | reg | ... | ... | DN | Draw | R9 |
| | | | | | ⌐Region | |

**Command:** region
**Select objects:** *(Select one or more closed objects.)*
**Select objects:** *(Press **Enter** to end object selection.)*
**1 loop extracted.**
**1 region created.**

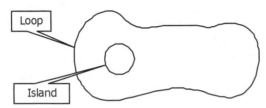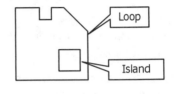

## COMMAND LINE OPTION

**Select objects** selects objects to convert to a region; AutoCAD discards unsuitable objects.

## RELATED COMMANDS

*All drawing commands.*

## RELATED SYSTEM VARIABLE

**DelObj** toggles whether objects are deleted during the conversion by the **Region** command.

## TIPS

- The **Region** command converts closed line sets, closed 2D and planar 3D polylines, and closed curves.

- The **Region** command rejects open objects, intersections, and self-intersecting curves.

- The resulting region is unpredictable when more than two curves share an endpoint.

- Polylines with width lose their width when converted to a region.

- An island can be considered a "hole" in the region.

## DEFINITIONS

*Curve* — an object made of circles, ellipses, splines, and joined circular and elliptical arcs.

*Island* — a closed shape fully within (*not touching or intersecting*) another closed shape.

*Loop* — a closed shape made of closed polylines, closed lines, and curves.

*Region* — a 2D closed area defined as a ShapeManager object.

# Reinit

Reinitializes digitizers and input-output ports, and reloads the *acad.pgp* file (*short for REINITialize*).

| Command | Alias | Ctrl+ | F-key | Alt+ | Menu Bar | Tablet |
|---------|-------|-------|-------|------|----------|--------|
| reinit | ... | ... | ... | ... | ... | ... |

**Command:** reinit

*Displays dialog box:*

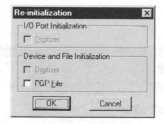

## DIALOG BOX OPTIONS

*I/O Port Initialization option*
**Digitizer** reinitializes ports connected to the digitizer; grayed out if no digitizer is configured.

*Device and File Initialization options:*
**Digitizer** reinitializes the digitizer driver; grayed out if no digitizer is configured.
**PGP File** reloads the *acad.pgp* file.

## RELATED COMMAND
**MenuLoad** reloads menu files.

## RELATED SYSTEM VARIABLE
**Re-Init** reinitializes via system variable settings.

## RELATED FILES
*acad.pgp* is the program parameters file in \*autocad 2004\support* folder.
***.hdi** are device drivers specific to AutoCAD.

## TIPS
- AutoCAD allows you to connect both the digitizer and the plotter to the same port, since you do not need the digitizer during plotting; use the **Reinit** command to reinitialize the digitizer after plotting.

- AutoCAD reinitializes all ports and reloads the *acad.pgp* file each time another drawing is loaded.

# Rename

<u>V . 2.1</u>   Changes the names of blocks, dimension styles, layers, linetypes, plot styles, text styles, UCS names, views, plotstyles, and viewports.

| Commands | Aliases | Ctrl+ | F-key | Alt+ | Menu Bar | Tablet |
|----------|---------|-------|-------|------|----------|--------|
| rename   | ren     | ...   | ...   | OR   | Format ⸂Rename | V1 |
| -rename  | -ren    |       |       |      |          |        |

## Command: rename

*Displays dialog box:*

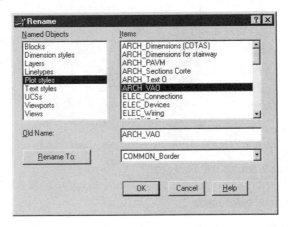

## DIALOG BOX OPTIONS

**Named Objects** lists the named objects in the drawing.

**Items** lists the names of named objects in the current drawing.

**Old Name** displays the current name of an object to be renamed.

**Rename to** allows you to enter a new name for the object.

## -RENAME Command

**Command:** -rename

**Enter object type to rename**

**[Block/Dimstyle/LAyer/LType/Plotstyle/Style/Ucs/VIew/VPort]:** *(Enter an option.)*

*Example usage:*

**[Block/Dimstyle/LAyer/LType/Plotstyle/Style/Ucs/VIew/VPort]:** b

**Enter old block name:** diode-20

**Enter new block name:** diode-02

### COMMAND LINE OPTIONS

**Block** changes the names of blocks.

**Dimstyle** changes the names of dimension styles.

**LAyer** changes the names of layers.

**LType** changes the names of linetypes.

**Plotstyle** changes the names of plotstyles.

**Style** changes the names of text styles.

**Ucs** changes the names of UCS configurations.

**VIew** changes the names of view configurations.

**VPort** changes the names of viewport configurations.

### RELATED SYSTEM VARIABLES

**CeLType** specifies the name of the current linetype.

**CLayer** specifies the name of the current layer.

**DimStyle** specifies the name of the current dimension style.

**InsName** specifies the name of the current block.

**TextStyle** specifies the name of the current text style.

**UcsName** specifies the name of the current UCS view.

### TIPS

* You cannot rename layer "0", dimstyle "Standard," anonymous blocks, groups, or linetype "Continuous."

* To rename a group of similar names, use **\*** (the wildcard for "all") and **?** (the wildcard for a single character).

* Names can be up to 255 characters in length.

* The **Properties** command does *not* allow you to rename blocks.

* The **DdRename** command no longer works in AutoCAD 2000; use the **Rename** command instead.

* The **PlotStyle** option is available only when plotstyles are attached to the drawing.

* In AutoCAD 2004, the **textStyle** option was renamed **Style**.

* You cannot use this command during **RefEdit**.

# Render

**Rel.12** Creates renderings of 3D objects.

| Command | Alias | Ctrl+ | F-key | Alt+ | Menu Bar | Tablet |
|---------|-------|-------|-------|------|----------|--------|
| render | rr | ... | ... | VER | View | M1 |
| | | | | | ⮡Render | |
| | | | | | ⮡Render | |

**Command:** render

*Displays dialog box:*

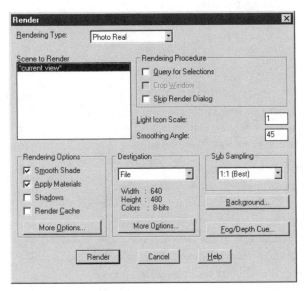

## DIALOG BOX OPTIONS

**Rendering Type** selects between basic Render, Photo Real, or Photo Raytrace; also lists installed third-party renderers.

**Scene to Render**   lists the names of scenes defined by the Scene command; default = *Current view*.

## Rendering Procedure options

**Query For Selections**

☑ prompts you to select the objects to render; unselected objects appear in wireframe in the rendering only when you select the Merge option from the Background option.

☐ renders all objects in the current viewport.

**Crop Window**

☑ prompts you to select a windowed area to render.

☐ renders the entire current viewport.

**Skip Render Dialog** does not display the Render dialog box the next time you use the Render command.

**Light icon scale** sizes light blocks Overhead, Direct, and Sh_Spot.

**Smoothing angle** converts edges to smooth curves. For example, when the angle between two surfaces is greater than the default of 45 degrees, AutoCAD renders an edge; when it is less than 45 degrees, the edge is smoothed to a curve.

Rendering Options options

**Smooth shade** smooths the edges of multifaced surfaces.

**Apply materials** applies surface materials defined by the RMat command.

**Shadows** generates shadows when Photo Real and Photo Raytrace rendering modes are selected.

**Render cache** caches the objects to help speed rendering.

**More options** displays the Render Options dialog box, which varies according to the type of rendering selected.

Destination options

- **Viewport** displays the rendering in the current viewport.
- **Render Window** displays the rendering in a separate window.
- **File** saves the rendering to a file on disk, does not display the rendering on screen.

**More Options** when File is selected, displays File Output Configuration dialog box.

**Sub Sampling** renders a fraction of pixels; ranges from 1:1 for best quality (default) to 8:1 for fastest.

**Background** displays the Background dialog box; see the Background command.

**Fog/Depth Cue** displays the Fog dialog box; see the **Fog** command.

**Render** renders the scene.

**Render Options** dialog box

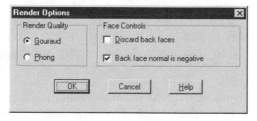

Render Quality options

- **Gouraud** calculates light intensity at each vertex; faster speed.
- **Phong** calculates light intensity at each pixel; higher quality.

Face Controls options

**Discard back faces** speeds up rendering by ignoring the backs of objects.

**Back face normal is negative** may create odd looking objects at times; turn off.

**Photo Real Render Options** dialog box

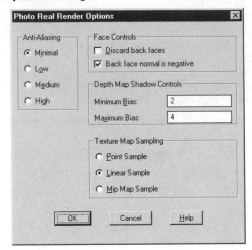

Anti-Aliasing options
- **Minimal** renders with analytical horizontal anti-aliasing; fastest rendering.
- **Low** renders with four samples per pixel.
- **Medium** renders with nine samples per pixel.
- **High** renders with 16 samples per pixel; best quality.

Face Controls options

**Discard back faces** speeds up rendering by ignoring the backs of objects.

**Back face normal is negative** may create odd looking objects at times; turn off.

Depth Map Shadow Controls options

**Minimum bias** adjusts the shadow map bias to prevent self-shadows and detached shadows; default = 2.0; ranges from 2.0 to 20.0.

**Maximum bias** limits the shadow map bias to 10 more than minimum bias; default = 4.0.

Texture Map Sampling options
- **Point sample** renders the nearest pixel within a bitmap.
- **Linear sample** averages the four neighbor pixels pyramidically (default).
- **Mip map sample** averages pixels with the *mip* method, which pyramidically averages a square sample area.

**Photo Raytrace Render Options** dialog box

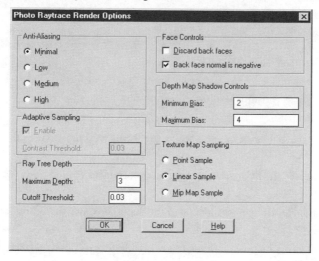

Anti-Aliasing options
- **Minimal** renders with analytical horizontal anti-aliasing; fastest rendering but lowest quality.
- **Low** renders with four samples per pixel.
- **Medium** renders with nine samples per pixel.
- **High** renders with 16 samples per pixel; best quality but slowest rendering.

Adaptive Sampling options

**Enable** toggles adaptive sampling; available when minimal anti-aliasing is turned off.

**Contrast Threshold** specifies the sensitivity of adaptive sampling; larger values increase rendering speed, but might reduce image quality; default = 0.03; ranges from 0.0 to 1.0.

Ray Tree Depth options

**Maximum Depth** limits the ray tree depth to track reflected and refracted rays; default = 3.

**Cutoff Threshold** defines the percentage bounce cutoff; default = 0.03 means 3%.

Face Controls options

**Discard back faces** speeds up rendering by ignoring the backs of objects.

**Back face normal is negative** may create odd looking objects at times; turn off.

Depth Map Shadow Controls options

**Minimum bias** Adjusts shadow map bias to prevent self-shadows and detached shadows; default = 2.0; ranges from 2.0 to 20.0.

**Maximum bias** limits shadow map bias to 10 more than minimum bias; default = 4.0.

Texture Map Sampling options
- **Point sample** renders the nearest pixel within a bitmap.
- **Linear sample** averages the four neighbor pixels (default).
- **Mip map sample** averages pixels with the *mip* method, which pyramidically averages a square sample area.

## RELATED COMMANDS

*All rendering-related commands.*

**Hide** removes hidden lines from wireframe view.

**Replay** displays a BMP, TIFF, or Targa raster (bitmap) file in the current viewport.

**RPref** sets up options for the Render command.

**SaveImg** saves the image in the current viewport to a raster file.

**ShadEdge** performs real-time shading of 3D objects.

## TIPS

- If you do not place a light or define a scene, the **Render** command uses the current view and ambient light.

- If you do not select a light or scene, the **Render** command renders all objects using all lights and the current view.

- If you set up the **Render** command to skip the dialog box, use the **RPref** command to set rendering options.

- When outputting to a file, you have the following file format options: BMP, TGA, PCX, PostScript, and TIFF.

- You cannot create a rendering in paper space mode.

## Removed Command

**RenderUnload** was removed from Release 14. Use **Arx** to unload *Render.Arx* instead.

# Your First Rendering

■ *Basic Rendering*

**Step 1**

Create a 3D drawing or select a 3D sample drawing.

**Step 2**

Start the **Render** command, click the **Render** button, and wait a few seconds.

■ *Advanced Rendering*

**Step 1**

Create a 3D drawing or select a 3D sample drawing.

**Step 2**

Use the **MatLib** command to load material definitions into drawing.

**Step 3**

Using the **RMat** command, assign materials to colors, layers, and/or objects.

**Step 4**

Use the **Light** command to place and aim lights: point, spot, and distant.

**Step 5**

Use the **Scene** command to collect lights and a named view into a named object.

**Step 6**

Render the named scene with the **Render** command.

**Step 7**

Use the **SaveImg** command to save the rendering to a TIFF, Targa, or *.gif* file on disk.

**Step 8**

View the saved rendering file with the **Replay** command.

## RENDERING EFFECTS

Wireframe drawing:

Basic rendering; most options turned off:

**Smooth Shading** on:    **Attach Materials** on; requires Photo Real or Raytrace mode:

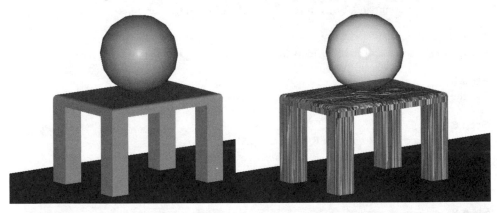

**Background** set to **Gradient**:    **Background** set to **Image**:

**Fog** set to white:

**Fog** set to black:

**Shadows** turned on:

**Shadow Volumes** turned off:

**Sub Sampling** sct to **4:1**:

# RendScr

<u>Rel.12</u>  Redisplays the most-recent rendering *(short for RENDer SCReen)*.

| Command | Alias | Ctrl+ | F-key | Alt+ | Menu Bar | Tablet |
|---------|-------|-------|-------|------|----------|--------|
| rendscr | ... | ... | ... | ... | ... | ... |

**Command:** rendscr

## COMMAND LINE OPTIONS
*None.*

## RELATED COMMAND
**Render** creates a rendering of the current 3D viewport.

## RELATED SYSTEM VARIABLES
*None.*

# Replay

Rel.12  Displays BMP, TIFF, and Targa files as bitmaps.

| Command | Alias | Ctrl+ | F-key | Alt+ | Menu Bar | Tablet |
|---------|-------|-------|-------|------|----------|--------|
| replay | ... | ... | ... | TDV | Tools | V8 |
| | | | | | ⬐Display Image | |
| | | | | | ⬐View | |

**Command:** replay

*Displays the **Select File** dialog box. Select file, and then click **Open**. Displays dialog box:*

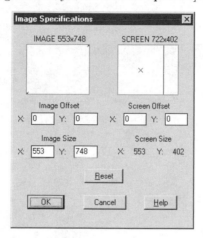

## DIALOG BOX OPTIONS

**Image** selects the displayed area by clicking on the image tile.

**Image Offset** specifies the x,y coordinates of the image's lower-left corner.

**Image Size** specifies the size of the image in pixels.

**Screen** specifies the maximum size of the image.

**Screen Offset** specifies the x,y coordinates of the image's lower-left corner.

**Reset** restores the default values.

## RELATED COMMANDS

**Import** displays a dialog box to loading some raster and vector files.

**SaveImg** saves a rendering as a BMP, TIFF, or Targa raster file.

## RELATED FILES

***.bmp** are Windows bitmap files.

***.tif** are RGBA TIFF files, up to 32-bits in color depth.

***.tga** are RGBA Targa v2.0 files, up to 32-bits in color depth.

*In folder \autocad 2004\textures:*

***.tga** contains many Targa-format images.

# 'Resume

<u>**V. 2.0**</u>    Resumes script files after being paused by the BACKSPACE key.

| Command | Alias | Ctrl+ | F-key | Alt+ | Menu Bar | Tablet |
|---------|-------|-------|-------|------|----------|--------|
| 'resume | ... | ... | ... | ... | ... | ... |

**Command:** resume

## COMMAND LINE OPTIONS

BACKSPACE pauses the script file.

ESC stops the script file.

## RELATED COMMANDS

**RScript** reruns the current script file.

**Script** loads and runs a script file.

 # RevCloud

**2004** Draws revision clouds, and converts objects into revision clouds.

| Command | Alias | Ctrl+ | F-key | Alt+ | Menu Bar | Tablet |
|---------|-------|-------|-------|------|----------|--------|
| revcloud | ... | ... | ... | DU | Draw<br>↳Revision Cloud | ... |

**Command:** revcloud
**Minimum arc length: 0.5000  Maximum arc length: 0.5000**
**Specify start point or [Arc length/Object] <Object>:** *(Pick a point, or enter an option.)*
**Guide crosshairs along cloud path...** *(Move cursor to create cloud.)*
**Revision cloud finished.**

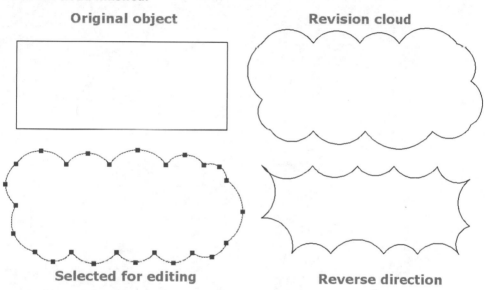

Original object

Revision cloud

Selected for editing

Reverse direction

## COMMAND LINE OPTIONS

**Specify start point** specifies the starting point of the cloud.

**Guide crosshairs along cloud path** indicates the outline of the cloud.

Arc Length options
**Specify minimum length of arc <0.5000>:** *(Enter a minimum value.)*
**Specify maximum length of arc <1.0000>:** *(Enter a maximum value.)*

**Specify minimum length of arc** specifies the minimum arc length.

**Specify maximum length of arc** specifies the maximum arc length.

Object options
**Select object:** *(Select one object to convert to a revision cloud.)*
**Reverse direction [Yes/No] <No>:** *(Type **Y** or **N**.)*

**Select object** selects the object to convert into a revision cloud.

**Reverse direction** turns the revision cloud inside-out.

. . . . . . . . . . . . . . . . . . . . . . . . . . . . . . . . . . . . . . . . . . . .

**RELATED SYSTEM VARIABLE**

**DimScale** affects the size of the arcs.

**TIPS**

- When the cursor reaches the start point, the revision cloud closes automatically.

- The revision cloud is drawn as a polyline.

- To edit the revision cloud, select it, and then move the grips.

- The arc length is not available from a system variable because it is stored in the Windows registry.

- The arc length is multiplied by the value stored in the **DimScale** system variable.

# Revolve

**Rel.11** Creates 3D solid objects by revolving closed 2D objects about an axis.

| Command | Alias | Ctrl+ | F-key | Alt+ | Menu Bar | Tablet |
|---------|-------|-------|-------|------|----------|--------|
| revolve | rev | ... | ... | DIR | Draw | Q7 |
| | | | | | ⮑ Solids | |
| | | | | | ⮑ Revolve | |

**Command:** revolve
**Current wire frame density: ISOLINES=4**
**Select objects:** *(Select one or more closed objects.)*
**Select objects:** *(Press **Enter** to end object selection.)*
**Specify start point for axis of revolution or define axis by [Object/X (axis)/Y (axis)]:** *(Pick a point, or enter an option.)*
**Specify angle of revolution <360>:** *(Enter a value.)*

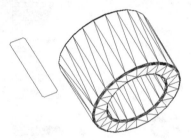

*The polyline rectangle (left) was used to create the solid object (right).*

## COMMAND LINE OPTIONS

**Select objects** selects a closed object to revolve: closed polyline, circle, ellipse, donut, polygon, closed spline, or a region.

Axis of revolution options

**Specify start point for axis of revolution** indicates the axis of revolution; you must specify the endpoints.

**Object** selects the object that determines the axis of revolution.

**X** uses the positive x axis as the axis of revolution.

**Y** uses the positive y axis as the axis of revolution.

**Specify angle of rotation** specifies the amount of rotation; full circle = 360 degrees.

## RELATED COMMANDS

**Extrude** extrudes 2D objects into a 3D solid model.

**Rotate** rotates open and closed objects, forming a 3D surface.

## TIPS

• **Revolve** works with just one object at a time.

• This command does not work with open objects or self-intersecting polylines.

 # RevSurf

**Rel.10** Generates 3D surfaces of revolution defined by a path curve and an axis (*short for REVolved SURFace*).

| Command | Alias | Ctrl+ | F-key | Alt+ | Menu Bar | Tablet |
|---------|-------|-------|-------|------|----------|--------|
| revsurf | ... | ... | ... | DFS | Draw | O8 |
| | | | | | ⌇Surfaces | |
| | | | | | ⌇Revolved Surface | |

**Command:** revsurf
**Current wire frame density: SURFTAB1=6 SURFTAB2=6**
**Select object to revolve:** *(Select an object.)*
**Select object that defines the axis of revolution:** *(Select an object.)*
**Specify start angle <0>:** *(Enter a value.)*
**Specify included angle (+=ccw, -=cw) <360>:** *(Enter a value.)*

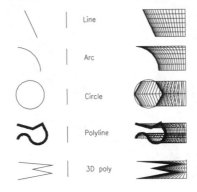

*Path Curve   +   Axis   =   Resulting Revolved Surface*

## COMMAND LINE OPTIONS

**Select object to revolve** selects the single object that will be revolved about an axis.
**Select object that defines the axis of revolution** selects the axis object.
**Start angle** specifies the starting angle.
**Included angle** specifies the angle to rotate about the axis.

## RELATED COMMANDS

**PEdit** edits revolved surfaces.
**Revolve** revolves a 2D closed object into a 3D solid.

## RELATED SYSTEM VARIABLES

**SurfTab1** specifies the mesh density in the m-direction (default = 6).
**SurfTab2** specifies the mesh density in the n-direction (default = 6).

## TIPS

- Unlike the **Revolve** command, **RevSurf** works with open and closed objects.

- If a multi-segment polyline is the axis of revolution, the rotation axis is defined as the vector pointing from the first vertex to the last vertex, ignoring the intermediate vertices.

. . . . . . . . . . . . . . . . . . . . . . . . . . . . . . . . . . . . . . . . . . . . . . . . . . .

# RMat

**Rel.13** Applies material definitions to colors, layers, and objects; used by the **Render** command (*short for Render MATerials*).

| Command | Alias | Ctrl+ | F-key | Alt+ | Menu Bar | Tablet |
|---------|-------|-------|-------|------|----------|--------|
| rmat | ... | ... | ... | VEM | View | P1 |
| | | | | | ⮩Render | |
| | | | | | ⮩Materials | |

**Command:** rmat
*Displays dialog box:*

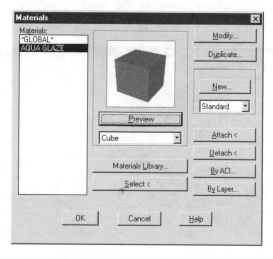

## DIALOG BOX OPTIONS

**Materials** lists the names of materials loaded into drawing by the MatLib command.

**Preview** previews the material mapped to a sphere or cube.

**Materials Library** displays the Materials Library dialog box; see the MatLib command.

**Select** selects the objects to which to attach the material definition.

**Modify** edits a material definition; displays a different Modify Material dialog box for standard, granite, marble and wood-based materials; see the New option.

**Duplicate** duplicates a material definition so that you can edit it; displays a different Modify Material dialog box for standard, granite, marble and wood-based materials; see the New option.

**New** creates a new material definition; displays a different dialog box for the four types of materials: standard, granite, marble, wood.

**Attach** selects the objects to which to attach the material definition.

**Detach** selects the objects from which to detach the material definition.

**By ACI** attaches material via ACI number; displays dialog box:

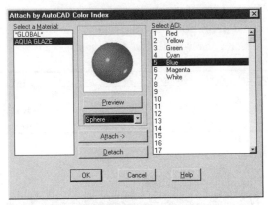

**By Layer** attaches material to a layer name; displays dialog box:

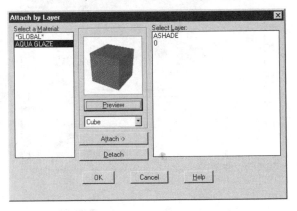

*New Material options*

**Material Name** names the material.

**Value** adjusts the value of selected Attributes between 0 and a larger number.

*Color options*

**By ACI** makes the material's base color the same as the object's.

**Lock** locks the ambient and reflective colors to the base color.

**Mirror** allows mirrored reflection.

**Color System** selects RGB or HLS color system:

- **HLS** specifies color by levels of Hue, Luminescence, and Saturation.
- **RGB** specifies color by levels of Red, Green, Blue.

**Bitmap Blend** determines the degree the bitmap is rendered.

**File Name** specifies the bitmap's filename.

**Adjust Bitmap** displays the Adjust Material Bitmap Placement dialog box.

**Find File** selects the bitmap file; displays the Bitmap File dialog box.

**New Standard Material** dialog box

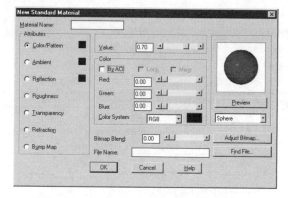

Attributes options
- **Color/Pattern** sets the base color of the material.
- **Ambient** sets the ambient color shown in shadowed areas.
- **Reflection** sets the highlight color of the material.
- **Roughness** sets the size of the highlighted area; the higher the value of the roughness, the larger the highlight area.
- **Transparency** sets the amount of transparency of the material:

| Transparency | Meaning |
| --- | --- |
| **0.0** | No transparency. |
| **0.1** *through* **9.9** | Increasing transparency with edge fall-off. |
| **1.0** | Perfect transparency. |

- **Refraction** controls refraction of the material; Photo Raytrace rendering only.
- **Bump Map** attaches a bitmap to the material.

**New Granite Material** dialog box

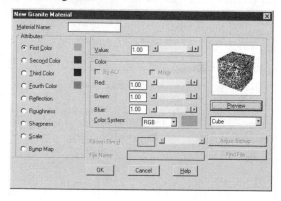

Attributes options
- **First Color** sets the base color of the granite material.
- **Second Color** sets the second color of the granite material.

- **Third Color** sets the third color of the granite material.
- **Fourth Color** sets the fourth color of the granite material.
- **Reflection** sets the highlight color of the material.
- **Roughness** sets the size of the highlighted area; the higher the value of the roughness, the larger the highlighted area.
- **Sharpness** sets the sharpness of the four colors of the material:

| Sharpness | Meaning |
| --- | --- |
| **0.0** | Complete blurring of colors. |
| **0.1** *through* **9.9** | Increasing sharpness of colors. |
| **1.0** | All four colors perfectly distinct. |

- **Scale** sizes the material relative to its attached object.
- **Bump Map** attaches a bitmap to the material.

**New Marble Material** dialog box

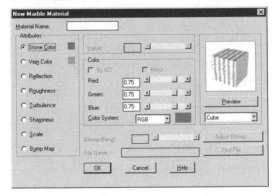

*Attributes options*
- **Stone Color** sets the base color of the marble material.
- **Vein Color** sets the secondary color of the marble material.
- **Reflection** sets the highlight color of the material.
- **Roughness** sets the size of the highlighted area; the higher the value of the roughness, the larger the highlighted area.
- **Turbulence** determines the swirling of the vein color.
- **Sharpness** sets the sharpness of the four colors of the material:

| Sharpness | Meaning |
| --- | --- |
| **0.0** | Complete blurring of colors. |
| **0.1** *through* **9.9** | Increasing sharpness of colors |
| **1.0** | All colors perfectly distinct. |

**New Wood Material** dialog box

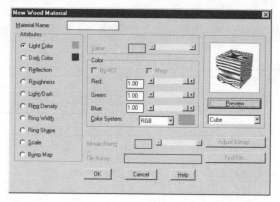

Attributes options
- **Light Color** sets the base color of the wood material.
- **Dark Color** sets the secondary color of the wood material.
- **Reflection** sets the highlight color of the material.
- **Roughness** sets the size of the highlighted area; the higher the value of the roughness, the larger the highlight area.
- **Light/Dark** sets the amount of contrast between the two colors of the material:

| Light/Dark | Meaning |
|---|---|
| **0.0** | Dark color. |
| **0.1** *through* **9.9** | Increasing lightness of color. |
| **1.0** | Light color. |

- **Ring Density** specifies the number of rings in the material definition.
- **Ring Width** specifies the variation in ring width; 0 = narrow rings; 1.0 = wide rings.
- **Ring Shape** specifies 0 = circular rings; 1.0 = irregular rings.
- **Scale** sizes the material relative to its attached object.
- **Bump Map** attaches a bitmap to the material.

## RELATED COMMANDS

**MatLib** loads material definitions into the drawing.

**Render** renders the drawing using material definitions.

## TIPS
- You must use the commands in this order:

    1. **MatLib** loads materials into the drawing.

    2. **RLib** attaches materials to objects.

    3. **Render** with **Apply Materials** turned on.

- The **By ACI** option lets you attach a material definition to all objects of one color.

- The **By Layer** option lets you attach a material definition to all objects on one layer.

# RMLin

**2000i** Inserts redline markup files from Autodesk's Volo products (*short for Redline Markup Language INput*).

| Commands | Alias | Ctrl+ | F-key | Alt+ | Menu Bar | Tablet |
|----------|-------|-------|-------|------|----------|--------|
| rmlin | ... | ... | ... | IU | Insert | ... |
| | | | | | ⌐Markup | |

**-rmlin**

**Command:** rmlin
  *Displays **Insert Markup** dialog box. Select a .rml file, and then click **OK**.*

- - - - - - - - - - - - - - - - - - - - - - - - - - - - - - - - - - - - - - - - - - - - - - - - - - - -

## -RMLIN Command
**Command:** -rmlin
**Enter name of RML file:** *(Enter the name of file.)*

### COMMAND LINE OPTION
**Enter name of RML file** specifies the name of the redline markup file to import.

### RELATED COMMAND
**DxfIn** inserts a *.dxf* file, which could contain redline markups from non-Autodesk viewer software.

### TIPS
- When an *.rml* file is inserted in the drawing, AutoCAD places it on a new layer called "_Markup_." This layer has the following properties that differ from the defaults:

| Property | Value |
|----------|-------|
| Name | Markup. |
| Color | Red. |
| Locked/unlocked | Locked. |

- Redlines or markup are comments added to drawings; redlines are not normally part of the drawing. Redlines consists of notes, circles, arrows, hyperlinks, and so on.

- This command is meant for use with Autodesk's own viewing and redlining software, called Volo. It saves redlines in *.rml* files.

- - - - - - - - - - - - - - - - - - - - - - - - - - - - - - - - - - - - - - - - - - - - - - - - - - - -

 # Rotate

**V. 2.5**   Rotates objects about base points in the 2D plane.

| Command | Alias | Ctrl+ | F-key | Alt+ | Menu Bar | Tablet |
|---------|-------|-------|-------|------|----------|--------|
| rotate | ro | ... | ... | MR | Modify<br>↳ Rotate | V20 |

**Command:** rotate
**Current positive angle in UCS: ANGDIR=counterclockwise ANGBASE=0**
**Select objects:** *(Select one or more objects.)*
**Select objects:** *(Press **Enter** to end object selection.)*
**Specify base point:** *(Pick a point.)*
**Specify rotation angle or [Reference]:** *(Enter a value, or type **R**.)*

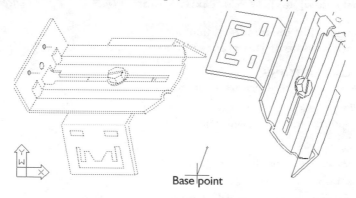

Base point

## COMMAND LINE OPTIONS

**Select objects** selects the objects to be rotated.

**Specify base point** picks the point about which the objects will be rotated.

**Specify rotation angle** specifies the angle by which the objects will be rotated.

**Reference** allows you to specify the current rotation angle and new rotation angle.

## RELATED COMMANDS

**Rotate3D** rotates objects in 3D space.

**UCS** rotates the coordinate system.

**Snap** rotates the cursor.

## TIPS

- AutoCAD rotates the selected object(s) about the base point.

- At the 'Specify rotation angle' prompt, you can show the rotation by moving the cursor. AutoCAD dynamically displays the new rotated position as you move the cursor.

- Use object snap modes, such as **INTersection**, to position the base point and rotation angle(s) accurately.

# Rotate3D

<u>Rel.11</u>  Rotates objects about axes in 3D space.

| Command | Alias | Ctrl+ | F-key | Alt+ | Menu Bar | Tablet |
|---------|-------|-------|-------|------|----------|--------|
| rotate3d | ... | ... | ... | M3R | Modify | W22 |
| | | | | | ⬁3D Operation | |
| | | | | | ⬁Rotate 3D | |

**Command:** rotate3d
**Current positive angle: ANGDIR=counterclockwise ANGBASE=0**
**Select objects:** *(Select one or more objects.)*
**Select objects:** *(Press **Enter** to end object selection.)*

**Specify first point on axis or define axis by [Object/Last/View/Xaxis/Yaxis/Zaxis/2points]:** *(Pick a point, or enter an option.)*
**Specify second point on axis:** *(Pick another point.)*
**Specify rotation angle or [Reference]:** *(Enter a value, or type **R**.)*

## COMMAND LINE OPTIONS

**Select objects** selects the objects to be rotated.

**Specify rotation angle** rotates objects by a specified angle: relative rotation.

**Reference** specifies the starting and ending angle: absolute rotation.

Define Axis By options

**Object** selects object to specify the rotation axis.

**Last** selects the previous axis.

**View** specifies that the current view direction is the rotation axis.

**Xaxis** specifies that the x axis is the rotation axis.

**Yaxis** specifies that the y axis is the rotation axis.

**Zaxis** specifies that the z axis is the rotation axis.

**2 points** defines two points on the rotation axis.

## RELATED COMMANDS

**Align** rotates, moves, and scales objects in 3D space.

**Mirror3d** mirrors objects in 3D space.

**Rotate** rotates objects in 2D space.

# RPref

**Rel.12**    Specifies options for the **Render** command (*short for Render PREFerences*).

| Command | Alias | Ctrl+ | F-key | Alt+ | Menu Bar | Tablet |
|---------|-------|-------|-------|------|----------|--------|
| rpref | rpr | ... | ... | VEP | View | R2 |
| | | | | | ⍦Render | |
| | | | | | ⍦Preferences | |

**Command:** rpref

*Displays dialog box::*

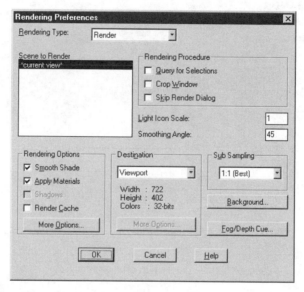

## DIALOG BOX OPTIONS
*Options are identical to the **Render** command, with the exception of the **OK** button replacing the **Render** button.*

## RELATED COMMANDS
*All rendering-related commands.*

**Hide** removes hidden lines from wireframe view.

**Render** renders the scene.

**ShadEdge** produces real-time shading of 3D objects.

**3dOrbit** creates perspective view.

## TIP
- If you set up the **Render** command to skip the dialog box, use the **RPref** command to set rendering options.

# 'RScript

<u>**V. 2.0**</u>  Repeats script files *(short for Repeat SCRIPT)*.

| Command | Alias | Ctrl+ | F-key | Alt+ | Menu Bar | Tablet |
|---------|-------|-------|-------|------|----------|--------|
| 'rscript | ... | ... | ... | ... | ... | ... |

**Command:** rscript

## COMMAND LINE OPTIONS
*None.*

## RELATED COMMANDS
**Resume** resumes a script file after being interrupted.

**Script** loads and runs a script file.

# RuleSurf

**Rel.10** Draws 3D ruled surfaces between two objects (*short for RULEd SURFace*).

| Command | Alias | Ctrl+ | F-key | Alt+ | Menu Bar | Tablet |
|---------|-------|-------|-------|------|----------|--------|
| rulesurf | ... | ... | ... | DFR | Draw | Q8 |
| | | | | | ⅋Surfaces | |
| | | | | | ⅋Ruled Surface | |

**Command:** rulesurf
**Select first defining curve:** *(Pick an object.)*
**Select second defining curve:** *(Pick another object.)*

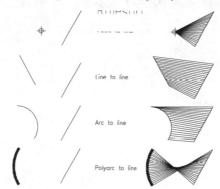

## COMMAND LINE OPTIONS

**Select first defining curve** selects the first object for the ruled surface.

**Select second defining curve** selects the second object for the ruled surface.

## RELATED COMMANDS

**EdgeSurf** draws a 3D surface bounded by four edges.

**RevSurf** draws a 3D surface of revolution.

**TabSurf** draws a 3D tabulated surface.

## RELATED SYSTEM VARIABLE

**SurfTab1** determines the number of rules drawn.

## TIPS

- **RuleSurf** uses objects as the boundary curve: points, lines, arcs, circles, polylines, and 3D polylines.

- Both boundaries must either be closed or open; the exception is using the point object.

- The **RuleSurf** command begins drawing its mesh as follows:

| Object | RuleSurf |
|--------|----------|
| **Open object** | From the object's endpoint closest to your pick. |
| **Circle** | From the zero-degree quadrant. |
| **Closed polyline** | From the last vertex. |

- **RuleSurf** draws with a circle in the opposite direction of a closed polyline.

# Save, SaveAs

**V. 2.0** Saves the current drawing to disk, after prompting for a filename.

| Commands | Alias | Ctrl+ | F-key | Alt+ | Menu Bar | Tablet |
|----------|-------|-------|-------|------|----------|--------|
| save | saveurl | ... | ... | ... | ... | ... |
| saveas | ... | Shift+S ... | | FA | File<br>⤷Save As | V24 |

**Command:** save *or* saveas

*Displays dialog box.*

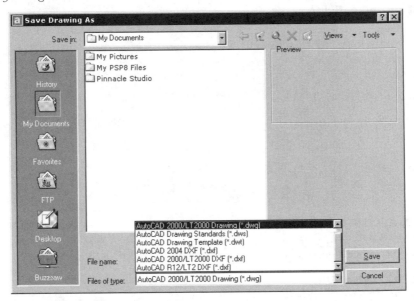

## DIALOG BOX OPTIONS

**Save in** selects the folder, hard drive, or network drive in which to save the drawing.

**File name** names the drawing; maximum = 255 characters; default=*drawing1.dwg*.

**Files of type** saves the drawing in a variety of formats:

| Save as type | Meaning |
|--------------|---------|
| **AutoCAD 2004 DWG** | Saves the drawing in native format. |
| **AutoCAD 2000/LT 2000** | Saves the drawing in AutoCAD 2000/2000i/ 2002 format. |
| **Standard DWS** | Saves the drawing as a CAD standard. |
| **Template DWT** | Saves the drawing in the \\*template* folder. |
| **AutoCAD 2004 DXF** | Exports the drawing in 2004 DXF format. |
| **AutoCAD 2000/LT 2000 DXF** | Exports the drawing in 2000/2000i/2002 DXF format. |
| **R12, LT 2 DXF** | Exports the drawing in R12 DXF format. |

**Save** saves the drawing, and then returns to AutoCAD.

**Cancel** does not save the drawing.

*When a drawing of the same name already exists in the same folder, displays dialog box:*

### Tools | Options dialog box
*From the toolbar, select* **Tools | Options:**

### DWG Options tab

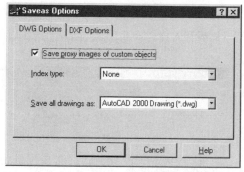

**Save proxy images of custom objects**

☑ saves an image of the custom objects in the drawing file.

☐ saves a frame around each custom object.

**Index type** saves indices with drawing; useful only for xrefed and partially-loaded drawings:

| Index type | Meaning |
|------------|---------|
| None | Creates no indices (default). |
| Layer | Loads layers that are on and thawed. |
| Spatial | Loads only the visible portion of a clipped xref. |
| Layer and Spatial | Combines the two options immediately above. |

**Save all drawings as** selects the default format for saving drawings.

### DXF Options tab

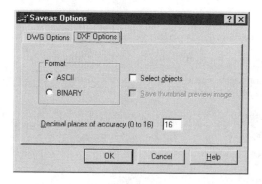

## Format options
**ASCII** saves the drawing in ASCII format, which can be read by humans – as well as common text editors – but takes up more disk space.

**BINARY** saves the drawing in binary format, which takes up less disk space, but cannot be read by some software programs.

## Additional options
**Select objects** allows you to save selected objects to the DXF file; AutoCAD prompts you: "Select objects:"

**Save thumbnail preview image** saves a thumbnail image of the drawing.

**Decimal places of accuracy** specifies the number of decimal places for real numbers: 0 to 16.

## **Template Description** dialog box
*Displayed when drawing is saved as a .dwt template file:*

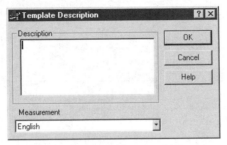

**Description** describes the template file; stored in *wizard.ini*.

**Measurement** selects English or metric as the default measurement system.

## RELATED COMMANDS
**Export** saves the drawing in a variety of raster and vector formats.

**Plot** saves the drawing in even more raster and vector formats.

**Quit** exits AutoCAD without saving the drawing.

**QSave** saves the drawing without prompting for a name.

## RELATED SYSTEM VARIABLES
**DBMod** indicates that the drawing was modified since being opened.

**DwgName** specifies the name of the drawing; *drawing1.dwg* when unnamed.

## TIPS
* The **Save** and **SaveAs** commands perform exactly the same function.

* Use these commands to save the drawing by another name.

* When a drawing is opened as read-only, use these commands to save the drawing to another filename.

* To save a drawing without seeing the **Save Drawing As** dialog box, use the **QSave** command.

* As of AutoCAD 2004, you can use CTRL+SHIFT+S to display the **Save Drawing As** dialog box.

. . . . . . . . . . . . . . . . . . . . . . . . . . . . . . . . . . . . . . . . . . . . . . . . . . . . .

## Removed Command
**SaveAsR12** was removed from AutoCAD Release 14. Use **SaveAs** instead, which saves drawings in earlier formats.

. . . . . . . . . . . . . . . . . . . . . . . . . . . . . . . . . . . . . . . . . . . . . . . . . . . . .

# SaveImg

**Rel.12** Saves the current viewport image — raster and vector — as BMP, TIFF, or Targa files on disk (*short for Save Image*).

| Command | Alias | Ctrl+ | F-key | Alt+ | Menu Bar | Tablet |
|---------|-------|-------|-------|------|----------|--------|
| saveimg | ... | ... | ... | TIS | Tools | V7 |
| | | | | | ⌐Display Image | |
| | | | | | ⌐Save | |

**Command:** saveimg

*Displays dialog box:*

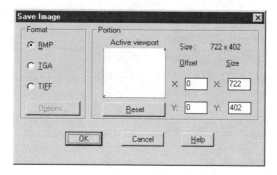

## DIALOG BOX OPTIONS

Format options
- **BMP** saves in Windows bitmap format.
- **TGA** saves in Targa format.
- **TIFF** saves in Tagged image file format.

**Options** displays the Options dialog box.

Portion options

**Active viewport** selects the area of the current viewport to save.

**Offset** specifies the lower-left x,y coordinates of the image area (default = 0,0).

**Size** specifies the upper-right x,y coordinates of the image area.

**TGA Options** dialog box

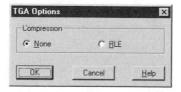

Compression options
- **None** saves *.tga* files with no file compression.
- **RLE** saves *.tga* files with run-length encoded compression.

**TIFF Options** dialog box

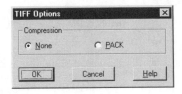

Compression options
* **None** saves TIFF files with no compression.
* **PACK** saves TIFF files with pack bits compression.

## RELATED COMMANDS

**JpgOut** saves the drawing in JPEG raster format.

**PngOut** saves the drawing in PNG raster format.

**TifOut** saves the drawing in TIFF raster format.

**WmfOut** saves the drawing in WMF vector format.

**Replay** displays a raster image in the current viewport.

**PRT SCR** saves the entire screen to the Clipboard.

## TIPS

* The **SaveImg** command saves files in these formats:

| Format | Specification |
| --- | --- |
| **BMP** | 24-bit Windows bitmap. |
| **TGA** | 32-bit RGBA TrueVision v2.0. |
| **TIFF** | 32-bit RGBA tagged image file. |

* The *.gif* format, which is used to display images on the Internet and was found in AutoCAD Release 12 and 13, was replaced by the less useful BMP format in AutoCAD Release 14.

## Removed Command

**SaveUrl** was removed from AutoCAD 2000; it was replaced by **Save** and **SaveAs**.

# Scale

**V. 2.5**  Changes the size of selected objects, to make them smaller or larger.

| Command | Alias | Ctrl+ | F-key | Alt+ | Menu Bar | Tablet |
|---------|-------|-------|-------|------|----------|--------|
| scale | sc | ... | ... | ML | Modify | V21 |
| | | | | | ⬐ Scale | |

**Command:** scale
**Select objects:** *(Select one or more objects.)*
**Select objects:** *(Press **Enter** to end object selection.)*
**Specify base point:** *(Pick a point.)*
**Specify a scale factor or [Reference]:** *(Enter scale factor, or type **R**.)*

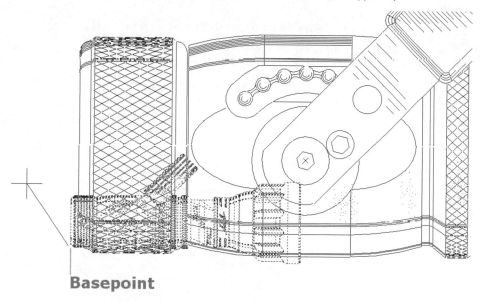

**Basepoint**

## COMMAND LINE OPTIONS

**Base point** specifies the point from which scaling takes place.

**Reference** supplies a reference value, followed by a new value.

**Scale factor** indicates the scale factor, which applies equally in the x and y directions.

| Scale Factor | Meaning |
|--------------|---------|
| **> 1.0** | Enlarges the size of the object(s). |
| **1.0** | Makes no change. |
| **> 0.0** *and* **< 1.0** | Reduces the size of the object(s). |
| **0.0** *or negative* | Illegal values. |

Reference options
**Specify reference length <1>:** *(Enter a base length.)*
**Specify new length:** *(Enter a new length.)*

Specify reference length specifies a baseline length; you can enter a value, or pick two points.

Specify new length specifies the new length that results in determining the scale factor.

## RELATED COMMANDS

Align scales an object in 3D space.

Insert allows a block to be scaled independently in the x, y, and z directions.

Plot allows a drawing to be plotted at any scale.

## TIPS

• You can interactively scale the object by moving the cursor to "grow" the object larger and smaller.

• Scale factors larger than 1.0 grow the object; smaller than 1.0 shrink the object.

• This command changes the size in all dimensions (x,y,z) equally; to change an object in just one dimension, use the **Stretch** command.

• The **Reference** option is useful for scaling a raster image to match the size of a drawing.

# ScaleText

**2002** Changes the height of text relative to its insertion point..

| Command | Alias | Ctrl+ | F-key | Alt+ | Menu Bar | Tablet |
|---------|-------|-------|-------|------|----------|--------|
| scaletext | ... | ... | ... | MOTS | Modify | ... |
| | | | | | ↳Object | |
| | | | | | ↳Text | |
| | | | | | ↳Scale | |

**Command:** scaletext
**Select objects:** *(Select one or more objects.)*
**Select objects:** *(Press **Enter** to end object selection.)*
**Enter a base point option for scaling**
**[Existing/Left/Center/Middle/Right/TL/TC/TR/ML/MC/MR/BL/BC/BR]**
**<Existing>:** *(Pick a point, or enter an option.)*
**Specify new height or [Match object/Scale factor] <0.2000>:** *(Enter a value, or an option.)*

## COMMAND LINE OPTIONS

**Enter a base point option for scaling** specifies the point from which scaling takes place.

**Existing** uses the existing insertion point.

**Specify new height** specifies the new height of the text.

**Match object** matches the height of another text object.

**Scale factor** scales the text by a factor:

| Scale Factor | Meaning |
|--------------|---------|
| > 1.0 | Enlarges the size of the object(s). |
| 1.0 | Makes no change in size. |
| > 0.0 *and* < 1.0 | Reduces the size of the object(s). |
| 0.0 *or negative* | Illegal values. |

Match Object options
**Select a text object with the desired height:** *(Pick another text object.)*
**Height=***nnn*

**Select a text object with the desired height** selects another text object to match its height.

Scale Factor options
**Specify scale factor or [Reference] <2.0000>:** *(Enter a value, or type **R**.)*

**Specify scale factor** scales the text by a factor; see Scale command.

**Reference** supplies a reference value, followed by a new value.

## RELATED COMMANDS

**Justify** changes the justification of the text.

**Properties** changes all aspects of the text.

**Style** creates text styles.

## TIP

• *Caution!* Scaling text larger may make it overlap nearby text.

**Scene**

<u>**Rel.12**</u>  Collects lights and a viewpoint into a named scene prior to rendering.

| Command | Alias | Ctrl+ | F-key | Alt+ | Menu Bar | Tablet |
|---------|-------|-------|-------|------|----------|--------|
| scene | ... | ... | ... | VES | View | N1 |
| | | | | | ⓈRender | |
| | | | | | ⓈScene | |

**Command:** scene

*Displays dialog box:*

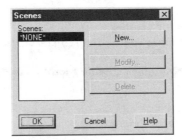

### DIALOG BOX OPTIONS

**New** creates new named scenes; displays the New Scene dialog box.

**Modify** changes existing scene definitions; displays the Modify Scene dialog box.

**Delete** deletes scenes from drawings; displays dialog box:

**New Scene** dialog box

**Scene Name** specifies a name for the scene; maximum = 8 characters.

**Views** selects one named view.

**Lights** selects one or more lights.

**Modify Scene** dialog box

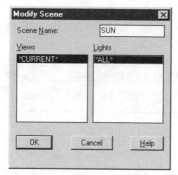

**Scene Name** changes the name of the scene; maximum = 8 characters.

**Views** selects a named view.

**Lights** selects one or more lights.

## RELATED COMMANDS

**Light** places lights in the drawing for the Scene command.

**Render** uses scenes to create renderings.

**RPref** specifies options for renderings.

**View** creates named views for the Scene command.

## TIPS

- Before you can use the **Scene** command, you must create at least one named view (with the **View** command), or place at least one light (with the **Light** command) in the drawing. Otherwise, there is no need to use the **Scene** command.

- If you select no lights for a scene, **Render** uses ambient light.

- Scene parameters are stored as attribute definitions in a block.

# 'Script

**V. 1.4**

Runs ASCII files containing sequences of AutoCAD commands and options.

| Command | Alias | Ctrl+ | F-key | Alt+ | Menu Bar | Tablet |
|---------|-------|-------|-------|------|----------|--------|
| 'script | scr | ... | ... | TR | Tools<br>⤷Run Script | V9 |

**Command:** script

*Displays the **Select Script File** dialog box. Select an .scr file, and then click **Open**.*

*Script file begins running as soon as it is loaded.*

## COMMAND LINE OPTIONS

BACKSPACE interrupts the script.

ESC stops the script.

## RELATED COMMANDS

**Delay** specifies the delay in milliseconds; pauses the script before executing the next command.

**Resume** resumes a script after it is interrupted.

**RScript** repeats script files.

## TIPS

• Since the **Script** command is transparent, it can be used during another command.

• Prefix the **VSlide** command with **\*** to preload it; this results in a faster slide show:

  \*vslide

• You can make a script file more flexible — such as pausing for user input, branching with conditionals — by inserting AutoLISP functions.

• Dialog boxes, toolbars, menus, and tool palettes cannot be controlled by scripts. Toolbar and menu macros can contain scripts, however.

## Quick Start Tutorial
# Writing Your First Script File

A script file consist of the exact keystrokes you enter for any command.

### Step 1
Open a text editor. The script file must be plain ASCII text. Write the script using a text editor, such as Notepad, not a word processor.

### Step 2
Write the script. Here is an example script that places a door symbol in the drawing:

```
; Inserts DOOR2436 symbol at x,y = (76,100)
; x-scale = 0.5, y-scale = 1.0, rotation = 90 degrees
insert door2436 76,100 0.5 1.0 90
```

In the script, these characters have special meaning:

| Character | Meaning |
|---|---|
|  | (*Space or end-of-line*) Equivalent to pressing the SPACEBAR or ENTER key. |
| ; | (*Semicolon*) Include a comment in the script file. |
| * | (*Asterisk*) Prefix the **VSlide** command to preload .*sld* files. |

### Step 3
Save script files with any filename and the .*scr* extension. For this example, use *insertdoor.scr*.

### Step 4
Return to AutoCAD, and run the script with the **Script** command:

```
Command: script
Script file: insertdoorblock.scr
Command: Insert
Block name (or ?): door2436
Insertion point: 76,100
X scale factor <1>/Corner/XYZ: 0.5
Y scale factor (default = X): 1.0
Rotation angle <0>: 90
```

### Step 5
If required, you can rerun the script with the **RScript** command.

# Section

__Rel.11__ Creates 2D region objects from the intersection of a plane and a 3D solid.

| Command | Alias | Ctrl+ | F-key | Alt+ | Menu Bar | Tablet |
|---------|-------|-------|-------|------|----------|--------|
| section | sec | ... | ... | DIE | Draw | ... |
| | | | | | ⮡Solid | |
| | | | | | ⮡Section | |

**Command:** section
**Select objects:** *(Select one or more objects.)*
**Select objects:** *(Press* **Enter** *to end object selection.)*
**Specify first point on section plane by [Object/Zaxis/View/XY/YZ/ZX/ 3points]:** *(Enter an option.)*

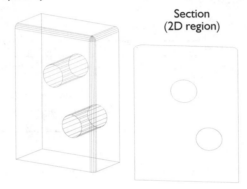

Section
(2D region)

## COMMAND LINE OPTIONS

**Select objects** selects the 3D solid objects to be sectioned.

**Object** aligns the section plane with an object: circle, ellipse, arc, elliptical arc, 2D spline, or 2D polyline.

**Zaxis** specifies the normal (*z axis*) to the section plane.

**View** uses the current view plane as the section plane.

**XY** uses the x,y plane of the current view.

**YZ** uses the y,z plane of the current view.

**ZX** uses the z,x plane of the current view.

**3 points** picks three points to specify the section plane.

## RELATED COMMAND

**Slice** cuts a slice out of a solid model, creating another 3D solid.

## TIPS

• Section blocks are placed on the current layer, not the object's layer.

• Regions are ignored.

• One cutting plane is required for each selected solid.

# SecurityOptions

<u>2004</u>  Sets passwords and digital signatures for securing drawings against unauthorized access.

| Command | Alias | Ctrl+ | F-key | Alt+ | Menu Bar | Tablet |
|---|---|---|---|---|---|---|
| securityoptions | ... | ... | ... | ... | ... | ... |

**Command:** securityoptions

*Displays dialog box:*

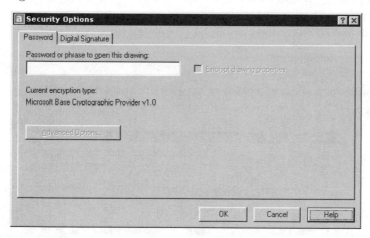

## DIALOG BOX OPTIONS

### Password tab

**Password or phrase to open this drawing** assigns a password to lock the drawing.

**Encrypt drawing properties** encrypts drawing properties data; see the DwgProps command.

**Advanced Options** displays the Advanced Options dialog box; available only after a password is entered.

**OK** displays the Confirm Password dialog box:

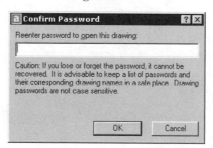

**Digital Signature** tab

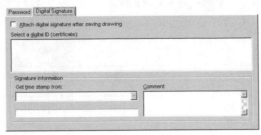

**Attach digital signature after saving drawing** attaches the digital signature the next time the drawing is saved.

**Select a digital ID (certificate)** lists digital signature services.

Signature information options

**Get time stamp from** lists time servers.

**Comment** provides additional comments to be included with the digital signature.

**Advanced Options** dialog box

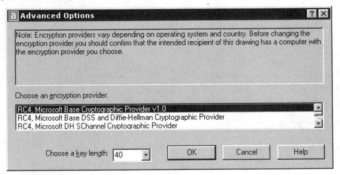

**Choose an encryption provider** selects from several encryption types.

**Choose a key length** selects 40, 48, or 56 bits of encryption.

**RELATED COMMAND**

**SigValidate** displays information about the digital signatures attached to drawings.

**TIPS**

- When opening drawings protected by passwords, AutoCAD displays the **Password** dialog box for entering the password:

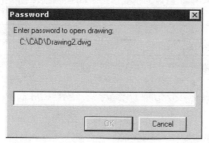

- When the password entered is correct, AutoCAD opens the drawing.

- When the password entered is incorrect, AutoCAD complains:

- To remove the password, in the **Security Options** dialog box, remove the password, and then click **OK**.

- The first time you assign a digital signature to the drawing, you are prompted to obtain one:

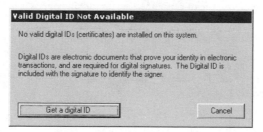

- *Caution!* If the password is lost, the drawing is not recoverable. Autodesk recommends that you (1) save a backup copy of the drawing not protected with any password; and (2) maintain a list of password names stored in a secure location.

- *Caution!* The encryption provider must be the same on the receiving computers; otherwise the drawings cannot be opened.

- To remove this feature from AutoCAD, install with the Custom option, and then unselect Drawing Encryption.

 **Select**

**V. 2.5**   Creates a selection set of objects before executing a command.

| Command | Alias | Ctrl+ | F-key | Alt+ | Menu Bar | Tablet |
|---------|-------|-------|-------|------|----------|--------|
| select  | ...   | A     | ...   | ...  | ...      | ...    |

**Command:** select
**Select objects:** *(Select one or more objects, or enter an option.)*
**Select objects:** *(Press **Enter** to end object selection.)*

### COMMAND LINE & TOOLBAR OPTIONS

*pick* selects a single object.

**AU** switches from pick mode to C or W mode, depending on whether an object is found at the initial pick point (short for AUtomatic).

**ALL** selects all objects in the drawing; press CTRL+A to select all objects.

**BOX** goes into C or W mode, depending on how the cursor moves.

**C** selects objects in and crossing the selection box (Crossing).

**CP** selects all objects inside and crossing the selection polygon (Crossing Polygon).

**F** selects all objects crossing a selection polyline (Fence).

**G** selects objects contained in a named group (Group).

**M** sakes multiple selections before AutoCAD scans the drawing; saves time in a large drawing (Multiple).

**SI** selects only a single set of objects before terminating the Select command (SIngle).

**W** selects all objects inside the selection box (Window).

**WP** selects objects inside the selection polygon (Windowed Polygon).

**L** selects the last-drawn object still visible on the screen (Last).

**P** selects previously-selected objects (Previous).

**R** removes objects from the selection set (Remove).

**U** removes the most-recently added selected objects (Undo).

**A** continues to add objects after using the R option (Add).

ENTER exits the Select command.

ESC aborts the Select command.

### RELATED COMMANDS
*All commands that prompt 'Select objects'.*

**Filter** specifies objects to be added to the selection set.

**QSelect** selects objects via a dialog box interface.

## RELATED SYSTEM VARIABLES

**PickAdd** controls how objects are added to a selection set.

**PickAuto** controls automatic windowing at the 'Select objects' prompt.

**PickDrag** controls the method of creating a selection box.

**PickFirst** controls the command-object selection order.

## TIPS

- Pressing CTRL+A selects all objects in the drawing without needing the **Select** command.

- **Select All** selects all objects in the drawing, except those on frozen and locked layers:

- **Select Crossing** selects objects within and crossing the selection rectangle.

- **Select Window** selects objects within the selection rectangle.

- **Select Fence** selects objects crossing the selection polyline.

- **Select CPolygon** selects objects within and crossing the selection polygon.

- **Select WPolygon** selects objects within the selection polygon.

# SelectURL

<u>Rel.14</u> Highlights all objects and areas that have hyperlinks attached to them
*(undocumented command)*.

| Command | Alias | Ctrl+ | F-key | Alt+ | Menu Bar | Tablet |
|---------|-------|-------|-------|------|----------|--------|
| selecturl | ... | ... | ... | ... | ... | ... |

**Command:** selecturl

*Highlights all objects and areas with a hyperlink.*

*If there are no hyperlinks in the drawing, AutoCAD reports:*

**No objects with hyperlinks found.**

## COMMAND LINE OPTIONS

*None.*

## RELATED COMMANDS

**AttachURL** attaches URL to objects and areas.

**Hyperlink** attaches URLs to objects via a dialog box.

## TIPS

* Examples of URLs include:

  http://www.autodeskpress.com   Autodesk Press Web site.

  news://adesknews.autodesk.com Autodesk news server.

  ftp://ftp.autodesk.com   Autodesk FTP server.

  http://www.upfrontezine.com   Author Ralph Grabowski's Web site.

* Do not delete layer URLLAYER.

* The URL is stored as follows:

| Attachment | URL |
|------------|-----|
| **One object** | Stored as xdata (extended entity data). |
| **Multiple objects** | Stored as xdata in each object. |
| **Area** | Stored as xdata in a rectangle object on layer URLLAYER. |

## DEFINITIONS

*DWF* — short for "drawing Web format," Autodesk's file format for displaying drawings on the Internet.

*URL* — short for "uniform resource locator," the universal file naming convention.

# 'SetIDropHandler

<u>2004</u>   Specifies how iDrop objects should be handed in drawings.

| Command | Alias | Ctrl+ | F-key | Alt+ | Menu Bar | Tablet |
|---------|-------|-------|-------|------|----------|--------|
| 'setidrophandler | ... | ... | ... | ... | ... | ... |

**Command:** setidrophandler

*Displays dialog box:*

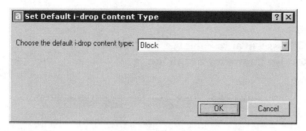

## DIALOG BOX OPTION

**Choose the default i-drop content type** specifies how iDrop objects are handled when placed in drawings.

## TIPS

• When iDrop content is dragged from a Web page into a drawing, AutoCAD treats the objects as a block.

• At time of writing this book, the only option is "Block."

• For more information on iDrop technology, see www.autodesk.com/developidrop.

 # SetUV

**Rel.14** Maps materials onto objects for rendering.

| Command | Alias | Ctrl+ | F-key | Alt+ | Menu Bar | Tablet |
|---------|-------|-------|-------|------|----------|--------|
| setuv | ... | ... | ... | VEA | View | R1 |
| | | | | | ⤷Render | |
| | | | | | ⤷Mapping | |

**Command:** setuv
**Select objects:** *(Select one or more objects.)*
**Select objects:** *(Press Enter to end object selection.)*
**Updating the Render geometry database...**
*Displays dialog box::*

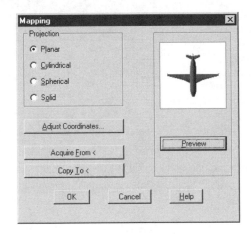

## DIALOG BOX OPTIONS

Projection options
- **Planar** specifies a plane for projecting the bitmap onto the selected object.
- **Cylindrical** specifies an axis of the cylindrical coordinate system and wrap line for projecting the bitmap onto the selected object.
- **Spherical** specifies the polar axis of the spherical coordinate system and wrap line for projecting the bitmap onto the selected object.
- **Solid** adjusts the coordinates to shift marble, granite, or wood materials.

Additional options
**Adjust Coordinates** displays different dialog boxes, depending on the setting of Projection.
**Acquire From** dismisses the dialog box temporarily, allowing you to select a mapping object already in the drawing.
**Copy To** dismisses the dialog box temporarily, allowing you to select the objects to which to apply the mapping.

**Adjust Planar Coordinates** dialog box

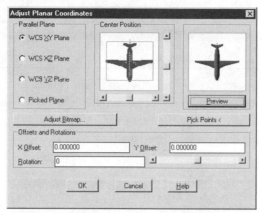

**Parallel Plane**    selects a WCS reference, or picks a plane with the **Pick Points** radio button.

**Center Position** shows a parallel projection of the selected object's mesh onto the current parallel plane:

| Position | Meaning |
|---|---|
| **Blue** | Current projection square. |
| **Blue tick mark** | Top of the projection square. |
| **Green** | Projection square's left edge. |

**Adjust Bitmap** displays the Adjust Object Bitmap Placement dialog box.

**Pick Points** specifies a projection plane by picking points in the drawing.

**Offset** changes the x and y offset of the map.

**Rotation** changes the rotation angle of the map.

**Adjust Cylindrical Coordinates** dialog box

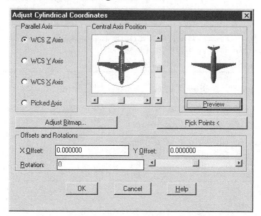

**Parallel Axis** selects a WCS reference axis of the WCS, or picks an axis with the Pick Points radio button.

**Central Axis Position** displays a parallel projection of the object's mesh onto a plane perpendicular to the current axis.

| Position | Meaning |
|---|---|
| **Blue circle** | Projection axis. |
| **Green radius** | Wrap line. |

**Adjust Bitmap** displays the Adjust Object Bitmap Placement dialog box.

**Pick Points** specifies a projection axis by picking points in the drawing.

**Offset** changes the x and y offset of the map.

**Rotation** changes the rotation angle of the map.

### Adjust Spherical Coordinates dialog box

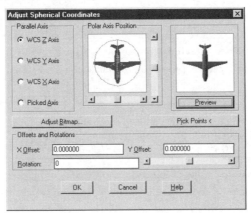

**Parallel Axis** selects one of the three perpendicular axes of the WCS, or picks an axis with the Pick Points button.

**Polar Axis Position** displays a parallel projection of the object's mesh onto a plane perpendicular to the current axis.

| Position | Meaning |
|---|---|
| **Blue circle** | Projection axis. |
| **Green radius** | Wrap line. |

**Adjust Bitmap** displays the Adjust Object Bitmap Placement dialog box.

**Pick Points** specifies a projection axis by picking points in the drawing.

**Offset** changes the x and y offset of the map.

**Rotation** changes the rotation angle of the map.

**Adjust Solid Coordinates** dialog box

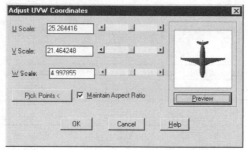

**U Scale, V Scale, W Scale** sets 3D projection coordinates.

**Pick Points** specifies a projection axis by picking points in the drawing.

**Adjust Object Bitmap Placement** dialog box

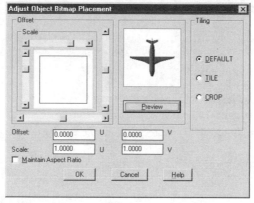

**Offset** sets the u and v offset distances via slider bars interactively.

**Scale** sets the u and v scale of the bitmap interactively.

Tiling options

**Tile** tiles the bitmap.

**Crop** does not tile the bitmap.

**TIPS**

• The upper-left corner pixel of the bitmap defines the transparent color; to have no transparent colors in the bitmap, deliberately make that pixel a unique color.

• **Crop** tiling turns on transparent mode: all pixels have the same color as the upper-left pixel, which is treated as transparent.

• UV (and sometimes W) are equivalent to the x,y,z coordinates; the letters U, V, W are used, however, because they are independent of the x,y,z coordinates.

# 'SetVar

**V. 2.5** Lists the settings of system variables; allows you to change variables that are not read-only (*short for SET VARiable*).

| Command | Alias | Ctrl+ | F-key | Alt+ | Menu Bar | Tablet |
|---------|-------|-------|-------|------|----------|--------|
| 'setvar | set | ... | ... | TQV | Tools | U10 |
| | | | | | ⇘Inquiry | |
| | | | | | ⇘Set Variable | |

**Command:** setvar
**Enter variable name or [?]:** *(Enter a name, or type* **?**.*)*

## COMMAND LINE OPTIONS

**Enter variable name** indicates the system variable name you want to access.

**?** lists the names and settings of system variables.

## TIPS

- See *Appendix A* for the complete list of all system variables found in AutoCAD 2004.

- Example usage of this command:

    **Command:** setvar
    **Enter variable name or [?]:** visretain
    **Enter new value for VISRETAIN <0>:** 1

- Almost all system variables can be entered without the **SetVar** command. For example:

    **Command:** visretain
    **New value for VISRETAIN <0>:** 1

## Removed Command

**Shade** was removed from AutoCAD 2000; it was replaced by **ShadeMode**.

# ShadeMode

<u>2000</u>  Generates renderings of 3D models quickly in a variety of modes *(replaces the Shade command)*.

| Command | Alias | Ctrl+ | F-key | Alt+ | Menu Bar | Tablet |
|---------|-------|-------|-------|------|----------|--------|
| shademode | shade ... | | ... | VS | View | N2 |
| | | | | | ⤷Shade | |

**Command:** shademode
**Current mode: 2D wireframe**
**Enter option [2D wireframe/3D wireframe/Hidden/Flat/Gouraud/ fLat+edges/gOuraud+edges] <2D wireframe>:** *(Enter an option.)*

## COMMAND LINE OPTIONS

**2D wireframe**  displays wireframe models in 2D space.

**3D wireframe** displays wireframe models in 3D space.

**Hidden** removes hidden faces.

**Flat** displays flat shaded faces.

**fLat+edges** displays flat shaded faces, with outlined faces of the background color.

**Gouraud** displays smooth shaded faces.

**gOuraud+edges** displays smooth shaded faces, with outlined faces of the background color.

## RELATED COMMANDS

**Hide** performs a true hidden-line removal of 3D drawings.

**MSlide** saves a rendered view as an SLD-format slide file.

**MView** performs a hidden-line view of individual viewports during plotting.

**Plot** performs a hidden-line view during plotting.

**Render** performs a more realistic rendering.

**3dOrbit** performs hidden-line removal of perspective views.

## RELATED SYSTEM VARIABLES

*None.*

## TIPS

- As an alternative to the **ShadeMode** command, the **Render** module does high-quality renderings of 3D drawings, but takes somewhat longer to do so.

- The smaller the viewport, the faster the shading.

This drawing is displayed in **ShadeMode**'s default **2D wireframe** mode; notice the standard 2D UCS icon.

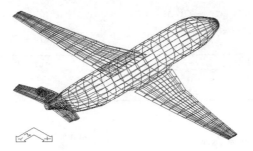

This drawing is displayed in **3D wireframe** mode; notice the 3D UCS icon.

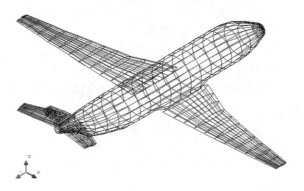

This drawing is displayed in **Hidden** mode; notice that faces hidden by other faces are not displayed:

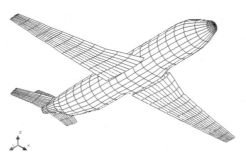

This drawing is displayed in **Flat** mode; notice that each face is filled with a shade of gray.

 This drawing is displayed **Gouraud** mode; notice that the faces are smoothed.

 This drawing is displayed in **fLat+edges** mode; notice that each face is outlined by the background color (*white*).

 This drawing is displayed in **gOuraud+edges** mode; notice the outlining of each face.

# Shape

<u>**V. 1.0**</u>   Inserts predefined shapes into drawings.

| Command | Alias | Ctrl+ | F-key | Alt+ | Menu Bar | Tablet |
|---------|-------|-------|-------|------|----------|--------|
| shape | ... | ... | ... | ... | ... | ... |

**Command:** shape
**Enter shape name or [?]:** *(Enter the name, or type* **?***.)*
**Specify insertion point:** *(Pick a point.)*
**Specify height <1>:** *(Specify the height.)*
**Specify rotation angle <0>:** *(Specify the angle.)*

## COMMAND LINE OPTIONS

**Enter shape name** indicates the name of the shape to insert.

**?** lists the names of currently-loaded shapes.

**Specify insertion point** indicates the insertion point of the shape.

**Specify height** specifies the height of the shape.

**Specify rotation angle** specifies the rotation angle of the shape.

## RELATED COMMANDS

**Load** loads an SHX-format shape file into the drawing.

**Insert** inserts a block into the drawing.

**Style** loads *.shx* font files into the drawing.

## RELATED SYSTEM VARIABLE

**ShpName** specifies the current *.shp* file name.

## TIPS

• Shapes are defined by *.shp* files, which must be compiled into *.shx* files before they can be loaded by the **Load** command.

• In addition, shapes must be loaded by the **Load** command before they can be inserted with the **Shape** command.

• Compile an *.shx* file into an *shp* file with the **Compile** command.

• Shapes are used to define the text and symbols found in complex linetypes.

*Some electronic shapes.*

# Shell

**V. 2.5**    Exits AutoCAD temporarily to the operating system's command mode; runs external commands defined in *acad.pgp*.

| Command | Alias | Ctrl+ | F-key | Alt+ | Menu Bar | Tablet |
|---------|-------|-------|-------|------|----------|--------|
| shell | sh | ... | ... | ... | ... | ... |

**Command:** shell
**OS Command:** *(Enter a command.)*

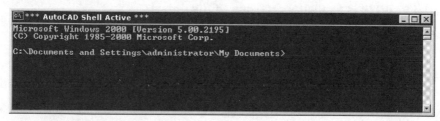

## COMMAND LINE OPTIONS

**OS Command** specifies an operating system command.

ENTER remains in the operating system's command mode for more than one command.

**Exit** returns to AutoCAD from the OS command mode.

## RELATED COMMAND

**Quit** exits AutoCAD back to Windows.

## RELATED FILE

*acad.pgp* is the external command definition file.

## TIP

• To view and edit the *acad.pgp* file, use the **Tools | Customize | Edit Custom Files | Program Parameters (acad.pgp)** command. AutoCAD displays the *acad.pgp* file in Notepad.

# ShowMat

Rel.13 Lists the rendering material attached to objects *(short for SHOW MATerial).*

| Command | Alias | Ctrl+ | F-key | Alt+ | Menu Bar | Tablet |
|---------|-------|-------|-------|------|----------|--------|
| showmat | ... | ... | ... | ... | ... | ... |

**Command:** showmat
**Select object:** *(Select one object.)*

*Sample response:*
**Material BRONZE is explicitly attached to the object.**

## COMMAND LINE OPTION
**Select object** selects the object to examine.

## RELATED COMMANDS
**MatLib** loads the material definitions into the drawing.

**RMat** attaches the materials to objects, colors, and layers.

## TIP
- When no material is attached to the model, AutoCAD reports, "Material *GLOBAL* is attached by default or by block."

# SigValidate

**2004**  Displays digital signature information stored in drawings.

| Command | Alias | Ctrl+ | F-key | Alt+ | Menu Bar | Tablet |
|---------|-------|-------|-------|------|----------|--------|
| sigvalidate | ... | ... | ... | ... | ... | ... |

**Command:** sigvalidate

*Displays dialog box::*

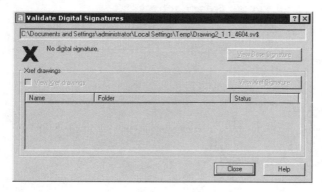

## DIALOG BOX OPTIONS

**View Base Signature** displays the Digital Signature Contents dialog box.

Xref Drawings options

**View Xref Drawings**

☑ lists xref drawings attached to this drawing.

☐ does not list xrefs.

**View Xref Signature** displays the Digital Signatures Contents dialog box for the selected xref; available only when the selected xref has a digital signature.

## RELATED COMMAND

**SecurityOptions** attaches digital signatures to drawings.

## TIPS

- Digital signatures validate the authenticity of drawings.

- Digital signatures indicates whether the drawing was changed since being signed.

- Autodesk notes that digital signatures become invalid for these reasons: the drawing was corrupted when the digital signature was attached; the drawing was corrupted in transit; and the digital ID is no longer valid.

- This command requires that the drawing be saved first; if not, AutoCAD reports, "This drawing has been modified. The last saved version will be validated."

- If this command does not work, it was removed from AutoCAD; re-install with the Custom option, and then select Drawing Encryption.

# Sketch

**V. 1.4**  Allows freehand drawing as a series of lines or polylines.

| Command | Alias | Ctrl+ | F-key | Alt+ | Menu Bar | Tablet |
|---------|-------|-------|-------|------|----------|--------|
| sketch  | ...   | ...   | ...   | ...  | ...      | ...    |

**Command:** sketch
**Record increment <0.1000>:** *(Enter a value.)*
**Sketch. Pen eXit Quit Record Erase Connect .**

*Click pick button to begin drawing:*
**<Pen down>**

*Click pick button again to stop drawing:*
**<Pen up>**

*Press* ENTER *to record sketching and exit* **Sketch:**
*nnn* **lines recorded.**

## COMMAND LINE OPTIONS

*Commands can be invoked by mouse and digitizer buttons:*

**Connect** connects to the last drawing segment (*as an alternative, press button #6*).

**Erase** erases temporary segments as the cursor moves over them (*button #5*).

**eXit** records the temporary segments, and exits the Sketch command (*button #3 or* SPACEBAR *or* ENTER).

**Pen** lifts and lowers the pen (*pick button #1*).

**Quit** discards temporary segments, and exits the Sketch command (*button #4 or* ESC).

**Record** records the temporary segments as permanent (*button #2*).

**.** (*Period*) connects the last segment to the current point (*button #1*).

## RELATED COMMANDS

**Line** draws line segments.

**PLine** draws polyline and polyline arc segments.

## RELATED SYSTEM VARIABLES

**SketchInc** specifies the current recording increment for the Sketch command (default = 0.1).

**SKPoly** controls the type of sketches recorded:

| SkPoly | Meaning |
|---|---|
| 0 | Record sketches as lines (default). |
| 1 | Record sketches as polylines. |

## TIPS

• During the **Sketch** command, the definitions of the pointing device's buttons change to:

| Button | Meaning | Keystroke |
|---|---|---|
| 0 | Raises and lowers the *pen* | P |
| 1 | Draws line to the current *point* | . |
| 2 | Records the sketch | R |
| 3 | Records the sketch and e*X*its | X *or* ENTER *or* SPACEBAR |
| 4 | Discards the sketch and *Q*uits | Q *or* ESC |
| 5 | *E*rases the sketch | E |
| 6 | *C*onnects to last-drawn segment | C |

• Only the first several button commands are available on a mouse.

• Pull-down menus are unavailable during the **Sketch** command.

 # Slice

**Rel.11**   Cuts 3D solids with planes, creating two 3D solids.

| Command | Alias | Ctrl+ | F-key | Alt+ | Menu Bar | Tablet |
|---------|-------|-------|-------|------|----------|--------|
| slice | sl | ... | ... | DIL | Draw | ... |
| | | | | | ⮡ Solids | |
| | | | | | ⮡ Slice | |

**Command:** slice
**Select objects:** *(Select one or more solid objects.)*
**Select objects:** *(Press Enter to end object selection.)*
**Specify first point on slicing plane by [Object/Zaxis/View/XY/YZ/ZX/3points] <3points>:** *(Pick a point, or enter an option.)*
**Specify a point on desired side of the plane or [keep Both sides]:** *(Pick a point, or type B.)*

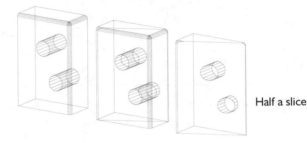

Half a slice

## COMMAND LINE OPTIONS
**Select objects** selects the 3D solid model to slice.

Slicing plane options
**Object** aligns the cutting plane with a circle, ellipse, arc, elliptical arc, 2D spline, or 2D polyline.
**View** aligns the cutting plane with the viewing plane.
**XY** aligns the cutting plane with the x,y plane of the current UCS.
**YZ** aligns the cutting plane with the y,x plane of the current UCS.
**Zaxis** aligns the cutting plane with two normal points.
**ZX** aligns the cutting plane with the z,x plane of the current UCS.
**3 points** aligns the cutting plane with three points.

Additional options
**keep Both sides** retains both halves of the cut solid model.
**Specify a point on the desired side of the plane** retains either half of the cut solid model.

## TIP
• This command cannot slice a 2D region, a 3D wireframe model, or other 2D shapes.

# 'Snap

<u>V. 1.0</u>   Sets the drawing "resolution," grid and hatch origin, isometric mode, and angle of grid, hatch, and ortho.

| Command | Alias | Ctrl+ | F-key | Alt+ | Menu Bar | Tablet |
|---------|-------|-------|-------|------|----------|--------|
| 'snap   | sn    | B     | F9    | ...  | ...      | ...    |

**Command:** snap
**Specify snap spacing or [ON/OFF/Aspect/Rotate/Style/Type]<0.5>:** *(Enter a value, or an option.)*

## COMMAND LINE OPTIONS

**Snap spacings** sets the snap increment.

**Aspect** sets separate x and y increments.

**OFF** turns off snap.

**ON** turns on snap.

**Rotate** rotates the crosshairs for snap and grid.

**Style** switches between standard and isometric style.

**Type** switches between grid or polar snap.

Aspect options
**Specify horizontal spacing <0.5>:** *(Enter a value.)*
**Specify vertical spacing <0.5>:** *(Enter a value.)*

**Specify horizontal spacing** specifies the spacing between snap points along the x-direction.

**Specify vertical spacing** specifies the spacing between snap points along the y-direciton.

Rotate options
**Specify base point <0.0,0.0>:** *(Enter a value.)*
**Specify rotation angle <0>:** *(Enter a value.)*

**Specify base point** specifies the point from which snap increments are determined.

**Specify rotation angle** rotates the snap about the basepoint.

Style options
**Enter snap grid style [Standard/Isometric] <S>:** *(Type **S** or **I**.)*
**Specify vertical spacing <0.5>:** *(Enter a value.)*

**Standard** specifies rectangular snap.

**Isometric** specifies isometric snap.

**Specify vertical spacing** specifies the spacing between snap points along the y-direciton.

Type options
**Enter snap type [Polar/Grid] <Grid>:** *(Type **P** or **G**.)*

**Polar** specifies polar snap.

**Grid** specifies rectangular snap.

## STATUS BAR OPTION

*Click* **SNAP** *to turn snap on and off; right-click for shortcut menu:*

**Polar Snap On** turns on polar snap.

**Grid Snap On** turns on snap.

**Off** turns off snap.

**Settings** displays the Snap and Grid tab of the Drafting Settings dialog box.

## RELATED COMMANDS

**DSettings** sets snap values via a dialog box.

**Grid** turns on the grid.

**Isoplane** switches to a different isometric drawing plane.

## RELATED SYSTEM VARIABLES

**SnapAng** specifies the current angle of the snap rotation.

**SnapBase** specifies the base point of the snap rotation.

**SnapIsoPair** specifies the current isometric plane setting.

**SnapMode** determines whether snap is on.

**SnapStyl** determines the style of snap.

**SnapUnit** specifies the current snap increment in the x and y directions.

## TIPS

- Setting the snap is setting the cursor resolution. For example, setting a snap distance of 0.1 means that when you move the cursor, it jumps in 0.1 increments. You can, however, still type in numerical values of greater resolution, such as 0.1234.

- There is no snap in the z-direction.

- The **Aspect** option is not available when the **Style** option is set to **Isometric**; you may, however, rotate the isometric grid.

- The options of the **Snap** command affect several other commands:

| Command | Snap Option | Effect |
|---------|-------------|--------|
| **Ellipse** | Style | Adds **Isocircle** option to the **Ellipse** command. |
| **Grid** | Rotate | Rotates the grid display. |
| **Hatch** | Rotate | Rotates the hatching angle. |
| **Hatch** | SnapBase | Relocates the origin of the hatch pattern. |
| **Ortho** | Rotate | Rotates the ortho angle. |

- You can toggle snap mode by double-clicking the word **SNAP** on the status bar.

# SolDraw

**Rel.13**  Creates profiles and sections in viewports created with the SolView command (*short for SOLids DRAWing*).

| Command | Alias | Ctrl+ | F-key | Alt+ | Menu Bar | Tablet |
|---------|-------|-------|-------|------|----------|--------|
| soldraw | ... | ... | ... | DIUD | Draw | ... |
| | | | | | ⮑**Solids** | |
| | | | | | ⮑**Setup** | |
| | | | | | ⮑**Drawing** | |

**Command:** soldraw
**Select viewports to draw ..**
**Select objects:** *(Select one or more viewports.)*
**Select objects:** *(Press **Enter** to end object selection.)*

## COMMAND LINE OPTION
**Select objects** selects a viewport; must be a floating viewport in model space (Tilemode=0).

## RELATED COMMANDS
**SolProf** creates profile images of 3D solids.
**SolView** creates floating viewports.

## TIPS
- The **SolView** command must be used before this command.
- This command performs the following actions:

  1. Creates visible and hidden lines representing the silhouette and edges of solids in the viewport.
  2. Projects to a plane perpendicular to the viewing direction.
  3. Generates silhouettes and edges for all 3D solids and portions of solids behind the cutting plane.
  4. Crosshatches sectional views.

- Existing profiles and sections in the selected viewport are erased.
- All layers — except the ones needed to display the profile or section — are frozen in each viewport.
- The following layers are used by **SolDraw**, **SolProf**, and **SolView**: *viewname*-**VIS**, *viewname*-**HID**, and *viewname*-**HAT**.
- Hatching uses the values set in system variables **HpName**, **HpScale**, and **HpAng**.

 # Solid

**V. 1.0** Draws solid-filled triangles and quadrilaterals; does *not* create 3D solids.

| Command | Alias | Ctrl+ | F-key | Alt+ | Menu Bar | Tablet |
|---------|-------|-------|-------|------|----------|--------|
| solid | so | ... | ... | DF2 | Draw | L8 |
| | | | | | ⮑Surfaces | |
| | | | | | ⮑2D Solid | |

**Command:** solid
**Specify first point:** *(Pick a point.)*
**Specify second point:** *(Pick a point.)*
**Specify third point:** *(Pick a point.)*
**Specify fourth point or <exit>:** *(Pick a point, or press **Enter** to draw triangle.)*

*Three-point solid.*          *Pick order makes a difference for four-point solids.*

## COMMAND LINE OPTIONS

**First point** picks the first corner.

**Second point** picks the second corner.

**Third point** picks the third corner.

**Fourth point** picks the fourth corner; alternatively, press ENTER to draw triangle.

ENTER draws quadilaterials; ends the **Solid** command.

## RELATED COMMANDS

**Fill** turns object fill off and on.

**BHatch** fills any shape with a solid fill pattern.

**Trace** draws lines with width.

**PLine** draws polylines and polyline arcs with width.

## RELATED SYSTEM VARIABLE

**FillMode** determines whether solids are displayed filled or outlined.

# SolidEdit

**2000**    Edits the faces and edges of 3D solids *(short for SOLids EDITor).*

| Command | Alias | Ctrl+ | F-key | Alt+ | Menu Bar | Tablet |
|---------|-------|-------|-------|------|----------|--------|
| solidedit | ... | ... | ... | MN | Modify<br>⤷Solids Editing | ... |

**Command:** solidedit
**Solids editing automatic checking: SOLIDCHECK=1**
**Enter a solids editing option [Face/Edge/Body/Undo/eXit] <eXit>:** *(Enter an option.)*

## COMMAND LINE OPTIONS

**Undo** undoes the last editing actions, one at a time, up to the start of the **SolidEdit** command.

**eXit** exits body mode.

Face options
**Enter a face editing option**
**[Extrude/Move/Rotate/Offset/Taper/Delete/Copy/coLor/Undo/eXit] <eXit>:**
*(Enter an option.)*

 **Extrude** extrudes one or more faces to the specified distance, or along a path; a positive value extrudes the face in the direction of its normal.

 **Move** moves one or more faces the specified distance.

 **Rotate** rotates one or more faces about an axis by a specified angle.

 **Offset** offsets one or more faces by the specified distance, or through a specified point; a positive value increases the size of the solid, while a negative value decreases the size.

 **Taper** tapers one or more faces by a specified angle; a positive angle tapers in, while

negative angle tapers out.

 **Delete** removes the selected faces; also removes attached chamfers and fillets.

 **Copy** copies the selected faces as a 3D region or a 3D body object.

 **Color** changes the color of the selected faces.

Edge options
**Enter an edge editing option [Copy/coLor/Undo/eXit] <eXit>:** *(Enter an option.)*

**Copy** copies the selected 3D edges as a line, arc, circle, ellipse, or spline.

**coLor** changes the color of the selected edges.

Body options
**Enter a body editing option**
**[Imprint/seParate solids/Shell/cLean/Check/Undo/eXit] <eXit>:** *(Enter an option.)*

**Imprint** imprints a selection set of arcs, circles, lines, 2D and 3D polylines, ellipses, splines, regions, bodies, and 3D solids on the face of a 3D solid.

**seParate solids** separates 3D solids into independent 3D solid objects; the solid objects must have disjointed volumes; this option does *not* separate 3D solids that were joined by a Boolean editing command, such as Intersect, Subtract, and Union.

**Shell** creates a hollow, thin-walled solid of specified thickness; a positive thickness creates a shell toward the inside of the solid, while a negative value creates the shell on the outside of the solid.

**cLean** removes redundant edges and vertices, imprints, unused geometry, shared edges, and shared vertices.

**Check** checks whether the object is a 3D solid; duplicates the function of the SolidCheck system variable.

## RELATED SYSTEM VARIABLE
**SolidCheck** toggles solid validation on and off (default = on).

## RELATED COMMANDS
*All commands related to creating and editing 3D solid models.*

## TIP
• When working with the **SolidEdit** command, you can select a face, an edge, and an internal point on a face, or use the **CP** *(crossing polygon)*, **CW** *(crossing window)*, **F** *(fence)* object selection options.

 # SolProf

**Rel.13** Creates profile images of 3D solids (*short for SOLid PROFile*).

| Command | Alias | Ctrl+ | F-key | Alt+ | Menu Bar | Tablet |
|---------|-------|-------|-------|------|----------|--------|
| solprof | ... | ... | ... | DIUP | Draw | ... |
| | | | | | ⳩Solids | |
| | | | | | ⳩Setup | |
| | | | | | ⳩Profile | |

**Command:** solprof
**Select objects:** *(Select one or more objects.)*
**Select objects:** *(Press Enter to end object selection.)*
**Display hidden profile lines on separate layer? [Yes/No] <Y>:** *(Type Y or N.)*
**Project profile lines onto a plane? [Yes/No] <Y>:** *(Type Y or N.)*
**Delete tangential edges? [Yes/No] <Y>:** *(Type Y or N.)*
*n* **solids selected.**

## COMMAND LINE OPTIONS

**Select objects** selects the objects to profile.

**Display hidden profile lines on separate layer?**

**No** specifies that all profile lines are visible; a block is created for the profile lines for every selected solid.

**Yes** generates just two blocks: one for visible lines and one for hidden lines.

**Project profile lines onto a plane?**

**No** creates profile lines with 3D objects.

**Yes** creates profile lines with 2D objects.

**Delete tangential edges?**

**No** does not display *tangential edges*, the transition line between two tangent faces.

**Yes** displays tangential edges.

## RELATED COMMANDS

**SolDraw** creates profiles and sections in viewports.

**SolView** creates floating viewports.

## TIPS

- The **SolView** command must be used before the **SolProf** command.

- Solids that share a common volume can produce dangling edges, if you generate profiles with hidden lines. To avoid this, first use the **Union** command.

- AutoCAD must be in a layout and a modelspace viewport must be active before you can use the **SolProf** command.

# SolView

Rel.13 Creates floating viewports in preparation for the SolDraw and SolProf commands (*short for SOLid VIEWs*).

| Command | Alias | Ctrl+ | F-key | Alt+ | Menu Bar | Tablet |
|---------|-------|-------|-------|------|----------|--------|
| solview | ... | ... | ... | DIUV | Draw | ... |
| | | | | | ⤷Solids | |
| | | | | | ⤷Setup | |
| | | | | | ⤷View | |

**Command:** solview
**Enter an option [Ucs/Ortho/Auxiliary/Section]:** *(Enter an option.)*

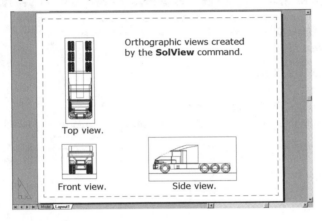

Orthographic views created by the **SolView** command.

Top view.

Front view.      Side view.

## COMMAND LINE OPTIONS

Ucs options
**Enter an option [Named/World/?/Current] <Current>:** *(Enter an option.)*

**Named** creates the profile view using the x,y plane of a named UCS.

**World** creates the profile view using the x,y plane of the WCS

**?** lists the names of existing UCSs.

**Current** creates the profile view using the x,y-plane of the current UCS.

Ortho options
**Specify side of viewport to project:** *(Pick a point.)*
**Specify view center:** *(Pick a point.)*
**Enter view name:** *(Enter a name.)*

**Pick side of viewport to project** selects the edge of one viewport.

**View center** picks the center of the view.

**Clip** picks two corners for a clipped view.

**View name** names the view.

Auxiliary options

**Specify first point of inclined plane:** *(Pick a point.)*
**Specify second point of inclined plane:** *(Pick a point.)*
**Specify side to view from:** *(Pick a point.)*
**Specify view center:** *(Pick a point.)*
**Enter view name:** *(Enter a name.)*

**Inclined plane's 1st point** picks the first point.

**Inclined plane's 2nd point** picks the second point.

**Side to view from** determines the view side.

Section options

**Specify first point of cutting plane:** *(Pick a point.)*
**Specify second point of cutting plane:** *(Pick a point.)*
**Specify side to view from:** *(Pick a point.)*
**Enter view scale <5.9759>:** *(Pick a point.)*
**Specify view center <specify viewport>:** *(Pick a point.)*
**Specify first corner of viewport:** *(Pick a point.)*
**Specify opposite corner of viewport:** *(Pick a point.)*
**Enter view name:** *(Enter a name.)*

**Cutting plane 1st point** picks the first point.

**Cutting plane 2nd point** picks the second point.

**Side to view from** determines the view side.

**Viewscale** specifies the scale of the new view.

**eXit** exits the command.

## RELATED COMMANDS

**SolDraw** creates profiles and sections in viewports.

**SolProf** creates profile images of 3D solids.

## TIPS

- This command creates orthographic, auxiliary, and sectional views of 3D solids only.

- This command must be used before the **SolDraw** command, because it creates that layers required by **SolDraw**.

- The layers created by this command have the following prefixes:

| Layer Name | View |
|---|---|
| *viewname*-VIS | Visible lines. |
| *viewname*-HID | Hidden lines. |
| *viewname*-DIM | Dimensions. |
| *viewname*-HAT | Hatch patterns for sectional views. |

- *Caution!* Autodesk warns that "The information stored on these layers is deleted and updated when you run **SolDraw**. Do not place permanent drawing information on these layers."

**'SpaceTrans**

**2002** Converts distances between model and space units (*short for SPACE TRANSlation*).

| Command | Alias | Ctrl+ | F-key | Alt+ | Menu Bar | Tablet |
|---------|-------|-------|-------|------|----------|--------|
| 'spacetrans | ... | ... | ... | ... | ... | ... |

**Command:** spacetrans

*This command does not operate in Model tab. In model view of a layout tab:*

**Specify paper space distance <1.000>:** *(Enter a value.)*

*In paper space of a layout tab:*

**Select a viewport:** *(Pick a point.)*

**Specify model space distance <1.000>:** *(Enter a value.)*

## COMMAND LINE OPTIONS

**Specify paper space distance** specifies the paper space length to be converted to its model space equivalent, usually the scale factor.

**Select a viewport** selects a paper space viewport.

**Specify model space distance** specifies the model space length to be converted to its paper space equivalent, usually the scale factor.

## RELATED COMMANDS

**Text** places text in the drawing.

**SolProf** creates profile images of 3D solids.

## TIPS

- This command is meant to be used transparently during another command, and not at the 'Command:' prompt.

- The purpose of this command is to converts lengths from model or paper space to an equivalent in the other space.

- Autodesk recommends using this command for converting text heights to make text legible.

# 'Spell

**Rel.13** Checks the spelling of text in the drawing.

| Command | Alias | Ctrl+ | F-key | Alt+ | Menu Bar | Tablet |
|---------|-------|-------|-------|------|----------|--------|
| 'spell | sp | ... | ... | TE | Tools ⌐Spelling | T10 |

**Command:** spell
**Select objects:** *(Select one or more text objects.)*
**Select objects:** *(Press **Enter** to end object selection.)*

*When unrecognized text is found, displays dialog box:*

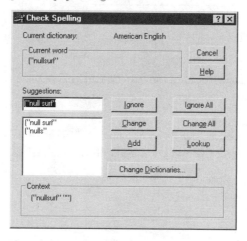

*When selected text is recognized, or when spelling check is complete, displays dialog box:*

## DIALOG BOX OPTIONS

**Ignore** ignores the word, and goes on to the next word.

**Ignore All** ignores all words with this spelling.

**Change** changes the word to the suggested spelling.

**Change All** changes all words with this spelling.

**Add** adds the word to the user (custom) dictionary.

**Lookup** checks spelling of the word in the Suggestions box.

**Change dictionaries** selects a different dictionary; displays the Change Dictionaries dialog box.

**Change Dictionaries** dialog box

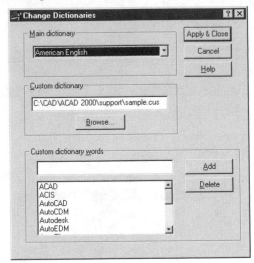

Main dictionary selects a language for the dictionary.

Custom dictionary options

**Directory** specifies the drive, folder, and filename of the custom dictionary.

**Browse** displays the **Select Custom Dictionary** dialog box.

Custom Dictionary Words options:

**Add** adds a word.

**Delete** removes a custom word from the dictionary.

**RELATED COMMANDS**

**DdEdit** edits text.

**MText** places paragraph text.

**Text** places lines of text in the drawing.

**RELATED SYSTEM VARIABLES**

**DctCust** specifies the name of the custom spelling dictionary.

**DctMain** specifies the name of the main spelling dictionary.

**RELATED FILES**

*enu.dct* is the dictionary word file.

*\*.cus* are the custom dictionary files.

**TIPS**

- A spell checker does *not* check your spelling; words that are spelled correctly but used incorrectly (such as *its* and *it's*) are not flagged. Rather, a spell checker looks for words that it does not recognize, which are words not in its dictionary file.

- As of AutoCAD 2002, the **Spell** command also checks words in blocks.

# Sphere

**Rel.11** Draws a 3D sphere as a solid model.

| Command | Alias | Ctrl+ | F-key | Alt+ | Menu Bar | Tablet |
|---------|-------|-------|-------|------|----------|--------|
| sphere | ... | ... | ... | DIS | Draw | K7 |
| | | | | | ⇘Solids | |
| | | | | | ⇘Sphere | |

**Command:** sphere
**Current wire frame density: ISOLINES=4**
**Specify center of sphere <0,0,0>:** *(Pick a point.)*
**Specify radius of sphere or [Diameter]:** *(Enter a value, or type **D**.)*

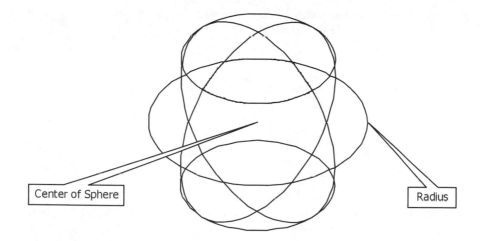

Center of Sphere

Radius

## COMMAND LINE OPTIONS

**Specify center of sphere** locates the center point of the sphere.

**Diameter** specifies the diameter of the sphere.

**Radius** specifies the radius of the sphere.

## RELATED COMMANDS

**Box** draws solid boxes.

**Cone** draws solid cones.

**Cylinder** draws solid cylinders.

**Torus** draws solid tori.

**Wedge** draws solid wedges.

**Ai_Sphere** draws surface meshed spheres.

## RELATED SYSTEM VARIABLES

**DispSilh** specifies the silhouette display of 3D solids:

| DispSilh | Meaning |
|----------|---------|
| 0 | Off. |
| 1 | On. |

**IsoLines** specifies the number of tessellation lines that define the surface of the 3D solid:

| IsoLines | Meaning |
|----------|---------|
| 0 | Minimum. |
| 4 | Default. |
| 2047 | Maximum. |

## TIPS

• The **Sphere** command places the sphere's central axis parallel to the z axis of the current UCS, with the latitudinal isolines parallel to the x,y plane.

• Notice the effect of the **DispSilh** system variable, which toggles the silhouette display of 3D solids, after executing the **Hide** command:

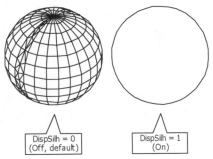

| DispSilh = 0 (Off, default) | DispSilh = 1 (On) |

• Notice the effect of the **IsoLines** system variable, which controls the number of tessellation lines used to define the surface of a 3D solid:

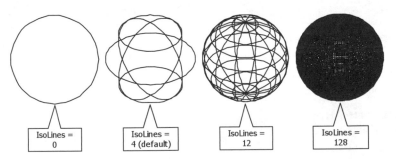

| IsoLines = 0 | IsoLines = 4 (default) | IsoLines = 12 | IsoLines = 128 |

• You must use the **Regen** command after changing the **DispSilh** and **IsoLines** system variables.

 # Spline

**Rel.13** Draws NURBS (*Non-Uniform Rational Bezier Spline*) curves.

| Command | Alias | Ctrl+ | F-key | Alt+ | Menu Bar | Tablet |
|---------|-------|-------|-------|------|----------|--------|
| spline | spl | ... | ... | DS | Draw<br>↳Spline | L9 |

**Command:** spline
**Specify first point or [Object]:** *(Pick a point, or type **O**.)*
**Specify next point:** *(Pick a point.)*
**Specify next point or [Close/Fit tolerance] <start tangent>:** *(Pick a point, or enter an option.)*
**Specify next point or [Close/Fit tolerance] <start tangent>:** *(Press **Enter** to define tangency points.)*
**Specify start tangent:** *(Pick a point.)*
**Specify end tangent:** *(Pick a point.)*

Open spline                    Closed spline

## COMMAND LINE OPTIONS

**Specify first point** picks the starting point of the spline.
**Object** converts 2D and 3D splined polylines into a NURBS spline.
**Specify next point** picks the next tangent point.
**Close** closes the spline at the start point.
**Fit** changes the spline tolerance; 0 = curve passes through fit points.
**Specify start tangent** specifies the tangency of the starting point of the spline.
**Specify end tangent** specifies the tangency of the endpoint of the spline.

## RELATED COMMANDS

**PEdit** edits splined polylines.
**PLine** draws splined polylines.
**SplinEdit** edits NURBS splines.

## RELATED SYSTEM VARIABLE

**DelObj** toggles whether the original polyline is deleted with the **Object** option.

## TIP

• A closed spline has the same start and end tangent.

# SplinEdit

**Rel.13**  Edits NURBS splines.

| Command | Alias | Ctrl+ | F-key | Alt+ | Menu Bar | Tablet |
|---------|-------|-------|-------|------|----------|--------|
| splinedit | spe | ... | ... | ... | Modify | Y18 |
| | | | | | ↳Spline | |

**Command:** splinedit
**Select spline:** *(Select one spline object.)*
**Enter an option [Fit data/Close/Move vertex/Refine/rEverse/Undo]:** *(Enter an option.)*

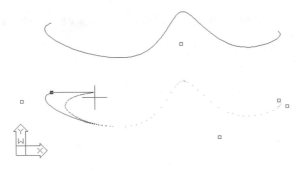

## COMMAND LINE OPTIONS

**Close** closes the spline, if open.

**Move vertex** moves a control vertex.

**Open** opens the spline, if closed.

**Refine** adds a control point; changes the spline's order or weight.

**rEverse** reverses the spline's direction.

**Undo** undoes the most-recent edit change.

**eXit** exits the command.

Fit Data options
**Enter a fit data option**
**[Add/Close/Delete/Move/Purge/Tangents/toLerance/eXit] <eXit>:** *(Enter an option.)*

**Add** adds fit points.

**Close** closes the spline, if open.

**Delete** removes fit points.

**Move** moves fit points.

**Open** opens the spline, if closed.

**Purge** removes fit point data from the drawing.

**Tangents** edits the start and end tangents.

**toLerance** refits the spline with the new tolerance value.

**eXit** exits suboptions.

## RELATED COMMANDS

**PEdit** edits a splined polyline.

**PLine** creates polylines, including splined polylines.

**Spline** draws a NURBS spline.

## TIPS

- The spline loses its fit data when you use these **SplinEdit** command options:

    **Refine**

    **Fit Purge**

    **Fit Tolerance** followed by **Fit Move**

    **Fit Tolerance** followed by **Fit Open** or **Fit Close**.

- The maximum order for a spline is 26; once the order has been elevated, it cannot be reduced.

- The larger the weight, the closer the spline is to the control point.

- This command automatically converts a spline-fit polyline into a spline object, even if you do not edit the polyline.

# Standards

**2002** Loads standards into the current drawing.

| Command | Alias | Ctrl+ | F-key | Alt+ | Menu Bar | Tablet |
|---------|-------|-------|-------|------|----------|--------|
| standards | ... | ... | ... | TSC | Tools | ... |
| | | | | | ⌐CAD Standards | |
| | | | | | ⌐Configure | |

**Command:** standards

*Displays tabbed dialog box:*

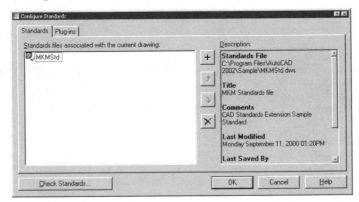

## DIALOG BOX OPTIONS

**+** adds a standards file (**F3**); displays the Select standards file dialog box.

**Move up** moves a standards file higher in the list.

**Move down** moves a standards file lower in the list.

**X** remove the selected standards file (**DEL**).

**Check Standards** displays the Check Standards dialog box; see the CheckStandards command.

## Plug-ins tab

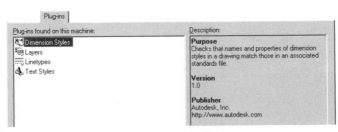

*Displays information about each item in the CAD standard.*

## RELATED COMMAND

**CheckStandards** checks the drawing against standards loaded by the Standards command.

## RELATED FILE

***.dws** are the CAD Standards files, stored in DWG format.

 **Stats**

**Rel.12** Lists statistics of the most-recent rendering (*short for STATisticS*).

| Command | Alias | Ctrl+ | F-key | Alt+ | Menu Bar | Tablet |
|---------|-------|-------|-------|------|----------|--------|
| stats | ... | ... | ... | VET | View | ... |
| | | | | | ⏷Render | |
| | | | | | ⏷Statistics | |

**Command:** stats

*Displays dialog box:*

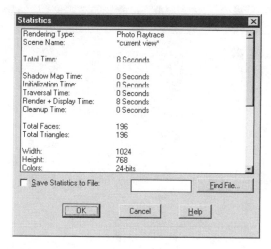

### DIALOG BOX OPTION

**Save Statistics to File** saves the rendering statistics to the file specified in the adjacent text entry box.

### RELATED AUTOCAD COMMAND

**Render** creates renderings.

### TIP

• The dialog box is blank if no rendering has taken place in the drawing.

### DEFINITIONS

*Scene Name* — name of the currently-selected scene; when no scene is current, displays "(none)."

*Last Rendering Type* — name of the currently-selected renderer; default is AutoCAD Render.

*Rendering Time* — time required to the create most recent rendering; reported in HH:MM:SS (hours, minutes, seconds) format.

*Total Faces* — number of faces processed in the most recent rendering; a single 3D object consists of many faces.

*Total Triangles* — number of triangles processed in the most recent rendering; a rectangular face is typically divided into two triangles.

# 'Status

**V. 1.0**  Displays information about the current drawing and environment.

| Command | Alias | Ctrl+ | F-key | Alt+ | Menu Bar | Tablet |
|---------|-------|-------|-------|------|----------|--------|
| 'status | ... | ... | ... | TYS | Tools | ... |
| | | | | | ↳ Inquiry | |
| | | | | | ↳ Status | |

**Command:** status

*Example output:*

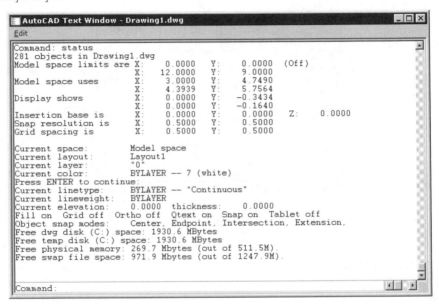

## COMMAND LINE OPTIONS

ENTER continues the listing.

F2 returns to the graphics screen.

## RELATED COMMANDS

**DbList** lists information about all the objects in the drawing.

**List** lists information about the selected objects.

**Stats** lists information about the most recent rendering.

## DEFINITIONS

*Model Space limits, Paper Space limits* — the x,y coordinates stored in the **LimMin** and **LimMax** system variables; 'Off' indicates limits checking is turned off (**LimCheck**).

*Model Space uses, Paper Space uses* — the x,y coordinates of the lower-left and upper-right extents of objects in the drawing; 'Over' indicates drawing extents exceed the drawing limits.

*Display shows* — the x,y coordinates of the lower-left and upper-right corners of the current display.

*Insertion base is* — the x,y,z coordinates stored in system variable **InsBase**.

*Snap resolution is, Grid spacing is* — the snap and grid settings, as stored in the **SnapUnit** and **GridUnit** system variables.

*Current space* — the indication whether whether model space or paper space is current.

*Current layout* — the indication of the name of the current layout.

*Current layer, Current color, Current linetype, Current lineweight, Current plot style, Current elevation, Thickness* — the current values for the layer name, color, linetype name, elevation, and thickness, as stored in system variables **CLayer, CeColor, CeLType, CeLweight, CPlotSytle, Elevation,** and **Thickness**.

*Fill, Grid, Ortho, Qtext, Snap, Tablet* — the current settings for the fill, grid, ortho, qtext, snap, and tablet modes, as stored in the system variables **FillMode, GridMode, OrthoMode, TextMode, SnapMode,** and **TabMode**.

*Object Snap modes* — the currently-set object modes, as stored in the system variable **OsMode**.

*Free disk (dwg + temp = C)* — the amount of free disk space on the drive storing AutoCAD's temporary files, as held by by system variable **TempPrefix**.

*Free physical memory* — the amount of free RAM.

*Free swap file space* — the amount of free space in AutoCAD's swap file on disk.

# StlOut

**Rel.12** Exports 3D solids and bodies in binary or ASCII SLA format (*short for STereoLithography OUTput*).

| Command | Alias | Ctrl+ | F-key | Alt+ | Menu Bar | Tablet |
|---------|-------|-------|-------|------|----------|--------|
| stlout | ... | ... | ... | FE | File | ... |
| | | | | ⌂STL | ⌂Export | |
| | | | | | ⌂Lithography | |

**Command:** stlout
**Select a single solid for STL output...**
**Select objects:** *(Select one or more solid objects.)*
**Select objects:** *(Press **Enter** to end object selection.)*
**Create a binary STL file? [Yes/No] <Y>:** *(Type **Y** or **N**.)*
  *Displays the **Create STL File** dialog box; enter a name, and then click **Save**.*

## COMMAND LINE OPTIONS

**Select objects** selects a single 3D solid object.

**Y** creates a binary-format *.sla* file.

**N** creates an ASCII-format *.sla* file.

## RELATED COMMANDS

*All solid modeling commands.*

**AcisOut** exports 3D solid models to an ASCII SAT-format ACIS file.

**AmeConvert** converts AME v2.x solid models into ACIS models.

## RELATED SYSTEM VARIABLE

**FaceTRes** determines the resolution of triangulating solid models.

## RELATED FILE

***.stl** is the the SLA-compatible file with STL extension created by this command.

## TIPS

- The solid model must lie in the positive x,y,z-octant of the WCS.

- The **StlOut** command exports a single 3D solid; it does not export regions or any other AutoCAD object.

- Even though this command prompts you twice to 'Select objects', selecting more than one solid causes AutoCAD to complain, "Only one solid per file permitted."

- The resulting *.stl* file cannot be imported back into AutoCAD.

## DEFINITIONS

*STL* — stereolithography data file, which consists of a faceted representation of the model.

*SLA* — stereoLithography Apparatus.

# Stretch

**V. 2.5** Stretches objects to lengthen, shorten, or distort them.

| Command | Alias | Ctrl+ | F-key | Alt+ | Menu Bar | Tablet |
|---------|-------|-------|-------|------|----------|--------|
| stretch | s | ... | ... | MH | Modify | V22 |
| | | | | | ⤷Stretch | |

**Command:** stretch
**Select objects to stretch by crossing-window or crossing-polygon...**
**Select objects:** *(Type C or CP.)*
**Specify first corner:** *(Pick a point.)*
**Specify opposite corner:** *(Pick a point.)*
**Select objects:** *(Press Enter to end object selection.)*
**Specify base point or displacement:** *(Pick a point.)*
**Specify second point of displacement or <use first point as displacement>:** *(Pick a point.)*

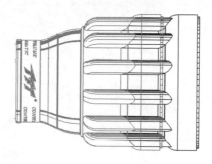

## COMMAND LINE OPTIONS

**First corner** selects object; must be CPolygon or Crossing object selection.

**Select objects** selects other objects using any selection mode.

**Base point** indicates the starting point for stretching.

**Second point** stretches the object larger or smaller.

## RELATED COMMANDS

**Change** changes the size of lines, circles, text, blocks, and arcs.

**Scale** increases or decreases the size of any object.

## TIPS

- The effect of the **Stretch** command is not always obvious; be prepared to use the **Undo** command.

- The first time you select objects for the **Stretch** command, you must use **Crossing** or **CPolygon** object selection; objects entirely within the selection window are moved.

- The **Stretch** command will not move a hatch pattern unless the hatch's origin is included in the selection set.

- Use the **Stretch** command to update associative dimensions automatically by including the dimension's endpoints in the selection set.

# 'Style

**V. 2.0**  Creates and modifies text styles, which define the properties of a font.

| Commands | Aliases | Ctrl+ | F-key | Alt+ | Menu Bar | Tablet |
|----------|---------|-------|-------|------|----------|--------|
| 'style | st | ... | ... | OS | Format | U2 |
|  | ddstyle |  |  |  | ⇘Text Style |  |
| -style |  |  |  |  |  |  |

**Command:** style

*Displays dialog box:*

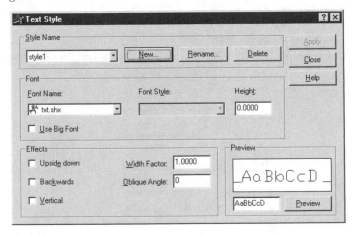

## DIALOG BOX OPTIONS

Style Name options

**Style Name** selects an existing text style.

**New** creates new text styles; displays dialog box:

**Rename** renames existing text styles; displays dialog box:

**Delete** deletes text styles.

## Font options

**Font Name** specifies the names of AutoCAD SHX and TrueType TTF fonts.

**Font Style** selects from available font styles, such as **Bold**, *Italic*, and ***Bold Italic***.

**Height** specifies the text height.

**Use Big Font** specifies the use of a big font file, typically for Asian alphabets.

## Effects options (not available for all fonts)

**Upside Down** draws text upside down:

**Backwards** draws text backwards:

**Vertical** draws text vertically.

**Width Factor** changes the width of characters.

*Width factor = 0.5 (left) and 2.0 (right):*

**Oblique Angle** slants characters forward or backward:

*Oblique angle = 30 (left) and -30 (right):*

**Preview** previews the effects on the style.

## Additional options

**Apply** applies the changes to the style.

**Close** closes the dialog box; in some cases, you can click the Close button before clicking the Apply button; then AutoCAD displays the following warning dialog box:

## -STYLE Command
**Command:** -style
**Enter name of text style or [?] <STANDARD>:** *(Enter a name, or type **?**.)*
**Specify full font name or font filename (TTF or SHX) <txt>:** *(Enter a name.)*
**Specify height of text <0.0000>:** *(Enter a value.)*
**Specify width factor <1.0000>:** *(Enter a value.)*
**Specify obliquing angle <0>:** *(Enter a value.)*
**Display text backwards? [Yes/No] <N>:** *(Type **Y** or **N**.)*
**Display text upside-down? [Yes/No] <N>:** *(Type **Y** or **N**.)*
**Vertical? <N>** *(Type **Y** or **N**.)*
**"STANDARD" is now the current text style.**

### COMMAND LINE OPTIONS

**Enter name of text style** names the text style; maximum = 31 characters (default = "STANDARD").

**?** lists the names of styles already defined in the drawing.

**Specify full font name or font filename** names the font file (SHX or TTF) from which the style is defined (default = *txt.shx*).

**Specify height of text** specifies the height of the text (default = 0 units).

**Specify width factor** specifies the width factor of the text (default = 1.00).

**Specify obliquing angle** specifies the obliquing angle or slant of the text (default = 0 degrees).

**Display text backwards**

　**Yes** prints text backwards — mirror writing.

　**No** prints text forwards.

**Display text upside-down**

　**Yes** prints text upside-down.

　**No** prints text rightside-up (default).

**Vertical**

　**Yes** prints text vertically; not available for all fonts.

　**No** prints text horizontally (default).

### RELATED COMMANDS

**Change** changes the style assigned to selected text.

**Purge** removes any unused text style definitions.

**Rename** renames a text style name.

**MText** places paragraph text.

**Text** places a single line of text.

### RELATED SYSTEM VARIABLES

**TextStyle** specifies the current text style.

**TextSize** specifies the current text height.

## RELATED FILES

*\*.shp* is Autodesk's format for vector source fonts.

*\*.shx is* Autodesk's format for compiled vector fonts; stored in \\*autocad 2004*\\*fonts* folder.

*\*.ttf* are TrueType font files; stored in \\*windows*\\*fonts* subdirectory.

## TIPS

- A **Width Factor** of 0.85 fits in 15% more text without sacrificing legibility.

- An **Obliquing Angle** of +15% can sometimes enhance the look of a font.

- The **Obliquing Angle** can be positive (forward slanting) or negative (backward slanting).

- To use PostScript fonts in the drawing, use the **Compile** command to convert PostScript *.pfb* files into *.shx* format.

- You can use any TrueType font with AutoCAD.

- The **Style** command affects the font used with the **Text**, **DText**, **MText**, and dimensioning commands, including **Leader** and **Tolerance**.

- Some of the fonts included with AutoCAD 2004:

- A text height of 0 lets you specify the text height while you are adding text with the **Text**, **DText**, and **MText** commands. If you specify a text height other than 0 in the **Style** command, that height is always used with that particular style name.

 **StylesManager**

Displays the Plot Styles window.

| Command | Alias | Ctrl+ | F-key | Alt+ | Menu Bar | Tablet |
|---------|-------|-------|-------|------|----------|--------|
| stylesmanager | ... | ... | ... | FY | File | Y24 |
| | | | | | ⌖Plot Style Manager | |

**Command:** stylesmanager

*You can also access this command through the Windows Control Panel: **Autodesk Plot Style Manager**.*

*Displays window:*

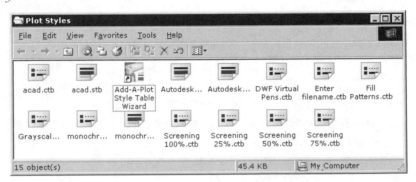

## WINDOW OPTIONS

acad.ctb

***.ctb*** (color-table based) opens the Plot Style Table Editor dialog box.

acad.stb

*** .stb*** (style-table based) opens the Plot Style Table Editor dialog box, as well.

**Add-A-Plot Style Table Wizard** opens Add Plot Style Table wizard; see the R14PenWizard command.

## DEFINITIONS

*StylesManager* — the program which modifies plot styles stored in plot style tables.

*Plot styles* — when drawings have a plot style table attached to their model and layout tabs, any changes to the plot style changes the object using the plot style. Plot styles can be assigned by object or by layer. See **PlotStyle** and **Layer** commands.

*Color-table based (.cbt)* — assigns colors to objects and layers, as used by older releases of AutoCAD. For example, the color of the object specifies the width of pen. This style of controlling the plot is now called "color dependent."

*Style-table based (.stb)* — newer releases of AutoCAD can control every aspect of the plot through "plot styles." By changing the *.stb* file attached to a layout tab, you immediately change the plot style for all objects and layers in the layout. This, for example, allows a quick change from monochrome to color plotting. A single drawing file can contain multiple plot styles.

## DIALOG BOX OPTIONS

**General** tab

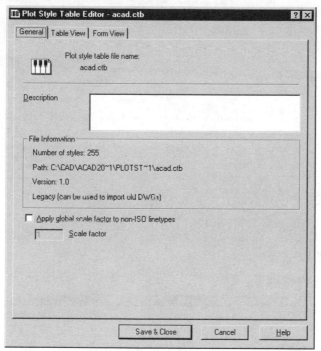

**Description** allows you to describe the plot style.

**Apply global scale factor to non-ISO linetypes and fill patterns** applies the scale factor to all non-ISO linetypes and hatch patterns in the plot.

**Scale factor** specifies the scale factor.

**Save and Close** saves the changes, and closes the dialog box.

**Cancel** cancels the changes, and closes the dialog box.

## CBT File Table View tab

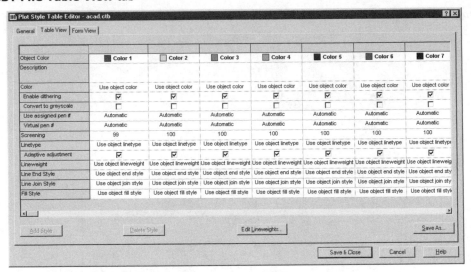

*See the following pages for options.*

## CBT File Form View tab

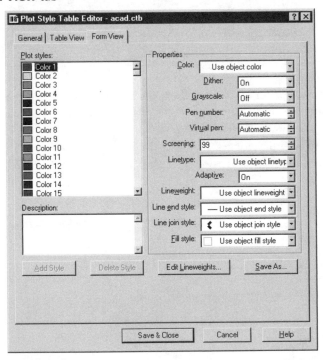

*See the following pages for options.*

## STB File Table View tab

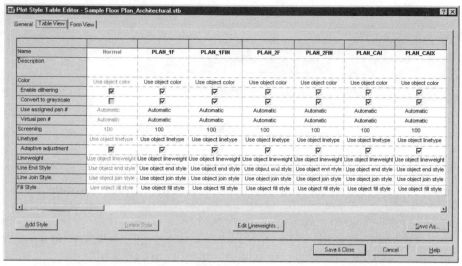

*See the following pages for options.*

## STB File Form View tab

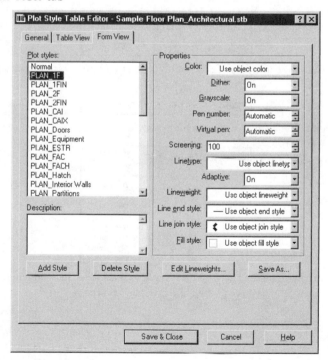

*See the following pages for options.*

Table View and Form View options are identical

**Object color** specifies the color of the object.

**Description** allows you to describe the plot style.

**Color** specifies the plotted color for the objects.

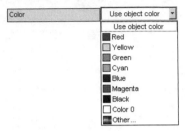

**Enable dithering** toggles dithering, if the plotter is supports dithering, to generate more colors than the plotter is capable of; this setting should be turned off for plotting vectors, and turned on for plotting renderings.

**Convert to grayscale** converts colors to shades of gray, if the plotter supports gray scale.

**Use assigned pen #** specifies the pen number of pen plotters; range is 1 to 32.

**Virtual pen number** specifies the virtual pen number (default = Automatic); range is 1 to 255. A value of 0 (or Automatic) tells AutoCAD to assign virtual pens from ACI (AutoCAD Color Index); this setting is meant for non-pen plotters that can make use of virtual pens.

**Screening** specifies the intensity of plotted objects; range is 0 (plotted "white") to 100 (full density):

**Linetype** specifies the linetype with which to plot the objects:

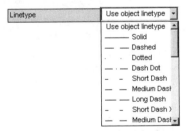

**Adaptive adjustment** adjusts linetype scale to prevent the linetype from ending in the middle of its pattern; keep off when the plotted linetype scale is crucial.

**Lineweight** specifies how wide lines are plotted, in millimeters:

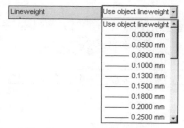

**Line end style** specifies how the ends of lines are plotted:

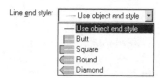

**Line join style** specifies how the intersections of lines are plotted:

**Fill style** specifies how objects are filled:

**Add Style** adds a plot style.

**Delete Style** removes a plot style.

**Edit Lineweights** displays the **Edit Lineweights** dialog box.

**Save As** displays the **Save As** dialog box.

## TIPS

* To configure a printer or plotter for *virtual pens*:

    1. In the **Device and Document Settings** tab, open the **PC3 Editor** dialog box.

    2. In the **Vector Graphics** section, select **255 Virtual Pens** from **Color Depth**.

* **CTB** is short for "color-dependent based" style table, which is compatible with earlier versions of AutoCAD.

* **STB** is short for "style-table based."

* You can attach a different *.stb* file to each layout and to the model tab in a drawing.

# Subtract

Rel.11 Removes the volume of one set of 3D models or 2D regions from another.

| Command | Alias | Ctrl+ | F-key | Alt+ | Menu Bar | Tablet |
|---------|-------|-------|-------|------|----------|--------|
| subtract | su | ... | ... | MNS | Modify | X16 |
| | | | | | ⮡Solids Editing | |
| | | | | | ⮡Subtract | |

**Command:** subtract
**Select objects:** *(Select one or more solid objects.)*
**Select objects:** *(Press **Enter** to end object selection.)*
**1 solid selected.**
**Objects to subtract from them...**
**Select objects:** *(Select one or more solid objects.)*
**Select objects:** *(Press **Enter** to end object selection.)*
**1 solid selected.**

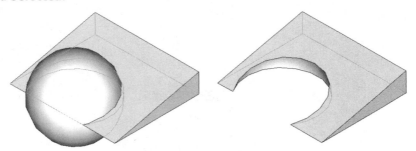

*Original objects (left); sphere subtracted from wedge (right).*

## COMMAND LINE OPTION

**Select objects** selects the objects to be subtracted.

## RELATED COMMANDS

**Intersect** removes all but the intersection of two solid volumes.

**Union** joins two solids.

## TIPS

- AutoCAD subtracts the objects you select *second* from the objects you select *first*.

- You can use this commands on 3D solids and 2D regions.

- When subtracting regions, they must lie in the same plane.

- When one solid is fully inside another, AutoCAD performs the subtraction, but reports, "Null solid created — deleted."

# SysWindows

**Rel.13**    Controls multiple windows (*short for SYStem WINDOWS*).

| Command | Alias | Ctrl+ | F-key | Alt+ | Menu Bar | Tablet |
|---|---|---|---|---|---|---|
| syswindows | ... | F6 | ... | W | Windows | ... |
| | | | | | ⤷*varies* | |

**Command:** syswindows
**Enter an option [Cascade/tile Horizontal/tile Vertical/Arrange icons]:** *(Enter an option.)*

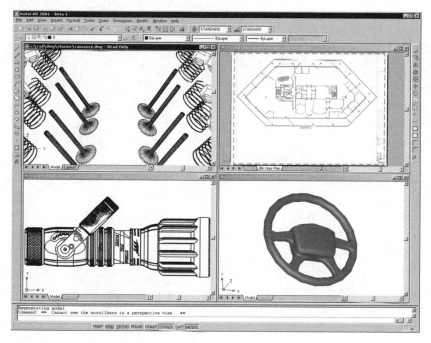

## COMMAND LINE OPTIONS

**Arrange icons** arranges icons in an orderly fashion.

    **Cascade** cascades the window.

    **tileHorizontal** tiles the window horizontally.

    **tileVertical** tiles the window vertically.

**TITLE BAR OPTIONS**

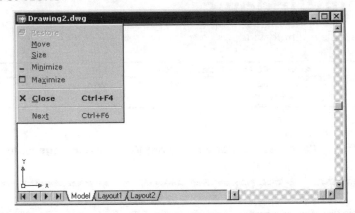

**Restore** restores the window to its "windowized" size.

**Move** moves the window.

**Size** resizes the window.

**Minimize** minimizes the window.

**Maximize** maximizes the window.

**Close** closes the window.

**Next** switches the focus to the next window.

**RELATED COMMANDS**

**Close** closes the current window.

**CloseAll** closes all windows.

**Open** opens one or more drawings, each in its own window.

**MView** creates paper space viewports in a window.

**Vports** creates model space viewports in a window.

**TIPS**

• The **SysWindows** command had no practical effect until AutoCAD 2000, since AutoCAD Release 13 and 14 supported a single window only.

• Window control icons:

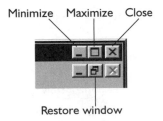

• Press **CTRL+F6** to switch quickly between currently-loaded drawings.

• Press **CTRL+F4** to close the currentl drawing.

# 'Tablet

V. 1.0 Configures and calibrates digitizing tablets, and toggles tablet mode.

| Command | Alias | Ctrl+ | F-key | Alt+ | Menu Bar | Tablet |
|---------|-------|-------|-------|------|----------|--------|
| 'tablet | ta | T | F4 | TT | Tools | X7 |
|  |  |  |  |  | ⌕Tablet |  |

**Command:** tablet

*When no tablet is configured, AutoCAD complains, "Your pointing device cannot be used as a tablet."*

*When a digitizing tablet is configured, AutoCAD continues:*

**Enter an option [ON/OFF/CAL/CFG]:** *(Enter an option.)*

## COMMAND LINE OPTIONS

**CAL** calibrates the coordinates for the tablet.

**CFG** configures the menu areas on the tablet.

**OFF** turns off the tablet's digitizing mode.

**ON** turns on the tablet's digitizing mode.

## RELATED SYSTEM VARIABLE

**TabMode** toggles use of the tablet:

| TabMode | Meaning |
|---------|---------|
| 0 | Tablet mode disabled. |
| 1 | Tablet mode enabled. |

## RELATED FILES

*acad.mnu* is the menu source code that defines the functions of tablet menu areas.

*tablet.dwg* is an AutoCAD drawing of the printed template overlay.

## TIPS

- To change the tablet overlay, edit the *tablet.dwg* file, and then plot it to fit your digitizer.

- **Tablet** does not work if a digitizer has not been configured with the **Options** command.

- AutoCAD supports up to four independent menu areas; macros are specified by the ***TABLET1 through ***TABLET4 sections of the *acad.mnu* menu file.

- Menu areas may be skewed, but corners must form a right angle.

- Projective transformation is a limited form of "rubber sheeting": straight lines remain straight, but not necessarily parallel.

## DEFINITIONS

*Affine Transformation* — requires three pick points; sets an arbitrary linear 2D transformation with independent x,y scaling and skewing.

*Orthogonal Transformation* — requires two pick points; sets the translation; the scaling and rotation angles remain uniform.

*Residual Error* — is largest where mapping is least accurate; second largest is where mapping is second-least accurate.

*Outcome of Fit* — reports on the results of transformation types:

| Outcome | Meaning |
| --- | --- |
| **Exact** | Enough points to transform data. |
| **Success** | More than enough points to transform data. |
| **Impossible** | Not enough points to transform data. |
| **Failure** | Too many colinear and coincident points. |
| **Cancelled** | Fitting cancelled during projective transformation. |

*Projective Transformation* — maps a perspective projection from one plane to another plane.

*RMS Error* — specifies root mean square error; smaller is better; measures closeness of fit.

*Standard Deviation* —when near zero, residual error at each point is roughly the same.

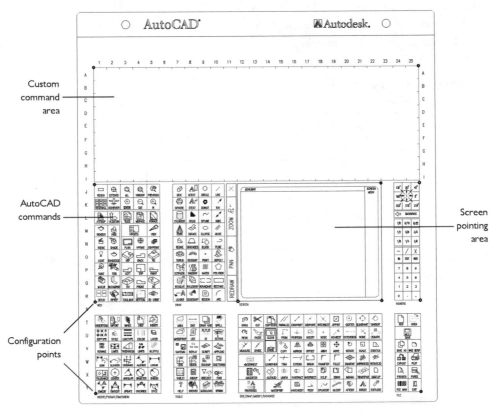

# TabSurf

**Rel.10** Draws tabulated surfaces as 3D meshes, defined by path curves and direction vectors (*short for TABulated SURFace*).

| Command | Alias | Ctrl+ | F-key | Alt+ | Menu Bar | Tablet |
|---------|-------|-------|-------|------|----------|--------|
| tabsurf | ... | ... | ... | DFT | Draw<br>⸂Surfaces<br>⸂Tabulated Surface | P8 |

**Command:** tabsurf
**Select object for path curve:** *(Select an object.)*
**Select object for direction vector:** *(Select an object.)*

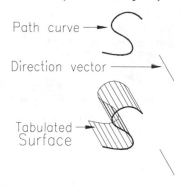

## COMMAND LINE OPTIONS

**Select object for path curve** selects the object that defines the tabulation path.

**Select object for direction vector** selects the vector that defines the tabulation direction.

## RELATED COMMANDS

**Edge** changes the visibility of 3D face edges.

**Explode** reduces a tabulated surface to 3D faces.

**PEdit** edits a 3D mesh, such as a tabulated surface.

**EdgeSurf** draws a 3D mesh surface between boundaries.

**RevSurf** draws a revolved 3D mesh surface around an axis.

**RuleSurf** draws a 3D mesh surface between open or closed boundaries.

## RELATED SYSTEM VARIABLE

**SurfTab1** defines the number of tabulations drawn by TabSurf in the *n*-direction.

## TIPS

• The path curve can be an open or closed object: line, 2D polyline, 3D polyline, arc, circle, or ellipse.

• The direction vector defines the direction and length of the extrusion.

• The number of *m*-direction tabulations is always 2, and lies along the direction vector.

• The number of *n*-direction tabulations is determined by system variable **SurfTab1** (default = 6) along curves only.

· · · · · · · · · · · · · · · · · · · · · · · · · · · · · · · · · · · · · · · · · · · · · · · · ·

 **Text**

**V. 1.0**   Places text, one line at a time, in drawings.

| Command | Alias | Ctrl+ | F-key | Alt+ | Menu Bar | Tablet |
|---------|-------|-------|-------|------|----------|--------|
| text | dtext | ... | ... | ... | Draw | K8 |
| | dt | | | | ⮑ Text | |
| | | | | | ⮑ Single Line Text | |

**Command:** text
**Current text style: "Standard" Text height: 0.2000**
**Specify start point of text or [Justify/Style]:** *(Pick a point, or enter an option.)*
**Specify height <0.2000>:** *(Enter a value.)*
**Specify rotation angle of text <0>:** *(Enter a value.)*
**Enter text:** *(Enter text, and then press **Enter**.)*
**Enter text:** *(Press **Enter** to end the command.)*

**COMMAND LINE OPTIONS**

**Specify start point of text** indicates the starting point of the text.

ENTER continues text one line below the previously-placed text.

**Specify height** indicates the height of the text; this prompt does not appear if the style has set the height to a value other than 0.

**Specify rotation angle of text** indicates the rotation angle of the text.

**Enter text** specifies the text; press **Enter** twice to end the command.

Justify options
**Enter an option [Align/Fit/Center/Middle/Right/TL/TC/TR/ML/MC/MR/BL/ BC/BR]:** *(Enter an option.)*

**Align** aligns the text between two points with adjusted text height.

**Fit** fits the text between two points with fixed text height.

**Center** centers the text along the baseline.

**Middle** centers the text horizontally and vertically.

**Right** right-justifies the text.

**TL** top-left justification.

**TC** top-center justification.

**TR** top-right justification.

**ML** middle-left justification.

**MC** middle-center justification.

**MR** middle-right justification.

**BL** bottom-left justification.

**BC** bottom-center justification.

**BR** bottom-right justification.

Style options
**Enter style name or [?] <Standard>:** *(Enter a name, or type **?**.)*

**Style name** indicates a different style name.

**?** lists the currently-loaded styles.

## SPECIAL SYMBOLS

**%%c** draws diameter symbol: Ø.

**%%d** draws degree symbol: °.

**%%o** starts and stops overlining.

**%%p** draws the plus-minus symbol: ±.

**%%u** starts and stops underlining.

**%%%** draws the percent symbol: %.

## RELATED COMMANDS

**DdEdit** edits text.

**Change** changes the text height, rotation, style, and content.

**Properties** changes all aspects of text.

**Style** creates new text styles.

**MText** places paragraph text in drawings.

## RELATED SYSTEM VARIABLES

**TextSize** is the current height of text.

**TextStyle** is the current style of text.

**ShpName** is the default shape name

## TIPS

- Use the **Text** command to place text easily in many locations in the drawing. It displays text on screen as you type.

- You can erase text by pressing the **BACKSPACE** key while at the 'Text' prompt.

- *Warning*: the spacing between lines of text does not match the current snap spacing.

- Transparent commands do not work during the **Text** command.

- You can enter any justification mode at the 'Start point' prompt.

- The 'Enter text' prompt repeats until cancelled with **ENTER**.

The dot indicates the text insertion point.

# 'TextScr

**V. 2.1**  Switches from the AutoCAD window to the Text window (*short for TEXT SCReen*).

| Command | Alias | Ctrl+ | F-key | Alt+ | Menu Bar | Tablet |
|---------|-------|-------|-------|------|----------|--------|
| 'textscr | ... | ... | F2 | VLT | View | ... |
| | | | | | ⌐Display | |
| | | | | | ⌐Text Window | |

**Command:** textscr

*Displays the **Text** window:*

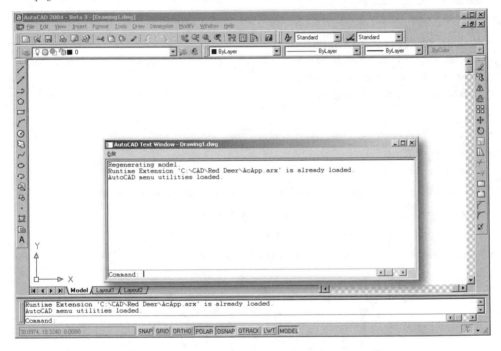

## EDIT MENU OPTIONS

**Paste to CmdLine** pastes text from the Clipboard to the command line; available only when the Clipboard contains text.

**Copy** copies selected text to the Clipboard.

**Copy History** copies all text to the Clipboard.

**Paste** pastes text from the Clipboard into the Text window; available only when the Clipboard contains text.

**Options** displays Options dialog box; see the Options command.

## COMMAND LINE OPTIONS

*Command window navigation:*

| Key | Meaning |
|-----|---------|
| ← | Moves the cursor left by one character. |
| → | Moves the cursor right by one character. |
| ↑ | Displays the previous line in the command history. |
| ↓ | Displays the next line in the command history. |
| Page Up | Displays the previous screen of text. |
| Page Down | Moves to the next screen of text. |
| Home | Moves the cursor to the start of the line. |
| End | Moves the cursor to the end of the line. |
| Insert | Toggles insert mode. |
| Delete | Deletes the character to the right of the cursor. |
| BACKSPACE | Deletes the character to the left of the cursor. |

## SHORTCUT MENU OPTIONS

*Right-click the **Text** window:*

**Recent Command** displays a list of recently-used commands.

**Copy** copies selected text to the Clipboard.

**Copy History** copies all text to the Clipboard.

**Paste** pastes text from the Clipboard into the Text window; available only when Clipboard contains text.

**Paste to CmdLine** pastes text from the Clipboard to the command line; available only when the Clipboard contains text.

**Options** displays the **Options** dialog box; see the **Options** command.

## RELATED COMMAND

**GraphScr** switches from the Text window to the AutoCAD drawing window.

## RELATED SYSTEM VARIABLE

**ScreenMode** reports whether the screen is in text or graphics mode:

| ScreenMode | Meaning |
|------------|---------|
| 0 | Text screen. |
| 1 | Graphics screen. |
| 2 | Dual screen displaying both text and graphics. |

## Removed Command

**TiffIn** was removed from AutoCAD Release 14. Use the **ImageAttach** command instead.

# TifOut

Exports the current viewports in TIFF raster format.

| Command | Alias | Ctrl+ | F-key | Alt+ | Menu Bar | Tablet |
|---------|-------|-------|-------|------|----------|--------|
| tifout | ... | ... | ... | ... | ... | ... |

**Command:** tifout

*Displays* **Create Raster File** *dialog box. Specify a filename, and the click* **Save.**

**Select objects or <all objects and viewports>:** *(Select objects, or press* **Enter** *to select all objects and viewports.)*

## COMMAND LINE OPTIONS

**Select objects** selects specific objects.

**All objects and viewports** selects all objects and all viewports, whether in model space or in layout mode.

## RELATED COMMANDS

**BmpOut** exports drawings in BMP (bitmap) format.

**Image** places raster images in the drawing.

**JpgOut** exports drawings in JPEG (joint photographic expert group) format.

**PngOut** exports drawings in PNG (portable network graphics) format.

## TIPS

- The rendering effects of the **ShadEdge** command are preserved, but not of the **Render** command.

- TIFF files are commonly used in desktop publishing.

- TIFF is short for "tagged image file format," and was developed by Aldus, the forerunner of Adobe.

/ The Illustrated AutoCAD 2004 Quick Reference

# 'Time

**V. 2.5**  Displays time-related information about the current drawing.

| Command | Alias | Ctrl+ | F-key | Alt+ | Menu Bar | Tablet |
|---------|-------|-------|-------|------|----------|--------|
| 'time | ... | ... | ... | TQT | Tools | ... |
| | | | | | ⮑Inquiry | |
| | | | | | ⮑Time | |

**Command:** time
**Display/ON/OFF/Reset:** *(Enter an option.)*

*Example output:*

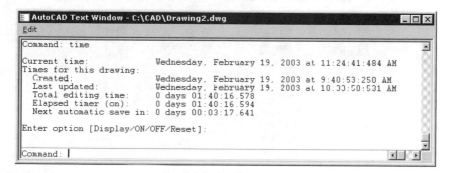

```
AutoCAD Text Window - C:\CAD\Drawing2.dwg
Edit

Command: time

Current time:            Wednesday, February 19, 2003 at 11:24:41:484 AM
Times for this drawing:
  Created:               Wednesday, February 19, 2003 at 9:40:53:250 AM
  Last updated:          Wednesday, February 19, 2003 at 10:33:50:531 AM
  Total editing time:    0 days 01:40:16.578
  Elapsed timer (on):    0 days 01:40:16.594
  Next automatic save in: 0 days 00:03:17.641

Enter option [Display/ON/OFF/Reset]:

Command:
```

## COMMAND LINE OPTIONS

**Display** displays the current time and date.

**OFF** turns off the user timer.

**ON** turns on the user timer.

**Reset** resets the user timer.

## RELATED COMMANDS

**Status** displays information about the current drawing and environment.

**Preferences**  sets the automatic backup time.

## RELATED SYSTEM VARIABLES

**CDate** is the current date and time.

**Date** is the current date and time in Julian format.

**SaveTime** is the interval for automatic drawing saves.

**TDCreate** is the date and time the drawing was created.

**TDInDwg** is the time the drawing spent in AutoCAD.

**TDUpdate** is the last date and time the drawing was changed.

**TDUsrTimer** is the current user timer setting.

## TIP

- The time displayed by the **Time** command is only as accurate as your computer's clock; unfortunately, the clock in some personal computers can stray by several minutes per week.

## Removed Command

**Today** was removed from AutoCAD 2004; it was replaced by the Communication Center.

# Tolerance

**Rel.13**   Places geometric tolerancing symbols and text.

| Command | Alias | Ctrl+ | F-key | Alt+ | Menu Bar | Tablet |
|---------|-------|-------|-------|------|----------|--------|
| tolerance | tol | ... | ... | NT | Dimension | X1 |
| | | | | | ⤷ Tolerance | |

**Command:** tolerance

*Displays dialog box.*

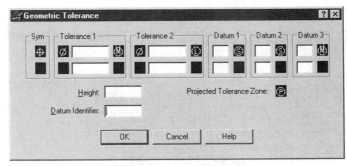

*Clicking a symbol displays this dialog box:*

**Enter tolerance location:** *(Pick a point.)*

## DIALOG BOX OPTIONS

**Sym** specifies the geometric characteristic symbol.

**Tolerance** specifies the first tolerance value.

**Dia** specifies the places optional Ø (diameter) symbol.

**Value** specifies the tolerance value.

**Datum** specifies the datum reference.

**Height** specifies the projected tolerance zone value.

**Projected Tolerance Zone** places the projected tolerance zone symbol.

**Datum Identifier** creates the datum identifier symbol, such as -A-.

**MC** displays Material Condition dialog box.

## Material Condition dialog box

(M)  specifies maximum material condition.

(L)  specifies least material condition.

(S)  specifies regardless of feature size.

## RELATED FILES

**gdt.shp** is the tolerance symbol definition source file.

**gdt.shx** is the compiled tolerance symbol file.

## TIP

- You can use the **DdEdit** command to edit tolerance symbols and feature control frames.

## DEFINITIONS

*Datum* — a theoretically-exact geometric reference that establishes the tolerance zone for the feature. These objects can be used as a datum: point, line, plane, cylinder, and other geometry.

*Material Condition* — symbols that modify the geometric characteristics and tolerance values (modifiers for features that vary in size).

*Projected Tolerance Zone* — the height of the fixed perpendicular part's extended portion; changes the tolerance to positional tolerance.

*Tolerance* — the amount of variance from perfect form.

### Orientation Symbols

⊕  Position.

◎  Concentricity and coaxiality.

⚌  Symmetry.

//  Parallelism.

⊥  Perpendicularity.

∠  Angularity.

### Form Symbols

⌭  Cylindricity.

▱  Flatness.

○  Circularity and roundness.

—  Straightness.

### Profile Symbols

⌓  Profile of the surface.

⌒  Profile of the line.

↗  Circular runout.

↗↗  Total runout.

# -Toolbar

<u>Rel.13</u>  Displays and hides toolbars via the command line.

| Command | Alias | Ctrl+ | F-key | Alt+ | Menu Bar | Tablet |
|---------|-------|-------|-------|------|----------|--------|
| -toolbar | ... | ... | ... | ... | ... | ... |

*Note:* **Toolbar** *and* **TbConfig** *are aliases for the* **Customize** *command.*

**Command:** -toolbar
**Enter toolbar name or [ALL]:** *(Enter a name, or type* **ALL**.*)*
**Enter an option [Show/Hide]:** *(Type* **S** *or* **H**.*)*

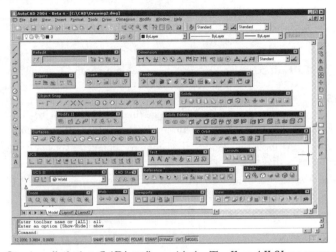

*Opening all of AutoCAD's toolbars with the* **-Toolbar All Show** *command.*

## COMMAND LINE OPTIONS

**Toolbar name** specifies the name of the toolbar.

**ALL** applies the command to all toolbars; must be entered in all capital letters.

**Show** displays the toolbar.

**Hide** dismisses the toolbar.

## RELATED COMMANDS

**Customize** customizes toolbars via a dialog box.

**MenuLoad** loads a partial menu file, including toolbar definitions.

**Tablet** configures the tablet.

## RELATED SYSTEM VARIABLE

**ToolTips** toggles the display of tooltips.

## RELATED FILES

*\*.mnc* are compiled menu files.

*\*.mns* are AutoCAD source menu files.

*\*.bmp* are bitmap files, that define custom icon buttons.

# ToolPalettes

**2004** Displays the Tool Palettes window.

| Command | Alias | Ctrl+ | F-key | Alt+ | Menu Bar | Tablet |
|---------|-------|-------|-------|------|----------|--------|
| toolpalettes | ... | 4 | ... | TP | Tools | ... |
| | | | | | ⤷Tool Palettes Window | |

**Command:** toolpalettes

*Displays window:*

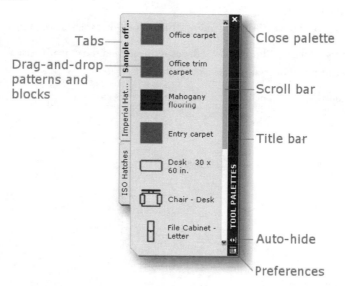

## WINDOW OPTIONS

**Tabs** selects various sets of palettes.

**x** closes the palette window; alternatively, press CTRL+4.

**Title bar** drags around the window.

**Auto-hide** collapses the window when the cursor moves away.

**Preferences** displays a shortcut menu.

## PREFERENCES MENU

*Right-click the palette, or click the **Preferences** button:*

**Move** moves the window.

**Size** changes the size of the window.

**Close** closes the window.

**Allow Docking** toggles whether the window can be docked at the side of AutoCAD.

**Auto-hide** toggles whether the window reduces its size when the cursor is elsewhere.

**Transparency** displays the Transparency dialog box.

**New Tool Palette** creates a new tab, and then prompts you for a name:

**Rename** renames the tab.

**Customize** displays the Tool Palettes tabs of the Customize dialog box; see Customize command.

## SHORTCUT MENU

*Right-click the tabs:*

**Move up** moves the tab up one position.

**Move down** moves the tab down one position.

**View Options** displays the View Options dialog box.

**New Tool Palette** creates a new tab, and then prompts you for a name.

**Delete Tool Palette** removes the tab, after prompting you for confirmation.

**Rename Tool Palette** renames the tab.

**Transparency** dialog box

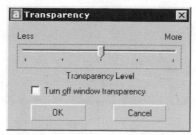

**Transparency Level** changes the transparency of the palette from none (Less) to More; see figure below for example of maximum transparency.

**Turn off window transparency** makes the palette opaque.

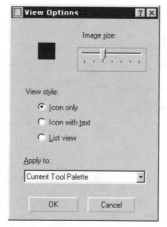

*View Options* dialog box

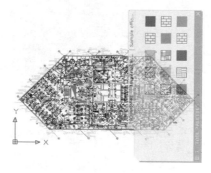

**Image Size** changes the size of icons from small (14 pixels square) to large (54 pixels).

**View style** toggles the display of icons and text:

- **Icon only** displays icons only.
- **Icon with text** displays icons and text.
- **List view** displays small icons with text.

**Apply to** determines if changes apply to the current palette only, or to all palettes.

## RELATED COMMANDS

**Customize** exports and imports Tool Palettes.

**DesignCenter** displays the content of drawings.

**ToolPalettesClose** closes the Tool Palette window.

## RELATED SYSTEM VARIABLES

**PaletteOpaque** determines whether the Tool Palette window can be transparent.

**TpState** notes whether the Tool Palettes window is open.

## RELATED FILES

*.xtp* stores the content of Tool Palette windows.

## TIPS

- To share the content of Tool Palettes:

    1. From the menu bar, select **Tools | Customize**.

    2. In the **Tool Palette** tab, click **Export** to save your tool palettes to disk.

    3. Alternatively, click **Import** to access Tool Palettes created by others.

- To bring content from the DesignCenter into the Tool Palette window:

    1. In **DesignCenter**, right-click an item.

    2. From the shortcut menu, select **Create Tool Palette**.

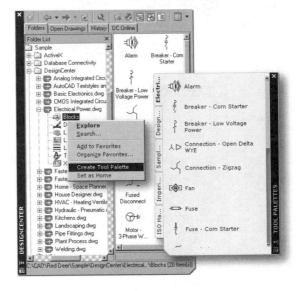

- The *.xtp* files store content in XML (extended markup language) format.

# ToolPalettesClose

**2004** Closes the Tool Palettes window.

| Command | Alias | Ctrl+ | F-key | Alt+ | Menu Bar | Tablet |
|---|---|---|---|---|---|---|
| toolpalettesclose | ... | 4 | ... | TP | Tools | ... |
| | | | | | ⬦Tool Palettes Window | |

**Command:** toolpalettesclose

*Closes the Tool Palette Window.*

## COMMAND LINE OPTIONS
*None.*

## TIP
- To close the Tool Palettes window, you can also click the **x** in the upper right corner.

# Torus

**Rel.11** Draws 3D tori as solid models.

| Command | Alias | Ctrl+ | F-key | Alt+ | Menu Bar | Tablet |
|---------|-------|-------|-------|------|----------|--------|
| torus | tor | ... | ... | DIT | Draw | O7 |
| | | | | | ⮡Solids | |
| | | | | | ⮡Torus | |

**Command:** torus
**Current wire frame density: ISOLINES=4**
**Specify center of torus <0,0,0>:** *(Pick a point.)*
**Specify radius of torus or [Diameter]:** *(Specify the radius, or type D.)*
**Specify radius of tube or [Diameter]:** *(Specify the radius, or type D.)*

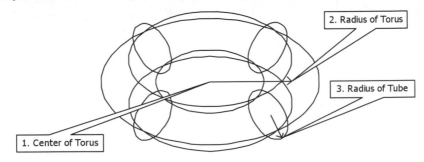

1. Center of Torus
2. Radius of Torus
3. Radius of Tube

## COMMAND LINE OPTIONS
**Center of torus** indicates the center of the torus.

**Diameter** indicates the diameter of the torus and the tube.

**Radius** indicates the radius of the torus and the tube.

## RELATED COMMANDS
**Ai_Torus** creates a torus from 3D polyfaces.

**Cone** draws solid cones.

**Cylinder** draws solid cylinders.

**Sphere** draws solid spheres.

## TIPS
- When the torus radius is negative, the tube radius must be a larger positive number. A negative torus radius creates a football shape. Specify a tube diameter larger than the torus diameter to create a hole-less torus.

*Football:*          *Hole-less torus:*

# Trace

**V. 1.0**  Draws lines with width.

| Command | Alias | Ctrl+ | F-key | Alt+ | Menu Bar | Tablet |
|---------|-------|-------|-------|------|----------|--------|
| trace | ... | ... | ... | ... | ... | ... |

**Command:** trace
**Specify trace width <0.050>:** *(Enter a value.)*
**Specify start point:** *(Pick a point.)*
**Specify next point:** *(Pick another point.)*
**Specify next point:** *(Pres **Enter** to end the command.)*

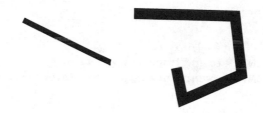

## COMMAND LINE OPTIONS

**Trace width** specifies the width of the trace.
**Start point** picks the starting point.
**Next point** picks the next vertex.
ENTER exits the Trace command.

## RELATED COMMANDS

**Line** draws lines with zero width.
**MLine** draws up to 16 parallel lines.
**PLine** draws polylines and polyline arcs with varying widths.
**LWeight** gives every object a width.

## RELATED SYSTEM VARIABLES

**FillMode** toggles display of fill or outline traces (default = 1, on).
**TraceWid** specifies the current width of the trace (default = 0.05).

## TIPS

* Traces are drawn along the centerline of the pick points.

* Display of a trace segment is delayed by one pick point.

* During the drawing of traces, you cannot back up, because an **Undo** option is missing; if you require this feature, draw wide lines with the **PLine** command, setting the **Width** option.

* There is no option for controlling joints (always bevelled) or endcapping (always square); if you require these features, draw wide lines with the **MLine** command, setting the solid fill, endcap, and joint options with the **MlStyle** command.

# *Tracking*

Rel.14 Locates x and y points visually, relative to other points in the command sequence; *not* a command, but a command modifier.

| Modifer | Alias | Ctrl+ | F-key | Alt+ | Menu Bar | Tablet |
|---------|-------|-------|-------|------|----------|--------|
| tacking | tk | ... | ... | ... | ... | T15 |
|         | track |       |       |      |          |        |

*Example usage:*

**Command:** line

**Specify first point:** *(Pick a point.)*

**Specify next point or [Undo]:** tk

*Enters tracking mode:*

**First tracking point:** *(Pick a point.)*

**Next point (Press ENTER to end tracking):** *(Pick a point.)*

**Next point (Press ENTER to end tracking):** *(Press* **Enter** *to end tracking)*

*Exits tracking mode:*

**Specify next point or [Undo]:** *(Pick a point.)*

## COMMAND LINE OPTIONS

**First tracking point** picks the first tracking point.

**Next point** picks the next tracking point.

ENTER exits tracking mode.

## RELATED COMMANDS

*Any command that prompts for a point, such as 'Specify first point' and 'Specify next point.'*

## TIPS

- **Tracking** is not a command, but a command option modifier.

- **Tracking** can be used in conjunction with direct distance entry.

- In tracking mode, AutoCAD automatically turns on **Ortho** mode to constrain the cursor vertically and horizontally.

- If you start tracking in the x direction, the next tracking direction is y, and vice versa.

- You can use tracking as many times as you need to at the 'Specify first point' and 'Specify next point' prompts.

# Transparency

<u>Rel.14</u>  Toggles the transparency of background pixels in raster images.

| Command | Alias | Ctrl+ | F-key | Alt+ | Menu Bar | Tablet |
|---------|-------|-------|-------|------|----------|--------|
| transparency | ... | ... | ... | MOIT | Modify | X21 |
| | | | | | ↳Object | |
| | | | | | ↳Image | |
| | | | | | ↳Transparency | |

**Command:** transparency
**Select image(s):** *(Select one or more images inserted with the* **Image** *command.)*
**Select image(s):** *(Press* **Enter** *to end object selection.)*
**Enter transparency mode [ON/OFF] <OFF>:** *(Type* **ON** *or* **OFF***.)*

## COMMAND LINE OPTIONS

**Select image(s)** selects the objects whose transparency to change.

**ON** makes the background pixels transparent.

**OFF** makes the background pixels opaque.

## RELATED COMMANDS

**ImageAttach** attaches a raster image as an externally-referenced file.

**ImageAdjust** changes the brightness, contrast, and fading of a raster image.

## TIPS

• This command works only with raster images placed in drawings with the **Image** command.

• This command is meant for use with raster file formats that allow transparent pixels.

• When on, transparent pixels allow graphics under the image to show through.

# TraySettings

<u>2004</u>  Specifies settings for the Communications Center, located at the right end of the status bar (called the "tray").

| Command | Alias | Ctrl+ | F-key | Alt+ | Menu Bar | Tablet |
|---------|-------|-------|-------|------|----------|--------|
| traysettings | ... | ... | ... | ... | ... | ... |

**Command:** traysettings

*Displays dialog box:*

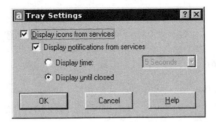

## DIALOG BOX OPTIONS

**Display icons from services:**

☑ displays "tray" at right end of the status line; see below.

☐ turns off the tray.

**Display notification from services:**

☑ displays balloon notifications from a variety of services; see below.

☐ turns off notifications:

- **Display time** specifies the duration that a notification balloon is displayed.
- **Display until closed** specifies that the notification balloon is displayed until closed by user.

## RELATED COMMANDS

*None.*

## RELATED SYSTEM VARIABLES

**TrayIcons** toggles the display of the tray on the status bar.

**TrayNotify** toggles whether notification balloons are displayed.

**TrayTimeOut** determines the length of time that notification balloons are displayed

## TIP

- Examples of notification balloons: click <u>underlined</u> links for more information, or click **x** to dismiss.

# 'TreeStat

**Rel.12** Displays the status of the drawing's spatial index, including the number and depth of nodes (*short for TREE STATistics*).

| Command | Alias | Ctrl+ | F-key | Alt+ | Menu Bar | Tablet |
|---|---|---|---|---|---|---|
| 'treestat | ... | ... | ... | ... | ... | ... |

**Command:** treestat

*Sample output:*

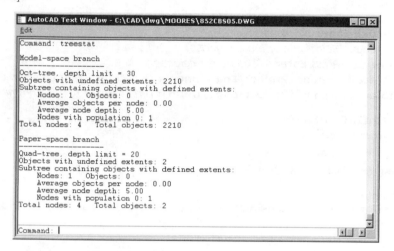

## RELATED SYSTEM VARIABLES

**TreeDepth** specifies the size of the tree-structured spatial index in *xxyy* format:

| Depth | Meaning |
|---|---|
| xx | Number of model space nodes (default = 30). |
| yy | Number of paper space nodes (default = 20). |
| -xx | 2D drawing. |
| +xx | 3D drawing (default). |

**TreeMax** is the maximum number of nodes (default = 10,000,000).

## TIPS

* Better performance occurs with fewer objects per oct-tree node. When redraws and object selection seem slow, increase the value of **TreeDepth**.

* Each node consumes 80 bytes of memory.

## DEFINITIONS

*Oct Tree* — the model space branch of the spatial index, where all objects are either 2D or 3D. *Oct* comes from the eight volumes in the x,y,z coordinate system of 3D space.

*Quad Tree* — the paper space branch of the spatial index, where all objects are two-dimensional. *Quad* comes from the four areas in the x,y coordinate system of 2D space.

*Spatial Index* — objects indexed by oct-region to record their position in 3D space; has a tree structure with two primary branches: oct tree and quad tree. Objects are attached to *nodes*; each node is a branch of the *tree*.

# Trim

**V. 2.5** Trims lines, arcs, circles, and 2D polylines back to real or projected cutting lines and views.

| Command | Alias | Ctrl+ | F-key | Alt+ | Menu Bar | Tablet |
|---|---|---|---|---|---|---|
| trim | tr | ... | ... | MT | Modify ↳Trim | W15 |

**Command:** trim
**Current settings: Projection=UCS Edge=None**
**Select cutting edges ...**
**Select objects:** *(Select one or more objects.)*
**Select objects:** *(Press* **Enter** *to end object selection.)*
**Select object to trim or [Project/Edge/Undo]:** *(Select object, or enter an option.)*
**Select object to trim or [Project/Edge/Undo]:** *(Press* **Enter** *to end command.)*

## COMMAND LINE OPTIONS

**Select objects** selects the cutting edges.

**Select object to trim** picks the object at the trim end.

**Undo** untrims the last trim action.

### Edge options

**Enter an implied edge extension mode [Extend/No extend] <No extend>:** *(Type* **E** *or* **N**.*)*

**Extend** extends the cutting edge to trim object.

**No extend** trims only at an actual cutting edge.

### Project options

**Enter a projection option [None/Ucs/View] <Ucs>:** *(Enter an option.)*

**None** uses only objects as cutting edge.

**Ucs** trims at the x,y plane of current UCS.

**View** trims at the current view plane.

## RELATED COMMANDS

**Change** changes the size of lines, arcs and circles.

**Extend** lengthens lines, arcs and polylines.

**Lengthen** lengthens open objects.

# U

**V. 2.5**  Undoes the most recent AutoCAD command (*short for Undo*).

| Command | Alias | Ctrl+ | F-key | Alt+ | Menu Bar | Tablet |
|---------|-------|-------|-------|------|----------|--------|
| u | ... | Z | ... | EU | Edit | T12 |
| | | | | | ↳Undo | |

**Command:** u

## COMMAND LINE OPTIONS
*None.*

## RELATED COMMANDS
**Oops** unerases the most-recently erased object.

**Quit** exits the drawing, undoing all changes.

**Redo** reverses the most-recent undo, if U or Undo was the previous command.

**Undo** allows more sophisticated control over undo.

## RELATED SYSTEM VARIABLE
*None.*

## TIPS
- The **U** command is convenient for stepping back through the design process, undoing one command at a time.

- The **U** command is the same as the **Undo 1** command; for greater control over the undo process, use the **Undo** command.

- The **Redo** command redoes the most-recent undo only; use **MRedo** otherwise.

- The **Quit** command, followed by the **Open** command, restores the drawing to its original state, if not already saved.

- Because the undo mechanism creates a mirror drawing file on disk, disable the **Undo** command with system variable **UndoCtl** (set to 0) when your computer is low on disk space.

- Commands that involve writing to file, plotting, and some display functions (such as **Render**, **Shade**, and **Hide**) are not undone.

# Ucs

Rel.10 Defines new coordinate planes, and restores existing UCSs (*short for User-defined Coordinate System*).

| Command | Alias | Ctrl+ | F-key | Alt+ | Menu Bar | Tablet |
|---------|-------|-------|-------|------|----------|--------|
| ucs | ... | ... | ... | TW | Tools | W7 |
| | | | | | ⮡New UCS | |
| | | | | TH | Tools | ... |
| | | | | | ⮡Orthographic UCS | |

**Command:** ucs
**Current ucs name: \*TOP\***
**Enter an option [New/Move/orthoGraphic/Prev/Restore/Save/Del/Apply/?/World] <World>:** *(Enter an option.)*

## COMMAND LINE OPTIONS

**New** creates a new user-defined coordinate system.

**Move** moves the UCS along the z axis.

**orthoGraphic** selects a standard orthographic UCS: top, bottom, front, back, left, and right.

**Prev** restores the previous UCS orientation.

**Restore** restores a named UCS.

**Save** saves the current UCS by name.

**Del** deletes the name of a saved UCS.

**Apply** applies the UCS setting to a selected viewport, or all active viewports.

**?** lists the names of saved UCS orientations.

**World** aligns the UCS with the WCS.

New options
**Specify origin of new UCS or [ZAxis/3point/OBject/Face/View/X/Y/Z] <0,0,0>:** *(Enter an option.)*

*For compatibility with earlier versions of AutoCAD, you may enter any of these options at the earlier 'Enter an option' prompt.*

**Specify origin of new UCS** moves the UCS to a new origin point.

**ZAxis** aligns the UCS with a new origin and z axis.

**3point** aligns the UCS with a point on the positive x-axis and positive x,y plane.

**OBject** aligns the UCS with a selected object.

**Face** aligns the UCS with the face of a 3D solid object.

**View** aligns the UCS with the current view.

**X** rotates the UCS about the x axis.

**Y** rotates the UCS about the y axis.

**Z** rotates the UCS about the z axis.

## RELATED TOOLBARS

**UCS** toolbar

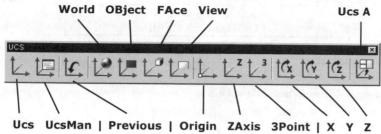

**UCS II** toolbar

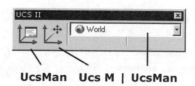

## RELATED SYSTEM VARIABLES

**UcsAxisAng** specifies the default rotation angle when the UCS is rotated around an axis using the X, Y, or Z options of this command

**UcsBase** specifies the UCS that defines the origin and orientation of orthographic UCS settings.

**UcsFollow** shows the plan view in the new UCS automatically:

| UcsFollow | Meaning |
|-----------|---------|
| 0 | No change in view (default). |
| 1 | Displays plan view of new UCS. |

**UcsIcon**      determines visibility and location of the UCS icon:

| UcsIcon | Meaning |
|---------|---------|
| 0 | UCS icon not displayed. |
| 1 | UCS icon displayed in lower-right corner. |
| 2 | UCS icon displayed at the UCS origin, when possible. |
| 3 | UCS icon displayed at UCS always (default). |

**UcsOrg** specifies the WCS coordinates of UCS icon (default = 0,0,0).

**UcsOrtho** specifies whether the related UCS is automatically displayed when an orthographic view is restored.

**UcsView** specifies whether the current UCS is saved when a view is created with the View command.

**UcsVp** specifies that the UCS reflects the UCS of the currently-active viewport.

**UcsXdir** specifies the X direction of the current UCS (default = 1,0,0).

**UcsYdir** specifies the Y direction of the current UCS (default = 0,1,0).

**WorldUcs** correlates the WCS to the UCS:

| WorldUcs | Meaning |
| --- | --- |
| 0 | Current UCS is WCS. |
| 1 | UCS is same as WCS (default). |

## RELATED COMMANDS

**UcsMan** modifies the UCS via a dialog box.

**UcsIcon** controls the visibility of the UCS icon.

**Plan** changes the view to the plan view of the current UCS.

## TIPS

- Use the **UCS** command to draw objects at odd angles in 3D space.

- Although you can create UCSes in paper space, you cannot use 3D viewing commands in paper space.

- UCSes can be aligned with these objects: points, lines, traces, 2D polylines, solids, arcs, circles, texts, shapes, dimensions, attribute definitions, 3D faces, and block references.

- UCSes do *not* align with these objects: mlines, rays, xlines, 3D polylines, splines, ellipses, leaders, viewports, 3D solids, 3D meshes, and regions.

## DEFINITIONS

*UCS* — user-defined 2D coordinate system oriented in 3D space; sets a working plane, orients 2D objects, defines the extrusion direction, and the axis of rotation.

*WCS* — world coordinate system; the default 3D x,y,z coordinate system.

# UcsIcon

<u>Rel.10</u>   Controls the location and display of the UCS icon.

| Command | Alias | Ctrl+ | F-key | Alt+ | Menu Bar | Tablet |
|---------|-------|-------|-------|------|----------|--------|
| ucsicon | ... | ... | ... | VLU | View | L2 |
| | | | | | ⮡Display | |
| | | | | | ⮡UCS Icon | |

**Command:** ucsicon
**Enter an option [ON/OFF/All/Noorigin/ORigin/Properties] <ON>:** *(Enter an option.)*

## COMMAND LINE OPTIONS

**All** forces the changes of this command effective in all viewports.

**Noorigin** displays the UCS icon always in lower-left corner.

**OFF** turns off the display of the UCS icon.

**ON** turns on the display of the UCS icon.

**ORigin** displays the UCS icon at the current UCS origin.

**Properties** displays the UCS Icon dialog box.

## DIALOG BOX OPTIONS

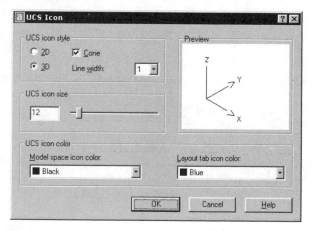

UCS icon style options

* **2D** displays flat UCS icon.
* **3D** displays tripod UCS icon.

**Cone** option (available only with 3D style):

☑ displays arrowheads as 3D cones.

☐ displays plain arrowheads.

**Line width** changes the line width from 1 to 2 to 3 pixels.

### UCS icon size option
*Slide bar* changes the icon size from 5 to 95 pixels.

### UCS icon color
**Model space icon color** selects the icon's color in model space.

**Layout tab icon color** selects the icon's color in layout (paper space).

## RELATED SYSTEM VARIABLE
**UcsIcon** determines the visibility and location of UCS icon:

| UcsIcon | Meaning |
|---|---|
| 0 | UCS icon not displayed. |
| 1 | UCS icon displayed in lower-right corner. |
| 2 | UCS icon displayed at the UCS origin, when possible. |
| 3 | UCS icon displayed at UCS always (default). |

## RELATED COMMAND
**UCS** creates and controls user-defined coordinate systems.

## TIPS
- The UCS icon varies, depending on the current viewpoint relative to the active UCS:

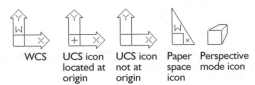

WCS    UCS icon located at origin    UCS icon not at origin    Paper space icon    Perspective mode icon

- When you cannot reliably draw or edit, AutoCAD displays the "broken pencil" icon (below):

- When AutoCAD switches from 2D wireframe mode to one of the **ShadeMode** command's options, such as 3D wireframe and flat rendered, the UCS icon changes to a 3D icon (below).

- There is generally no need for the UCS icon in 2D drafting, and it can be safely turned off.

# UcsMan

**2000** Displays the UCS dialog box.

| Commands | Aliases Ctrl+ | F-key | Alt+ | Menu Bar | Tablet |
|----------|---------------|-------|------|----------|--------|
| ucsman | dducs ... | ... | ... | **Tools** | **W8** |
| | | | | ⮡**Named UCS** | |
| | dducsp | | **Tools** | **W9** | |
| | | | | ⮡**Orthographic UCS** | |
| | | | | ⮡**Preset** | |
| **+ucsman** | | | | | |

**Command:** ucsman

*Displays dialog box.*

## DIALOG BOX OPTIONS

**Named UCSs** tab

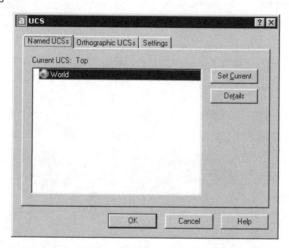

**Named UCSs** lists the names of the AutoCAD-generated and user-defined coordinate systems of the current viewport in the active drawing; the arrowhead points to the current UCS.

**Set Current** restores the selected UCS.

**Details** displays the UCS Details dialog box:

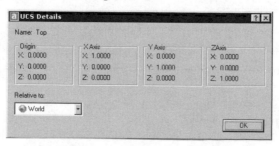

**Orthographic UCSs** tab

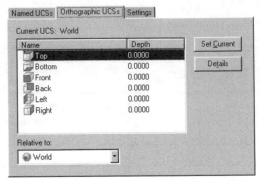

**Name** lists the six standard orthographic UCS views: top, bottom, front, back, left, and right.

**Depth** specifies the height of the UCS above the x,y plane.

**Relative to** specifies the orientation of the selected UCS relative to WCS or to a customized UCS.

**Set Current** activates the selected UCS.

**Details** displays the UCS Details dialog box.

**Settings** tab

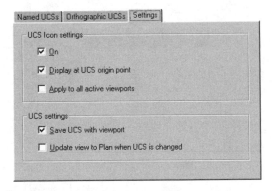

UCS icon settings options

**On** displays the UCS icon in the current viewport; each viewport can display the UCS icon independently.

**Display at UCS origin point:**

☑ displays the UCS icon at the origin of the current UCS.

☐ displays the UCS icon at the lower-left corner of the viewport.

**Apply to all active viewports:**

☑ applies these UCS icon settings to all active viewports in the current drawing.

☐ applies to the current viewport only.

UCS settings options

**Save UCS with viewport:**

☑ saves the UCS setting with the viewport.

☐ current viewport determines UCS settings.

**Update view to Plan when UCS is changed** restores plan view when the UCS changes.

## SHORTCUT MENU OPTIONS

*Right-click the list in the **Named UCSs** tab:*

**Set Current** sets the selected UCS as active.

**Rename** renames the selected UCS; you cannot rename the World UCS.

**Delete** erases the selected UCS; you cannot delete the World UCS.

**Details** displays the UCS Details dialog box.

*Right-click the list in the **Orthographic UCS** tab:*

**Set Current** sets the selected UCS as active.

**Reset** restores the origin of the selected UCS.

**Depth** moves the UCS in the z direction.

**Details** displays the UCS Details dialog box.

. . . . . . . . . . . . . . . . . . . . . . . . . . . . . . . . . . . . . . . . . . . . . . . . . . .

# +UCSMAN Command

**Command:** +ucsman

**Tab index <0>:** *(Type **1**, **2**, or **3**.)*

## COMMAND LINE OPTION

**Tab index** displays the tab related to the tab number:

| Index | Meaning |
|-------|---------|
| 0 | **Named UCS** tab. |
| 1 | **Orthographic UCS** tab. |
| 2 | **Settings** tab. |

## RELATED SYSTEM VARIABLES

*See **UCS** command.*

## RELATED COMMANDS

**UCS** displays the UCS options at the command line.

**UcsIcon** changes the display of the UCS icon.

**Plan** displays the plan view of the WCS or a UCS.

## TIPS

• The functions of this command were formerly carried out by the **DdUcs** and **DdUcsP** commands.

• **Unnamed** is the first entry, when the current UCS is unnamed.

• **World** is the default for new drawings; it cannot be renamed or deleted.

• **Previous** is the previous UCS; you can move back through several previous UCSs.

. . . . . . . . . . . . . . . . . . . . . . . . . . . . . . . . . . . . . . . . . . . . . . . . . . .

# Undefine

**Rel. 9** Makes an AutoCAD command unavailable.

| Command | Alias | Ctrl+ | F-key | Alt+ | Menu Bar | Tablet |
|---------|-------|-------|-------|------|----------|--------|
| undefine | ... | ... | ... | ... | | |

**Command:** undefine
**Enter command name:** *(Enter name.)*

*Example usage:*
**Command:** undefine
**Enter command name:** line
**Command:** line
**Unknown command. Type ? for list of commands.**
**Command:** .line
**From point:**

## COMMAND OPTIONS

**Enter command name** specifies the name of the command to make unavailable.

**.** *(period)* is the prefix for undefined commands to redefine them temporarily.

## RELATED COMMAND

**Redefine** redefines an AutoCAD command.

## TIPS

• Commands created by programs cannot be undefined, including the following programming interfaces:

> AutoLISP and Visual LISP.
> ObjectARx.
> Visual Basic for Applications.
> External commands.
> Aliases.

• In menu macros written with international language versions of AutoCAD, precede command names with an underscore character ( _ ) to translate the command name into English automatically.

 # Undo

**V. 2.5**  Undoes the effect of the previous command(s).

| Command | Alias | Ctrl+ | F-key | Alt+ | Menu Bar | Tablet |
|---------|-------|-------|-------|------|----------|--------|
| undo | ... | ... | ... | ... | ... | ... |

**Command:** undo

**Enter the number of operations to undo or [Auto/Control/BEgin/End/Mark/Back] <1>:** *(Enter a number, or an option.)*

## COMMAND LINE OPTIONS

**Auto** treats a menu macro as a single command.

**Control** limits the options of the Undo command.

**BEgin** groups a sequence of operations (formerly the Group option).

**End** ends the group option.

**Mark** sets a marker.

**Back** undoes back to the marker.

*number* indicates the number of commands to undo.

Control options

**Enter an UNDO control option [All/None/One] <All>:**

**All** toggles on full undo.

**None** turns off the undo feature.

**One** limits the Undo command to a single undo.

## RELATED COMMANDS

**Oops** unerases the most-recently erased object.

**Quit** leaves the drawing without saving changes.

**Redo** undoes the most recent undo.

**MRedo** undoes multiple undoes.

**U** undoes a single step.

## RELATED SYSTEM VARIABLES

**UndoCtl** determines the state of undo control:

| UndoCtrl | Meaning |
|----------|---------|
| 0 | Undo disabled. |
| 1 | Undo enabled. |
| 2 | Undo limited to one command. |
| 4 | Auto group mode. |
| 8 | Group currently active. |

**UndoMarks** specifies the number of undo marks placed in the **Undo** control stream.

## TIPS

- Since the undo mechanism creates a mirror drawing file on disk, disable the **Undo** command with system variable **UndoCtl** (set it to 0) when your computer is low on disk space.

- There are some commands that cannot be undone, such as **Save** and **Plot**.

# Union

<u>2000</u>  Joins two or more solids and regions together into a single body.

| Command | Alias | Ctrl+ | F-key | Alt+ | Menu Bar | Tablet |
|---------|-------|-------|-------|------|----------|--------|
| union | uni | ... | ... | MNU | Modify | X15 |
| | | | | | ⮑ Solids Editing | |
| | | | | | ⮑ Union | |

**Command:** union
**Select objects:** *(Select one or more solid objects.)*
**Select objects:** *(Select one or more solid objects.)*
**Select objects:** *(Press **Enter** to end command.)*

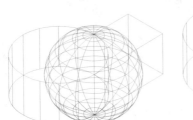
*Before Union*

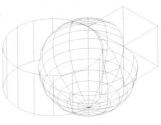
*After Union*

## COMMAND LINE OPTION

**Select objects** selects the objects to join into a single object; you must select at least two solid objects.

## RELATED COMMANDS

**Intersect** creates a solid model from the intersection of two objects.

**Subtract** creates a solid model by subtracting one object from another.

## TIPS

• You must select at least two solid or coplanar region objects.

• The two objects need not overlap for this c*ommand to operate.

# 'Units

**V. 1.4**   Controls the display and format of coordinates and angles.

| Commands | Aliases Ctrl+ | F-key | Alt+ | Menu Bar | Tablet |
|----------|---------------|-------|------|----------|--------|
| 'units   | un  ···       | ···   | OU   | Format   | V4     |
|          | ddunits       |       |      | ⬚Units   |        |
| -units   | -un           |       |      |          |        |

**command:** units

*Displays dialog box:*

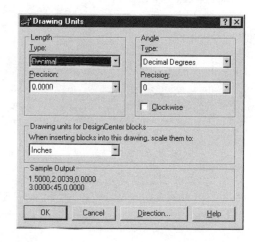

## DIALOG BOX OPTIONS

Length options

**Type** sets the format for units of linear measure displayed by AutoCAD: Architectural, Decimal, Engineering, Fractional, or Scientific.

**Precision** specifies the number of decimal places or fractional accuracy.

Angle options

**Type** sets the current angle format.

**Precision** sets the precision for the current angle format.

**Clockwise** calculates positive angles in the clockwise direction.

**Drawing units for DesignCenter blocks** specifies the units when blocks are inserted from the DesignCenter.

**Direction** displays the Direction Control dialog box.

**Direction Control** dialog box

Base Angle options

**East** sets the base angle to 0 degrees (default).

**North** sets the base angle to 90 degrees.

**West** sets the base angle to 180 degrees.

**South** sets the base angle to 270 degrees.

**Other** turns on the Angle option.

**Angle** sets the base angle to any direction.

**Pick an angle** dismisses the dialog box temporarily, and allows you to define the base angle by picking two points in the drawing; AutoCAD prompts you 'Pick angle' and 'Specify second point.'

**TIPS**

- Since **'Units** is a transparent command, you can change units during another command.

- The 'Direction Angle' prompt lets AutoCAD start the angle measurement from any direction.

- AutoCAD accepts the following notations for angle input:

| Notation | Meaning |
|----------|---------|
| < | Specify an angle based on current units setting. |
| << | Bypass angle translation set by **Units** command to use 0-angle-is-east direction and decimal degrees. |
| <<< | Bypass angle translation; use angle units set by the **Units** command and 0-angle-is-east direction. |

- The system variable **UnitMode** forces AutoCAD to display units in the same manner that you enter them.

- Do not use a suffix — such as 'r' or 'g' — for angles entered as radians or grads; instead, use the **Units** command to set angle measurement to radians and grads.

- The **Drawing units for DesignCenter blocks** option is for inserting blocks from AutoCAD DesignCenter, and especially when the block was created in other units.

- To not scale a block when dragged from the DesignCenter window, select **Unitless**.

## -UNITS Command

**Command:** -units

| Report formats: | (Examples) |
|---|---|
| 1. Scientific | 1.55E+01 |
| 2. Decimal | 15.50 |
| 3. Engineering | 1'-3.50" |
| 4. Architectural | 1'-3 1/2" |
| 5. Fractional | 15 1/2 |

With the exception of Engineering and Architectural formats, these formats can be used with any basic unit of measurement. For example, Decimal mode is perfect for metric units as well as decimal English units.

**Enter choice, 1 to 5 <2>:** *(Enter a value.)*

**Enter number of digits to right of decimal point (0 to 8) <4>:** *(Enter a value.)*

| Systems of angle measure: | (Examples) |
|---|---|
| 1. Decimal degrees | 45.0000 |
| 2. Degrees/minutes/seconds | 45d0'0" |
| 3. Grads | 50.0000g |
| 4. Radians | 0.7854r |
| 5. Surveyor's units | N 45d0'0" E |

**Enter choice, 1 to 5 <1>:** *(Enter a value.)*

**Enter number of fractional places for display of angles (0 to 8) <0>:** *(Enter a value.)*

**Direction for angle 0:**
 East  3 o'clock = 0
 North 12 o'clock = 90
 West  9 o'clock = 180
 South  6 o'clock = 270
**Enter direction for angle 0 <0>:** *(Enter a value.)*

**Measure angles clockwise? [Yes/No] <N>** *(Enter **Y** or **N**.)*

## COMMAND LINE OPTIONS

**Report formats** selects scientific, decimal, engineering, architectural, or fractional format for length display.

**Number of digits to right of decimal point** specifies the number of decimal places between 0 and 8.

**Systems of angle measure** selects decimal degrees, degrees/minutes/seconds, grads, radians, or surveyor's units for angle display.

**Denominator of smallest fraction to display** specifies the denominator of faction displays, such as 1/2 or 1/128.

**Number of fractional places for display of angles** specifies the number of decimal places between 0 and 8.

**Direction for angle 0** selects the direction for 0 degrees as east, north, west, or south.

**Do you want angles measured clockwise?**

 **Yes** measures angles clockwise.

 **No** measures angles counterclockwise.

## RELATED SYSTEM VARIABLES

**AngBase** specifies the direction of zero degrees.

**AngDir** specifies the direction of angle measurement.

**AUnits** specifies the units of angles.

**AuPrec** specifies the displayed precision of angles.

**InsUnits** specifies the drawing units for blocks dragged from the DesignCenter:

| InsUnits | Meaning |
|----------|---------|
| 0 | Unitless |
| 1 | Inches |
| 2 | Feet |
| 3 | Miles |
| 4 | Millimeters |
| 5 | Centimeters |
| 6 | Meters |
| 7 | Kilometers |
| 8 | Microinches |
| 9 | Mils |
| 10 | Yards; 3 feet |
| 11 | Angstroms; 0.1 nanometers |
| 12 | Nanometers; 10E-9 meters |
| 13 | Microns; 10E-6 meters |
| 14 | Decimeters; 0.1 meter |
| 15 | Decameters; 10 meters |
| 16 | Hectometers; 100 meters |
| 17 | Gigameters; 10E9 meters |
| 18 | Astronomical Units; 149.597E8 kilometers |
| 19 | Light Years; 9.4605E9 kilometers |
| 20 | Parsecs; 3.26 light years |

**InsUnitsDefSource** specifies source units that should be used.

**InsUnitsDefTarget** specifies target units that should be used.

**LUnits** specifies the units of measurement.

**LuPrec** specifies the displayed precision of coordinates.

**UnitMode** toggles the type of display units.

## RELATED COMMAND

**New** sets up drawings with Imperial or metric units.

# VbaIde

<u>2000</u>  Displays the Visual Basic window (*short for Visual Basic for Applications Integrated Development Environment*).

| Command | Alias | Alt+ | F-key | Alt+ | Menu Bar | Tablet |
|---------|-------|------|-------|------|----------|--------|
| vbaide | ... | F11 | ... | TMB | Tools | ... |
| | | | | | ⤷Macro | |
| | | | | | ⤷Visual Basic Editor | |

**Command:** vbaide

*Displays window:*

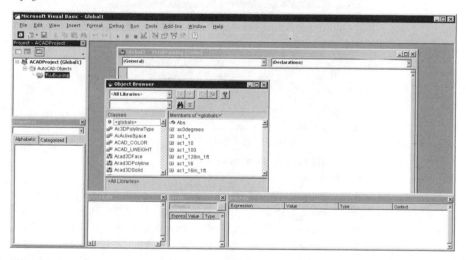

## MENU BAR OPTION

*Select* **Help | Microsoft Visual Basic Help** *for assistance in using this VBA IDE window.*

## RELATED COMMANDS

**VbaLoad** loads a VBA project; displays the Open VBA Project dialog box.

**VbaMan** displays the VBA Manager dialog box.

**VbaRun** displays the Macros dialog box.

**VbaStmt** executes a VBA expression at the command line.

**VbaUnload** unloads a VBA project.

## RELATED SYSTEM VARIABLES

*None.*

**TIPS**

- VBA is short for "Visual Basic for Applications," a macro programming language common to a number of Windows applications; it is based on the Visual Basic programming language.

- AutoCAD 2004 contains sample VBA projects in the \*autocad 2004\sample\vba* folder.

- For more information about Visual Basic for Applications, read the *ActiveX and VBA Developer's Guide* included with AutoCAD.

- Since many viruses can be spread via VBA macros, I strongly recommend that your computer have real-time virus protection to prevent infection.

- Since loading a macro from other sources into AutoCAD can also expose your computer to malicious viruses, AutoCAD displays the following warning:

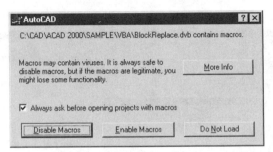

**More Info** displays AutoCAD's on-line help window.

**Always ask before opening projects with macros:**

☑ displays this dialog box.

☐ prevents this dialog box from being displayed; to turn back on, use the VbaRun command's Options dialog box; see VbaRun command.

**Disable Macros** loads a VBA project file, but disables macros; you can view, edit, and save the macros. To renable the macros, close the project, and then open it again with Enable Macros.

**Enable Macros** loads the project file with macros enabled.

**Do Not Load** prevents the project file from being loaded.

# VbaLoad

2000 Loads a VBA project into AutoCAD (*short for Visual Basic for Applications LOAD*).

| Commands | Alias | Ctrl+ | F-key | Alt+ | Menu Bar | Tablet |
|----------|-------|-------|-------|------|----------|--------|
| vbaload | ... | ... | ... | TML | Tools | ... |
| | | | | | ⌂Macro | |
| | | | | | ⌂Load Project | |
| -vbaload | | | | | | |

**Command:** vbaload

*Displays the **Open VBA Project** dialog box; select a .dvd file, and then click **Open**.*

*When the project contains macros, the **AutoCAD** dialog box is displayed; see the **Vbalde** command.*

. . . . . . . . . . . . . . . . . . . . . . . . . . . . . . . . . . . . . . . .

## -VBALOAD Command
**Command:** -vbaload
**Open VBA project <*projectname*>:** *(Enter project name.)*

### COMMAND LINE OPTION
**Open VBA project** specifies the project path and filename.

### RELATED COMMANDS
**Vbaide** displays the Visual Basic for Applications development environment window.

**VbaMan** displays the VBA Manager dialog box.

**VbaRun** displays the Macros dialog box.

**VbaStmt** executes a VBA expression at the command line.

**VbaUnload** unloads a VBA project.

### RELATED SYSTEM VARIABLES
*None.*

### TIPS
- You may load one or more VBA projects; there is no practical limit to the number.

- To unload a VBA project, use the **VbaUnload** command.

- This command does not load embedded VBA projects; these projects are automatically loaded with the drawing.

- When the project contains macros, AutoCAD displays the AutoCAD dialog box to warn you about protection against macro viruses; see the **Vbaide** command.

- For more information about Visual Basic for Applications, read the *ActiveX and VBA Developer's Guide* included with AutoCAD.

# VbaMan

<u>2000</u>  Displays the VBA Manager dialog box (*short for Visual Basic for Applications MANager*).

| Command | Alias | Ctrl+ | F-key | Alt+ | Menu Bar | Tablet |
|---------|-------|-------|-------|------|----------|--------|
| vbaman | ... | ... | ... | TMV | Tools | ... |
| | | | | | ⤷Macro | |
| | | | | | ⤷VBA Manager | |

**Command:** vbaman

*Displays dialog box:*

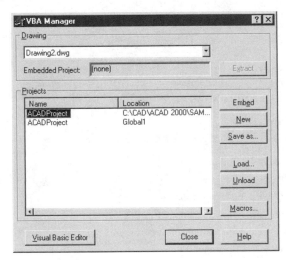

## DIALOG BOX OPTIONS

Drawing options
**Drawing** lists the names of drawings currently loaded in AutoCAD.

**Embedded Project** specifies the name of the embedded project.

**Extract** moves the embedded project from the drawing to a global project file.

Projects options
**Embed** embeds the project in the drawing.

**New** creates a new project; default name = Global *n*.

**Save as** saves a global project.

**Load** displays the Open VBA Project dialog box; see the VbaLoad command.

**Unload** unloads the global project.

**Macros** displays the Macros dialog box; see the VbaRun command.

Additional options
**Visual Basic Editor** displays the Visual Basic Editor; see the VbaIde command.

# VbaRun

<u>2000</u> Displays the Macros dialog box (*short for Visual Basic for Applications RUN*).

| Command | Alias | Alt+ | F-key | Alt+ | Menu Bar | Tablet |
|---------|-------|------|-------|------|----------|--------|
| vbarun | ... | F8 | ... | TMM | Tools | ... |
| | | | | | ⌐Macro | |
| | | | | | ⌐Macros | |

-vbarun

**Command:** vbarun

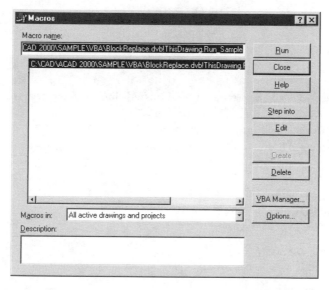

## DIALOG BOX OPTIONS

**Macro name** specifies the name of the macro; enter a name or select one from the list.

**Macros in** specifies the projects and drawings containing macros from all active drawings and projects, all active drawings, all active projects, and any single drawing or project currently open in AutoCAD.

**Description** describes the macro; you may modify the description.

Buttons

**Run** runs the macro.

**Close** closes the dialog box.

**Help** displays context-sensitive on-line help.

**Step into** displays the Visual Basic Editor and executes the macro, pausing at the first executable line of code.

**Edit** displays the Visual Basic Editor with the macro; see the Vbaide command.

**Create** displays the Visual Basic Editor with an empty procedure.

**Delete** erases the selected macro.

**VBA Manager** displays the VBA Manager dialog box; see the VbaMan command.

**Options** displays the VBA Options dialog box.

**VBA Options** *dialog box*

**Enable auto embedding** creates an embedded VBA project for all drawings when you open the drawing:

**Allow Break on errors:**

☑ stops the macro, and displays the **Visual Basic Editor** with the code, showing the error in the macro.

☐ displays an error message, and stops the macro.

**Enable macro virus protection** enables virus protection, which displays a dialog box when VBA macros are loaded; see the VbaIde command.

## -VBARUN Command
**Command:** -vbarun
**Macro name:** *(Enter macro name.)*

### COMMAND LINE OPTION
**Macro name** treats a menu macro as a single command.

### RELATED COMMANDS
**Vbaide** displays the Visual Basic for Applications integrated development environment window.

**VbaLoad** loads a VBA project; displays the Open VBA Project dialog box.

**VbaMan** displays the VBA Manager dialog box.

**VbaStmt** executes a VBA expression at the command line.

**VbaUnload** unloads a VBA project.

### RELATED SYSTEM VARIABLES
*None.*

### TIPS
• A *macro* is an executable subroutine; each project can contain one or more macros.

• When the macro's name is not unique among loaded projects, include the module's project and names in this format: **Project.Module.Macro**

• When the macro is not yet loaded, include the *.dvb* file name using this format: **Filenamedvb!Project.Module.Macro**

# VbaStmt

**2000** Executes a single line VBA expression at the command prompt (*short for Visual Basic for Applications StaTeMenT*).

| Command | Alias | Ctrl+ | F-key | Alt+ | Menu Bar | Tablet |
|---------|-------|-------|-------|------|----------|--------|
| vbastmt | ... | ... | ... | ... | ... | ... |

**Command:** vbastmt
**Statement:** *(Enter VBA statement.)*

## COMMAND LINE OPTION

**Statement** specifies the VBA statement for AutoCAD to execute.

## RELATED COMMANDS

**Vbaide** displays the Visual Basic for Applications integrated development environment window.

**VbaLoad** loads a VBA project; displays the Open VBA Project dialog box.

**VbaMan** displays the VBA Manager dialog box.

**VbaRun** displays the Macros dialog box.

**VbaUnload**   unloads a VBA project.

## RELATED SYSTEM VARIABLES

*None.*

## TIPS

* A VBA *statement* is a complete instruction containing keywords, operators, variables, constants, and expressions.

* A VBA *macro* is an executable subroutine.

* At the 'Statement' prompt, enter a single line of code; use the colon ( : ) to separate multiple statements on the single line.

# VbaUnload

**2000** Unloads a VBA project (*short for Visual Basic for Applications UNLOAD*).

| Command | Alias | Ctrl+ | F-key | Alt+ | Menu Bar | Tablet |
|---------|-------|-------|-------|------|----------|--------|
| vbaunload | ... | ... | ... | ... | ... | ... |

**Command:** vbaunload
**Unload VBA Project:** *(Enter project name.)*

## COMMAND LINE OPTION
**Unload VBA Project** specifies the name of the VBA project to unload.

ENTER unloads the active global project.

## RELATED COMMANDS
**VbaIde** displays the Visual Basic for Applications integrated development environment window.

**VbaLoad** loads a VBA project; displays the Open VBA Project dialog box.

**VbaMan** displays the VBA Manager dialog box.

**VbaRun** displays the Macros dialog box.

**VbaStmt** executes a VBA expression at the command line.

## RELATED SYSTEM VARIABLES
*None.*

## TIPS
• When you do not enter a project name, AutoCAD unloads the active global project.

• To load a VBA project, use the **VbaLoad** command.

# View

**V. 2.0** Saves and displays view by name in the current viewport.

| Commands | Aliases Ctrl+ | F-key | Alt+ | Menu Bar | Tablet |
|----------|---------------|-------|------|----------|--------|
| view | v    ... | ... | VN | View<br>⤷**Named Views** | M5 |
| +view | | | | | |
| -view | -v | | V3 | View<br>⤷**3D Views** | O3-Q5 |

**Command:** view

*Displays dialog box.*

## DIALOG BOX OPTIONS

**Named Views** tab

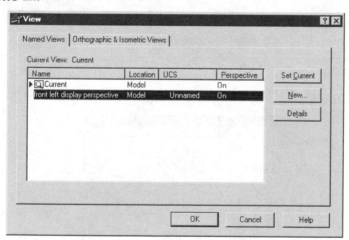

**Name** lists the names of saved views in the current drawing.

**Location** locates the view in Model or a Layout.

**UCS** names the UCS saved with the view.

**Perspective** specifies whether the view was saved in perspective view, was clipped, or neither.

Buttons

**Set Current** restores the named view.

**New** displays the **New View** dialog box.

**Details** displays the **View Details** dialog box.

**Orthographic & Isometric Views** tab

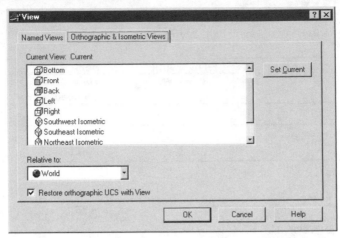

**Names** lists the names of standard orthographic and isometric views; an arrowhead points to the current view.

**Relative to** sets the selected view relative to the WCS or a named UCS.

**Restore orthographic UCS with View** restores the associated UCS.

**Set Current** sets the selected view; after clicking OK, AutoCAD automatically zooms to the extents of the view.

*New View* dialog box

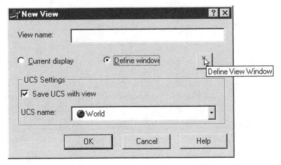

**View name** specifies the view name; up to 255 characters long.

**Current display** stores the current viewport as the named view.

**Define window** stores a windowed area as the named view.

**Define View Window** dismisses the dialog box temporarily, and prompts you to pick two corners that define the view.

UCS Settings options

**Save UCS with view** toggles the option to save a UCS with the named view.

**UCS name** selects the name of a UCS to store with the named view.

**View Details** dialog box

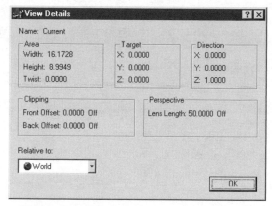

Relative to sets the selected view relative to the WCS or a named UCS.

## SHORTCUT MENU OPTIONS
*Right-click the list in the Named Views tab:*

**Set Current** sets the selected view as active.

**Rename** renames the selected view; you cannot rename the Current view.

**Delete** erases the selected view; you cannot delete the Current view.

**Details** displays the View Details dialog box.

## +VIEW Command
**Command:** +view
**Tab index <0>:** *(Type 0 or 1.)*

## COMMAND LINE OPTIONS
**Tab index** specifies the tab of the **View** dialog box to display:

| Index | Meaning |
|-------|---------|
| 0 | Named Views tab (default). |
| 1 | Orthographic & Isometric Views tab. |

## -VIEW Command

**Command:** -view

**Enter an option [?/Orthographic/Delete/Restore/Save/Ucs/Window]:** *(Enter an option.)*

### COMMAND LINE OPTIONS

**?** lists the names of views saved in the current drawing.

**Delete** deletes a named view.

**Restore** restores a named view.

**Save** saves the current view with a name.

**Ucs** saves the current UCS with the view.

**Window** saves a windowed view with a name.

Orthographic options

**Enter an option [Top/Bottom/Front/BAck/Left/Right] <Top>:** *(Enter an option.)*

**Restoring Model space view.**

**Select Viewport for view:** *(Pick a viewport.)*

**Regenerating model.**

**Enter an option** selects a standard orthographic view for the current viewport: Top, Bottom, Front, BAck, Left, or Right.

**Select Viewport for view** selects the viewport — in either Model or Layout tab — in which to apply the orthographic view.

### RELATED COMMANDS

**Rename** changes the names of views via a dialog box.

**UCS** creates and displays user-defined coordinate systems.

**PartialLoad** loads portions of drawings based on view names.

**Open** opens drawings and optionally starts with a named view.

**Plot** plots named views.

### RELATED SYSTEM VARIABLES

**ViewCtr** specifies the coordinates of the center of the view.

**ViewSize** specifies the height of the view.

### TIPS

• Name views in your drawing to move quickly from one detail to another.

• The **Plot** command plots named views of a drawing.

• Objects outside of the window created by the **Window** option may be displayed, but are not plotted.

• You create separate views in model and paper space; when listing named views with ?, AutoCAD indicates an 'M' or 'P' next to the view name.

# ViewRes

**V. 2.5** Controls the roundness of curved objects; determines whether zooms and pans are performed as redraws or regens (*short for VIEW RESolution*).

| Command | Alias | Ctrl+ | F-key | Alt+ | Menu Bar | Tablet |
|---------|-------|-------|-------|------|----------|--------|
| viewres | ... | ... | ... | ... | ... | ... |

**Command:** viewres
**Do you want fast zooms? [Yes/No] <Y>:** *(Type Y or N.)*
**Enter circle zoom percent (1-20000) <100>:** *(Enter a value.)*

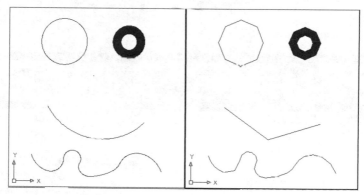

*Circle zoom percent = 100 (at left) and 1 (at right).*

## COMMAND LINE OPTIONS

**Do you want fast zooms?**

**Yes** AutoCAD tries to make every zoom and pan a redraw (faster).

**No** every zoom and pan causes a regeneration (slower).

**Enter circle zoom percent** specifies that smaller values display faster, but makes circles look less round (see figure); default = 100.

## RELATED SYSTEM VARIABLE

**WhipArc** toggles the display of circles and arcs as vectors or as true, rounded objects.

## RELATED COMMAND

**RegenAuto** determines whether AutoCAD uses redraws or regens.

· · · · · · · · · · · · · · · · · · · · · · · · · · · · · · · · · · · · · · · · · · · · · · · · · ·

## Removed Command

**VIConv** was removed from AutoCAD Release 14; use **3dsIn** and **3dsOut** instead.

· · · · · · · · · · · · · · · · · · · · · · · · · · · · · · · · · · · · · · · · · · · · · · · · · ·

# VLisp

**2000** Opens the VLisp integrated development environment (*short for Visual LISP*).

| Commands | Alias | Ctrl+ | F-key | Alt+ | Menu Bar | Tablet |
|----------|-------|-------|-------|------|----------|--------|
| vlisp | vlide | ... | ... | TSV | Tools | ... |
| | | | | | ↳AutoLISP | |
| | | | | | ↳Visual LISP Editor | |

vlide

**Command:** vlisp

*Displays window:*

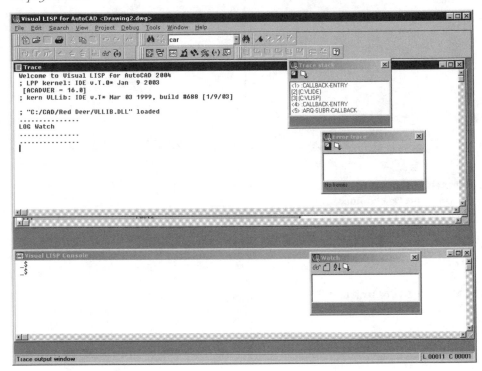

## MENU BAR OPTION

*Select* **Help | Visual LISP Help Topics** *for assistance in using this VLISP window.*

## RELATED COMMAND

**AppLoad** loads Visual LISP applications, as well as programs written in AutoLISP and other APIs.

## TIP

• Sample VLisp code can be found in the \*autocad 2004*\*sample*\*vlisp* folder.

 # VpClip

**2000** Clips a layout viewport *(short for ViewPort CLIPping)*.

| Command | Alias | Ctrl+ | F-key | Alt+ | Menu Bar | Tablet |
|---------|-------|-------|-------|------|----------|--------|
| vpclip | ... | ... | ... | ... | Modify | ... |
| | | | | | ⌐Clip | |
| | | | | | ⌐Viewport | |

**Command:** vpclip
**Select viewport to clip:** *(Pick a viewport.)*
**Select clipping object or [Polygonal] <Polygonal>:** *(Select an object, or type **P**.)*
*The selected viewport disappears, and is replaced by the new clipped viewport.*

## COMMAND LINE OPTIONS

**Select viewport to clip** selects the viewport that will be clipped.

**Select clipping object** selects the object that defines the clipping boundary: closed polyline, circle, ellipse, closed spline, or region.

Polygonal options
**Specify start point:** *(Pick a point.)*
**Specify next point or [Arc/Close/Length/Undo]:** *(Pick a point, or enter an option.)*
**Specify next point or [Arc/Close/Length/Undo]:** *(Type **C** to close.)*

**Specify start point** specifies the starting point for the polygon.

**Arc** draws an arc segment; see the Arc command.

**Close** closes the polygon.

**Length** draws a straight segment of specified length.

**Undo** undoes the previous polygon segment.

## RELATED COMMAND

**Mview** creates rectangular and polygonal viewports in paper space.

## TIPS

- This command does not operate in Model tab.

- An example of clipped viewports:

# VpLayer

**V. 2.1** Controls the visibility of layers in viewports, when TileMode is turned off (*short for ViewPort LAYER*).

| Command | Alias | Ctrl+ | F-key | Alt+ | Menu Bar | Tablet |
|---------|-------|-------|-------|------|----------|--------|
| vplayer | ... | ... | ... | ... | ... | ... |

**Command:** vplayer
**Enter an option [?/Freeze/Thaw/Reset/Newfrz/Vpvisdflt]:** *(Enter an option.)*
**Select a viewport:** *(Pick a viewport.)*

## COMMAND LINE OPTIONS

**Freeze** indicates the names of layers to freeze in this viewport.

**Newfrz** creates new layers that are frozen in all newly-created viewports (short for NEW FReeZe).

**Reset** resets the state of layers based on the Vpvisdflt settings.

**Thaw** indicates the names of layers to thaw in this viewport.

**Vpvisdflt** determines which layers will be frozen in a newly-created viewport and default visibility in existing viewports (short for ViewPort VISibility DeFauLT).

**?** lists the layers frozen in the current viewport.

## RELATED COMMANDS

**Layer** creates and controls layers in all viewports.

**MView** creates and joins viewports when tilemode is off.

## RELATED SYSTEM VARIABLE

**TileMode** controls whether viewports are tiled or overlapping.

# VPoint

**Rel.10** Changes the viewpoint of 3D drawings (*short for ViewPOINT*).

| Command | Alias | Ctrl+ | F-key | Alt+ | Menu Bar | Tablet |
|---------|-------|-------|-------|------|----------|--------|
| vpoint | -vp | ... | ... | V3V | View | N4 |
| | | | | | ↳3D Views | |
| | | | | | ↳VPOINT | |

**Command:** vpoint
**Current view direction: VIEWDIR=0.0000,0.0000,1.0000**
**Specify a view point or [Rotate] <display compass and tripod>:** *(Enter an option, or press **Enter** for the compass tripod.)*

## COMMAND LINE OPTIONS

**Specify a view point** indicates the new 3D viewpoint by coordinates.

**Rotate** indicates the new 3D viewpoint by angle.

ENTER brings up visual guides (see figure).

## RELATED COMMANDS

**DdVpoint** adjusts the viewpoint via a dialog box.

**DView** changes the viewpoint of 3D objects, and allows perspective mode.

**3dOrbit** rotates the viewpoint in real-time.

## RELATED SYSTEM VARIABLES

**VpointX** is the x-coordinate of the current 3D view.

**VpointY** is the y-coordinate of the current 3D view.

**VpointZ** is the z-coordinate of the current 3D view.

**WorldView** determines whether VPoint coordinates are in WCS or UCS.

## TIPS

- The *compass* represents the globe, flattened to two dimensions:
  - The north pole $(0, 0, z)$ is in the center.
  - The equator $(x, y, 0)$ is the inner circle
  - The south pole $(0, 0, -z)$ is the outer circle.
- As the cursor is moved on the compass, the *axis tripod* rotates showing the 3D view direction.
- To select the view direction, pick a location on the globe and press the pick button.

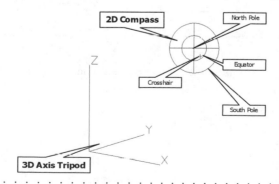

# VPorts

**V. 2.0** Creates viewports of the current drawing, when TileMode is on (*short for ViewPORTS*).

| Commands | Alias | Ctrl+ | F-key | Alt+ | Menu Bar | Tablet |
|----------|-------|-------|-------|------|----------|--------|
| vports | viewports | R | ... | VV | View<br>⤷Viewports | M3-4 |
| **+vports** | | | | | | |
| **-vports** | | | | VV1 | View<br>⤷Viewports<br>⤷1 Viewport | |

**Command:** vports

*Displays tabbed dialog box.*

## DIALOG BOX OPTIONS

**New Viewports** tab

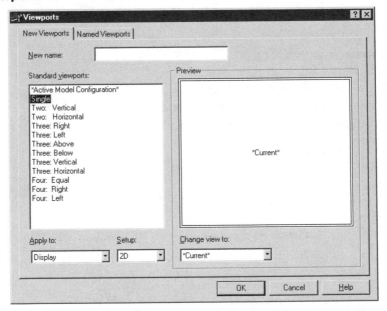

In model space

**New name** specifies the name for the viewport configuration; can be up to 255 characters long.

**Standard viewports** lists the available viewport configurations.

**Preview** displays a preview of the viewport configuration.

**Apply to** applies the viewport configuration to:

- Display.
- Current Viewport.

**Setup** selects 2D or 3D configuration; the 3D option applies orthogonal views, such as top, left, and front.

**Change view to** selects the type of view; in 3D mode, selects a standard orthoganal view.

In paper space

**Viewport spacing** specifies the spacing between the floating viewports; default = 0 units.

**Named Viewports** tab

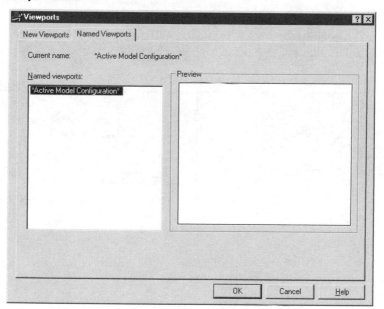

**Named viewports** lists the names of saved viewport configurations.

. . . . . . . . . . . . . . . . . . . . . . . . . . . . . . . . . . . . . . . . . . . . .

## +VPORTS Command
**Command:** +vports
**Tab index <0>:** *(Type* **0** *or* **1.***)*

**Tab index** specifies the tab to display:

| Index | Meaning |
|-------|---------|
| 0 | New Viewports tab (default). |
| 1 | Named Viewports tab. |

## -VPORTS Command

**Command:** -vports

*In layout mode, displays **MView** command prompts. In model space, prompts:*

**Enter an option [Save/Restore/Delete/Join/SIngle/?/2/3/4] <3>:** *(Enter an option.)*

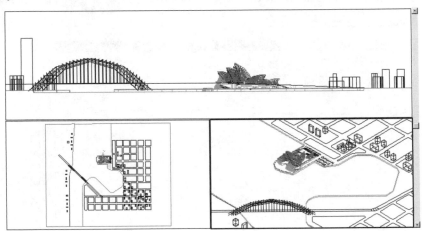

## COMMAND LINE OPTIONS

In model space

**Save** saves the settings of a viewport by name.

**Restore** restores a viewport definition.

**Delete** deletes a viewport definition.

**Join** joins two viewports together as one when they form a rectangle.

**SIngle** joins all viewports into a single viewport.

**?** lists the names of saved viewport configurations.

**4** divides the current viewport into four.

2 options

**Horizontal** creates one viewport over another.

**Vertical** creates one viewport beside another (default).

3 options

**Horizontal** creates three viewports over each other.

**Vertical** creates three viewports beside each other.

**Above** creates one viewport overtop of two viewports.

**Below** creates one viewport below two viewports.

**Left** creates one viewport left of two viewports.

**Right** creates one viewport right of two viewports (default).

In paper space

**Specify corner of viewport or [ON/OFF/Fit/Hideplot/Lock/Object/Polygonal/Restore/2/3/4]<Fit>:** *(Enter an option.)*

**ON** turns on the viewport; the objects in the viewport become visible.

**OFF** turns off the viewport; the objects in the viewport become invisible.

**Fit** creates one viewport that fills the display area.

**Hideplot** removes hidden lines when plotting in layout mode.

**Lock** locks the viewport, so that no editing can take place.

**Object** converts a closed polyline, ellipse, spline, region, or circle into a viewport.

**Polygonal** creates an non-rectangular viewport

*Other options are identical to those displayed in model tab.*

## RELATED COMMANDS

**MView** creates viewports in paper space.

**RedrawAll** redraws all viewports.

**RegenAll** regenerates all viewports.

**VpClip** clips a viewport.

## RELATED SYSTEM VARIABLES

**CvPort** is the current viewport number.

**MaxActVp** limits the maximum number of active viewports.

**TileMode** controls whether viewports can be overlapped or tiled.

## TIPS

- The **Join** option joins two viewports only when they form a rectangle.
- You can restore saved viewport arrangements in paper space using the **MView** command.
- Many display-related commands (such as **Redraw** or **Grid**) affect the current viewport only.

# VSlide

V. 2.0 Displays slide files in the current viewport (*short for View SLIDE*).

| Command | Alias | Ctrl+ | F-key | Alt+ | Menu Bar | Tablet |
|---------|-------|-------|-------|------|----------|--------|
| vslide  | ...   | ...   | ...   | ...  | ...      | ...    |

**Command:** vslide

*Displays **Select Slide File** dialog box. Select an .sld file, and then click **Open**.*

## COMMAND LINE OPTIONS
*None.*

## RELATED COMMANDS
**MSlide** creates slide files of the current viewport.

**Redraw** erases slides from the screen.

## RELATED AUTODESK PROGRAM
*slidelib.exe* creates an SLB-format library file of a group of slide files.

## RELATED AUTOCAD FILES
***.sld** stores individual slide files.

***.slb** stores a library of slide files.

## TIPS
* For faster viewing of a series of slides, placing an asterisk proceeding the **VSlide** command preloads the *.sld* slide file, as in:

  **Command:** *vslide filename

* Use the following format to display a specific slide stored in an SLB slide library file:

  **Command:** vslide
  **Slide file:** acad.slb(slidefilename)

# WBlock

**V. 1.4** Writes blocks or entire drawings to disk (*short for Write BLOCK*).

| Commands | Aliases | Ctrl+ | F-key | Alt+ | Menu Bar | Tablet |
|----------|---------|-------|-------|------|----------|--------|
| wblock | w | ... | ... | FE | File | ... |
|  |  |  |  | ⅁DWG | ⅁Export |  |
|  |  |  |  |  | ⅁Block |  |
| -wblock | -w |  |  |  |  |  |

**Command:** wblock

*Displays dialog box:*

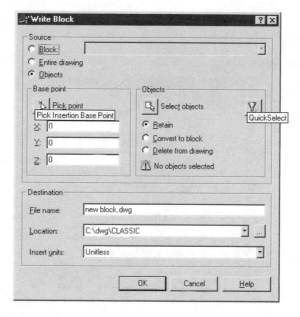

## DIALOG BOX OPTIONS

Source options

**Block** specifies the name of a block to save as a *.dwg* file.

**Entire drawing** selects the current drawing to save as a *.dwg* file.

**Objects** specifies the objects from the drawing to save as a *.dwg* file.

Base point options

**Pick Insertion Base Point** dismisses the dialog box temporarily to allow you to select the insertion base point.

**X** specifies the x coordinate of the insertion point.

**Y** specifies the y coordinate of the insertion point.

**Z** specifies the z coordinate of the insertion point.

Objects options

**Select Objects** dismisses the dialog box temporarily to allow you to select one or more objects.

**Quick Select** displays the Quick Select dialog box; see the QSelect command.

**Retain** retains the selected objects in the current drawing after saving them as a file.

**Convert to block** converts the selected objects to a block in the drawing, after saving them as a *.dwg* file; names the block under File name in the Destination section.

**Delete from drawing** deletes the selected objects from the drawing, after saving them as a *.dwg* file.

Destination options

**File name** specifies a file name for the block or objects.

**Location** specifies a path.

**...** displays the Browse for Folder dialog box.

**Insert units**   specifies the units when the *.dwg* file is inserted as a block.

***Browse for Folder*** *dialog box*

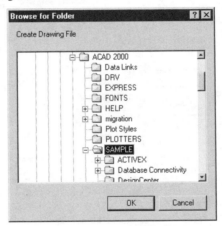

**Create Drawing File** selects the drive and folder in which to store the *.dwg* file extracted from the current drawing.

## -WBLOCK Command

**Command:** -wblock

*Displays the **Create Drawing File** dialog box. Name the file, and then click **Save**.*

**Enter name of existing block or**

**[= (block=output file)/* (whole drawing)] <define new drawing>:** *(Enter name, or use = and * options.)*

### COMMAND LINE OPTIONS

**Enter name of existing block** specifies the name of a current block in the drawing.

**=** *(equals)* writes block to a *.dwg* file, using block's name as file name.

**\*** *(asterisk)* writes the entire drawing to a *.dwg* file.

ENTER creates a block on disk of the selected objects.

SPACEBAR moves the selected objects to the specified drawing.

### RELATED COMMANDS

**Block** creates a block of a group of objects.

**Insert** inserts a block or another drawing into the drawing.

### RELATED SYSTEM VARIABLES

*None.*

### TIPS

- Use the **WBlock** command to extract blocks from the drawing and store them on a disk drive. This allows the creation of a block library.

- Support for the DesignXML (extended markup language) format was withdrawn from AutoCAD 2004.

# Wedge

<u>Rel.11</u>   Draws 3D wedges as solid models.

| Command | Alias | Ctrl+ | F-key | Alt+ | Menu Bar | Tablet |
|---------|-------|-------|-------|------|----------|--------|
| wedge | we | ... | ... | DIW | Draw | N7 |
| | | | | | ↳Solids | |
| | | | | | ↳Wedge | |

**Command:** wedge
**Specify first corner of wedge or [CEnter] <0,0,0>:** *(Pick a point, or type CE.)*
**Specify corner or [Cube/Length]:** *(Pick a point, or enter an option.)*
**Specify height:** *(Pick a point.)*

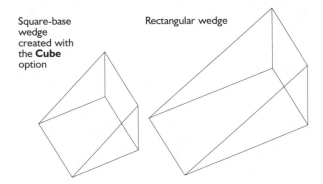

Square-base wedge created with the **Cube** option

Rectangular wedge

**COMMAND LINE OPTIONS**

**Corner** draws the wedge's base between two pick points.

**CEnter** draws the wedge's base about the center of the sloped face.

**Cube** draws a cubic wedge.

**Length** specifies the length, width, and height of the wedge.

CEnter options
**Specify center of wedge <0,0,0>:** *(Pick a point.)*
**Specify opposite corner or [Cube/Length]:** *(Pick a point, or enter an option.)*
**Specify height:** *(Pick a point.)*

**Specify center of wedge** indicates the midpoint of the wedge's base.

**Specify opposite corner** indicates the distance from the midpoint to one corner.

Cube options
**Specify corner or [Cube/Length]:** *(Type C.)*
**Specify length:** *(Specify the length.)*

**Specify length** indicates the length of all three sides.

Length options

**Specify corner or [Cube/Length]:** *(Type **L**.)*
**Specify length:** *(Specify the length.)*
**Specify width:** *(Specify the width.)*
**Specify height:** *(Specify the height.)*

> **Specify length** indicates the length along the x-axis.
>
> **Specify width** indicates the width along the y-axis.
>
> **Specify height** indicates the height along the z-axis.

## RELATED COMMANDS

**Ai_Wedge** draws a wedge as a 3D surface model.

**Box** draws solid boxes.

**Cone** draws solid cones.

**Cylinder** draws solid cylinders.

**Sphere** draws solid spheres.

**Torus** draws solid tori.

## RELATED SYSTEM VARIABLES

*None.*

## TIPS

- *Length* means size in the x-direction.

- *Width* means size in the y-direction.

- *Height* means size in the z-direction.

- Use negative values for length, width, and height to draw the wedge in the negative x, y, and z directions

- The **IsoLines** system variable has no effect on wedges.

# WhoHas

Determines which computer has drawings open.

| Command | Alias | Ctrl+ | F-key | Alt+ | Menu Bar | Tablet |
|---------|-------|-------|-------|------|----------|--------|
| whohas | ... | ... | ... | ... | ... | ... |

**Command:** whohas

*Displays the **Select Drawing to Query** dialog box. Select a drawing file, and then click **Open**.*

*When the drawing is open, reports:*

**Owner: ralphg**

**Computer's Name : HEATHER**

**Time Acessed : Monday, March 3, 2004 11:27:28 AM**

*When the drawing is not open, reports:*

**User: unknown.**

## COMMAND LINE OPTIONS

*None.*

## RELATED COMMANDS

**Open** opens drawings.

**XAttach** attaches drawings that can be opened by other users.

## TIP

- This command is meant for use over networks as a convenient way to find out which users are editing specific drawings.

/ The Illustrated AutoCAD 2004 Quick Reference

# WipeOut

2004 Fills areas with the background color to "wipe out" portions of drawings; meant for use underneath text.

| Command | Alias | Ctrl+ | F-key | Alt+ | Menu Bar | Tablet |
|---------|-------|-------|-------|------|----------|--------|
| wipeout | ... | ... | ... | ... | ... | ... |

**Command:** wipeout
**Specify first point or [Frames/Polyline] <Polyline>:** *(Pick a point, or enter an option.)*
**Specify next point:** *(Pick a point.)*
**Specify next point or [Undo]:** *(Pick a point, or type **U**.)*
**Specify next point or [Close/Undo]:** *(Pick a point, or type **C** or **U**.)*
**Specify next point or [Close/Undo]:** *(Type **C** to end the command.)*

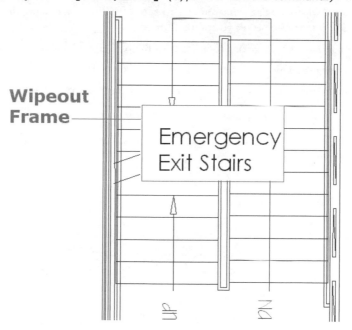

Wipeout Frame

Emergency Exit Stairs

## COMMAND LINE OPTIONS

**Specify first point** specifies the starting point of the polygon.

**Undo** undoes the last segment.

**Close** closes the polygon.

Frames options
**Enter mode [ON/OFF] <ON>:** *(Type **ON** or **OFF**.)*

**ON** turns on the wipeout boundary.

**OFF** turns off the boundary polygon.

Polyline options
**Select a closed polyline:** *(Pick a closed polyline.)*
**Erase polyline? [Yes/No] <No>:** *(Type **Y** or **N**.)*

Select a closed polyline picks a polyline that forms the wipeout boundary.

**Erase polyline?**

**Yes** erases the polyline.

**No** leaves the polyline in place.

## RELATED COMMAND

**DrawOrder** displays overlapping objects in a different order.

## TIPS

- Wipeout boundaries can be edited with grips editing.

- When the **Frames** option is turned off, wipeouts cannot be edited.

- The **Frames** option applies to all wipeouts in the drawing; you cannot turn frames on and off for individual wipeouts. The workaround is to turn off all frames, and use then the **Polyline** option for those wipeouts you prefer framed

- The wipeout consists of a image object drawn with the background color.

- To create a rectangular wipeout frame, turn on snap and ortho modes.

- To make text appear above the wipeout, use the **DrawOrder** command to move the text to the **Front**.

# WmfIn

**Rel.12** Imports WMF and CLP files (*short for Windows MetaFile IN*).

| Command | Alias | Ctrl+ | F-key | Alt+ | Menu Bar | Tablet |
|---------|-------|-------|-------|------|----------|--------|
| wmfin | ... | ... | ... | IW | Insert | ... |
| | | | | | ↳Windows Metafile | |

**Command:** wmfin

*Displays the* **Import WMF** *dialog box. Select a file, and the click* **Open.**

**Specify insertion point or [Scale/X/Y/Z/Rotate/PScale/PX/PY/PZ/PRotate]:** *(Pick a point, or enter an option.)*

**Enter X scale factor, specify opposite corner, or [Corner/XYZ] <1>:** *(Specify a value, pick a point, or enter an option.)*

**Enter Y scale factor <use X scale factor>:** *(Specify a value, or press* **Enter.***)*

**Specify rotation angle <0>:** *(Specify a value, or press* **Enter.***)*

## COMMAND LINE OPTIONS

**Insertion point** picks the insertion point of the lower-left corner of the WMF image.

**X scale factor** scales the WMF image in the x direction (default = 1).

**Corner** scales the WMF image in the x and y directions.

**XYZ** scales the image in the x, y, and z directions.

**Y scale factor** scales the image in the y direction (default = x scale).

**Rotation angle** rotates the image (default = 0).

## RELATED COMMANDS

**WmfOpts** controls the importation of WMF files.

**WmfOut** exports selected objects in WMF format.

## RELATED FILES

*\*.clp* are Windows Clipboard files.

*\*.wmf* are Windows Metafiles.

## TIPS

* The WMF image is placed as a block with the name **WMF0**; subsequent placements of WMF files increment the digit: **WMF1**, **WMF2**, and so on.

* Exploding the WMF*n* block results in polylines; even circles, arcs, and text are converted to polylines; solid-filled areas are exploded into solid triangles.

* The *.clp* file is created by the Windows Clipboard. After using **CTRL+C** to copy objects to the Clipboard, you can open the Clipboard Viewer, and then save the image as a *.clp* file.

* The *.clp* support is undocumented by Autodesk; the **WmfOpts** command has no effect on imported *.clp* files.

* The *.clp* file is pasted as a block; when exploded, constituent parts are 2D polylines.

# WmfOpts

**Rel.12** Controls the importation of WMF files (*short for Windows Meta File OPTionS*).

| Command | Alias | Ctrl+ | F-key | Alt+ | Menu Bar | Tablet |
|---------|-------|-------|-------|------|----------|--------|
| wmfopts | ... | ... | ... | IWP | Insert | ... |
| | | | | | ⤷Windows Metafile | |
| | | | | | ⤷Options | |

**Command:** wmfopts

*Displays dialog box:*

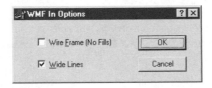

## DIALOG BOX OPTIONS

**Wire Frame**

☑ displays the *.wmf* file with lines only, no filled areas (default).

☐ displays area fills.

**Wide Lines**

☑ displays lines with width (default).

☐ displays lines with a width of zero.

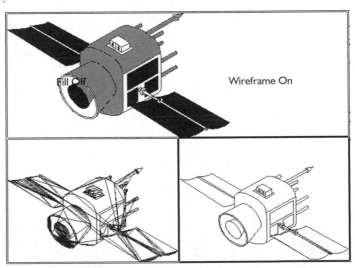

## RELATED COMMANDS

**WmfIn** imports *.wmf* files.

**WmfOut** exports selected objects in *.wmf* format.

# WmfOut

Rel.12 Exports selected objects as a WMF vector file *(short for Windows MetaFile OUTput)*.

| Command | Alias | Ctrl+ | F-key | Alt+ | Menu Bar | Tablet |
|---------|-------|-------|-------|------|----------|--------|
| wmfout | ... | ... | ... | FE | File | ... |
| | | | | ⌐WMF | ⌐Export | |
| | | | | | ⌐Metafile | |

**Command:** wmfout

*Displays the **Create WMF File** dialog box. Enter a file name, and then click **Save**.*
**Select objects:** *(Select one or more objects.)*
**Select objects:** *(Press **Enter** to end object selection.)*

## COMMAND LINE OPTION

**Select objects** selects the objects to export.

## RELATED SYSTEM VARIABLE

**WmfBkgnd** toggles the background color of an exported *.wmf* file:

| WmfBkgnd | Meaning |
|----------|---------|
| 0 | Transparent background. |
| 1 | AutoCAD background color. |

**WmfForegnd** switches the foreground and background colors of an exported *.wmf* file:

| WmfForegnd | Meaning |
|------------|---------|
| 0 | Foreground is darker than background color. |
| 1 | Background is darker than foreground color. |

## RELATED COMMANDS

**WmfOpts** controls the importation of *.wmf* files.

**WmfIn** imports files in *.wmf* format.

**CopyClip** copies selected objects to the Clipboard in several formats, including *.wmf* , also called "picture" format.

## TIPS

- *.wmf* files created by AutoCAD are resolution-dependent; small circles and arcs lose their roundness.

- The **All** selection does not select all objects in the drawing; instead, the **WmfOut** command selects all objects *visible* in the current viewport.

# XAttach

**Rel.14** Attaches externally-referenced drawings to the current drawing (*short for eXternal reference ATTACH*).

| Command | Alias | Ctrl+ | F-key | Alt+ | Menu Bar | Tablet |
|---------|-------|-------|-------|------|----------|--------|
| xattach | xa | ... | ... | IX | Insert | ... |
| | | | | | ⌐External Reference | |

**Command:** xattach

*Displays the **Select File to Attach** dialog box. Select a file, and click **Open**.*

*Displays dialog box:*

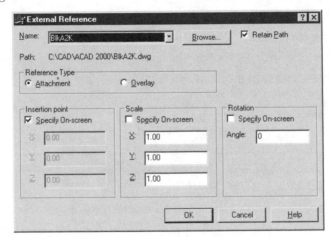

*After you click **OK**, AutoCAD confirms at the command line:*

**Attach Xref FILENAME: C:\filename.dwg**
**FILENAME loaded.**

## DIALOG BOX OPTIONS

**Name** specifies the file name of the external *.dwg* to be attached; the drop list shows the names of currently-attached xrefs (externally-referenced files).

**Browse** displays the **Select File To Attach** dialog box.

**Retain Path**

☑ saves the xref's filename and full path in the *.dwg* file.

☐ saves only the filename of the xref; when the xref cannot be found, AutoCAD searches the Support File Search Path and the ProjectName.

Reference Type options
- **Attachment** attaches the xref.
- **Overlay** overlays the xref.

## Insertion Point options

**Specify On-screen** specifies the insertion point of the xref in the drawing.

*After clicking OK to dismiss the dialog box; AutoCAD prompts you:*

**Specify insertion point or [Scale/X/Y/Z/Rotate/PScale/PX/PY/PZ/PRotate]:** *(Pick a point, or enter an option.)*

**Scale** sets the scale factor for the x, y, and z axes.

**X** sets the x-scale factor.

**Y** sets the y-scale factor.

**Z** sets the z-scale factor.

**Rotate** specifies the rotation angle.

**PScale** presets the scale factor for the x, y, and z axes.

**PX** presets the x-scale factor.

**PY** presets the y-scale factor.

**PZ** presets the z-scale factor.

**PRotate** presets the rotation angle.

## Scale options

**Specify On-screen** specifies the scale of the xref in the drawing.

*After clicking OK to dismiss the dialog box; AutoCAD prompts you:*

**Enter X scale factor, specify opposite corner, or [Corner/XYZ] <1>:** *(Enter a value, or enter an option.)*

**Enter Y scale factor <use X scale factor>:** *(Enter a value, or press* **Enter.***)*

**X scale factor** scales the xref in the x direction.

**Corner** indicates the x,y scale factor by picking two points of a rectangle.

**XYZ** specifies the scale factor in the x, y, and z directions.

**Y scale factor** scales the xref in the y direction.

## Rotation Angle options

**Specify On-screen** specifies the rotation of the xref in the drawing.

*After you click OK to dismiss the dialog box; AutoCAD prompts you:*

**Specify rotation angle <0>:** *(Enter a value, or press* **Enter.***)*

**Rotation angle** specifies the rotation angle of the xref.

## RELATED SYSTEM VARIABLES

**DemandLoad** specifies if and when AutoCAD demand-loads a third-party application when a drawing contains custom objects created by the application:

| DemandLoad | Meaning |
| --- | --- |
| 0 | Turns off demand loading. |
| 1 | Loads application when drawings contain proxy objects. |
| 2 | Loads application when the application's commands are invoked. |
| 3 | Loads application when drawings contains proxy objects, or when the application's commands are invoked. |

**IdxCtl** controls the creation of layer and spatial indices:

| IndexCtl | Meaning |
| --- | --- |
| 0 | Creates no indices (default). |
| 1 | Creates layer index. |
| 2 | Creates spatial index. |
| 3 | Creates both layer and spatial indices. |

**ProjectName** holds the project name for the current drawing (default = "").

**VisRetain** specifies how the layer settings — on-off, freeze-thaw, color, and linetype — in xref drawings are defined by the current drawing:

| VisRetain | Meaning |
| --- | --- |
| 0 | Xref layer definition in the current drawing takes precedence. |
| 1 | Settings for xref-dependent layers take precedence over xref layer definition in the current drawing. |

**XEdit** determines whether the drawing may be edited in-place, when being referenced by another drawing:

| XEdit | Meaning |
| --- | --- |
| 0 | Cannot be edited in-place. |
| 1 | Can be edited in-place. |

**XLoadCtl** controls the loading of xref drawings:

| XLoadCtl | Meaning |
| --- | --- |
| 0 | Loads the entire xref drawing. |
| 1 | Demand loading; xref is opened. |
| 2 | Demand loading; copy of the xref is opened. |

**XLoadPath** stores the path of temporary copies of demand-loaded xref drawings.

**XRefCtl** controls whether .*xlg* external reference log files are written:

| XRefCtl | Meaning |
| --- | --- |
| 0 | XLG file not written (default). |
| 1 | XLG log file written. |

## RELATED COMMANDS

**RefEdit** edits xref drawings.

**XBind** binds portions of xref drawings to the current drawing.

**XClip** clips the display of xrefs.

**XRef** attaches another drawing to the current drawing.

# XBind

__Rel.11__  Binds portions of externally-referenced drawings to the current drawing
(*short for eXternal BINDing*).

| Commands | Aliases | Ctrl+ | F-key | Alt+ | Menu Bar | Tablet |
|----------|---------|-------|-------|------|----------|--------|
| xbind | xb | ... | ... | MOEB | Modify<br>♦Object<br>  ♦External Reference<br>    ♦Bind | X19 |
| -xbind | -xb | | | | | |

**Command:** xbind

*Displays dialog box:*

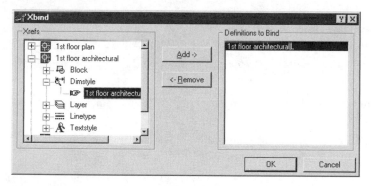

## DIALOG BOX OPTIONS

**Xrefs** lists xrefs, along with their bindable objects: blocks, dimension styles, layer names,
linetypes, and text styles.

**Definitions to Bind** lists definitions that will be bound.

Buttons

**Add** adds a definition to the binding list.

**Remove** removes a definition from the binding list.

## -XBIND Command

**Command:** -xbind

**Enter symbol type to bind [Block/Dimstyle/LAyer/LType/Style]:** *(Enter an option.)*

**Enter dependent name(s):** *(Enter one or more names, separated by commas.)*

### COMMAND LINE OPTIONS

**Block** binds blocks to the current drawing.

**Dimstyle** binds dimension styles to the current drawing.

**LAyer** binds layer names to the current drawing.

**LType** binds linetype definitions to the current drawing.

**Style** binds text styles to the current drawing.

**Enter dependent names** specifies the named objects to bind.

### RELATED SYSTEM VARIABLES

*None.*

### RELATED COMMANDS

**RefEdit** edits xref drawings.

**XRef** attaches another drawing to the current drawing.

### TIPS

- The **XBind** command lets you copy named objects from another drawing to the current drawing.

- Before you can use the **XBind** command, you must first use the **XAttach** command to attach an xref to the current drawing.

- Blocks, dimension styles, layer names, linetypes, and text styles are known as "dependent symbols."

- When a dependent symbol is part of an xrefed drawing, AutoCAD uses a vertical bar (|) to separate the xref name from the symbol name, as in *filename*|*layername*.

- After you use the **XBind** command, AutoCAD replaces the vertical bar with **$0$**, as in *filename***$0$***layername*. The second time you bind that layer from that drawing, **XBind** increments the digit, as in *filename***$1$***layername*.

- When the **XBind** command binds a layer with a linetype (other than Continuous), it automatically binds the linetype.

- When the **XBind** command binds a block — with a nested block, dimension style, layer, linetype, text style, and/or reference to another xref — it automatically binds those objects as well.

 # XClip

**Rel.12** Clips portions of blocks and externally-referenced drawings (*short for eXternal CLIP; formerly the **XRefClip** command*).

| Command | Alias | Ctrl+ | F-key | Alt+ | Menu Bar | Tablet |
|---------|-------|-------|-------|------|----------|--------|
| xclip | xc | ... | ... | ... | Modify | X18 |
| | | | | | ⃝Clip | |
| | | | | | ⃝Xref | |

**Command:** xclip
**Select objects:** *(Select one or more blocks or xrefs.)*
**Select objects:** *(Press **Enter** to end object selection.)*
**Enter clipping option**
**[ON/OFF/Clipdepth/Delete/generate Polyline/New boundary] <New>:** *(Enter an option.)*

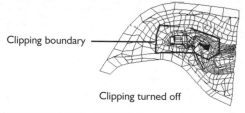

Clipping boundary

Clipping turned off

Clipping turned on

## COMMAND LINE OPTIONS

**Select objects** selects the xref or block, *not* the clipping polyline.

**ON** turns on clipped display.

**OFF** turns off clipped display; displays all of the xref or block.

**Clipdepth** sets front and back clipping planes for 3D xrefs and blocks.

**Delete** erases the clipping boundary.

**generate Polyline** draws a polyline over top of the clipping boundary.

**New boundary** places a new rectangular or irregular polygon clipping boundary, or creates an irregular clipping boundary from an existing polyline.

## RELATED COMMANDS

**XBind** bind parts of the xrefs drawing to the current drawing.

**Xref** displays an xref'ed drawing in the current drawing.

## RELATED SYSTEM VARIABLE

 **XClipFrame** toggles the display of the clipping boundary.

## TIPS

- While the old **XRefClip** command could not create an irregularly clipped xref, the new **XClip** command creates arbitrary clipping boundaries.

- The **XClip** command works for both blocks and xrefs.

- A spline-fit polyline results in a curved clip boundary, but a curve-fit polyline does not.

# XLine

**Rel.13**  Places infinitely long construction lines in drawings.

| Command | Alias | Ctrl+ | F-key | Alt+ | Menu Bar | Tablet |
|---------|-------|-------|-------|------|----------|--------|
| xline | xl | ... | ... | DT | Draw ⌐Construction Line | L10 |

**Command:** xline
**Specify a point or [Hor/Ver/Ang/Bisect/Offset]:** *(Pick a point, or enter an option.)*
**Through point:** *(Pick a point.)*
**Through point:** *(Press **Enter** to end the command.)*

### COMMAND LINE OPTIONS
**Specify a point** picks the midpoint for the xline.
**Through point** picks another point through which the xline passes.
**Ang** places the construction line at an angle.
**Bisect** bisects an angle with the construction line.
**From point** places the construction line through a point.
**Hor** places a horizontal construction line.
**Offset** places the construction line parallel to another object.
**Ver** places a vertical construction line.
ENTER exits the command.

Angle options
**Enter angle of xline (0) or [Reference]:** *(Enter an angle, or type **R**.)*
**Enter angle of xline** specifies the angle of the xline relative to the x-axis.
**Reference** specifies the angle relative to two points.

Bisect options
**Specify angle vertex point:** *(Pick a point.)*
**Specify angle start point:** *(Pick a point.)*
**Specify angle end point:** *(Pick a point.)*
**Specify angle vertex point** specifies the vertex of the angle.
**Specify angle start point** specifies the angle start point.
**Specify angle end point** specifies the angle end point.

Offset options

**Specify offset distance or [Through] <1.0000>:** *(Enter a distance, or type* **T***.)*
**Select a line object:** *(Select a line, xline, ray, or polyline.)*
**Specify side to offset:** *(Pick a point.)*

**Specify offset distance** specifies the distance between xlines.

**Through** picks a point through which the xline should pass.

**Select a line object** selects the line, xline, ray, or polyline to offset.

**Specify side to offset** specifies the offset side.

## RELATED COMMANDS

**Properties** modifies characteristics of xline and ray objects.

**Ray** places a semi-infinite construction line.

## RELATED SYSTEM VARIABLE

**OffsetDist** specifies the current offset distance.

## TIPS

• Use xlines to find the bisectors of triangles (using **MIDpoint** object snap), or to create intersection snap points (using the **INTersection** object snap).

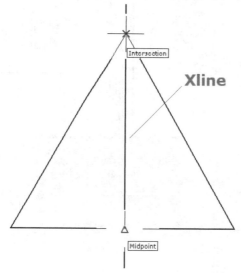

• Ray and xline construction lines are plotted; they do not affect the extents.

# XOpen

**2004** Opens externally-referenced drawings in new windows (*short for eXternal OPEN*).

| Command | Alias | Ctrl+ | F-key | Alt+ | Menu Bar | Tablet |
|---------|-------|-------|-------|------|----------|--------|
| xopen | ... | ... | ... | ... | ... | ... |

**Command:** xopen
**Select xref:** *(Select an externally-referenced drawing.)*

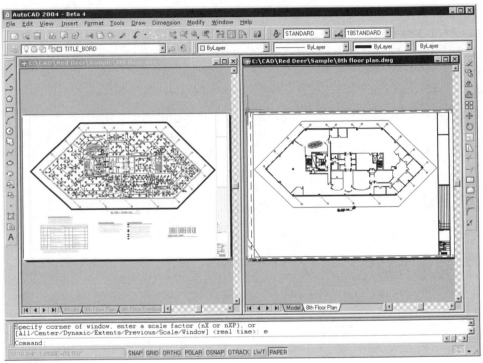

*Original drawing (left), and "xopened" drawing in separate window (right).*

## COMMAND LINE OPTION

**Select xrefs** selects the xref drawing to open; it must exist in the drawing.

## TIPS

• This command does not work with blocks.

• When you select a non-xref object, AutoCAD complains, "Object is not an Xref."

# Xplode

Rel.12 Explodes complex objects into simpler objects, with user control (*short for eXPLODE*).

| Command | Alias | Ctrl+ | F-key | Alt+ | Menu Bar | Tablet |
|---------|-------|-------|-------|------|----------|--------|
| xplode | xp | ... | ... | ... | ... | ... |

**Command:** xplode
**Select objects to XPlode.**
**Select objects:** *(Select one or more objects.)*
**Select objects:** *(Press **Enter** to end object selection.)*
**Enter an option [Individually/Globally] <Globally>:** *(Type **I** or **G**.)*
**Enter an option [All/Color/LAyer/LType/Inherit from parent block/Explode]**
**<Explode>:** *(Enter an option.)*

*Block and polyline (at left); exploded (at right).*

## COMMAND LINE OPTIONS

**Select objects** selects objects to be exploded.

**Individually** allows you to specify options for each selected object.

**Globally** applies options to all selected objects.

**All** displays all prompts individually with the Color, LAyer, and LType options.

**Color** specifies a single color for all objects after they are exploded: red, yellow, green, cyan, blue, magenta, white, bylayer, byblock, or any color number.

**LWeight** specifies a lineweight.

**LAyer** specifies the layer name for the exploded objects.

**LType** specifies any loaded linetype name for the exploded objects.

**Inherit from parent block** assigns the color, linetype, lineweight, and layer to the exploded objects, based on the original object.

**Explode** reduces the complex object into its components.

## RELATED COMMANDS

**Explode** explodes the object without options.

**U** reverses the explosion.

## RELATED SYSTEM VARIABLES

*None.*

**TIPS**

- Examples of complex objects include blocks and polylines; examples of simple objects include lines, circles, and arcs.

- Blocks with unequal scale factors cannot be exploded with this command; in its place, use the **Explode** command.

- Mirrored blocks can be exploded.

- The **LWeight** option is not displayed when the lineweight is off.

- The 'Enter an option [Individually/Globally]' option appears only when more than one valid object is selected for explosion.

- Specifying **BYLayer** for the color or linetype means that the exploded objects take on the color or linetype of the object's original layer.

- Specifying **BYBlock** for the color or linetype means that the exploded objects take on the color or linetype of the original object.

- The default layer is the current layer, not the exploded object's original layer.

- The **XPlode** command breaks down complex objects as follows:

| Object | Exploded into |
|---|---|
| **Block** | Component objects. |
| **Attributes** | Attribute values are deleted; displays attribute definitions. |
| **2D polyline** | Line and arc segments; width and tangency are lost. |
| **3D polyline** | Line segments. |
| **Leaders** | Line segments, splines, mtext, and tolerance objects; arrowheads become solids or blocks. |
| **Mtext** | Text. |
| **Multiline** | Line and arc segments. |
| **Polyface mesh** | Point, line, or 3D faces. |
| **Region** | Lines, arcs, and splines. |
| **3D solids** | Planar surfaces become regions; nonplanar surfaces become bodies. |
| **3D bodies** | Single-surface body, regions, or curves. |

- The **Inherit** option works only when the parts were originally drawn with color, linetype, and lineweight set to BYBLOCK, and drawn on layer 0.

 # XRef

**Rel.11** Attaches externally-referenced drawings to the current drawing (*short for eXternal REFerence*).

| Commands | Aliases | Ctrl+ | F-key | Alt+ | Menu Bar | Tablet |
|----------|---------|-------|-------|------|----------|--------|
| xref | xr | ... | ... | IR | Insert | T4 |
| | | | | | ⬘Xref Manager | |
| -xref | -xr | | | | | |

**Command:** xref

*Displays dialog box:*

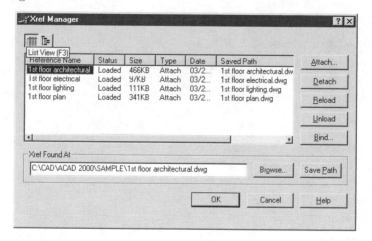

## DIALOG BOX OPTIONS

Buttons

**Attach** attaches a drawing as an xref (externally-referenced drawing); displays the Attach Xref dialog box; see the XAttach command.

**Detach** detaches xrefs.

**Reload** reloads and displays the most-recently saved version of the xref.

**Unload** unloads the xref; does not remove it permanently; rather it does not display the xref.

**Bind** binds named objects — blocks, dimension styles, layer names, linetypes, and text styles — to the current drawing; displays the Bind Xrefs dialog box; see the XBind command.

Additional options

**Xref Found At** displays the path to the xref file.

**Browse** selects a path or file name; displays the Select New Path dialog box.

**Save Path** saves the path displayed by the Xref Found At option.

## -XREF Command

**Command:** -xref

**Enter an option [?/Bind/Detach/Path/Unload/Reload/Overlay/Attach]
<Attach>:** *(Enter an option.)*

### COMMAND LINE OPTIONS

**?** lists the names of xref files.

**Bind** makes the xref drawing part of the current drawing.

**Detach** removes xref files.

**Path** respecifies paths to xref drawings.

**Unload** unloads xref files.

**Reload** updates the xref files.

**Overlay** overlays the xref files.

**Attach** attaches another drawing to the current drawing.

### RELATED TOOLBAR ICONS

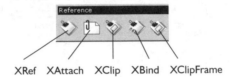

XRef    XAttach    XClip    XBind    XClipFrame

### RELATED COMMANDS

**Insert** adds another drawing to the current drawing.

**RefEdit** edits xref drawings.

**XBind** binds parts of xrefs to the current drawing.

**XClip** clip portions of xrefs.

### RELATED SYSTEM VARIABLES

*See the XAttach command.*

### TIPS

*   *Caution!* Nested xrefs cannot be unloaded.

*   Tree view shows how the xrefs are nested:

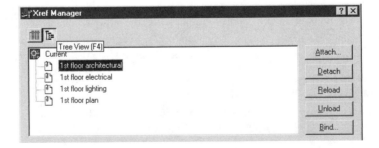

# 'Zoom

**V. 1.0** Makes drawings larger or smaller in the current viewport.

| Command | Aliases | Ctrl+ | F-key | Alt+ | Menu Bar | Tablet |
|---------|---------|-------|-------|------|----------|--------|
| 'zoom | z | ... | ... | VZ | View | K11 |
| | rtzoom | | | | ↳Zoom | |

**Command:** zoom
**Specify corner of window, enter a scale factor (nX or nXP), or**
**[All/Center/Dynamic/Extents/Previous/Scale/Window] <real time>:** *(Pick a point, enter an option, or press **Enter**.)*

## COMMAND LINE OPTIONS

*(pick a point)*   begins the Window option.

**realtime** press ENTER to start real-time zoom.

ENTER *or* ESC ends real-time zoom.

**All** displays the drawing limits or extents, whichever is greater.

**Dynamic** brings up the dynamic zoom view.

**Extents** displays the current drawing extents.

**Previous** displays the previous view generated by Pan, View, or Zoom.

**Vmax** displays the current virtual screen limits (short for Virtual MAXimum; undocumented option).

**Window** indicates the two corners of the new view.

· · · · · · · · · · · · · · · · · · · · · · · · · · · · ·

Center options
**Specify center point:** *(Pick a point.)*
**Enter magnification or height <>:** *(Enter a value.)*

**Center point** indicates the center point of the new view.

**Enter magnification or height** indicates a magnification value or height of view.

Left options (undocumented)
**Lower left corner point:** *(Pick a point.)*
**Enter magnification or height <>:** *(Enter a value.)*

**Lower left corner point** indicates the lower-left corner of the new view.

**Enter magnification or height** indicates a magnification value or height of view.

Scale(X/XP) options

*n***X** displays a new view as a factor of the current view.

*n***XP** displays a paper space view as a factor of model space.

## RIGHT-CLICK OPTIONS

*During real-time zoom, right-click in the drawing:*

Exit real-time zoom mode: ——— Exit
Real-time pan: ——— Pan
Real-time zoom: ——— ✔ Zoom
3D orbit mode: ——— 3D Orbit
Zoom window: ——— Zoom Window
Original view: ——— Zoom Original
Zoom to drawing extents: ——— Zoom Extents

## RELATED COMMANDS

**DsViewer** displays Aerial View window.

**Pan** moves the view to a different location.

**View** saves zoomed views by name.

**3dZoom** performs real-time zooms in perspective viewing mode.

## RELATED SYSTEM VARIABLES

**ViewCtr** specifies the coordinates of the current view's center point.

**ViewSize** specifies the height of the current view.

## TIPS

- A zoom factor of 1 displays the entire drawing as defined by the limits; a zoom factor of 2 enlarges objects (zooms in), while 0.5 makes objects smaller (zooms out).

- Transparent zoom is *not* possible during the **VPoint**, **Pan**, **DView**, and **View** commands.

# 3D

<u>Rel.11</u>  Draws 3D surface primitives with polymeshes *(short for three Dimensions)*.

| Command | Alias | Ctrl+ | F-key | Alt+ | Menu Bar | Tablet |
|---------|-------|-------|-------|------|----------|--------|
| 3d | ... | ... | ... | DF3 | Draw | N8 |
| | | | | | ⸂Surfaces | |
| | | | | | ⸂3D Surfaces | |

**Command:** 3d
**Enter an option**
**[Box/Cone/DIsh/DOme/Mesh/Pyramid/Sphere/Torus/Wedge]:** *(Enter an option.)*

See the **Ai_** *commands for details, such as* **Ai_Box** *and* **Ai_Wedge**.

Selecting **Draw | Surfaces | 3D Surfaces** *from the menu bar displays the dialog box:*

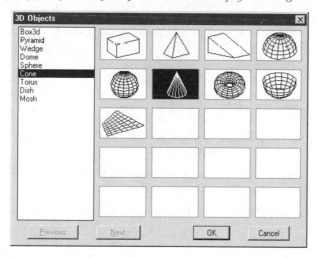

## ICON MENU OPTIONS

**Box** draws 3D boxes and cubes.

**Cone** draws cones.

**DIsh** draws dishes, the bottom half of spheres.

**Dome** draws domes, the top half of spheres.

**Mesh** draws a 3D mesh.

**Pyramid** draws a variety of pyramids.

**Sphere** draws spheres.

**Torus** draws tori (3D donut) shapes.

**Wedge** draws wedges.

## RELATED COMMANDS

**Ai_Box** draws 3D surface boxes and cubes.

**Ai_Cone** draws 3D surface cones.

**Ai_Dish** draws 3D surface dishes.

**Ai_Dome** draws 3D surface domes.

**Ai_Mesh** draws 3D meshes.

**Ai_Pyramid** draws 3D surface pyramids.

**Ai_Sphere** draws 3D surface spheres.

**Ai_Torus** draws 3D surface tori.

**Ai_Wedge** draws 3D surface wedges.

**Box** draws 3D solid boxes and cubes.

**Cone** draws 3D solid cones.

**Cylinder** draws 3D solid cylinders.

**Sphere** draws 3D solid spheres.

**Torus** draws 3D solid tori.

**Wedge** draws 3D solid wedges.

## TIPS

- The **3D** command creates 3D objects made of 3D meshes, and *not* of 3D solids.

- To draw a cylinder, apply thickness to a circle.

- You *cannot* perform Boolean operations on 3D surface models.

- To convert a 3D solid model to a 3D surface model, export the drawing with the **3dsOut** command, then import with the **3dsIn** command.

- Use the **Ucs** command to place 3D surface models in space; use the **VPoint** and **Dview** commands to view surface models from different 3D viewpoints.

- You can apply the **Hide**, **Shade,** and **Render** commands to 3D surface models.

# 3dArray

Rel.11 Creates 3D rectangular and polar arrays.

| Command | Alias | Ctrl+ | F-key | Alt+ | Menu Bar | Tablet |
|---------|-------|-------|-------|------|----------|--------|
| 3darray | ... | ... | ... | M33 | Modify | W20 |
| | | | | | ⤷ 3D Operation | |
| | | | | | ⤷ 3D Array | |

**Command:** 3darray
**Select objects:** *(Select one or more objects.)*
**Select objects:** *(Press Enter to end object selection.)*
**Enter the type of array [Rectangular/Polar] <R>:** *(Type R or P.)*

3D rectangular array

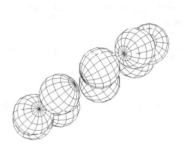

3D Polar array.

## COMMAND LINE OPTIONS

**Select objects** selects the objects to be arrayed.

**R** creates rectangular 3D arrays.

**P** creates a polar array in 3D space.

Rectangular Array options
**Enter the number of rows (---) <1>:** *(Enter a value.)*
**Enter the number of columns (|||) <1>:** *(Enter a value.)*
**Enter the number of levels (...) <1>:** *(Enter a value.)*
**Specify the distance between rows (---) <1>:** *(Enter a value.)*
**Specify the distance between columns (|||) <1>:** *(Enter a value.)*
**Specify the distance between levels (...) <1>:** *(Enter a value.)*

**Enter the number of rows** specifies the number of rows in the x direction.

**Enter the number of columns** specifies the number of columns in the y direction.

**Enter the number of levels** specifies the number of levels in the z direction.

**Specify the distance between rows** specifies the distance between objects in the x direction.

**Specify the distance between columns** specifies the distance between objects in the y direction.

**Specify the distance between levels** specifies the distance between objects in the z direction.

Polar Array options
**Enter the number of items in the array:** *(Enter a number.)*
**Specify the angle to fill (+=ccw, -=cw) <360>:** *(Enter an angle.)*
**Rotate arrayed objects? [Yes/No] <Y>:** *(Type **Y** or **N**.)*
**Specify center point of array:** *(Pick a point.)*
**Specify second point on axis of rotation:** *(Pick a point.)*

**Enter the number of items** specifies the number of objects to array.

**Specify the angle to fill** specifies the distance along the circumference that objects are arrayed (default = 360 degrees).

**Rotate arrayed objects?**

**Yes** objects rotate so that they face the central axis (default).

**No** objects rotate.

**Specify enter point of array** specifies the x,y,z coordinates for one end of the axis for the polar array.

**Specify second point on axis of rotation** specifies the x,y,z coordinates for the other end of the array axis.

ESC interrupts the drawing of arrays.

## RELATED COMMANDS

**Array** creates a rectangular or polar array in 2D space.

**Copy** creates one or more copies of the selected object.

**MInsert** creates a rectangular block-array of blocks.

# 3dClip

<u>2000</u>   Performs real-time front and back clipping (*short for three Dimensional CLIPping*).

| Command | Alias | Ctrl+ | F-key | Alt+ | Menu Bar | Tablet |
|---------|-------|-------|-------|------|----------|--------|
| 3dclip | ... | ... | ... | ... | ... | ... |

**Command:** 3dclip

*Displays window:*

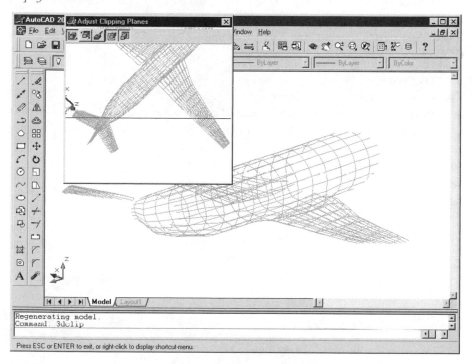

**TOOLBAR OPTIONS**

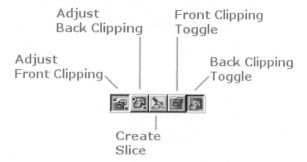

## SHORTCUT MENU OPTIONS

*Right-click Adjust Clipping Planes window:*

**Adjust Front Clipping** switches to front clipping mode.

**Adjust Back Clipping** switches to back clipping mode.

**Create Slice** switches to slicing — ganged front and back clipping — mode.

**Front Clipping On** toggles on and off front clipping.

**Back Clipping On** toggles on and off back clipping.

**Close** closes the Adjust Clipping Planes window.

## RELATED COMMANDS

**Vpoint** creates a static 3D viewpoint.

**3dCOrbit** places the drawing into real-time, 3D, continuous orbit mode.

**3dDistance** performs real-time 3D forward and backward panning.

**3dOrbit** provides real-time 3D viewing of the drawing.

**3dPan** performs real-time 3D sideways panning.

**3dSwivel** tilts the 3D view.

**3dZoom** performs real-time 3D zooming.

## TIP

- Use this command to remove objects in the front of a 3D scene, or to expose the interior of a 3D model, such as this airplane.

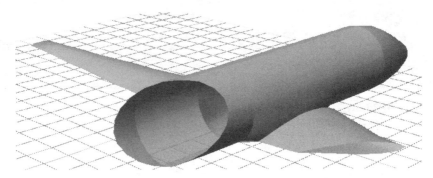

# 3dConfig

<u>2004</u> Configuration of 3D display characteristics.

| Command | Alias | Ctrl+ | F-key | Alt+ | Menu Bar | Tablet |
|---------|-------|-------|-------|------|----------|--------|
| 3dconfig | ... | ... | ... | ... | ... | ... |

**Command:** 3dconfig
**Configure: 3DCONFIG**
**Enter option [Adaptive degradation/Dynamic tessellation/Render options/ Geometry/acceLeration/eXit] <Adaptive degradation>:** *(Enter an option.)*

## COMMAND LINE OPTIONS

**Adaptive degradation** allows AutoCAD to switch to lower-quality rendering in order to maintain display speed.

**Dynamic tessellation** specifies the smoothness of faceted objects in 3D drawings.

**Render options** enhances the display of lights, materials, textures, and transparency in 3D views.

**Geometry** specifies the display of isolines and back faces in 3D drawings.

**acceLeration** specifies hardware or software acceleration.

**eXit** exits the command.

Adaptive Degradation options
**Enter mode [ON/OFF] <ON>:** *(Enter **ON** or **OFF**.)*

**ON** turns on adaptive degradation.

**OFF** turns off adaptive degradation.

*When on, the following options become available:*

**Current display options: Wireframe Bounding Box**
**Enter option [Flat shaded/Wireframe/Bounding box/Maintain speed fps/ eXit] <Flat shaded>:** *(Enter an option.)*

**Flat shaded** degrades the 3D display to flat shading, when on.

**Wireframe** degrades the 3D display to wireframe, when on.

**Bounding box** degrades the 3D display showing each object as a rectangular box, when on.

**Maintain speed fps** specifies the speed at which to display, in frames per second; range is 5fps to 60fps.

Dynamic tessellation options
**Enter option [Surface tessellation/Curve tessellation/Tessellations to cache/ eXit] <Surface tessellations>:** *(Enter an option.)*

**Surface tessellation** specifies the number of tessellation lines to display on surfaces; ranges from 0 to 100 (default = 88).

**Curve tessellation** specifies the number of tessellation lines to display on curved surfaces; ranges from 0 to 100 (default = 88).

**Tessellations to cache** specifies the number of tessellations to cache; ranges from 1 to 3 (default = 3).

Render options
**Enter mode [ON/OFF] <ON>:** *(Enter* **ON** *or* **OFF.***)*
**Enter option [Lights/Materials/eXit] <Lights>:** *(Enter an option.)*

ON turns on render options.

OFF turns off render options.

*When Render mode is on, the following options become available:*

**Lights** determines whether assigned lights (with the Light command) or the default global light are used.

**Materials** determines whether assigned materials (with the RMat command) or default global material are used.

*When Materials mode is on, the following options become available:*
**Configure: Textures**
**Enter mode [ON/OFF] <ON>:** *(Enter* **ON** *or* **OFF.***)*
**Configure: Transparency**
**Enter mode [Low/Medium/High] <Low>:** *(Enter an option.)*

**Textures** determines whether assigned textures (with the RMat and SetUV commands) are used.

**Transparency** determines the quality of transparency:

- **Low** applies screen-door effect for faster speed (default).
- **Medium** applies blending effect for moderate speed.
- **High** applies blending and extra processing for slower speed.

Geometry options
**Enter option [Isolines on top/Discard backfaces] <Isolines on top>:** *(Enter an option.)*

Isolines on top

On displays isolines on front and back faces in all shading modes (except hidden).

Off hides isolines on back faces.

Discard backfaces

On discards back faces to enhance performance.

Off displays back faces.

acceLeration options
**Enter option [Hardware/Software/eXit] <Hardware>:** *(Enter an option.)*

**Hardware** configures graphics card to perform 3D display rendering, the faster option.

**Software** configures 3D display through software, if the graphics board is unable to.

*When Hardware option selected:*
**Enter option [Driver name/Geometry acceleration/Antialias lines/eXit] <Driver name>:**

**Driver name** allows you to select a specific driver for the graphics board.

**Geometry acceleration** allows a more precise display, if turned on and if supported by the graphics board.

**Antialias lines** uses anti-aliasing to drawn smoother lines, if on.

**RELATED COMMAND**

3dOrbit provides real-time 3D viewing of the drawing.

# 3dCOrbit, Distance, Pan, Swivel, Zoom

<u>2000</u>   This collection of commands is a subset of 3dOrbit.

| Command | Alias | Ctrl+ | F-key | Alt+ | Menu Bar | Tablet |
|---------|-------|-------|-------|------|----------|--------|
| 3dcorbit | ... | ... | ... | ... | ... | ... |

 **Command:** 3dcorbit

Places the 3D view into continuous orbiting mode.

 **Command:** 3ddistance

Interactively changes the 3D viewing distance.

 **Command:** 3dpan

Interactively pans the 3D view.

 **Command:** 3dswivel

Interactively twists the 3D view.

 **Command:** 3dzoom

Interactively zooms the 3D view.

## COMMAND LINE OPTIONS
ESC exits the command.
ENTER exits the command.

## RELATED COMMANDS
**Vpoint** creates a static 3D viewpoint.
**3dClip** performs real-time 3D front and back clipping.
**3dOrbit** performs real-time 3D viewing of the drawing.
**3dViewCtr** specifies the center point for 3D orbiting views.

## TIPS
- To set the 3D model to continuous orbit mode, drag the cursor across the drawing.
- You must use **3dPan** and **3dZoom** commands when the current drawing is in perspective mode; when you try to use the regular **Pan** and **Zoom** command, AutoCAD complains, '** That command may not be invoked in a perspective view **.'

# 3dFace

**V. 2.6**  Draws 3D faces with three or four corners.

| Command | Alias | Ctrl+ | F-key | Alt+ | Menu Bar | Tablet |
|---------|-------|-------|-------|------|----------|--------|
| 3dface | 3f | ... | ... | DFF | Draw | M8 |
| | | | | | ⌖Surfaces | |
| | | | | | ⌖3D Face | |

**Command:** 3dface
**Specify first point or [Invisible]:** *(Pick a point, or type I followed by the point.)*
**Specify second point or [Invisible]:** *(Pick a point, or type I followed by the point.)*
**Specify third point or [Invisible] <exit>:** *(Pick a point, or type I followed by the point.)*
**Specify fourth point or [Invisible] <create three-sided face>:** *(Pick a point, or type I followed by the point.)*
**Specify third point or [Invisible] <exit>:** *(Press Enter to exit the command.)*

## COMMAND LINE OPTIONS

**First point** picks the first corner of the face.

**Second point** picks the second corner of the face.

**Third point** picks the third corner of the face.

**Fourth point** picks the fourth corner of the face; or press ENTER to create a triangular face.

**Invisible** makes the edge invisible.

## RELATED COMMANDS

**3D** draws a 3D object: box, cone, dome, dish, pyramid, sphere, torus, or wedge.

**Properties** modifies the 3d face, including visibility of edges.

**Edge** changes the visibility of the edges of 3D faces.

**EdgeSurf** draws 3D surfaces made of 3D meshes.

**PEdit** edits 3D meshes.

**PFace** draws generalized 3D meshes.

## RELATED SYSTEM VARIABLE

**SplFrame** controls the visibility of edges.

## TIPS

- A 3D face is the same as a 2D solid, except that each corner can have a different z-coordinate.

- Unlike the procedure for **Solid**, corner coordinates are entered in natural order.

- The **i** (short for invisible) suffix must be entered before object snap modes, point filters, and corner coordinates.

- Invisible 3D faces, where all four edges are invisible, do not appear in wireframe views; they hide objects behind them in hidden-line mode, and are rendered in shaded views.

- You can use the **Properties** and **Edge** commands to change the visibility of 3D faces.

- 3D faces cannot be extruded.

# 3dMesh

**Rel.10** Draws open 3D rectangular meshes made of 3D faces.

| Command | Alias | Ctrl+ | F-key | Alt+ | Menu Bar | Tablet |
|---------|-------|-------|-------|------|----------|--------|
| 3dmesh | ... | ... | ... | DFM | Draw | ... |
| | | | | | ⤷Surfaces | |
| | | | | | ⤷3D Mesh | |

**Command:** 3dmesh
**Enter size of mesh in M direction:** *(Enter a value.)*
**Enter size of mesh in N direction:** *(Enter a value.)*
**Specify location for vertex (0, 0):** *(Pick a point.)*
**Specify location for vertex (0, 1):** *(Pick a point.)*
**Specify location for vertex (1, 0):** *(Pick a point.)*
**Specify location for vertex (1, 1):** *(Pick a point.)*

## COMMAND LINE OPTIONS

**Enter size of mesh in M direction** specifies the m-direction mesh size (between 2 and 256).

**Enter size of mesh in N direction** specifies the n-direction mesh size (between 2 and 256).

**Specify location for vertex** (*m,n*) specifies a 2D or 3D coordinate for each vertex.

## RELATED COMMANDS

**3D** draws a variety of 3D objects.

**3dFace** draws a 3D face with three or four corners.

**Explode** explodes a 3D mesh into individual 3D faces.

**PEdit** edits a 3D mesh.

**PFace** draws a generalized 3D face.

**Xplode** explodes a group of 3D meshes.

## TIPS

- It is more convenient to use the **EdgeSurf**, **RevSurf**, **RuleSurf**, and **TabSurf** commands than the **3dMesh** command. The **3dMesh** command is meant for use by AutoLISP and other programs.

- The range of values for the m- and n-mesh size is 2 to 256.

- The number of vertices = **m** x **n**.

- The first vertex is at (0,0). The vertices can be any distance from each other.

- The coordinates for each vertex in row **m** must be entered before starting on vertices in row **m+1**.

- Use the **PEdit** command to close the mesh, since it is always created open.

- The **SurfU** and **SurfV** system variables do not affect the 3dmesh object.

 # 3dOrbit

2000 Provides real-time 3D viewing of the drawing.

| Command | Alias | Ctrl+ | F-key | Alt+ | Menu Bar | Tablet |
|---------|-------|-------|-------|------|----------|--------|
| 3dorbit | orbit | ... | ... | VB | View<br>3D Orbit | R5 |

**Command:** 3dorbit
**Press ESC or ENTER to exit, or right-click to display shortcut-menu.** *(Move the cursor to change the viewpoint, and then press **Enter** to exit the command.)*

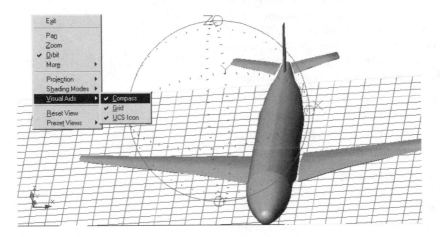

## COMMAND LINE OPTIONS

ESC exits the command.

ENTER exits the command.

## SHORTCUT MENU OPTIONS

*Right-click the drawing:*

**Exit** exits the command.

**Pan** pans the view in real-time; see the 3dPan command.

**Zoom** zooms the view in real-time; see the 3dZoom command.

**Orbit** rotates the view in real-time.

**More** displays a submenu with additional view controls.

**Projection** displays a submenu for toggling between parallel and perspective projection.

**Shading Modes** displays a submenu for selecting the type of wireframe or shaded modes; see the ShadeMode command.

**Visual Aids** displays a submenu of visual aids for navigating in 3D space.

**Reset View** resets to the original view when you first began the command.

**Preset Views** displays a submenu of standard views.

More options

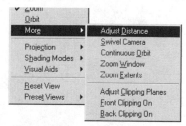

**Adjust Distance** moves the view closer and further away; see the 3dDistance command.

**Swivel Camera** rotates the view; see the 3dSwivel command.

**Continuous Orbit** continuously rotates the model; see the 3dCOrbit command.

**Zoom Window** performs a windowed zoom; see the Zoom command.

**Zoom Extents** displays the entire drawing.

**Adjust Clipping Planes** sets the front and back clipping planes; see the 3dClip command.

**Front Clipping On** toggles the front clipping plane.

**Back Clipping On** toggles the back clipping plane.

Projection options

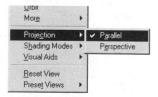

**Parallel** displays the view in orthogonal view.

**Perspective** displays the view in one-point perspective view; some commands do not work in perspective mode, such as the Pan and Zoom commands.

Shading Modes options (see ShadeMode command)

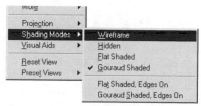

**Wireframe** displays the model in wireframe mode.

. . . . . . . . . . . . . . . . . . . . . . . . . . . . . . . . . . . . . . . . . . . . . . . . . .

**Hidden** removes hidden lines from the view.

**Flat Shaded** flat-shades the model.

**Gouraud Shaded** smooth-shades the model.

**Flat Shaded, Edges On** flat-shades the model, and outlines faces with the background color.

**Gouraud Shaded, Edges On** smooth-shades the model, and outlines faces with the background color.

Visual Aids options

**Compass** toggles the display of the compass to help you navigate in 3D space.

**Grid** toggles the display of the grid as lines, to help you see the x,y-plane; see the Grid command.

**UCS Icon** toggles the display of the 3D UCS icon; see the UcsIcon command.

Preset Views options

**Top, Bottom, Front, Back, Left, Right** displays the standard orthogonal views.

**SW Isometric, SE Isometric, NW Isometric, NW Isometric** displays the standard isometric views.

### RELATED COMMANDS

**Vpoint** creates a static 3D viewpoint.

**3dClip** performs real-time 3D front and back clipping.

**3dCOrbit** places the drawing in real-time, 3D, continuous orbit mode.

**3dDistance** performs real-time 3D forward and backward panning.

**3dPan** performs real-time 3D sideways panning.

**3dSwivel** tilts the 3D view.

**3dZoom** performs real-time 3D zooming.

### TIP

• This command is meant to replace the **DView** command.

# 3dOrbitCtr

2004 Specifies the center point for 3D orbiting views.

| Command | Alias | Ctrl+ | F-key | Alt+ | Menu Bar | Tablet |
|---------|-------|-------|-------|------|----------|--------|
| 3dorbitctr | ... | ... | ... | ... | ... | ... |

**Command:** 3dorbitctr
**Specify orbit center point:** *(Pick a point.)*

## COMMAND LINE OPTION

**Specify orbit center point** specifies the center of rotation.

## RELATED COMMANDS

**3dClip** performs real-time 3D front and back clipping.

**3dCOrbit** places the drawing in real-time, 3D, continuous orbit mode.

**3dDistance** performs real-time 3D forward and backward panning.

**3dOrbit** provides real-time 3D viewing of the drawing.

**3dSwivel** tilts the 3D view.

**3dZoom** performs real-time 3D zooming.

 **3dPoly**

**Rel.10** Draws 3D polylines *(short for 3D POLYline)*.

| Command | Alias | Ctrl+ | F-key | Alt+ | Menu Bar | Tablet |
|---------|-------|-------|-------|------|----------|--------|
| 3dpoly | ... | ... | ... | D3 | Draw | O10 |
| | | | | | ↳3D Polyline | |

**Command:** 3dpoly
**Specify start point of polyline:** *(Pick a point.)*
**Specify endpoint of line or [Undo]:** *(Pick a point, or type **U**.)*
**Specify endpoint of line or [Undo]:** *(Pick a point, or type **U**.)*
**Specify endpoint of line or [Close/Undo]:** *(Pick a point, or type **U** or **C**, or press **Enter** to end the command.)*

## COMMAND LINE OPTIONS

**Specify start point** indicates the starting point of the 3D polyline.

**Close** joins the last endpoint with the start point.

**Undo** erases the last-drawn segment.

**Specify endpoint of line** indicates the endpoint of the current segment.

**ENTER** ends the command.

## RELATED COMMANDS

**Explode** reduces a 3D polyline into lines and arcs.

**PEdit** edits 3D polylines.

**PLine** draws 2D polylines.

## TIPS

• Since 3D polylines are made of straight lines, use the **PEdit** command to spline the polyline as a curve.

• 3D polylines do not support linetypes and widths.

• You may use lineweights to fatten up a 3D polyline.

• Use the .xy point filter to place pick points, and then specify the z-coordinate.

# 3dsIn

**Rel.13** Imports 3DS files created by 3D Studio *(short for 3D Studio IN)*.

| Command | Alias | Ctrl+ | F-key | Alt+ | Menu Bar | Tablet |
|---------|-------|-------|-------|------|----------|--------|
| 3dsin | ... | ... | ... | I3 | Insert | ... |
| | | | | | ↳3D Studio | |

**Command:** 3dsin

*Displays the **3D Studio File Import** dialog box. Select a .3ds file, and then click **Open**.*
*AutoCAD displays dialog box:*

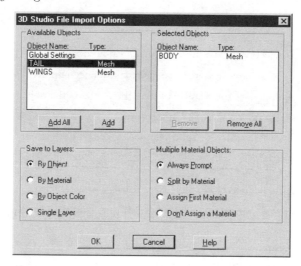

## DIALOG BOX OPTIONS

Available and Selected Objects options

**Object Name** names the object.

**Type** specifies the type of object.

**Add** adds the object to the Selected Objects list.

**Add All** adds all objects to the Selected Objects list.

**Remove** removes the object from Selected Objects list.

**Remove All** removes all objects from Selected Objects list.

Save to Layers options

**By Object** places each object on its own layer.

**By Material** places the objects on layers named after materials.

**By Object Color** places the objects on layers named "Color*nn*."

**Single Layer** places all objects on layer "AvLayer."

Multiple Material Objects options

**Always Prompt** prompts for each material.

**Split by Material** splits objects with more than one material into multiple objects, each with one material.

**Assign First Material** assigns the first material to the entire object.

**Don't Assign to a Material** removes all 3D Studio material definitions.

## RELATED COMMAND

**3dsOut** exports drawing as a 3DS file.

## RELATED FILES

*__*.3ds__* are 3D Studio files.

*__*.tga__* are converted bitmap and animation files.

## TIPS

- You are limited to selecting a maximum of 70 3D Studio objects.

- Conflicting object names are truncated and given a sequence number.

- The **By Object** option gives the AutoCAD layer the name of the object.

- The **By Object Color** option places all objects on layer "ColorNone" when no colors are defined in the 3DS file.

- 3D Studio assigns materials to faces, elements, and objects; AutoCAD assigns materials only to objects, colors, and layers.

- 3D Studio bitmaps are converted to *.tga* (Targa format) bitmaps.

- Only the first frame of animation files (CEL, CLI, FLC, and IFL) is converted to Targa bitmap files.

- Converted *.tga* files are saved to the 3DS file's folder.

- 3D Studio "ambient lights" lose their color.

- 3D Studio "omni lights" become point lights in AutoCAD.

- 3D Studio "cameras" become named views in AutoCAD.

# 3dsOut

**Rel.13** Exports AutoCAD drawings as 3D Studio files (*short for 3D Studio OUT*).

| Command | Alias | Ctrl+ | F-key | Alt+ | Menu Bar | Tablet |
|---------|-------|-------|-------|------|----------|--------|
| 3dsout | ... | ... | ... | FE | File | ... |
| | | | | Ⓢ3DS | ⓈExport | |
| | | | | | Ⓢ3D Studio | |

**Command:** 3dsout
**Select objects:** *(Select one or more objects.)*
**Select objects:** *(Press **Enter** to end object selection.)*

*Displays **3D Studio File Export** dialog box. Enter a file name, and then click **Save**.*
*AutoCAD displays dialog box:*

*After you select options, the **3dsOut** command exports selected objects:*
**Generating objects**
**Writing preamble**
**Converting and writing material definitions**
**UCSVIEW = 1 UCS will be saved with view**
**Collecting geometry**
**Converting object BODY**
**Unifying normals**
**Assigning smoothing**
**3D Studio file output completed**

## COMMAND LINE OPTION

**Select objects** selects objects to export; note that **3dsOut** exports only objects with a surface.

## DIALOG BOX OPTIONS

Derive 3D Studio Objects From options

**Layer** exports all objects on AutoCAD layers to single 3D Studio objects (default).

**ACI** exports all objects of an ACI color to single 3D Studio objects.

**Object Type** exports all objects of an AutoCAD object type to single 3D Studio objects.

AutoCAD Blocks option

**Override** specifies that each AutoCAD block becomes a single 3D Studio object; overrides the Derive 3D Studio Objects From options.

Smoothing options
**Auto-Smoothing**

☑ creates a 3D Studio smoothing group (default).

☐ no smoothing assigned to new 3D Studio objects.

**Degrees** smooths the face normals, when the angle between two face normals is equal to or less than this value (default = 30 degrees).

Welding options
**Auto-Welding**

☑ creates a 3D Studio welded vertex (default).

☐ vertices remain unwelded upon export.

**Threshold** welds two vertices into a single vertex when their interdistance is less than or equal to this value (default = 0.001).

## RELATED AUTOCAD COMMAND

**3dsIn** imports 3DS files into drawings.

## RELATED FILE

***.3ds** are 3D studio files.

## TIPS

- AutoCAD objects with 0 thickness are not exported, with the exception of circles, polygons, and polyface meshes.

- Solids and 3D faces must have at least three vertices.

- 3D solids and bodies are converted to meshes.

- AutoCAD blocks are exploded unless the **Override** option is turned on.

- The weld threshold distance ranges from 0.00 000 001 to 99,999,999.

- AutoCAD named views become 3D Studio "cameras."

- AutoCAD point lights become 3D Studio "omni lights."

# System Variables

AutoCAD stores information about its current state, the drawing and the operating system in over 400 *system variables*. The variables help programmers — who often work with menu macros and AutoLISP — determine the state of the AutoCAD system.

## CONVENTIONS

The following pages list all documented system variables, plus several more not documented by Autodesk. The listing uses the following conventions:

**Bold**            System variable is documented in AutoCAD 2004.

*Italicized*        System variable is not listed by the **SetVar** command or Autodesk documentation.

~~Strikethru Italic~~   System variable was removed from AutoCAD.

⌨               System variable must be accessed via the **SetVar** command.

🇦               System variable is new to AutoCAD 2004.

## COLUMN HEADINGS

**Default**         Default value, as set in the *acad.dwg* prototype drawing.

R/O             Read-only; cannot be changed by the user or by a program.

**Loc**             Location where the value of the system variable is saved:

| Location | Meaning |
| --- | --- |
| ACAD | Set by AutoCAD. |
| DWG | Saved in current drawing. |
| REG | Saved globally in Windows registry. |
| ... | Not saved. |

## TIPS

* The **SetVar** command lets you change the value of all variables, except those marked read-only (R/O).

* You can get a list of most system variables at the 'Command' prompt with the **?** option of the **SetVar** command, as follows:

> **Command:** setvar
> **Variable name or ?:** ?
> **Variable(s) to list <*>:** *(Press **Enter**.)*

* When a system variable is stored in the Windows registry, it affects all drawings.

* When a system variable is stored in the drawing, it affects the current drawing only.

* When a system variable is not stored, the variable is set when AutoCAD loads. The value of the variable is either read from the operating system, or set to a default value.

| Variable | Default | R/O | Loc | Meaning |
|---|---|---|---|---|
| ~~LInfo~~ | | | | *Removed from AutoCAD 2004.* |
| _PkSer | *varies* | R/O | *ACAD* | *Software package serial number, such as "117-69999999".* |
| _Server | 0 | R/O | REG | *Network authorization code.* |
| _VerNum | *varies* | R/O | REG | *Internal program build number, such as "T.0.98".* |

. . . . . . . . . . . . . . . . . . . . . . . . . . . . . . . . . . . . . .

**A**

| Variable | Default | R/O | Loc | Meaning |
|---|---|---|---|---|
| **AcadLspAsDoc** | 0 | ... | REG | *acad.lsp* is loaded into:<br>**0** Just the first drawing.<br>**1** Every drawing. |
| **AcadPrefix** | *varies* | R/O | ... | Path spec'd by ACAD environment variable, such as "d:\acad 2004\support; d:\acad 2004\fonts". |
| **AcadVer** | "15.06" | R/O | ... | AutoCAD version number. |
| **AcisOutVer** | 40 | R/O | ... | ACIS version number, such as 15, 16, 17, 18, 20, 21, 30, 40, or 70. |
| ▣ **AdcState** | 0 | R/O | ... | Specifies if **DesignCenter** is active. |
| **AFlags** | 0 | ... | ... | Attribute display code:<br>**0** No mode specified.<br>**1** Invisible.<br>**2** Constant.<br>**4** Verify.<br>**8** Preset. |
| **AngBase** | 0 | ... | DWG | Direction of zero degrees relative to UCS |
| **AngDir** | 0 | ... | DWG | Rotation of angles:<br>**0** Clockwise<br>**1** Counterclockwise |
| **ApBox** | 0 | ... | REG | AutoSnap aperture box cursor:<br>**0** Off.<br>**1** On. |
| **Aperture** | 10 | ... | REG | ⌨ Object snap aperture in pixels:<br>**1** Minimum size.<br>**50** Maximum size. |
| **Area** | 0.0000 | R/O | ... | ⌨ Area measured by the last **Area**, **List**, or **Dblist** commands. |
| **AttDia** | 0 | ... | DWG | Attribute entry interface:<br>**0** Command-line prompts.<br>**1** Dialog box. |
| **AttMode** | 1 | ... | DWG | Display of attributes:<br>**0** Off.<br>**1** Normal.<br>**2** On. |
| **AttReq** | 1 | ... | REG | Attribute values during insertion are:<br>**0** Default values.<br>**1** Prompt for values. |
| **AuditCtl** | 0 | ... | REG | Determines creation of *.adt* audit log file:<br>**0** File not created.<br>**1** *.adt* file created. |

. . . . . . . . . . . . . . . . . . . . . . . . . . . . . . . . . . . . . .

| Variable | Default | R/O | Loc | Meaning |
|----------|---------|-----|-----|---------|
| **AUnits** | 0 | ... | DWG | Mode of angular units:<br>**0** Decimal degrees.<br>**1** Degrees-minutes-seconds.<br>**2** Grads.<br>**3** Radians.<br>**4** Surveyor's units. |
| **AUPrec** | 0 | ... | DWG | Decimal places displayed by angles. |
| **AutoSnap** | 63 | ... | REG | Controls AutoSnap display:<br>**0** Turns off all AutoSnap features.<br>**1** Turns on marker.<br>**2** Turns on SnapTip.<br>**4** Turns on magnetic cursor.<br>**8** Turns on polar tracking .<br>**16** Turns on object snap tracking .<br>**32** Turns on tooltips for polar tracking and object snap tracking . |
| *AuxStat* | *0* | ... | DWG | *-32768  Minimum value.*<br>*32767  Maximum value.* |
| ~~*AxisMode*~~ | *0* | ... | DWG | *Removed from AutoCAD 2002.* |
| *AxisUnit* | *0.0000* | ... | DWG | *Obsolete system variable.* |

· · · · · · · · · · · · · · · · · · · · · · · · · · · · · · · · · · · · · · · · · ·

**B**

| Variable | Default | R/O | Loc | Meaning |
|----------|---------|-----|-----|---------|
| **BackZ** | 0.0000 | R/O | DWG | Back clipping plane offset. |
| **BindType** | 0 | ... | ... | When binding an xref or editing an xref, xref names are converted from:<br>**0** **xref\|name** to **xref$0$name**.<br>**1** **xrcf\|name** to **name**. |
| **BlipMode** | 0 | ... | DWG | ⌨ Display of blip marks:<br>**0** Off.<br>**1** On. |

· · · · · · · · · · · · · · · · · · · · · · · · · · · · · · · · · · · · · · · · · ·

**C**

| Variable | Default | R/O | Loc | Meaning |
|----------|---------|-----|-----|---------|
| **CDate** | *varies* | R/O | ... | Current date and time in the format YyyyMmDd.HhMmSsDd, such as 20010503.18082328 |
| **CeColor** | "BYLAYER" | ... | DWG | Current color. |
| **CeLtScale** | 1.0000 | ... | DWG | Current linetype scaling factor. |
| **CeLType** | "BYLAYER" | ... | DWG | Current linetype. |
| **CeLWeight** | -1 | ... | DWG | Current lineweight in millimeters; valid values are 0, 5, 9, 13, 15, 18, 20, 25, 30, 35, 40, 50, 53, 60, 70, 80, 90, 100, 106, 120, 140, 158, 200, and 211, plus the following:<br>**-1** BYLAYER.<br>**-2** BYBLOCK.<br>**-3** DEFAULT as defined by **LwDdefault** |
| **ChamferA** | 0.5000 | ... | DWG | First chamfer distance. |
| **ChamferB** | 0.5000 | ... | DWG | Second chamfer distance. |

· · · · · · · · · · · · · · · · · · · · · · · · · · · · · · · · · · · · · · · · · ·

| Variable | Default | R/O | Loc | Meaning |
|---|---|---|---|---|
| ChamferC | 1.0000 | ... | DWG | Chamfer length. |
| ChamferD | 0 | ... | DWG | Chamfer angle. |
| ChamMode | 0 | ... | ... | Chamfer input mode:<br>0 Chamfer by two lengths.<br>1 Chamfer by length and angle. |
| CircleRad | 0.0000 | ... | ... | Most-recent circle radius. |
| CLayer | "0" | ... | DWG | Current layer name. |
| CmdActive | 1 | R/O | ... | Type of current command:<br>1 Regular command.<br>2 Transparent command.<br>4 Script file.<br>8 Dialog box.<br>16 AutoLISP is active . |
| CmdDia | 1 | ... | REG | Formerly determined whether the **Plot** command displayed at the command line prompt or via a dialog box; no longer has an effect as of AutoCAD 2000; replaced by **PlQuiet**. |
| CmdEcho | 1 | ... | ... | AutoLISP command display:<br>0 No command echoing.<br>1 Command echoing. |
| CmdNames | *varies* | R/O | ... | Current command, such as "SETVAR". |
| CMLJust | 0 | ... | DWG | Multiline justification mode:<br>0 Top.<br>1 Middle.<br>2 Bottom. |
| CMLScale | 1.0000 | ... | DWG | Scales width of multiline:<br>-*n* Flips offsets of multiline.<br>0 Collapses to single line.<br>1 Default.<br>*n* Scales by a factor of *n*. |
| CMLStyle | "STANDARD" | ... | DWG | Current multiline style name. |
| Compass | 0 | ... | ... | Toggles display of the 3D compass:<br>0 Off.<br>1 On. |
| Coords | 1 | ... | DWG | Coordinate display style:<br>0 Updated by screen picks.<br>1 Continuous display.<br>2 Polar display upon request. |
| CPlotStyle | "ByColor" | ... | DWG | Current plot style; values defined by AutoCAD are:<br>"ByLayer"<br>"ByBlock"<br>"Normal"<br>"User Defined" |
| CProfile | "<<Unnamed Profile>>" | R/O | REG | Current profile. |
| CTab | "Model" | R/O | DWG | Current tab. |
| ~~CurrentProfile~~ | *"<<UnnamedProfile>>"*... | ... | | *Removed from AutoCAD 2000; replaced by **CProfile**.* |
| CursorSize | 5 | ... | REG | Cursor size, in percent of viewport: |

| Variable | Default | R/O | Loc | Meaning |
|---|---|---|---|---|
| | | | | **1** Minimum size. |
| | | | | **100** Full viewport. |
| **CVPort** | 2 | ... | DWG | Current viewport number: |
| | | | | **2** Minimum (default). |

**D**

| **Date** | *varies* | R/O | ... | Current date in Julian format, such as 2448860.54043252 |
|---|---|---|---|---|
| ~~DBGListAll~~ | *0* | ... | *ACAD* | *Removed from AutoCAD 2002.* |
| **DBMod** | 4 | R/O | ... | Drawing modified, as follows: |
| | | | | **0** No modification since last save. |
| | | | | **1** Object database modified. |
| | | | | **2** Symbol table modified. |
| | | | | **4** Database variable modified. |
| | | | | **8** Window modified. |
| | | | | **16** View modified. |
| **DbcState** 🔲 | 0 | R/O | DWG | Specifies if dbConnect Manager is active. |
| **DctCust** | "d:\acad 2004\support\sample.cus" | | | |
| | | ... | REG | Name of custom spelling dictionary. |
| **DctMain** | "enu" | ... | REG | Code for spelling dictionary: |
| | | | | **ca** Catalan. |
| | | | | **cs** Czech. |
| | | | | **da** Danish. |
| | | | | **de** German; sharp 's'. |
| | | | | **ded** German; double 's'. |
| | | | | **ena** English; Australian. |
| | | | | **ens** English; British 'ise'. |
| | | | | **enu** English; American. |
| | | | | **enz** English; British 'ize'. |
| | | | | **es** Spanish; unaccented capitals. |
| | | | | **esa** Spanish; accented capitals. |
| | | | | **fi** Finish. |
| | | | | **fr** French; unaccented capitals. |
| | | | | **fra** French; accented capitals. |
| | | | | **it** Italian. |
| | | | | **nl** Dutch; primary. |
| | | | | **nls** Dutch; secondary. |
| | | | | **no** Norwegian; Bokmal. |
| | | | | **non** Norwegian; Nynorsk. |
| | | | | **pt** Portuguese; Iberian. |
| | | | | **ptb** Portuguese; Brazilian. |
| | | | | **ru** Russian; infrequent 'io'. |
| | | | | **rui** Russian; frequent 'io'. |
| | | | | **sv** Swedish. |
| **DefLPlStyle** | "ByColor" | R/O | REG | Default plot style for new layers. |
| **DefPlStyle** | "ByColor" | R/O | REG | Default plot style for new objects. |

## Variable    Default         R/O Loc    Meaning

**DelObj**      1                   ...    REG    Toggle source objects deletion:
                                                  **0**  Objects deleted.
                                                  **1**  Objects retained.

**DemandLoad** 3                    ...    REG    When drawing contains proxy objects:
                                                  **0**  Demand loading turned off.
                                                  **1**  Load app when drawing opened.
                                                  **2**  Load app at first command.
                                                  **3**  Load app when drawing opened or at
                                                  first command.

**DiaStat**     1              R/O    ...          User exited dialog box by clicking:
                                                  **0**  **Cancel** button.
                                                  **1**  **OK** button.

## Dimension Variables

**DimADec**     0                   ...    DWG    Angular dimension precision:
                                                  **-1**  Use **DimDec** setting (default).
                                                  **0**  Zero decimal places (minimum).
                                                  **8**  Eight decimal places (maximum).

**DimAlt**      Off                 ...    DWG    Alternate units:
                                                  **On**  Enabled.
                                                  **Off**  Disabled.

**DimAltD**     2                   ...    DWG    Alternate unit decimal places.
**DimAltF**     25.4000             ...    DWG    Alternate unit scale factor.
**DimAltRnd**   0.0000              ...    DWG    Rounding factor of alternate units.
**DimAltTD**    2                   ...    DWG    Tolerance alternate unit decimal places.
**DimAltTZ**    0                   ...    DWG    Alternate tolerance units zeros:
                                                  **0**  Zeros not suppressed.
                                                  **1**  All zeros suppressed.
                                                  **2**  Include 0 feet, but suppress 0 inches .
                                                  **3**  Includes 0 inches, but suppress 0 feet.
                                                  **4**  Suppresses leading zeros .
                                                  **8**  Suppresses trailing zeros .

**DimAltU**     2                   ...    DWG    Alternate units:
                                                  **1**  Scientific.
                                                  **2**  Decimal.
                                                  **3**  Engineering.
                                                  **4**  Architectural; stacked.
                                                  **5**  Fractional; stacked.
                                                  **6**  Architectural.
                                                  **7**  Fractional.
                                                  **8**  Windows desktop units setting.

**DimAltZ**     0                          DWG    Zero suppression for alternate units:
                                                  **0**  Suppress 0 ft and 0 in.
                                                  **1**  Include 0 ft and 0 in.
                                                  **2**  Include 0 ft; suppress 0 in.
                                                  **3**  Suppress 0 ft; include 0 in.
                                                  **4**  Suppress leading 0 in dec dims.
                                                  **8**  Suppress trailing 0 in dec dims.
                                                  **12**  Suppress leading and trailing zeroes.

| Variable | Default | R/O | Loc | Meaning |
|---|---|---|---|---|
| DimAPost | "" | ... | DWG | Prefix and suffix for alternate text. |
| DimAso | On | ... | DWG | Toggle associative dimensions:<br>**On** Dimensions are created associative.<br>**Off** Dimensions are not associative. |
| DimAssoc | 2 | ... | DWG | ▦ Controls creation of dimensions:<br>**0** Dimension elements are exploded.<br>**1** Single dimension object, attached to defpoints.<br>**2** Single dimension object, attached to geometric objects. |
| DimASz | 0.1800 | ... | DWG | Arrowhead length. |
| DimAtFit | 3 | ... | DWG | When insufficient space between extension lines, dimension text and arrows are fitted:<br>**0** Text and arrows outside extension lines.<br>**1** Arrows first outside, then text.<br>**2** Text first outside, then arrows.<br>**3** Either text or arrows, whichever fits better. |
| DimAUnit | 0 | ... | DWG | Angular dimension format:<br>**0** Decimal degrees.<br>**1** Degrees.Minutes.Seconds.<br>**2** Grad.<br>**3** Radian.<br>**4** Surveyor units. |
| DimAZin | 0 | ... | DWG | Supress zeros in angular dimensions:<br>**0** Display all leading and trailing zeros.<br>**1** Suppress 0 in front of decimal.<br>**2** Suppress trailing zeros behind decimal.<br>**3** Suppress zeros in front and behind the decimal. |
| DimBlk | "" | R/O | DWG | Arrowhead block name:<br>Architectural tick: "Archtick"<br>Box filled: "Boxfilled"<br>Box: "Boxblank"<br>Closed blank: "Closedblank"<br>Closed filled: "" (default)<br>Closed: "Closed"<br>Datum triangle filled: "Datumfilled"<br>Datum triangle: "Datumblank"<br>Dot blanked: "Dotblank"<br>Dot small: "Dotsmall"<br>Dot: "Dot"<br>Integral: "Integral"<br>None: "None"<br>Oblique: "Oblique"<br>Open 30: "Open30"<br>Open: "Open"<br>Origin indication: "Origin"<br>Right-angle: "Open90" |

| Variable | Default | R/O | Loc | Meaning |
|---|---|---|---|---|
| **DimBlk1** | "" | R/O | DWG | Name of first arrowhead's block; uses same list of names as under **DimBlk**. |
| | | | | **.** No arrowhead. |
| **DimBlk2** | "" | R/O | DWG | Name of second arrowhead's block; uses same list of names as under **DimBlk**. |
| | | | | **.** No arrowhead. |
| **DimCen** | 0.0900 | ... | DWG | Center mark size: |
| | | | | *-n* Draws center lines. |
| | | | | **0** No center mark or lines drawn. |
| | | | | *+n* Draws center marks of length *n*. |
| **DimClrD** | 0 | ... | DWG | Dimension line color: |
| | | | | **0** BYBLOCK (default) |
| | | | | **1** Red. |
| | | | | ... |
| | | | | **255** Dark gray. |
| | | | | **256** BYLAYER. |
| **DimClrE** | 0 | ... | DWG | Extension line and leader color. |
| **DimClrT** | 0 | ... | DWG | Dimension text color. |
| **DimDec** | 4 | ... | DWG | Primary tolerance decimal places. |
| **DimDLE** | 0.0000 | ... | DWG | Dimension line extension. |
| **DimDLI** | 0.3800 | ... | DWG | Dimension line continuation increment. |
| **DimDSep** | "." | ... | DWG | Decimal separator (must be a single char.) |
| **DimExe** | 0.1800 | ... | DWG | Extension above dimension line. |
| **DimExO** | 0.0625 | ... | DWG | Extension line origin offset. |
| ~~DimFit~~ | *3* | ... | *DWG* | *Obsolete: Autodesk recommends use of* **DimATfit** *and* **DimTMove** *instead.* |
| **DimFrac** | 0 | ... | DWG | Fraction format when **DimLUnit** is set to 4 or 5: |
| | | | | **0** Horizontal. |
| | | | | **1** Diagonal. |
| | | | | **2** Not stacked. |
| **DimGap** | 0.0900 | ... | DWG | Gap from dimension line to text. |
| **DimJust** | 0 | ... | DWG | Horizontal text positioning: |
| | | | | **0** Center justify. |
| | | | | **1** Next to first extension line. |
| | | | | **2** Next to second extension line. |
| | | | | **3** Above first extension line. |
| | | | | **4** Above second extension line. |
| **DimLdrBlk** | "" | ... | DWG | Block name for leader arrowhead; uses same name as **DimBlock**. |
| | | | | **.** Supresses display of arrowhead. |
| **DimLFac** | 1.0000 | ... | DWG | Linear unit scale factor. |
| **DimLim** | Off | ... | DWG | Generate dimension limits. |

| Variable | Default | R/O | Loc | Meaning |
|---|---|---|---|---|
| **DimLUnit** | 2 | ... | DWG | Dimension units (except angular); replaces **DimUnit**:<br>**1** Scientific.<br>**2** Decimal.<br>**3** Engineering.<br>**4** Architectural.<br>**5** Fractional.<br>**6** Windows desktop. |
| **DimLwD** | -2 | ... | DWG | Dimension line lineweight; valid values are BYLAYER, BYBLOCK, or an integer multiple of 0.01mm. |
| **DimLwE** | -2 | ... | DWG | Extension lineweight; valid values are BYLAYER, BYBLOCK, or an integer multiple of 0.01mm. |
| **DimPost** | "" | ... | DWG | Default prefix or suffix for dimension text (maximum 13 characters):<br>"" No suffix.<br>**<>mm** Millimeter suffix.<br>**<>Å** Angstrom suffix. |
| **DimRnd** | 0.0000 | ... | DWG | Rounding value for dimension distances. |
| **DimSAh** | Off | ... | DWG | Separate arrowhead blocks:<br>**Off** Use arrowhead defined by **DimBlk**.<br>**On** Use arrowheads defined by **DimBlk1** and **DimBlk2**. |
| **DimScale** | 1.0000 | ... | DWG | Overall scale factor for dimensions:<br>**0** Value is computed from the scale between current modelspace viewport and paperspace.<br>**>0** Scales text and arrowheads. |
| **DimSD1** | Off | ... | DWG | Suppress first dimension line:<br>**On** First dimension line is suppressed.<br>**Off** Not suppressed. |
| **DimSD2** | Off | ... | DWG | Suppress second dimension line:<br>**On** Second dimension line is suppressed.<br>**Off** Not suppressed. |
| **DimSE1** | Off | ... | DWG | Suppress the first extension line:<br>**On** First extension line is suppressed.<br>**Off** Not suppressed. |
| **DimSE2** | Off | ... | DWG | Suppress the second extension line:<br>**On** Second extension line is suppressed.<br>**Off** Not suppressed. |
| **DimSho** | On | ... | DWG | Update dimensions while dragging:<br>**On** Dimensions are updated during drag.<br>**Off** Dimensions are updated after drag. |
| **DimSOXD** | Off | ... | DWG | Suppress dimension lines outside extension lines:<br>**On** Dimension lines not drawn outside extension lines.<br>**Off** Are drawn outside extension lines. |

| Variable | Default | R/O Loc | Meaning |
|---|---|---|---|
| DimStyle | "STANDARD" | R/O DWG | ⌨ Current dimension style. |
| DimTAD | 0 | ... DWG | Vertical position of dimension text:<br>**0** Centered between extension lines.<br>**1** Above dimension line, except when dimension line not horizontal and **DimTIH** = 1.<br>**2** On side of dimension line farthest from the defining points.<br>**3** Conforms to JIS. |
| DimTDec | 4 | ... DWG | Primary tolerance decimal places. |
| DimTFac | 1.0000 | ... DWG | Tolerance text height scaling factor. |
| DimTIH | On | ... DWG | Text inside extensions is horizontal:<br>**Off** Text aligned with dimension line.<br>**On** Text is horizontal. |
| DimTIX | Off | ... DWG | Place text inside extensions:<br>**Off** Text placed inside extension lines, if room.<br>**On** Force text between the extension lines. |
| DimTM | 0.0000 | ... DWG | Minus tolerance. |
| DimTMove | 0 | ... DWG | Determines how dimension text is moved:<br>**0** Dimension line moves with text.<br>**1** Adds a leader when text is moved.<br>**2** Text moves anywhere; no leader. |
| DimTOFL | Off | ... DWG | Force line inside extension lines:<br>**Off** Dimension lines not drawn when arrowheads are outside.<br>**On** Dimension lines drawn, even when arrowheads are outside. |
| DimTOH | On | ... DWG | Text outside extension lines:<br>**Off** Text aligned with dimension line.<br>**On** Text is horizontal. |
| DimTol | Off | ... DWG | Generate dimension tolerances:<br>**Off** Tolerances not drawn.<br>**On** Tolerances are drawn. |
| DimTolJ | 1 | ... DWG | Tolerance vertical justification:<br>**0** Bottom.<br>**1** Middle.<br>**2** Top. |
| DimTP | 0.0000 | ... DWG | Plus tolerance. |
| DimTSz | 0.0000 | ... DWG | Size of oblique tick strokes:<br>**0** Arrowheads.<br>**>0** Oblique strokes. |
| DimTVP | 0.0000 | ... DWG | Text vertical position when **DimTAD**=0:<br>**1** Turns **DimTAD** on.<br>**>-0.7** *or* **<0.7** Dimension line is split for text. |
| DimTxSty | "STANDARD" | ... DWG | Dimension text style. |
| DimTxt | 0.1800 | ... DWG | Text height. |

| Variable | Default | R/O | Loc | Meaning |
|---|---|---|---|---|
| DimTZin | 0 | ... | DWG | Tolerance zero suppression: |
| | | | | **0** Suppress 0 ft and 0 in. |
| | | | | **1** Include 0 ft and 0 in. |
| | | | | **2** Include 0 ft; suppress 0 in. |
| | | | | **3** Suppress 0 ft; include 0 in. |
| | | | | **4** Suppress leading 0 in decimal dim. |
| | | | | **8** Suppress trailing 0 in decimal dim. |
| | | | | **12** Suppress leading and trailing zeroes. |
| ~~DimUnit~~ | *2* | ... | *DWG* | *Obsolete; replaced by* **DimLUnit** *and* **DimFrac.** |
| DimUPT | Off | ... | DWG | User-positioned text: |
| | | | | **Off** Cursor positions dimension line |
| | | | | **On** Cursor also positions text |
| DimZIN | 0 | ... | DWG | Suppression of 0 in feet-inches units: |
| | | | | **0** Suppress 0 ft and 0 in. |
| | | | | **1** Include 0 ft and 0 in. |
| | | | | **2** Include 0 ft; suppress 0 in. |
| | | | | **3** Suppress 0 ft; include 0 in. |
| | | | | **4** Suppress leading 0 in decimal dim. |
| | | | | **8** Suppress trailing 0 in decimal dim. |
| | | | | **12** Suppress leading and trailing zeroes. |
| DispSilh | 0 | ... | DWG | Silhouette display of 3D solids: |
| | | | | **0** Off. |
| | | | | **1** On. |
| Distance | 0.0000 | R/O | ... | Distance measured by last **Dist** command. |
| ~~Dither~~ | | | | *Removed from Release 14.* |
| DonutId | 0.5000 | ... | ... | Inside radius of donut. |
| DonutOd | 1.0000 | ... | ... | Outside radius of donut. |
| DragMode | 2 | ... | REG | ▦ Drag mode: |
| | | | | **0** No drag. |
| | | | | **1** On if requested. |
| | | | | **2** Automatic. |
| DragP1 | 10 | ... | REG | Regen drag display. |
| DragP2 | 25 | ... | REG | Fast drag display. |
| DwgCheck | 0 | ... | REG | Toggles checking if drawing was edited by software other than AutoCAD: |
| | | | | **0** Supresses dialog box. |
| | | | | **1** Displays warning dialog box. |
| DwgCodePage | *varies* | R/O | DWG | Drawing code page, such as "ANSI_1252" |
| DwgName | *varies* | R/O | ... | Current drawing filename, such as "drawing1.dwg". |
| DwgPrefix | *varies* | R/O | ... | Drawing's drive and folder, such as "d:\acad 2004\". |
| DwgTitled | 0 | R/O | ... | Drawing has filename: |
| | | | | **0** "drawing1.dwg". |
| | | | | **1** User-assigned name. |
| ~~DwgWrite~~ | | | | *Removed from AuoCAD Release 14.* |

. . . . . . . . . . . . . . . . . . . . . . . . . . . . . . . . . . . . . . . . . . . . . . . . . . . . . . . . . . . .

| Variable | Default | R/O | Loc | Meaning |
|---|---|---|---|---|

**E**

**EdgeMode**  0  ...  REG  Toggle edge mode for **Trim** and **Extend** commands:
  **0**  No extension.
  **1**  Extends cutting edge.

**Elevation**  0.0000  ...  DWG  Current elevation, relative to current UCS.

*EntExts*  *1*  ...  ...  *Controls how drawing extents are calculated:*
  ***0***  *Extents calculated every time; slows down AutoCAD but uses less memory.*
  ***1***  *Extents of every object is cached as a two-byte value (default).*
  ***2***  *Extents of every object is cached as a four-byte value (fastest but uses more memory).*

*EntMods*  *0*  R/O  ...  *Increments by one each time an object is modified to indicate that an object has been modified since the drawing was opened; value ranges from 0 to 4.29497E9.*

**ErrNo**  0  ...  ...  Error number from AutoLISP, ADS, & Arx

~~*ExtDir*~~  *Removed from Release 14.*

**Expert**  0  ...  ...  Suppresses the displays of prompts:
  **0**  Normal prompts
  **1**  "About to regen, proceed?" and "Really want to turn the current layer off?"
  **2**  "Block already defined. Redefine it?" and "A drawing with this name already exists. Overwrite it?"
  **3**  **Linetype** command messages.
  **4**  **UCS Save** and **VPorts Save**.
  **5**  **DimStyle Save** and **DimOverride.**

**ExplMode**  1  ...  ...  Toggle whether **Explode** and **Xplode** commands explode non-uniformly scaled blocks:
  **0**  Does not explode.
  **1**  Explodes.

**ExtMax**  -1.0E+20, -1.0E+20, -1.0E+20
    R/O  DWG  Upper-right coordinate of drawing extents.

**ExtMin**  1.0E+20, 1.0E+20, 1.0E+20
    R/O  DWG  Lower-left coordinate of drawing extents.

**ExtNames**  1  ...  DWG  Format of named objects:
  **0**  Names are limited to 31 characters, and can include A - Z, 0 - 9, dollar ($), underscore (_), and hyphen (-).
  **1**  Names are limited to 255 characters, and can include A - Z, 0 - 9, spaces, and any characters not used by Microsoft Windows or AutoCAD for special purposes.

| Variable | Default | R/O | Loc | Meaning |
|---|---|---|---|---|
| **F** | | | | |
| **FaceTRatio** | 0 | ... | ... | Controls the aspect ratio of facets on rouunded 3D bodies:<br>**0** Creates an *n* by 1 mesh.<br>**1** Creates an *n* by *m* mesh. |
| **FaceTRres** | 0.5000 | ... | DWG | Adjusts smoothness of shaded and hidden-line objects:<br>**0.01** Minimum value.<br>**10.0** Maximum value. |
| *FfLimit* | ... | ... | ... | *Removed from AutoCAD Release 14.* |
| **FileDia** | 1 | ... | REG | User interface:<br>**0** Command-line prompts.<br>**1** Dialog boxes, when available. |
| **FilletRad** | 0.5000 | ... | DWG | Current fillet radius. |
| **FillMode** | 1 | ... | DWG | Fill of solid objects:<br>**0** Off.<br>**1** On. |
| *Flatland* | *0* | R/O | ... | *Obsolete system variable.* |
| **FontAlt** | "simplex.shx" | ... | REG | Name for substituted font. |
| **FontMap** | "acad.fmp" | ... | REG | Name of font mapping file. |
| *Force_Paging* | *0* | ... | ... | ***0*** *Minimum (default).*<br>***4.29497E9*** *Maximum.* |
| **FrontZ** | 0.0000 | R/O | DWG | Front clipping plane offset. |
| **FullOpen** | 1 | R/O | ... | Drawing is:<br>**0** Partially loaded.<br>**1** Fully open. |
| **G** | | | | |
| **GfAng** 🔲 | 0 | ... | ... | Angle of gradient fill; 0 to 360 degrees. |
| **GfClr1** 🔲 | "RGB 000,000,255" | ... | ... | First gradient color in RGB format. |
| **GfClr2** 🔲 | RGB 255,255,153" | ... | ... | Second gradient color in RGB format. |
| **GfClrLum** 🔲 | 1.0 | ... | ... | Level of gray in one-color gradients:<br>**0** Black<br>**1** White |
| **GfClrState** 🔲 | 1 | ... | ... | Specifies type of gradient fill:<br>**0** Two-color<br>**1** One-color |

| Variable | Default | R/o | Loc | Meaning |
|---|---|---|---|---|
| **GfName** ⓐ | 1 | ... | ... | Specifies style of gradient fill: |
| | | | | **1** Linear |
| | | | | **2** Cylindrical |
| | | | | **3** Inverted cylindrical |
| | | | | **4** Spherical |
| | | | | **5** Inverted spherical |
| | | | | **6** Hemispherical |
| | | | | **7** Inverted hemispherical |
| | | | | **8** Curved |
| | | | | **9** Inverted curved |
| **GfAShift** ⓐ | 0 | ... | ... | Specifies the origin of the gradient fill: |
| | | | | **0** Centered |
| | | | | **1** Shifted up and left |
| *GlobCheck* | *0* | *...* | *...* | *Reports statistics on dialog boxes:* |
| | | | | *-1 Turn off local language.* |
| | | | | *0 Turn off.* |
| | | | | *1 Warns if larger than 640x400.* |
| | | | | *2 Also reports size in pixels.* |
| | | | | *3 Additional info.* |
| **GridMode** | 0 | ... | DWG | Display of grid: |
| | | | | **0** Off. |
| | | | | **1** On. |
| **GridUnit** | 0.5000,0.5000 | ... | DWG | X,y-spacing of grid. |
| **GripBlock** | 0 | ... | REG | Display of grips in blocks: |
| | | | | **0** At insertion point. |
| | | | | **1** At all objects within block. |
| **GripColor** | 160 | ... | REG | Color of unselected grips: |
| | | | | **1** Minimum color number; red. |
| | | | | **160** Default color; blue. |
| | | | | **255** Maximum color number. |
| **GripHot** | 1 | ... | REG | Color of selected grips: |
| | | | | **1** Default color, red. |
| | | | | **255** Maximum color number. |
| **GripHover** ⓐ | 3 | ... | REG | Grip fill color when cursor hovers. |
| **GripObjLimit** ⓐ | | | | |
| | 100 | ... | REG | Grips not displayed when more than this number: |
| | | | | **1** Minimum |
| | | | | **32767** Maximum. |
| **Grips** | 1 | ... | REG | Display of grips: |
| | | | | **0** Off. |
| | | | | **1** On |
| **GripSize** | 3 | ... | REG | Size of grip box, in pixels: |
| | | | | **1** Minimum size. |
| | | | | **255** Maximum size. |

| Variable | Default | R/O | Loc | Meaning |
|---|---|---|---|---|
| GripTips 🔲 | 1 | ... | REG | Determines if grip tips are displayed when the cursor hovers over custom objects:<br>**0** Off.<br>**1** On |

**H**

| Variable | Default | R/O | Loc | Meaning |
|---|---|---|---|---|
| HaloGap | 0 | ... | DWG | Distance to shorten a haloed line; specified as the percentage of 1". |
| Handles | 1 | R/O | ... | 🔳 Obsolete system variable. |
| HidePrecision | 0 | ... | DWG | Controls the precision of hide calculations:<br>**0** Single precision, less accurate, faster.<br>**1** Double precision, more accurate, but slower. |
| HideText | 0 | ... | ... | Determines whether text is hidden during the **Hide** command:<br>**0** Text is not hidden nor hides other objects, unless text object has thickness.<br>**1** Text is hidden and hides other objects. |
| Highlight | 1 | ... | ... | Object selection highlighting:<br>**0** Disabled.<br>**1** Enabled. |
| HPAng | 0 | ... | ... | Current hatch pattern angle. |
| HPAssoc | 0 | ... | ... | Determines if hatches are associative:<br>**0** Not associative<br>**1** Associative |
| HPBound | 1 | ... | REG | Object created by **BHatch** and **Boundary** commands:<br>**0** Region.<br>**1** Polyline. |
| HPDouble | 0 | ... | ... | Double hatching:<br>**0** Disabled.<br>**1** Enabled. |
| HPName | "ANSI31" | ... | ... | Current hatch pattern name<br>**""** No default.<br>**.** Set no default. |
| HPScale | 1.0000 | ... | ... | Current hatch scale factor; cannot be zero. |
| HPSpace | 1.0000 | ... | ... | Current spacing of user-defined hatching; cannot be zero. |
| HyperlinkBase | "" | ... | DWG | Path for relative hyperlinks. |

**I**

| Variable | Default | R/O | Loc | Meaning |
|---|---|---|---|---|
| ImageHlt | 0 | ... | REG | When a raster image is selected:<br>**0** Image frame is highlighted.<br>**1** Entire image is highlighted. |

| Variable | Default | R/O | Loc | Meaning |
|---|---|---|---|---|
| IndexCtl | 0 | ... | DWG | Creates layer and spatial indices:<br>  **0**  No indices created.<br>  **1**  Layer index created.<br>  **2**  Spatial index created.<br>  **3**  Both indices created. |
| InetLocation | "www.autodesk.com" | ... | REG | Default browser URL. |
| InsBase | 0.0000,0.0000,0.0000 | ... | DWG | Insertion base point relative to the current UCS for **Insert** and **DdInsert**. |
| InsName | "" | ... | ... | Current block name:<br>  **.**  Set to no default. |
| InsUnits | 1 | ... | ... | Drawing units when a block is dragged into drawing from DesignCenter:<br>  **0**  Unitless.<br>  **1**  Inches.<br>  **2**  Feet.<br>  **3**  Miles.<br>  **4**  Millimeters.<br>  **5**  Centimeters.<br>  **6**  Meters.<br>  **7**  Kilometers.<br>  **8**  Microinches.<br>  **9**  Mils.<br>  **10**  Yards.<br>  **11**  Angstroms.<br>  **12**  Nanometers.<br>  **13**  Microns.<br>  **14**  Decimeters.<br>  **15**  Decameters.<br>  **16**  Hectometers.<br>  **17**  Gigameters.<br>  **18**  Astronomical Units.<br>  **19**  Light Years.<br>  **20**  Parsecs. |
| InsUnitsDefSource | 1 | ... | REG | Source drawing units value; ranges from 0 to 20; see above. |
| InsUnitsDefTarget | 1 | ... | REG | Target drawing units value; ranges from 0 to 20. |
| IntersectionDisplay | 0 | ... | DWG | Determines whether polylines are drawn by the **Hide** command at 3D surface intersections. |
| ISaveBak | 1 | ... | REG | Controls whether *.bak* file is created:<br>  **0**  No file created.<br>  **1**  *.bak* backup file created. |
| ISavePercent | 50 | ... | REG | Percentage of waste in *.dwg* file before cleanup occurs:<br>  **0**  Every save is a full save. |

| Variable | Default | R/O | Loc | Meaning |
|---|---|---|---|---|
| IsoLines | 4 | ... | DWG | Isolines on 3D solids:<br>**0** No isolines; minimum.<br>**16** Good-looking.<br>**2,047** Maximum. |

. . . . . . . . . . . . . . . . . . . . . . . . . . . . . . . . . . . . . . . . . . . . . . .

## L

| Variable | Default | R/O | Loc | Meaning |
|---|---|---|---|---|
| LastAngle | 0 | R/O | ... | Ending angle of last-drawn arc. |
| LastPoint | *varies* | ... | ... | Last-entered point, such as 15,9,56. |
| LastPrompt | "" | R/O | ... | Last string on the command line; includes user input. |
| *LazyLoad* | *0* | ... | ... | *Toggle: 0 or 1.* |
| LayoutRegenCtl | | | | |
| | 2 | ... | REG | Controls display list for layouts:<br>**0** Display list is regenerated with each tab change.<br>**1** Display list is saved for model tab and last layout tab.<br>**2** Display list is saved for all tabs. |
| LensLength | 50.0000 | R/O | DWG | Perspective view lens length, in mm. |
| LimCheck | 0 | ... | DWG | Drawing limits checking:<br>**0** Disabled.<br>**1** Enabled. |
| LimMax | 12.0000,9.0000 | ... | DWG | Upper right drawing limits. |
| LimMin | 0.0000,0.0000 | ... | DWG | Lower left drawing limits. |
| LispInit | 1 | ... | REG | AutoLISP functions and variables are:<br>**0** Preserved from drawing to drawing.<br>**1** Valid in current drawing only. |
| Locale | "enu" | R/O | ... | ISO language code. |
| LocalRootPrefix ⬚ | | | | |
| | "d:\documents and Settings\administrator\local settings\appli..." | R/O | REG | Path to folder holding local customizable files. |
| LogFileMode | 0 | ... | REG | Text window written to *.log* file:<br>**0** No.<br>**1** Yes. |
| LogFileName | "d:\acad 2004\Drawing1_1_1_0000.log" | R/O | DWG | Filename and path for *.log* file. |
| LogFilePath | "d:\acad 2004\" | ... | REG | Path for the *.log* file. |
| LogInName | "" | R/O | ... | User's login name; max = 30 chars. |
| ~~LongFName~~ | | | | *Removed from AutoCAD Release 14.* |
| LTScale | 1.0000 | ... | DWG | ▦ Current linetype scale factor; cannot be 0. |
| LUnits | 2 | ... | DWG | Linear units mode:<br>**1** Scientific.<br>**2** Decimal.<br>**3** Engineering.<br>**4** Architectural.<br>**5** Fractional. |

. . . . . . . . . . . . . . . . . . . . . . . . . . . . . . . . . . . . . . . . . . . . . . .

| Variable | Default | R/o | Loc | Meaning |
|----------|---------|-----|-----|---------|
| **LUPrec** | 4 | ... | DWG | Decimal places of linear units. |
| **LwDefault** | 25 | ... | REG | Default lineweight, in millimeters; must be one of the following values: 0, 5, 9, 13, 15, 18, 20, 25, 30, 35, 40, 50, 53, 60, 70, 80, 90, 100, 106, 120, 140, 158, 200, or 211. |
| **LwDisplay** | 0 | ... | DWG | Toggles whether lineweight is displayed; setting is saved separately for Model space and each layout tab.<br>**0** Not displayed.<br>**1** Displayed. |
| **LwUnits** | 1 | ... | REG | Determines units for lineweight:<br>**0** Inches.<br>**1** Millimeters. |

## M

| Variable | Default | R/o | Loc | Meaning |
|----------|---------|-----|-----|---------|
| *MacroTrace* | *0* | ... | ... | *Diesel debug mode:*<br>*0  Off.*<br>*1  On.* |
| **MaxActVP** | 64 | ... | ... | Maximum viewports to regenerate:<br>**2** Minimum.<br>**64** Maximum (increased from 48 in R14). |
| *MaxObjMem* | *0* | ... | ... | *Maximum number of objects in memory; object pager is turned off when value = 0, <0, or 2,147,483,647.* |
| **MaxSort** | 200 | ... | REG | Maximum names sorted alphabetically. |
| **MButtonPan** | 1 | ... | REG | Determines behavior of wheelmouse:<br>**0** As defined by AutoCAD menu file.<br>**1** Pans when dragging with wheel. |
| **MeasureInit** | 0 | ... | REG | Drawing units:<br>**0** English.<br>**1** Metric. |
| **Measurement** | 0 | ... | DWG | Drawing units (overrides **MeasureInit**):<br>**0** English.<br>**1** Metric. |
| **MenuCtl** | 1 | ... | REG | Submenu display:<br>**0** Only with menu picks.<br>**1** Also with keyboard entry. |
| **MenuEcho** | 0 | ... | | Menu and prompt echoing:<br>**0** Display all prompts.<br>**1** Suppress menu echoing.<br>**2** Suppress system prompts.<br>**4** Disable **^P** toggle.<br>**8** Display all input-output strings. |
| **MenuName** | "acad" | R/o | REG | Current menu filename. |
| **MirrText** | 1 | ... | DWG | Text handling during **Mirror** command:<br>**0** Retain text orientation.<br>**1** Mirror text. |

## Variable   Default          R/O Loc   Meaning

**ModeMacro**   ""           ...   ...      Invoke Diesel programming language.

**MTextEd**   "Internal"     ...   REG      Name of the **MText** editor:
         **.**  Use default editor.
         **0**  Cancel the editing operation.
        **-1**  Use the secondary editor.
        **"blank"**     MTEXT internal editor.
        **"Internal"**  MTEXT internal editor.
        **"Notepad"**  Windows Notepad editor.
        **":lisped"**     Built-in AutoLISP function.
        *string* Name of editor; must be less than 256
        characters long and use this syntax:
        *:AutoLISPtextEditorFunction#TextEditor*

**MTextFixed** 🖽 0     ...   REG      Specifies mtext editor appearence:
        **0**  Displays mtext editor and text at same
        size and position as object being edited.
        **1**  Displays mtext editor at the same
        location as last used; used fixed height text.

**MTJigStrings** 🖽
      "abc"     ...   REG      Sample text displayed by mtext editor;
        maximum 10 letters; enter . for no text.

**MyDocumentsPrefix** 🖽
      ""C:\Documents and Settings\administrator\My Documents"
             R/O   REG      Path to the \\*my documents* folder of the
        currently logged-in user.

- - - - - - - - - - - - - - - - - - - - - - - - - - - - - - - - - - - - - - - - - -

**N**

*NodeName*   *"AC$"*     R/O   REG   *Name of network node; range is one to three characters.*

**NoMutt**   0     ...   ...      Suppresses the display of message (a.k.a.
        muttering) during scripts, LISP, macros:
        **0**  Display prompt, as normal.
        **1**  Suppress muttering.

- - - - - - - - - - - - - - - - - - - - - - - - - - - - - - - - - - - - - - - - - -

**O**

**ObscureColor**  0     ...   DWG      Color of objects obscured by **Hide**
        command:
        **0**  Invisible
        **1 - 255** Color number.

| Variable | Default | R/o | Loc | Meaning |
|---|---|---|---|---|
| **ObscureLtype** | 0 | ... | DWG | Linetype of objects obscured by **Hide** command:<br>0 Invisible.<br>1 Solid.<br>2 Dashed.<br>3 Dotted.<br>4 Short dash.<br>5 Medium dash.<br>6 Long dash.<br>7 Double short dash.<br>8 Double medium dash.<br>9 Double long dash.<br>10 Medium long dash.<br>11 Sparse dot. |
| **OffsetDist** | 1.0000 | ... | ... | Current offset distance:<br><0 Offsets through a specified point.<br>>0 Default offset distance. |
| **OffsetGapType** | 0 | ... | REG | Determines how to reconnect polyline when individual segments are offset:<br>0 Extend segments to fill gap.<br>1 Fill gap with fillet (arc segment).<br>2 Fill gap with chamfer (line segment). |
| **OleHide** | 0 | ... | REG | Display and plotting of OLE objects:<br>0 All OLE objects visible.<br>1 Visible in paper space only.<br>2 Visible in model space only.<br>3 Not visible. |
| **OleQuality** | 1 | ... | REG | Quality of display and plotting of embedded OLE objects:<br>0 Line art quality.<br>1 Text quality.<br>2 Graphics quality.<br>3 Photograph quality.<br>4 High quality photograph. |
| **OleStartup** | 0 | ... | DWG | Loading OLE source application improves plot quality:<br>0 Do not load OLE source app.<br>1 Load OLE source app when plotting. |
| **OrthoMode** | 0 | ... | DWG | Orthographic mode:<br>0 Off.<br>1 On. |
| **OSMode** | 4133 | ... | REG | Current object snap mode:<br>0 NONe.<br>1 ENDpoint.<br>2 MIDpoint.<br>4 CENter.<br>8 NODe.<br>16 QUAdrant.<br>32 INTersection.<br>64 INSertion. |

| Variable | Default | | R/o | Loc | Meaning |
| --- | --- | --- | --- | --- | --- |
| | | | | | 128 PERpendicular. |
| | | | | | 256 TANgent. |
| | | | | | 512 NEARest. |
| | | | | | 1024 QUIck. |
| | | | | | 2048 APPint. |
| | | | | | 4096 EXTension. |
| | | | | | 8192 PARallel. |
| | | | | | 16383 All modes on. |
| | | | | | 16384 Object snap turned off via **OSNAP** on the status bar. |
| OSnapCoord | 2 | | ... | REG | Keyboard overrides object snap: |
| | | | | | 0 Object snap override keyboard. |
| | | | | | 1 Keyboard overrides object snap. |
| | | | | | 2 Keyboard overrides object snap, except in script. |

- - - - - - - - - - - - - - - - - - - - - - - - - - - - - - - - - - - - -

## P

**PaletteOpaque** [a]

| Variable | Default | | R/o | Loc | Meaning |
| --- | --- | --- | --- | --- | --- |
| | 0 | | ... | REG | Determines if palettes can be made transparent: |
| | | | | | 0 Turned off by user. |
| | | | | | 1 Turned on by user. |
| | | | | | 2 Unavailable, but turned on by user. |
| | | | | | 3 Unavailable, and turned off by user. |
| PaperUpdate | 0 | | ... | REG | Determines how AutoCAD plots a layout with paper size different from plotter's default: |
| | | | | | 0 Displays a warning dialog box. |
| | | | | | 1 Changes paper size to that of the plotter configuration file. |
| PDMode | 0 | | ... | DWG | Point display mode: |
| | | | | | 0 Dot. |
| | | | | | 1 No display. |
| | | | | | 2 +-symbol. |
| | | | | | 3 x-symbol. |
| | | | | | 4 Short line. |
| | | | | | 32 Circle. |
| | | | | | 64 Square. |

| 0 | 1 | 2 | 3 | 4 |
| --- | --- | --- | --- | --- |
| 32 | 33 | 34 | 35 | 36 |
| 64 | 65 | 66 | 67 | 68 |
| 96 | 97 | 98 | 99 | 100 |

| Variable | Default | | R/o | Loc | Meaning |
| --- | --- | --- | --- | --- | --- |
| PDSize | 0.0000 | | ... | DWG | Point display size, in pixels: |
| | | | | | >0 Absolute size. |
| | | | | | 0 5% of drawing area height. |
| | | | | | <0 Percentage of viewport size. |

- - - - - - - - - - - - - - - - - - - - - - - - - - - - - - - - - - - - -

| Variable | Default | R/O Loc | Meaning |
|---|---|---|---|

**PEdit Accept** ⬚

| | 0 | ... reg | Suppresses display of the **PEdit** command's "Object selected is not a polyline. Do you want to turn it into one? <Y>" prompt. |
|---|---|---|---|
| **PEllipse** | 0 | ... DWG | Toggle **Ellipse** creation: <br> **0** True ellipse. <br> **1** Polyline arcs. |
| **Perimeter** | 0.0000 | R/O ... | Perimeter calculated by the last **Area**, **DbList**, and **List** commands. |
| **PFaceVMax** | 4 | R/O ... | Maximum vertices per 3D face. |
| *PHandle* | *0* | ... ACAD | *Ranges from 0 to 4.29497E9.* |
| **PickAdd** | 1 | ... REG | Effect of **SHIFT** key on selection set: <br> **0** Adds to selection set. <br> **1** Removes from selection set. |
| **PickAuto** | 1 | ... REG | Selection set mode: <br> **0** Single pick mode. <br> **1** Automatic windowing and crossing. |
| **PickBox** | 3 | ... REG | Object selection pickbox size, in pixels: <br> **0** Minimum size. <br> **50** Maximum size. |
| **PickDrag** | 0 | ... REG | Selection window mode: <br> **0** Pick two corners. <br> **1** Pick a corner; drag to second corner. |
| **PickFirst** | 1 | ... REG | Command-selection mode: <br> **0** Enter command first. <br> **1** Select objects first. |
| **PickStyle** | 1 | ... REG | Include groups and associative hatches in selection: <br> **0** Neither included. <br> **1** Include groups. <br> **2** Include associative hatches. <br> **3** Include both. |
| **Platform** | "Microsoft Windows NT Version 5.0 (x86)" | | |
| | | R/O ... | AutoCAD platform (name of the operating system). |
| **PLineGen** | 0 | ... DWG | Polyline linetype generation: <br> **0** From vertex to vertex. <br> **1** From end to end. |
| **PLineType** | 2 | ... REG | Automatic conversion and creation of 2D polylines by **PLine**: <br> **0** Not converted; old-format polylines created. <br> **1** Not converted; optimized polylines created. <br> **2** Polylines in older drawings are converted on open; **PLine** creates optimized polylines with Lwpolyline object. |
| **PLineWid** | 0.0000 | ... DWG | Current polyline width. |

| Variable | Default | R/o | Loc | Meaning |
|---|---|---|---|---|
| PlotId | "" | ... | REG | Obsolete; has no effect in AutoCAD 2004. |
| PlotRotMode | 1 | ... | DWG | Orientation of plots:<br>**0** Lower left = 0,0.<br>**1** Lower left plotter area = lower left of media.<br>**2** X, y-origin offsets calculated relative to the rotated origin position. |
| Plotter | 0 | ... | REG | Obsolete; has no effect in AutoCAD 2004. |
| PlQuiet | 0 | ... | REG | Toggles display during batch plotting and scripts (replaces **CmdDia**):<br>**0** Plot dialog boxes and nonfatal errors are displayed.<br>**1** Nonfatal errors are logged; plot dialog boxes are not displayed. |
| PolarAddAng | "" | ... | REG | Contains a list of up to 10 user-defined polar angles; each angle can be up to 25 characters long, each separated with a semicolon (;). For example: 0;15;22.5;45. |
| PolarAng | 90 | ... | REG | Specifies the increment of polar angle; contrary to Autodesk documentation, you may specify any angle. |
| PolarDist | 0.000 | ... | REG | The polar snap increment when **SnapStyl** is set to 1 (isometric). |
| PolarMode | 0 | ... | REG | Settings for polar and object snap tracking:<br>**0** Measure polar angles based on current UCS (absolute), track orthogonally; don't use additional polar tracking angles; and acquire object tracking points automatically.<br>**1** Measure polar angles from selected objects (relative).<br>**2** Use polar tracking settings in object snap tracking.<br>**4** Use additional polar tracking angles (via **PolarAng**).<br>**8** Press SHIFT to acquire object snap tracking points. |
| PolySides | 4 | ... | ... | Current number of polygon sides:<br>**3** Minimum sides.<br>**1024** Maximum sides. |
| Popups | 1 | R/O | ... | Display driver support of AUI:<br>**0** Not available.<br>**1** Available. |
| Product | "AutoCAD" | R/O | ACAD | Name of the software. |
| Program | "acad" | R/O | ACAD | Name of the software's executable file. |
| ProjectName | "" | ... | DWG | Project name of the current drawing; searches for xref and image files. |

| Variable | Default | R/o | Loc | Meaning |
|---|---|---|---|---|
| **ProjMode** | 1 | ... | REG | Projection mode for **Trim** and **Extend** commands:<br>**0** No projection.<br>**1** Project to x,y-plane of current UCS.<br>**2** Project to view plane. |
| **ProxyGraphics** | 1 | ... | REG | Proxy image saved in the drawing:<br>**0** Not saved; displays bounding box.<br>**1** Image saved with drawing. |
| **ProxyNotice** | 1 | ... | REG | Display warning message:<br>**0** No.<br>**1** Yes. |
| **ProxyShow** | 1 | ... | REG | Display of proxy objects:<br>**0** Not displayed.<br>**1** All displayed.<br>**2** Bounding box displayed. |
| **ProxyWebSearch** | 0 | ... | REG | Object enablers are checked:<br>**0** AutoCAD does not check for object enablers.<br>**1** AutoCAD checks for object enablers if an Internet connection is present. |
| **PsLtScale** | 1 | ... | DWG | Paper space linetype scaling:<br>**0** Use model space scale factor.<br>**1** Use viewport scale factor. |
| *PsProlog* | *""* | ... | REG | *PostScript prologue filename.* |
| *PsQuality* | *75* | ... | REG | *Resolution of PostScript display, in pixels:*<br>*<0 Display as outlines; no fill.*<br>*0 Not displayed.*<br>*>0 Display filled.* |
| **PStyleMode** | 1 | ... | DWG | Toggles the plot color matching mode of the drawing:<br>**0** Use named plot style tables.<br>**1** Use color-dependent plot style tables. |
| **PStylePolicy** | 1 | ... | REG | Determines whether the object color is associated with its plot style:<br>**0** Color and plot style not associated.<br>**1** Object's plot style is associated with its color. |
| **PsVpScale** | 0 | ... | ... | Sets the view scale factor (the ratio of units in paper space to the units in newly created model space viewports) for all newly-created viewports:<br>**0** Scaled to fit. |
| **PUcsBase** | "" | ... | DWG | Name of UCS defining the origin and orientation of orthographic UCS settings in paper space only. |

| Variable | Default | R/O | Loc | Meaning |
|----------|---------|-----|-----|---------|

**Q**

*QAFlags*  0  ...  ...  *Quality assurance flags:*
    **0**  *Turned off.*
    **1**  *The ^C metacharacters in a menu macro cancels grips, just as if user pressed* **Esc***.*
    **2**  *Long text screen listings do not pause.*
    **4**  *Error and warning messages are displayed at the command line, instead of in dialog boxes.*
    **128**  *Screen picks are accepted via the AutoLISP (command) function.*

**QTextMode**  0  ...  DWG  Quick text mode:
    **0**  Off.
    **1**  On.

**R**

**RasterPreview**  1  R/O  REG  Preview image:
    **0**  None saved.
    **1**  Saved in *.bmp* format.

**RefEditName**""  ...  ...  The reference filename when it is in reference-editing mode.

**RegenMode**  1  ...  DWG  Regeneration mode:
    **0**  Regen with each view change.
    **1**  Regen only when required.

**Re-Init**  0  ...  ...  Reinitialize I/O devices:
    **1**  Digitizer port.
    *2*  *Plotter port.*
    **4**  Digitizer.
    *8*  *Plotter.*
    **16**  Reload PGP file.

**RememberFolders**

    1  ...  REG  Controls path search method:
    **0**  Path specified in desktop AutoCAD icon is default for file dialog boxes.
    **1**  Last path specified by each file dialog box is remembered.

**ReportError** 🗗 1  ...  REG  Determines if AutoCAD sends an error report to Autodesk:
    **0**  No error report created.
    **1**  Error report is generated and sent to Autodesk.

**RoamableRootPrefix** 🗗

    "d:\documents and settings\administrator\application aata\aut..."

    R/O  REG  Path to root folder where roamable customized files are located.

~~RIAspect~~  *Removed from AutoCAD Release 14.*
~~RIBackG~~  *Removed from AutoCAD Release 14.*
~~RIEdge~~  *Removed from AutoCAD Release 14.*

| Variable | Default | R/o | Loc | Meaning |
|---|---|---|---|---|
| ~~RIGamut~~ | | | | Removed from *AutoCAD Release 14.* |
| ~~RIGrey~~ | | | | Removed from *AutoCAD Release 14.* |
| ~~RIThresh~~ | | | | Removed from *AutoCAD Release 14.* |
| **RTDisplay** | 1 | | ... REG | Raster display during real-time zoom and pan: |
| | | | |   **0** Display the entire raster image. |
| | | | |   **1** Display raster outline only. |

. . . . . . . . . . . . . . . . . . . . . . . . . . . . . . . . . . . . . . .

## S

| Variable | Default | R/o | Loc | Meaning |
|---|---|---|---|---|
| **SaveFile** | "auto.sv$" | R/O | REG | Automatic save filename. |
| **SaveFilePath** | "d:\temp\" | ... | REG | Path for automatic save files. |
| **SaveName** | "" | R/O | ... | Drawing save-as filename. |
| **SaveTime** | 10 | ... | REG | Automatic save interval, in minutes: |
| | | | |   **0** Disable auto save. |
| **ScreenBoxes** | 0 | R/O | ACAD | Maximum number of menu items |
| | | | |   **0** Screen menu turned off. |
| **ScreenMode** | 3 | R/O | ... | State of AutoCAD display screen: |
| | | | |   **0** Text screen. |
| | | | |   **1** Graphics screen. |
| | | | |   **2** Dual-screen display. |
| **ScreenSize** | *varies* | R/O | ... | Current viewport size, in pixels, such as 719.0000,381.0000. |
| **SDI** | 0 | ... | REG | Toggles multiple-document interface (SDI is "single document interface"): |
| | | | |   **0** Turns on MDI. |
| | | | |   **1** Turns off MDI (only one drawing may be loaded into AutoCAD). |
| | | | |   **2** MDI disabled for apps that cannot support MDI; read-only. |
| | | | |   **3** MDI disabled for apps that cannot support MDI, even when **SDI** set to 1; R/O. |
| **ShadEdge** | 3 | ... | DWG | **Shade** style: |
| | | | |   **0** Shade faces; 256-color shading. |
| | | | |   **1** Shade faces; edges background color. |
| | | | |   **2** Hidden-line removal. |
| | | | |   **3** 16-color shading. |
| **ShadeDif** | 70 | ... | DWG | Percent of diffuse to ambient light: |
| | | | |   **0** Minimum. |
| | | | |   **100** Maximum. |
| **ShortcutMenu** | 11 | ... | REG | Toggles availability of shortcut menus: |
| | | | |   **0** Disables all default, edit, and command shortcut menus. |
| | | | |   **1** Enables default shortcut menus. |
| | | | |   **2** Enables edit shortcut menus. |
| | | | |   **4** Enables command shortcut menus whenever a command is active. |
| | | | |   **8** Enables command shortcut menus only when command options are available at the command line. |

. . . . . . . . . . . . . . . . . . . . . . . . . . . . . . . . . . . . . . .

| Variable | Default | R/o | Loc | Meaning |
|---|---|---|---|---|
| **ShpName** | "" | ... | ... | Current shape name: <br> • Set to no default. |
| **SigWarn** ⬚ | 1 | ... | REG | Determines whether a warning is displayed when a file is opened with a digital signature. |
| **SketchInc** | 0.1000 | ... | DWG | **Sketch** command's recording increment. |
| **SkPoly** | 0 | ... | DWG | Sketch line mode: <br> 0 Record as lines. <br> 1 Record as a polyline. |
| **SnapAng** | 0 | ... | DWG | Current rotation angle for snap and grid. |
| **SnapBase** | 0.0000,0.0000 | ... | DWG | Current origin for snap and grid. |
| **SnapIsoPair** | 0 | ... | DWG | Current isometric drawing plane: <br> 0 Left isoplane. <br> 1 Top isoplane. <br> 2 Right isoplane. |
| **SnapMode** | 0 | ... | DWG | Snap mode: <br> 0 Off. <br> 1 On. |
| **SnapStyl** | 0 | ... | DWG | Snap style: <br> 0 Normal. <br> 1 Isometric. |
| **SnapType** | 0 | ... | REG | Toggles between standard or polar snap for the current viewport: <br> 0 Standard snap. <br> 1 Polar snap. |
| **SnapUnit** | 0.5000,0.5000 | ... | DWG | X,y-spacing for snap. |
| **SolidCheck** | 1 | ... | ... | Toggles solid validation: <br> 0 Off. <br> 1 On. |
| **SortEnts** | 96 | ... | DWG | Object display sort order: <br> 0 Off. <br> 1 Object selection. <br> 2 Object snap. <br> 4 Redraw. <br> 8 Slide generation. <br> 16 Regeneration. <br> 32 Plot. <br> 64 PostScript output. |
| **SplFrame** | 0 | ... | DWG | Polyline and mesh display: <br> 0 Polyline control frame not displayed; display polygon fit mesh; 3D faces invisible edges not displayed <br> 1 Polyline control frame displayed; display polygon defining mesh; 3D faces invisible edges displayed. |
| **SplineSegs** | 8 | ... | DWG | Number of line segments that define a splined polyline. |

| Variable | Default | R/O | Loc | Meaning |
|---|---|---|---|---|
| **SplineType** | 6 | ... | DWG | Spline curve type:<br>**5** Quadratic Bezier spline.<br>**6** Cubic Bezier spline. |
| **StandardsViolation** 🔳 | | | | |
| | 2 | ... | REG | Determines whether alerts are displayed when CAD standards are violated:<br>**0** No alerts.<br>**1** Alert displayed when CAD standard violated.<br>**2** Displays icon on status bar when file is opened with CAD standards, and when non-standard objects are created. |
| **Startup** 🔳 | 0 | ... | reg | Determines which dialog box is displayed by the **New** and **QNew** commands:<br>**0** Displays **Select Template** dialog box.<br>**1** Displays **Startup** and **Create New Drawing** dialog box. |
| ~~StartupToday~~ | | | | ~~Removed from AutoCAD 2004.~~ |
| **SurfTab1** | 6 | ... | DWG | Density of surfaces and meshes:<br>**5** Minimum.<br>**32766**Maximum. |
| **SurfTab2** | 6 | ... | DWG | Density of surfaces and meshes:<br>**2** Minimum.<br>**32766**Maximum. |
| **SurfType** | 6 | ... | DWG | Pedit surface smoothing:<br>**5** Quadratic Bezier spline.<br>**6** Cubic Bezier spline.<br>**8** Bezier surface. |
| **SurfU** | 6 | ... | DWG | Surface density in m-direction:<br>**2** Minimum.<br>**200** Maximum. |
| **SurfV** | 6 | ... | DWG | Surface density in n-direction:<br>**2** Minimum.<br>**200** Maximum. |
| **SysCodePage** | "ANSI_1252" | R/O | ... | System code page. |

**T**

| Variable | Default | R/O | Loc | Meaning |
|---|---|---|---|---|
| **TabMode** | 0 | ... | ... | Tablet mode:<br>**0** Off.<br>**1** On. |
| **Target** | 0.0000,0.0000,0.0000 | R/O | DWG | Target in current viewport. |
| **TDCreate** | *varies* | R/O | DWG | Time and date drawing created, such as 2448860.54014699. |
| **TDInDwg** | *varies* | R/O | DWG | Duration drawing loaded, such as 0.00040625. |
| **TDuCreate** | *varies* | R/O | DWG | The universal time and date the drawing was created, such as 2451318.67772165. |

| Variable | Default | R/O | Loc | Meaning |
|---|---|---|---|---|
| TDUpdate | *varies* | R/O | DWG | Time and date of last update, such as 2448860.54014699. |
| TDUsrTimer | *varies* | R/O | DWG | Time elapsed by user-timer, such as 0.00040694. |
| TDuUpdate | *varies* | R/O | DWG | The universal time and date of the last save, such as 2451318.67772165. |
| TempPrefix | "d:\temp" | R/O | ... | Path for temporary files. |
| TextEval | 0 | ... | ... | Interpretation of text input:<br>**0** Literal text.<br>**1** Read ( and ! as AutoLISP code. |
| TextFill | 1 | ... | REG | Toggle fill of TrueType fonts:<br>**0** Outline text.<br>**1** Filled text. |
| TextQlty | 50 | ... | DWG | Resolution of TrueType fonts:<br>**0** Minimum resolution<br>**100** Maximum resolution. |
| TextSize | 0.2000 | ... | DWG | Current height of text. |
| TextStyle | "Standard" | ... | DWG | Current name of text style. |
| Thickness | 0.0000 | ... | DWG | Current object thickness. |
| TileMode | 1 | ... | DWG | View mode:<br>**0** Display layout tab.<br>**1** Display model tab. |
| ToolTips | 1 | ... | REG | Display tooltips:<br>**0** Off.<br>**1** On. |
| TpState 🔲 | 0 | R/O | ... | Determines if Tool Palettes is open. |
| TraceWid | 0.0500 | ... | DWG | Current width of traces. |
| TrackPath | 0 | ... | REG | Determines the display of polar and object snap tracking alignment paths:<br>**0** Display object snap tracking path across the entire viewport.<br>**1** Display object snap tracking path between the alignment point and "From point" to cursor location.<br>**2** Turn off polar tracking path.<br>**3** Turn off polar and object snap tracking paths. |
| TrayIcons 🔲 | 1 | ... | REG | Determines if the tray is displayed on the status bar. |
| TrayNotify 🔲 | 1 | ... | REG | Determines whether service notifications are displayed by the tray. |
| TrayTimeout 🔲 | 5 | ... | REG | Specifies the length of time (in seconds) that tray notificaitons are displayed:<br>**0** Minimium<br>**5** Default<br>**10** Maximum |

| Variable | Default | R/o | Loc | Meaning |
|---|---|---|---|---|
| TreeDepth | 3020 | ... | DWG | Maximum branch depth in *xxyy* format:<br>*xx*  Model-space nodes.<br>*yy*  Paper-space nodes.<br>*>0*  3D drawing.<br>*<0*  2D drawing. |
| TreeMax | 10000000 | ... | REG | Limits memory consumption during drawing regeneration. |
| TrimMode | 1 | ... | REG | Trim toggle for **Chamfer** and **Fillet** commands:<br>**0**  Leave selected edges in place.<br>**1**  Trim selected edges. |
| TSpaceFac | 1.0000 | ... | ... | Mtext line spacing distance; measured as a factor of text height; valid values range from 0.25 to 4.0. |
| TSpaceType | 1 | ... | ... | Type of mtext line spacing:<br>**1**  At Least: adjust line spacing based on the height of the tallest character in a line of mtext.<br>**2**  Exactly: use the specified line spacing; ignores character height. |
| TStackAlign | 1 | ... | DWG | Vertical alignment of stacked text.<br>**0**  Bottom aligned.<br>**1**  Center aligned.<br>**2**  Top aligned. |
| TStackSize | 70 | ... | DWG | Sizes stacked text as a percentage of the selected text height:<br>**1**  Minimum %.<br>**127**  Maximum %. |

. . . . . . . . . . . . . . . . . . . . . . . . . . . . . . . . . . . . . . . . . . . . .

**U**

| Variable | Default | R/o | Loc | Meaning |
|---|---|---|---|---|
| UcsAxisAng | 90 | ... | REG | Default angle for rotating the UCS around an axes (via the **UCS** command using the **X**, **Y**, or **Z** options; valid values are limited to: 5, 10, 15, 18, 22.5, 30, 45, 90, or 180. |
| UcsBase | "" | ... | DWG | Name of the UCS that defines the origin and orientation of orthographic UCS settings. |
| UcsFollow | 0 | ... | DWG | New UCS views:<br>**0**  No change.<br>**1**  Automatic display of plan view. |
| UcsIcon | 3 | ... | DWG | ▨ Display of UCS icon:<br>**0**  Off.<br>**1**  On.<br>**2**  Display at UCS origin, if possible.<br>**3**  On, and displayed at origin. |
| UcsName | "" | R/O | DWG | Name of current UCS view:<br>"" Current UCS is unnamed. |
| UcsOrg | 0.0000,0.0000,0.0000 | R/O | DWG | Origin of current UCS relative to WCS. |

. . . . . . . . . . . . . . . . . . . . . . . . . . . . . . . . . . . . . . . . . . . . .

| Variable | Default | R/o | Loc | Meaning |
|---|---|---|---|---|
| UcsOrtho | 1 | ... | REG | Determines whether the related ortho-graphic UCS setting is restored automatically:<br>**0** UCS setting remains unchanged when orthographic view is restored.<br>**1** Related orthographic UCS setting is restored automatically when an orthographic view is restored. |
| UcsView | 1 | ... | REG | Determines whether the current UCS is saved with a named view:<br>**0** Not saved.<br>**1** Saved. |
| UcsVp | 1 | ... | DWG | Determines whether the UCS in active viewports remains fixed (locked) or changes (unlocked) to match the UCS of the current viewport:<br>**0** Unlocked.<br>**1** Locked. |
| UcsXDir | 1.0000,0.0000,0.0000 | R/O | DWG | X-direction of current UCS relative to WCS. |
| UcsYDir | 0.0000,1.0000,0.0000 | R/O | DWG | Y-direction of current UCS relative to WCS. |
| UndoCtl | 5 | R/O | ... | State of undo:<br>**0** Undo disabled.<br>**1** Undo enabled.<br>**2** Undo limited to one command.<br>**4** Auto-group mode.<br>**8** Group currently active. |
| UndoMarks | 0 | R/O | ... | Current number of undo marks. |
| UnitMode | 0 | ... | DWG | Units display:<br>**0** As set by **Units** command.<br>**1** As entered by user. |
| UserI1 thru **UserI5** | 0 | ... | DWG | Five user-definable integer variables. |
| UserR1 thru **UserR5** | 0.0000 | ... | DWG | Five user-definable real variables. |
| UserS1 thru **UserS5** | "" | ... | ... | Five user-definable string variables. |

. . . . . . . . . . . . . . . . . . . . . . . . . . . . . . . . . . . . . .

**V**

| Variable | Default | R/o | Loc | Meaning |
|---|---|---|---|---|
| ViewCtr | *varies* | R/O | DWG | X,y,z-coordinate of center of current view, such as 6.2433,4.5000,0.0000. |
| ViewDir | *varies* | R/O | DWG | Current view direction relative to UCS, such as 0.0000,0.0000,1.0000. |
| ViewMode | 0 | R/O | DWG | Current view mode:<br>**0** Normal view.<br>**1** Perspective mode on.<br>**2** Front clipping on.<br>**4** Back clipping on.<br>**8** UCS-follow on. |

. . . . . . . . . . . . . . . . . . . . . . . . . . . . . . . . . . . . . .

## Variable　Default　　　　R/o Loc　Meaning

|  |  |  |  |  |
|---|---|---|---|---|
|  |  |  |  | **16** Front clip not at eye. |
| **ViewSize** | 9.0000 | R/O | DWG | Height of current view. |
| **ViewTwist** | 0 | R/O | DWG | Twist angle of current view. |
| **VisRetain** | 1 | ... | DWG | Determines xref drawing's layer settings — on-off, freeze-thaw, color, and linetype:<br>**0** Xref layer settings in the current drawing takes precedence for xref-dependent layers.<br>**1** Settings for xref-dependent layers take precedence over the xref layer definition in the current drawing. |
| **VSMax** | *varies* | R/O | DWG | Upper-right corner of virtual screen, such as 37.4600,27.0000,0.0000. |
| **VSMin** | *varies* | R/O | DWG | Lower-left corner of virtual screen, such as -24.9734,-18.0000,0.0000. |

. . . . . . . . . . . . . . . . . . . . . . . . . . . . . . . . . . . . . . . . . . . . . . . . . .

**W**

|  |  |  |  |  |
|---|---|---|---|---|
| **WhipArc** | 0 | ... | REG | Display of circlular objects:<br>**0** Displayed as connected vectors.<br>**1** Displayed as true circles and arcs. |
| **WhipThread** | 3 | ... | REG | Controls multithreaded processing on two CPUs (if present) during drawing redraw and regeneration:<br>**0** Single-threaded calculations.<br>**1** Regenerations multi-threaded.<br>**2** Redraws multi-threaded.<br>**3** Regens and redraws multi-threaded. |
| **WmfBkgnd** | 1 | ... | ... | Controls background of *.wmf* files:<br>**0** Background is transparent.<br>**1** Background has same as AutoCAD's background color. |
| **WmfForegnd** | 0 | ... | ... | Controls foreground colors of exported *.wmf* images:<br>**0** Foreground is darker than background.<br>**1** Foreground is lighter than background. |
| **WorldUcs** | 1 | R/O | ... | Matching of WCS with UCS:<br>**0** Current UCS is not WCS.<br>**1** UCS is WCS. |
| **WorldView** | 1 | ... | DWG | Display during **3dOrbit**, **DView**, and **VPoint** commands:<br>**0** Display UCS.<br>**1** Display WCS.<br>**2** UCS changes relative to the UCS specified by the **UcsBase** system variable. |
| **WriteStat** | 1 | R/O | ... | Indicates whether drawing file is read-only:<br>**0** Drawing cannot be written to.<br>**1** Drawing can be writen to. |

. . . . . . . . . . . . . . . . . . . . . . . . . . . . . . . . . . . . . . . . . . . . . . . . . .

| Variable | Default | R/O | Loc | Meaning |
|---|---|---|---|---|
| **X** | | | | |
| **XClipFrame** | 0 | ... | DWG | Visibility of xref clipping boundary: <br> **0** Not visible. <br> **1** Visible. |
| **XEdit** | 0 | ... | DWG | Toggles whether drawing can be edited in-place when referenced by another drawing: <br> **0** Cannot in-place refedit. <br> **1** Can in-place refedit. |
| **XFadeCtl** | 50 | ... | REG | Fades objects not being edited in-place: <br> **0** No fading; minimum value. <br> **90** 90% fading; maximum value. |
| **XLoadCtl** | 1 | ... | REG | Controls demand loading: <br> **0** Demand loading turned off; entire drawing is loaded. <br> **1** Demand loading turned on; xref file opened. <br> **2** Demand loading turned on; a *copy* of the xref file is opened. |
| **XLoadPath** | "d:\temp" | ... | REG | Path for loading xref file. |
| **XRefCtl** | 0 | ... | REG | Determines creation of *.xlg* xref log files: <br> **0** File not written. <br> **1** *.xlg* file written. |
| **XrefNotify** ⓐ 2 | | ... | REG | Determines if alert is displayed: <br> **0** No alert displayed. <br> **1** Icon indicates xrefs are attached; a yellow alert indicates missing xrefs. <br> **2** Also displays balloon messages when an xref is modified. |
| **Z** | | | | |
| **ZoomFactor** | 40 | ... | REG | Controls the zoom level via mouse wheel; valid values range between 3 and 100. |

# Obsolete & Removed Commands

The following commands have been removed from AutoCAD since v 2.5.

| Command | Introduced | Removed | Replacement | Reaction |
|---------|-----------|---------|-------------|----------|
| 3Dline | R9 | R11 | Line | "Line" |
| AmeLite | R11 | R12 | Region | "Unknown command" |
| AscText | R11 | R13 | MText | "Unknown command" |
| Ase... | R12 | R13 | ASE... | "Unknown command" |
| | *(Most R12 ASE commands were combined into ASE commands with R13.)* | | | |
| Ase... | R13 | 2000 | dbConnect | "Unknown command" |
| AseUnload | R12 | R14 | Arx Unload | "Unknown command" |
| Axis | v1.4 | R12 | *none* | "Discontinued command" |
| BMake | R12 | 2000 | Block | Displays **Block Definition** dialog. |
| CConfig | R13 | 2000 | PlotStyle | "Discontinued command" |
| Config | R12 | R14 | Options | Displays **Options** dialog. |
| DdAttDef | R12 | 2000 | AttDef | Displays **Attribute Def** dialog. |
| DdAttE | R9 | 2000 | AttEdit | Displays **Edit Attributes** dialog. |
| DdAttExt | R12 | 2000 | AttExt | Displays **Attribute Ext** dialog. |
| DdChProp | R12 | 2000 | Properties | Displays **Properties** window. |
| DdColor | R13 | 2000 | Color | Displays **Select Color** dialog. |
| DdEModes | R9 | R14 | Object Properties | "Discontinued command" |
| DdGrips | R12 | 2000 | Options | Displays **Options** dialog. |
| DDim | R12 | 2000 | DimStyle | Displays **Dim Style Mgr** dialog. |
| DdInsert | R12 | 2000 | Insert | Displays **Insert** dialog. |
| DdLModes | R9 | R14 | Layer | Displays **Layer Manager** dialog. |
| DdLType | R9 | R14 | Linetype | Displays **Linetype Mgr** dialog. |
| DdModify | R12 | 2000 | Properties | Displays **Properties** window. |
| DdOSnap | R12 | 2000 | DSettings | Displays **Drafting Settings** dialog. |
| DdRename | R12 | 2000 | Rename | "Unknown command" |
| DdRModes | R9 | 2000 | DSettings | Displays **Drafting Settings** dialog. |
| DdSelect | R12 | 2000 | Options | Displays **Options** dialog. |
| DdUcs | R10 | 2000 | UcsMan | Displays **UCS** dialog. |
| DdUcsP | R12 | 2000 | UcsMan | Displays **UCS** dialog. |
| DdUnits | R12 | 2000 | Units | Displays **Units** dialog. |
| DdView | R12 | 2000 | View | Displays **View** dialog. |
| DText | v2.5 | 2000 | Text | Executes **Text** command. |
| DL, DLine | R11 | R13 | MLine | "Unknown command" |
| DwfOut | R14 | 2004 | Publish | Executes **Plot** command. |
| DwfOutD | R14 | 2000 | DwfOut | "Unknown command" |

| Command | Introduced | Removed | Replacement | Reaction *(con't)* |
|---|---|---|---|---|
| End | R11 | R13 | Quit | "Discontinued command" |
| EndRep | v1.0 | v2.5 | Minsert | "Discontinued command" |
| EndSv | v2.0 | v2.5 | End | "Discontinued command" |
| EndToday | 2000i | 2004 | *none* | "Unknown command" |
| ExpressTools | 2000 | 2002 | *none* | "Unknown command" |
| Files | v1.4 | R14 | *Explorer* | "Discontinued command" |
| FilmRoll | v2.6 | R13 | *none* | "Unknown command" |
| FlatLand | R10 | R11 | *none* | "Cannot set Flatland to that value" |
| GifIn | R12 | R14 | ImageAttach | "No longer supported" |
| HpConfig | R12 | 2000 | PlotStyle | "Discontinued command" |
| IgesIn, IgesOut | v2.5 | R13 | *none* | "Discontinued command" |
| InetCfg | R14 | 2000 | *none* | "Unknown command" |
| InetHclp | R14 | 2000 | Help | "Unknown command" |
| InsertUrl | R14 | 2000 | Insert | Displays **Insert** dialog. |
| ListUrl | R14 | 2000 | QSelect | "Unknown command" |
| MakePreview | R13 | R14 | *none* | "Discontinued command" |
| MeetNow | 2000i | 2004 | *none* | "Unkown command" |
| OceConfig | R13 | 2000 | PlotStyle | "Discontinued command" |
| OpenUrl | R14 | 2000 | Open | Displays **Select File** dialog. |
| OSnap | v2.0 | 2000 | DSettings | Displays **Drafting Settings** dialog. |
| PcxIn | R12 | R14 | ImageAttach | "No longer supported" |
| PsDrag | R12 | 2000i | *none* | "Unknown command" |
| PsIn | R12 | 2000i | *none* | "Unknown command" |
| Preferences | R11 | 2000 | Options | Displays Options dialog. |
| PrPlot | v2.1 | R12 | Plot | "Discontinued command." |
| QPlot | v1.1 | v2.0 | SaveImg | "Unknown command" |
| RConfig | R12 | R14 | *none* | "Unknown command" |
| RenderUnload | R12 | R14 | Arx Unload | "Unknown command" |
| Repeat | v1.0 | v2.5 | Minsert | "Discontinued command" |
| SaveAsR12 | R13 | R14 | SaveAs | "Unknown command" |
| SaveUrl | R14 | 2000 | SaveAs | Displays **Save Drawing As** dialog. |
| Snapshot | v2.0 | v2.1 | Saveimg | "Unknown command" |
| Sol... | R11 | R13 | *(AME commands lost their SOL-prefix.)* | |
| TbConfig | R12 | 2000i | Customize | Displays **Customize** dialog box. |
| TiffIn | R12 | R14 | ImageAttach | "No longer supported" |
| Today | 2000i | 2004 | *none* | "Unknown command" |
| Toolbar | R13 | 2000i | Customize | Displays **Customize** dialog box. |
| VlConv | R13 | R14 | 3dsIn & 3dsOut | "Unknown command" |

# External Programs

. . . . . . . . . . . . . . . . . . . . . . . . . . . . . . . . . . . . . . . . . . . . . . .

 AcSignApply.exe

Applies digital signatures to drawings (*short for AutoCad SIGNnature APPLY*).

**Start** | **Programs** | **Autodesk** | **AutoCAD 2004** | **Attach Digital Signatures**

*Displays dialog box:*

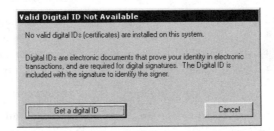

## DIALOG BOX OPTIONS

**Get a digital ID** attempts to contact servers to download digital IDs (identification).

**Cancel** cancels the program.

## RELATED AUTOCAD COMMANDS

**SecurityOptions** applies passwords and digital signatures to drawings.

**SigValidate** checks whether the drawing has a valid digital signature.

## TIPS

• Digital signatures validate the authenticity of drawings.

• Digital signatures indicate whether the drawing was changed since signed.

• Autodesk notes that digital signatures become invalid for these reasons: the drawing was corrupted when the digital signature was attached, the drawing was corrupted in transit, and the digital ID is no longer valid.

. . . . . . . . . . . . . . . . . . . . . . . . . . . . . . . . . . . . . . . . . . . . . . .

# AddPlWiz.exe

Runs the Add Plotter wizard (*short for ADD PLotter WIZARD*).

**Start** | **Settings** | **Control Panel** | **Autodesk Plotter Manager**

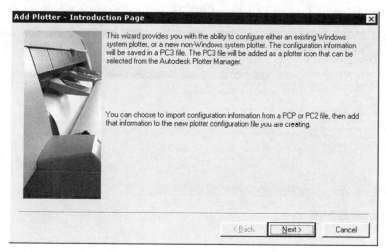

### DIALOG BOX OPTIONS

**Next** goes to the next dialog box.

**Cancel** cancels the program.

### RELATED AUTOCAD COMMANDS

**PlotterManager** opens the Plotters window, which includes the Add-a-Plotter wizard.

### RELATED FILES

*\*.pc3* are AutoCAD plotter configuration files, third generation.

*\*.pmp* are AutoCAD plotter model parameter files.

### TIPS

- See the **PlotterManager** command for details on using this program.

- You can create and edit *.pc3* plotter configuration files without AutoCAD. From the Start button on the Windows toolbar, select **Settings | Control Panel | Autodesk Plotter Manager**.

# AdRefMan.exe

Allows you to change the path to external files, such as text fonts, images, plot configurations, and xrefs (*short for AutoDesk REFerence MANager*).

**Start** | **Programs | Autodesk | AutoCAD 2004 | Reference Manager**

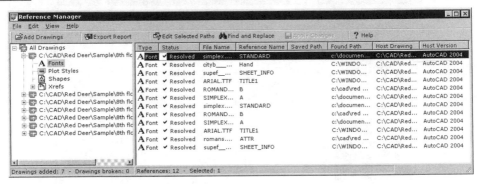

## MENU BAR OPTIONS

**File** menu

**Add Drawings** (CTRL+O) adds one or more drawings to the list.

**Remove Drawings** removes selected drawings from the list.

**Export Report** creates a report in *.csv* (comma separated values) format.

**Apply Changes** applies the changed paths to drawings.

**Exit** (ALT+F4) exits the program.

**Edit** menu

**Select All** (CTRL+A) selects all drawings.

**Unselect All** unselects all drawings.

**Invert Selection** reverses the selection: unselected drawings become selected.

**Find Selected Paths** (F2) displays the Edit Selected Paths dialog box.

**Find and Replace** (F3) displays the Find and Replace Selected Paths dialog box.

**View** menu

**List by Drawing** (CTRL+D) lists by files: drawings and xrefs.

**List by Reference Type** (CTRL+R) lists by objects: fonts, plot styles, shapes, and xrefs.

**Options** displays the Options dialog box.

**Help** menu

**Help** (F1) displays online help for this program.

**About** displays the About Autodesk Reference Manager dialog box.

## TOOLBAR OPTIONS

**Add Drawings** selects the drawings to be checked; displays the Add Xrefs dialog box when a drawing contains xrefs.

**Export Report** creates a report in *.csv* (comma separated values) format of the findings by this program; displays the Export Report dialog box.

**Edit Selected Paths** displays the Edit Selected Paths dialog box; enter a new path, and click OK.

**Find and Replace** displays the Find and Replace Selected Paths dialog box.

**Apply Changes** applies the changed paths to drawings; displays the Summary dialog box.

**Help** displays online help for this program.

## Options dialog box

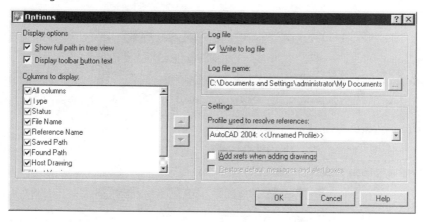

### Display Options options

**Show full path in tree view** determines whether the full path is displayed or just the file name.

**Display toolbar button text** toggles the display of text next to toolbar buttons.

**Columns to display** determines which data is displayed for each drawing.

### Log File options

**Write to log file** determines whether the results of this session are written to a *.txt* file.

**Log file name** specifies the name of the log file.

**...** displays the Browse for Folder dialog box.

### Settings options

**Profile used to resolve references** selects a profile (created with the Profile tab in AutoCAD's Options command), which specifies paths.

**Add xrefs when adding drawings** loads all nested xref drawing files.

**Restore default messages and alert boxes** turns on all dialog boxes that have been turned off with the "Don't display this message again" option.

**Add Xrefs** dialog box

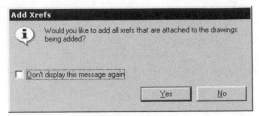

**Don't display this message again** suppresses further displays of this dialog box.

**Yes** loads all externally-referenced drawings (recommended).

**No** does not load xrefs.

**Edit Selected Paths** dialog box

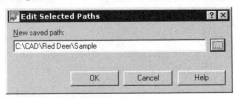

**New saved path** specifies the path to the object.

... displays the Browse for Folder dialog box.

**Find and Replace Selected Paths** dialog box

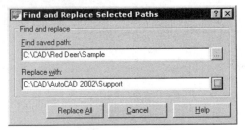

**Find saved path** specifies the current path to the object.

**Replace with** specifies the new path to the object.

... displays the Browse for Folder dialog box.

**Replace All** replaces the old path with the new path.

**Summary** dialog box

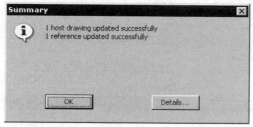

OK dismisses dialog box.

Details displays the Details dialog box:

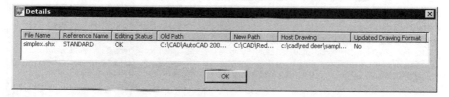

## RELATED AUTOCAD COMMANDS

**Image** attaches raster images to drawings.

**Options** specifies the default paths to external files.

**Style** defines the look of *.shx* and *.ttf* fonts.

**Shapes** loads *.shx* shape files into drawings.

**XRef** attaches drawings as external references.

## TIPS

- When drawings are moved, they may need to reference different folders and disk drives, which means that saved paths need to be updated. This program allows you to modify the paths without opening each drawing file AutoCAD.

- This program does not work with the following references:

    Text fonts not associated with a text style.

    OLE links.

    Hyperlinks.

    Database file links.

    Xrefs linked to URLs on the Web.

- This program cannot work with drawings open in AutoCAD, or have file attributes set to read-only.

# BatchPlot.exe

Launches AutoCAD to batch plot drawings.

 | **BatchPlot.exe**

*Displays program:*

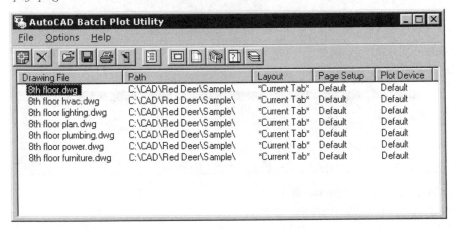

## MENU BAR & TOOLBAR OPTIONS

**File** menu

 **Add Drawing** adds drawings to be plotted; displays the Add Drawing File dialog box.

 **Remove** removes selected drawings from the list.

**New List** clears drawings from the list.

 **Open List** (CTRTL+O) opens *.bp3* files, which specify lists of drawings to plot.

**Append List** prompts you to select additional *.bp3* files, which adds drawings to the list.

 **Save List** (CTRTL+S) saves the list of drawings in *.bp3* files; displays the Save Batch Plot List File dialog box.

 **Plot** plots the drawings using AutoCAD.

 **Plot Test** loads (but does not plot) drawings to check for missing xrefs, fonts, and fonts.

 **Logging** displays the Logging dialog box.

## Options menu

 **Layouts** displays the Layouts dialog box.

 **Page Setups** displays the Page Setups dialog box.

 **Plot Devices** displays the Plot Devices dialog box.

 **Plot Settings** displays the Plot Settings tab of the Plot Settings dialog box.

**Layers** displays the Layers tab of the Plot Settings dialog box.

## Help menu

**Help** displays online help for this program.

**About** displays the About Batch Plot dialog box.

## Logging dialog box

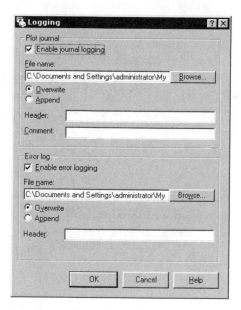

Plot Journal options

**Enable Journal Logging** turns on plot logging.

**File Name** specifies the name of the log file.

• **Overwrite** overwrites the previous plot file.

• **Append** adds to the end of an existing plot file.

**Header** specifies a line of text to be added to the beginning of the file.

**Comment** specifies additional text to be added to the beginning of the file.

Error Log options
**Enable Error Logging** turns on error logging; records problems that occur during batch plotting.

**File Name** specifies the name of the error log file.

- **Overwrite** overwrites the previous error log file.

- **Append** adds error message to the end of a the error log file.

**Header** specifies a line of text to be added to the beginning of the file.

## Layouts dialog box

- **Plot All Layouts** instructs AutoCAD to plot all the layouts in the selected drawing.
- **Plot Selected Layouts** instructs AutoCAD to plot only selected layouts , as listed below.
  **\*Current Tab\*** plots the tab current when the drawing was saved.
  **Model Tab** plots the model tab.
  **Last Active Layout Tab** plots the last layout tab active (other than a model tab).
**Show All Layouts** shows all layout names in the drawing.

## Page Setups dialog box

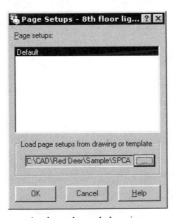

**Page Setups** lists the page setups in the selected drawing.

Load Page Setups from Drawing or Template options
... displays the Select Drawing dialog box.

**Plot Devices** dialog box

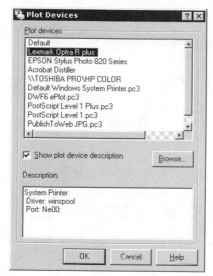

**Plot Devices** lists the printers and plotters supported by Window and AutoCAD.

**Show Plot Device Description** displays a description of the selected device.

**Browse** displays the Open Plot Configuration File dialog box; locate a plotter configuration file, and then click Open.

**Plot Settings** dialog box

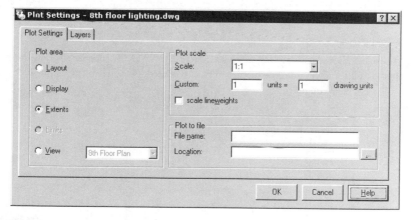

**Plot Settings** tab

Plot Area options

- **Layout** plots a specified layout.
- **Display** plots the view stored when the drawing was last saved.
- **Extents** plots the drawing extents.
- **Limits** plots the drawing limits.
- **Plots** the named view.

## Plot Scale options

**Plot Scale** selects a scale factor:

    Layout tabs default is 1:1 scale.

    Model tab default is Scaled to Fit.

**Custom** allows you to set your own scale factor.

**Scale Lineweights** makes lineweights thinner or thicker, according to the scale factor.

## Plot to File options

**File Name** specifies the name of the plot file.

**Location** selects the path and drive for the plot file.

## Layers tab

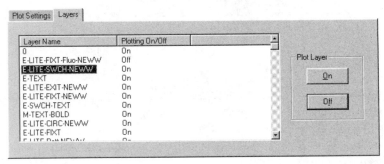

## Plot Layer options

**On** turns on the selected layer for plotting.

**Off** turns off the selected layer for plotting.

## TIP

• After you select **Plot**, each drawing is loaded into AutoCAD for plotting. When the drawing plots successfully, a check mark is displayed next to the drawing name. When the drawing fails to plot, an "x" is displayed.

# DwgCheckStandards.exe

Checks if drawing elements meet prescribed standards (*short for DraWinG CHECK STANDARDS*).

🏁Start | **Programs | Autodesk | AutoCAD 2004 | Batch Standards Checker**

## MENU BAR & TOOLBAR OPTIONS

**File** menu

New Check File (CTRL+N) clears the settings.

Open Check File (CTRL+O) displays the File Open dialog box; select *.chx* file, click Open.

Save Check File (CTRL+S) saves the parameters of this program as *.chx* files.

Save As (ALT+S) displays the File Save dialog box; name the file, and click Save.

Exit (ALT+F4) exits the program.

**Check** menu

Start Check (ALT+T) examines the drawing for violations of CAD standards.

Stop Check (ALT+P) halts the standards checking process.

**View Report** (ALT+V) displays reports in Web browsers.

 **Export Report** (ALT+E) saves reports in HTML format.

**Help** menu

**Help** (F1) displays online help for this command.

**About** displays the About Batch Standards Checker dialog box.

## DIALOG BOX OPTIONS

**Drawings** tab

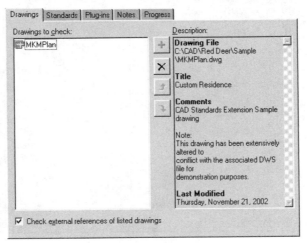

 **Add Drawing** (F3) opens additional drawing files; displays the File Open dialog box.

**Remove Drawing** (DEL) removes selected drawings from the list.

**Move Up** (F4) moves the selected drawings higher.

**Move Down** (F5) moves the selected drawings lower. Drawings are checked in the order in which they appear.

**Check external references of listed drawings** causes this program to check the standards of all attached xrefs. Drawings that cannot be located have an exclamation (!) prefix.

**Standards** tab

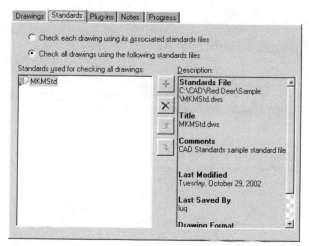

- **Check each drawing using its associated standards files** checks each drawing against its associated standard.
- **Check all drawings using the following standards files** checks all drawings against a single standard.

**Add Standards File** (F3) adds additional *.dws* standards list to the list.

**Remove Standards File** (DEL) removes selected standards from the list.

**Move Up** (F4) moves the selected standards higher.

**Move Down** (F5) moves the selected standards lower.

When two or more standards are loaded, sometimes they conflict. In this case, the settings in the higher standards take precedence over lower standards.

**Plug-ins** tab

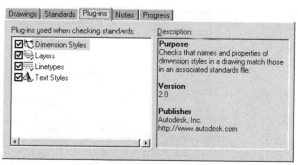

**Plug-ins used when checking standards** selects the styles to check.

☑ style is checked.

☐ style is not checked.

**Notes** tab

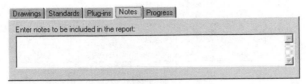

**Enter notes to be included in the report** provides an area for you to enter text.

**Progress** tab

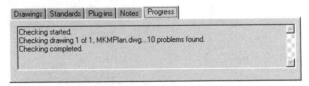

Displays the progress of the standards checking.

## RELATED AUTOCAD COMMANDS

**CheckStandards** checks whether drawings match CAD standards.

**Standards** applies CAD standards to drawings.

## RELATED FILES

***.chx*** are CAD standards files stored in XML format.

***.dws*** are CAD Standards files, stored in DWG format.

## TIP

• This program produces reports in HTML format, which can be viewed by Web browsers:

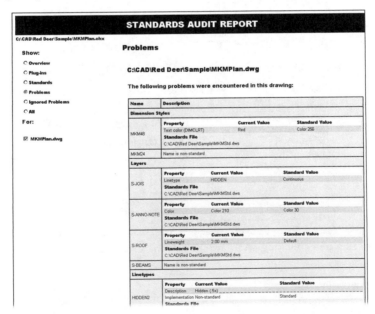

# ExpressViewer.exe

Displays and plots DWF files.

**Start** | **Programs | Autodesk | Autodesk Express Viewer**

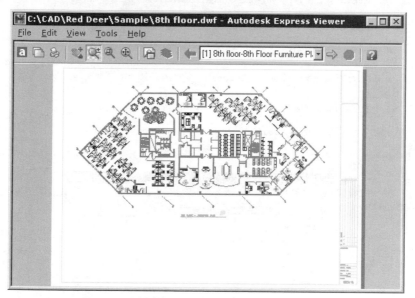

## MENU BAR & TOOLBAR OPTIONS

**File** menu

**Open** (CTRL+O) opens *.dwf* files; displays the Open File dialog box.

 **Print** (CTRL+P) displays the Print dialog box.

**Exit** (ALT+F4) exits this program.

**Edit** menu

 **Copy** (CTRL+C) copies the image to the Clipboard as a bitmap.

**View** menu

 **Layers** (l) displays the Layers window; allows layers to be turned on and off.

**Sheets** (s) displays the Sheets windows; allows a specific layout to be displayed.

**Views** (v) displays the Views window; allows a specific view to be displayed.

**Show** displays additional viewing options:

**Hyperlinks** (CTRL+H) toggles the display of hyperlinks embedded in the drawing.

**Page Tiles** (CTRL+T) illustrates sheets of paper, when Fit to Paper Size option is turned off in the Print dialog box.

**Paper Background** (CTRL+B) toggles the background color between white and gray.

**Toolbar** toggles the display of the toolbar.

## Tools menu

**Pan** (*arrow keys*) moves the view.

**Zoom** (+ *and* -) enlarges and reduces the drawing.

**Zoom Rectangle** (CTRL+R) enlarges a windowed area of the drawing.

**Fit in Window** (HOME) changes the view of the drawing to fit to the window.

## Help menu

**Contents** (F1) displays online help for this program.

**Check for Viewer Updates** requires an Internet connection.

**About** displays About Autodesk Express Viewer dialog box.

## Print dialog box

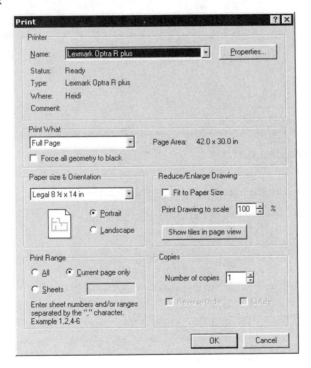

Printer options

**Name** selects a Windows system printer; Express view does not support AutoCAD plotter drivers.

**Properties** displays the Windows printer Properties dialog box.

Print What options
- **Full page** prints the entire drawing.
- **Current View** prints the zoomed in view.

**Force all geometry to black** prints all elements in black; useful for monochrome printers.

Paper Size & Orientation options

**Paper Size** selects the size of paper, of those supported by the selected printer.

**Portrait** orients the page vertically.

**Landscape** orients the page horizontally.

Reduce/Enlarge Drawing options

**Fit to Paper Size**

☑ fits the drawing to the paper size; ignores scale.

☐ tiles the print over several pages.

**Print Drawing to Scale** plots the drawing to a scale relative to its full size.

**Show tiles in page view** displays the viewer with dashed blue lines indicating the margins of pages.

Print Range options

**All** prints all sheets in the drawing.

**Current page only** prints the currently-displayed page only.

**Sheets** prints a range of sheets.

Copies options

**Number of copies** prints one or more copies.

**Reverse order** prints multiple sheets in reverse order.

**Collate** prints multiple sheets together.

**Layers** window

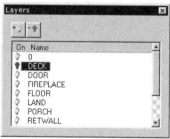

**On** turns on the selected layer.

**Off** turns off the selected layer.

**x** dismisses the window.

**Sheets** window

✓ displays the selected sheet.

**x** dismisses the window.

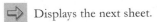 Displays the previous sheet (layout).

 Selects the name of a sheet.

⇨ Displays the next sheet.

Stops display of a sheet.

**Views** window

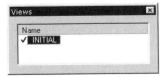

✓ displays the selected view.

**x** dismisses the window.

**RELATED AUTOCAD COMMANDS**

**Publish** creates multi-sheet *.dwf* files.

## TIPS

- Express Viewer displays (and prints) just one *.dwf* file at a time.

- This viewer does not display *.dwg* or *.dxf* drawing files.

- In the Print dialog box:

    **Reverse order** option is useful when your printer produces pages face up, meaning they are printed in reverse order. Turn on this option to reverse the reverse order, producing a print set in correct order.

    **Collate** option is useful when you have more than one set of drawings to produce. Turn on this option to print sheets together; the drawback is that printing takes longer.

- *.dwf* files cannot be imported into AutoCAD, without help from a third-party translator.

- When **Page Tiles** is turned on, blue dashed lines show the edges of paper:

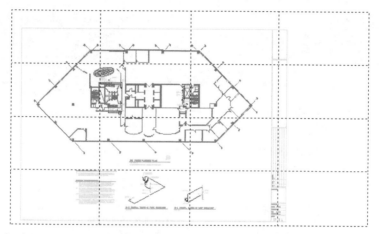

- Use a wheelmouse to zoom in and out by rolling the wheel forward and backward; to pan, hold down the wheel while moving the mouse.

# HpSetup.exe

Displays information on using merge control with all brands of plotters
(*obsolete program; short for Hewlett-Packard plotter SETUP*).

 | **HpSetup.exe**

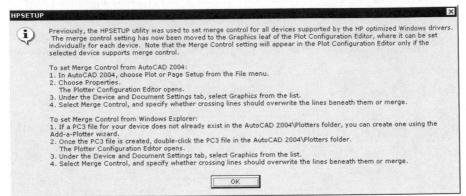

**HPSETUP**

> Previously, the HPSETUP utility was used to set merge control for all devices supported by the HP optimized Windows drivers. The merge control setting has now been moved to the Graphics leaf of the Plot Configuration Editor, where it can be set individually for each device. Note that the Merge Control setting will appear in the Plot Configuration Editor only if the selected device supports merge control.
>
> To set Merge Control from AutoCAD 2004:
> 1. In AutoCAD 2004, choose Plot or Page Setup from the File menu.
> 2. Choose Properties.
>    The Plotter Configuration Editor opens.
> 3. Under the Device and Document Settings tab, select Graphics from the list.
> 4. Select Merge Control, and specify whether crossing lines should overwrite the lines beneath them or merge.
>
> To set Merge Control from Windows Explorer:
> 1. If a PC3 file for your device does not already exist in the AutoCAD 2004\Plotters folder, you can create one using the Add-a-Plotter wizard.
> 2. Once the PC3 file is created, double-click the PC3 file in the AutoCAD 2004\Plotters folder.
>    The Plotter Configuration Editor opens.
> 3. Under the Device and Document Settings tab, select Graphics from the list.
> 4. Select Merge Control, and specify whether crossing lines should overwrite the lines beneath them or merge.

OK

## DIALOG BOX OPTIONS

**OK** exits the dialog box.

## RELATED AUTOCAD COMMANDS

**Plot** sets up drawings with plotters.

## TIPS

- In older releases of AutoCAD, only plotters compatible with HP had additional control over how drawings were plotted. This program, **HpSetup.exe** acted like an extension to the **Plot** command. As of AutoCAD 2000, all plotters benefit from the same degree of control by **Plot**.

- For compatibility reasons, this program continues to exist, merely displaying the dialog box shown above.

- This program was replaced by AutoCAD's **Plot** command enhanced in AutoCAD 2000.

# StyShWiz.exe

Runs the Add Plot Style Table wizard *(short for STYle SHeet WIZard.)*

**AutoCAD** | **StyShWiz.exe**

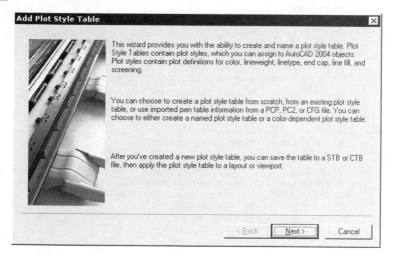

## DIALOG BOX OPTIONS

**Next** goes to the next dialog box.

**Cancel** cancels the program.

## RELATED AUTOCAD COMMANDS

**StylesManager** opens the Plot Styles window, which includes the Add-a-Plot Style Table wizard.

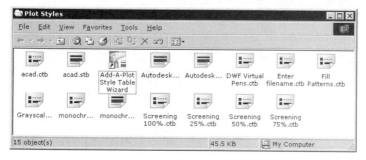

## RELATED FILES

**\*.ctb** are AutoCAD color-table based files.

**\*.stb** are AutoCAD style-table based files.

## TIPS

- See the **StylesManager** command for details on using this program.

- You can create and edit plotter style files without AutoCAD. From the **Start** button on the Windows toolbar, select **Settings | Control Panel | Autodesk Plot Style Manager**.

## APPENDIX D
# Express Tools

The following commands are included with the Express Tools package:

| Command | Description |
| --- | --- |
| AliasEdit | Edits the aliases stored in *acad.pgp*. |
| AlignSpace | Aligns model space objects, whether in different viewports or with objects in paper space. |
| ArcText | Places text along an arc. |
| AttIn | Imports attribute data. |
| AttOut | Quickly extracts attributes in tab-delimited format. |
| | |
| BExtend | Extends open objects to objects in blocks and xrefs. |
| BlockReplace | Replaces all inserts of one block with another. |
| BlockToXref | Convert blocks to xrefs. |
| BreakLine | Creates the break-line symbol. |
| BScale | Scales blocks from their insertion points. |
| BTrim | Trims to objects nested in blocks and external references. |
| Burst | Explodes blocks, converts attributes to text. |
| | |
| ChSpace | Moves objects between model and paper space. |
| ChUrls | **DdEdit**-like editor for hyperlinks (URL addresses). |
| ClipIt | Adds arcs, circles, and polylines to the **XClip** command. |
| CopyM | **Copy** command with repeat, divide, measure, and array options. |
| CopyToLayer | Copies objects to other layers. |
| | |
| DimEx | Exports dimension styles to an ASCII file. |
| DimIm | Imports dimension style files created with **DimEx**. |
| | |
| EtBug | Sends bug reports to Autodesk. |
| ExOffset | Adds options to the **Offset** command. |
| ExPlan | Adds options to the **Plan** command. |
| | |
| FS | Selects objects that touch the selected object. |
| FullScreen | Toggles between full-screen and regular window. |
| | |
| GetSel | Selects objects based on layer and type. |
| | |
| ImageApp | Specifies the external image editor. |
| ImageEdit | Launches the image editor to edit selected images. |
| | |
| LayCur | Changes the layer of selected objects to the current layer. |
| LayDel | Deletes layers from drawings — permanently. |
| LayFrz | Freezes the layers of selected objects. |
| LayIso | Isolates layers of selected objects (all other layers are frozen). |
| LayLck | Locks the layers of selected objects. |
| LayMch | Changes the layer of selected objects to that of a selected object. |
| LayMrg | Merges two layers; removes the first layer from the drawing. |
| LayOff | Turns off layers of selected objects. |
| LayOn | Turns on all layers. |
| LayoutMerge | Places objects from layouts onto one layout. |
| LayThw | Thaws all layers. |
| LayUlk | Unlocks layer of selected object. |
| LayVpi | Isolates object's layer in viewport. |

| Command | Description |
|---|---|
| LayWalk | Isolates each layer in sequential order. |
| LMan | Saves and restores layer settings. |
| Lsp | AutoLISP function searching utility. |
| LspSurf | LISP file viewer. |
| MkLtype | Creates linetypes from selected objects. |
| MkShape | Creates shapes from selected objects. |
| MoCoRo | Moves, copies, rotates, and scales objects. |
| MoveBak | Moves .bak files to specified folders. |
| MStretch | Stretches with multiple selection windows. |
| NCopy | Copies objects nested inside blocks and xrefs. |
| Plt2Dwg | Imports HPGL files into the drawings. |
| Propulate | Updates, lists, and clears drawing properties. |
| PsBScale | Sets and updates the scale of blocks relative to paper space. |
| PsTScale | Sets text height relative to paper space. |
| QlAttach | Associate leaders to annotation objects. |
| QlAttachSet | Associates leaders with annotations. |
| QlDetachSet | Dissassociates leaders from annotations. |
| QQuit | Closes all drawings, and then exits AutoCAD. |
| ReDir | Changes paths for xrefs, images, shapes, and fonts. |
| RepUrls | Replaces hyperlinks. |
| Revert | Closes the drawing, and re-opens the original. |
| RText | Inserts and edits remote text objects. |
| RtUcs | Changes UCSs in real time. |
| SaveAll | Saves all drawings. |
| ShowUrls | Lists URLs in a dialog box. |
| Shp2blk | Converts from a shape definition to a block definition. |
| SuperHatch | Uses images, blocks, external references, or wipeouts as hatch patterns. |
| SysvDlg | Launches an editor for system variables. |
| TCase | Changes text between Sentence, lower, UPPER, Title, and tOGGLE cASE. |
| TCircle | Surrounds text and multiline text with circles, slots, and rectangles. |
| TCount | Prefixes text with sequential numbers. |
| TextFit | Fits text between points. |
| TextMask | Places masks behind selected text. |
| TextUnmask | Removes masks from behind text. |
| TFrames | Toggles the frames surrounding images and wipeouts. |
| TJust | Justifies text created with the **MText** and **AttDef** commands. |
| TOrient | Re-orients text, multiline text, and block attributes. |
| TScale | Scales text, multiline text, attributes, and attribute definitions. |
| TSpaceInvaders | Finds and selects text with overlapping objects. |
| Txt2Mtxt | Converts single-line to multiline text. |
| TxtExp | Explodes selected text into polylines. |
| VpScale | Lists the scale of selected viewports. |
| VpSynch | Synchronizes viewports with a master viewport. |
| XData | Attaches xdata to objects. |
| XdList | Lists xdata attached to objects. |
| XList | Displays properties of objects nested in blocks and xref. |

# AutoCAD 2004 Keystroke Shortcuts

## Object Snap

*Modes:*

| | |
|---|---|
| **APP** | Apparent intersection. |
| **CEN** | Center. |
| **END** | Endpoint. |
| **EXT** | Extension. |
| **FROM** | From. |
| **INS** | Insertion point. |
| **INT** | Intersection. |
| **MID** | Midpoint. |
| **NEA** | Nearest. |
| **NOD** | Node (point). |
| **PAR** | Parallel. |
| **PER** | Perpendicular. |
| **QUA** | Quadrant. |
| **TAN** | Tangent. |

*Settings:*

| | |
|---|---|
| **NON** | None. |
| **OFF** | Turn off object snap. |
| **QUI** | Quick mode. |

## Selection Sets

| | |
|---|---|
| **(Pick)** | Selects one object. |
| **ALL** | Selects all objects. |
| **AU** | AUtomatic: *(pick)* or BOX. |
| **BOX** | Left to right = Crossing; Right to left = Window. |
| **C** | Crossing. |
| **CP** | Crossing polygon. |
| **F** | Fence. |
| **G** | Group. |
| **L** | Last. |
| **M** | Multiple (no highlighting). |
| **P** | Previous. |
| **SI** | Single selection. |
| **W** | Window. |
| **WP** | Window polygon. |

*Selection modes:*

| | |
|---|---|
| **A** | Add to selection set (default). |
| **R** | Remove from selection set. |
| **U** | Undo change to selection set. |

## Special Text Characters

*For **Text** command:*

| | | |
|---|---|---|
| **%%c** | Diameter symbol | Ø |
| **%%d** | Degree symbol | ° |
| **%%o** | Overline | |
| **%%%** | Percent symbol | % |
| **%%p** | Plus-minus symbol | ± |
| **%%u** | Underline | |
| **%%nnn** | ASCII character *nnn* | |

*For **MText** command:*

| | |
|---|---|
| **\~** | Nonbreaking space. |
| **\\** | Backslash. |
| **\{** | Opening brace. |
| **\}** | Closing brace. |
| **\C***n* | Sets color *n*. |
| **\F***x;* | Changes to font file name *x*. |
| **\H***n;* | Changes text height to *n* units. |
| **\L** | Underline. |
| **\l** | Turns off underline. |
| **\M+***n* | Multibyte shape number *n*. |
| **\O** | Overline. |
| **\o** | Turns off overline. |
| **\P** | End of paragraph. |
| **\Q***n;* | Changes obliquing angle to *n*. |
| **\S***n^m* | Stacks character *n* over *m*. |
| **\T***n;* | Changes tracking to *n*. |
| **\U+***n* | Places Unicode character *n*. |
| **\W***n;* | Changes width factor to *n*. |

## Color Numbers & Names

| | | |
|---|---|---|
| 0 | ... | Background color. |
| 1 | R | Red. |
| 2 | Y | Yellow. |
| 3 | G | Green. |
| 4 | C | Cyan. |
| 5 | B | Blue. |
| 6 | M | Magenta. |
| 7 | W | White. |
| 8-249 | ... | Other colors. |
| 250-255 | ... | Shades of grey. |
| BYLAYER | ... | Color from layer. |
| BYBLOCK | ... | Color from block. |

## Coordinate Input

*Default Directions:*
  *X positive = Right.*
  *Y positive = Up.*
  *Z positive = Right-hand rule.*

| | | |
|---|---|---|
| **x,y** | 2D cartesian coordinates: | |
| | From point: 2,3 | |
| **x,y,z** | 3D cartesian coordinates: | |
| | From point: 2,3,4 | |
| **d<a** | 2D polar coordinates: | |
| | From point: 2<45 | |
| **@d<a** | Relative coordinates: | |
| | From point: @2<45 | |
| **d<a,z** | 3D cylindical coordinates: | |
| | From point: 2<45,5 | |
| **@d<a,z** | Relative cylindrical coordinates: | |
| | From point: @2<45,5 | |
| **d<a<a** | 3D spherical coordinates: | |
| | From point: 5<45<30 | |

## Option Modifiers

| | |
|---|---|
| **tt** | Tracking. |
| **from** | Offsets from temporary ref point. |
| 🖱 | Direct distance entry: move mouse, and enter a distance. |

## Filter Operators

*Logical operators:*

| | |
|---|---|
| **\*** | Equal to any value. |
| **=** | Equal to. |
| **!=** | Not equal to. |
| **<** | Less than. |
| **>** | Greater than. |
| **<=** | Less than or equal to. |
| **>=** | Greater than or equal to. |

*Grouping operators:*

| | |
|---|---|
| **\*\*BEGIN** | Begin group. |
| **AND** | Intersection. |
| **OR** | Union. |
| **NOT** | Exclude. |
| **XOR** | Exclusive Or. |
| **\*\*END** | End of group. |

## Distance Modifiers

| | |
|---|---|
| **,** | Distance separator: |
| | From point: 2,3 |
| **'** | Feet (required). |
| **"** | Inches (optional): |
| | From point: 2'3,5'4" |
| *— and /* | Fractional inches: |
| | From point: 2'3-3/4,5'4" |
| **@** | Relative distance. |
| **\*** | Force use of WCS coordinates: |
| | From line: \*3,4,2 |
| **@\*** | Force relative WCS coordinates: |
| | From line: @\*3<45 |

## Angle Modifiers

*Angle defaults:*
  *0 degrees = East (3 o'clock).*
  *Positive direction = Counterclockwise.*

| | |
|---|---|
| **'** | Minutes. |
| **"** | Seconds. |
| **.** | Decimal of a degree or seconds. |
| **d** | Degrees. |
| **r** | Radian. |
| **g** | Grad. |
| **<** | Angle. |
| **<<** | Force 0 degrees = east, and decimal degrees. |
| **<<<** | Force 0=east, but uses current format. |
| **+** | Counterclockwise. |
| **—** | Clockwise. |

## Surveyor Units

| | | |
|---|---|---|
| **E** | East | 0 degrees. |
| **N** | North | 90 degrees. |
| **W** | West | 180 degrees. |
| **S** | South | 270 degrees. |

## Point Filters

| | |
|---|---|
| **.x** | Pick x, and enter y,z. |
| **.y** | Pick y, and enter x,z. |
| **.z** | Pick z, and enter x,y. |
| **.xy** | Pick x,y, and enter z. |
| **.xz** | Pick x,z, and enter y. |
| **.yz** | Pick y,z, and enter x. |